Statistics for Industry and Technology

Theory and Applications of Recent Robust Methods

Mia Hubert
Greet Pison
Anja Struyf
Stefan Van Aelst
Editors

Springer Basel AG

Mia Hubert
Department of Mathematics
Katholieke Universiteit Leuven
W. de Croylaan 54
3001 Leuven
Belgium
E-mail: Mia.Hubert@wis.kuleuven.ac.be

Greet Pison
Department of Applied Mathematics
and Computer Science
Ghent University
Krijgslaan 281 S9
9000 Gent
Belgium
E-mail: Greet.Pison@UGent.be

Anja Struyf
Department of Mathematics
and Computer Science
University of Antwerp
Middelheimlaan 1
2020 Antwerp
Belgium
E-mail: Anja.Struyf@ua.ac.be

Stefan Van Aelst
Department of Applied Mathematics
and Computer Science
Ghent University
Krijgslaan 281 S9
9000 Gent
Belgium
E-mail: Stefan.VanAelst@UGent.be

2000 Mathematics Subject Classification 62F35, 62G35

A CIP catalogue record for this book is available from the Library of Congress,
Washington D.C., USA

Bibliographic information published by Die Deutsche Bibliothek
Die Deutsche Bibliothek lists this publication in the Deutsche Nationalbibliografie;
detailed bibliographic data is available in the Internet at <http://dnb.ddb.de>.

ISBN 978-3-0348-9636-8 ISBN 978-3-0348-7958-3 (eBook)
DOI 10.1007/978-3-0348-7958-3

Originally published by Birkhäuser Verlag in 2004
Softcover reprint of the hardcover 1st edition 2004
Printed on acid-free paper produced of chlorine-free pulp. TCF ∞

ISBN 978-3-0348-9636-8

9 8 7 6 5 4 3 2 1 www.birkhauser-science. com

Contents

Preface ix

List of Reviewers xi

J.G. Adrover and V.J. Yohai
Bias Behavior of the Minimum Volume Ellipsoid Estimate 1

L. Aucremanne, G. Brys, M. Hubert, P.J. Rousseeuw and A. Struyf
A Study of Belgian Inflation, Relative Prices and Nominal Rigidities using New Robust Measures of Skewness and Tail Weight 13

G. Bassett, G. Gerber and P. Rocco
Robust Strategies for Quantitative Investment Management 27

C. Chen
An Adaptive Algorithm for Quantile Regression 39

A. Christmann
On Properties of Support Vector Machines for Pattern Recognition in Finite Samples 49

P. Čížek
Smoothed Local L-Estimation With an Application 59

S. Copt and M.-P. Victoria-Feser
Fast Algorithms for Computing High Breakdown Covariance Matrices with Missing Data 71

R.B. Dimova and N.M. Neykov
Generalized d-fullness Technique for Breakdown Point Study of the Trimmed Likelihood Estimator with Application 83

A. Durio and E.D. Isaia
On Robustness to Outliers of Parametric $\boldsymbol{L_2}$ Estimate Criterion in the Case of Bivariate Normal Mixtures: a Simulation Study 93

S. Engelen, M. Hubert, K. Vanden Branden and S. Verboven
Robust PCR and Robust PLSR: a Comparative Study 105

P.J. Hingley
Analytic Estimator Densities for Common Parameters under Misspecified Models 119

K. Joossens and C. Croux
Empirical Comparison of the Classification Performance of Robust Linear and Quadratic Discriminant Analysis 131

J. Jurečková and J. Picek
Estimates of the Tail Index Based on Nonparametric Tests 141

A. Kankainen, S. Taskinen and H. Oja
On Mardia's Tests of Multinormality 153

A. Kharin
Robustness in Sequential Discrimination of Markov Chains under "Contamination" 165

A. Marazzi and V.J. Yohai
Robust Box-Cox Transformations for Simple Regression 173

L. Mašíček
Consistency of the Least Weighted Squares Regression Estimator 183

S. Morgenthaler, R.E. Welsch and A. Zenide
Algorithms for Robust Model Selection in Linear Regression 195

D.M. Mount, N.S. Netanyahu and E. Zuck
Analyzing the Number of Samples Required for an Approximate Monte-Carlo LMS Line Estimator 207

D.J. Olive
Visualizing 1D Regression 221

M.R. Oliveira, J.A. Branco, C. Croux and P. Filzmoser
Robust Redundancy Analysis by Alternating Regression 235

E. Ollila and V. Koivunen
Robust ML-estimation of the Transmitter Location 247

J.F. Ortega
A Family of Scale Estimators by Means of Trimming 259

C. Ortelli and F. Trojani
Robust Efficient Method of Moments Estimation 271

E. Rafalin and D.L. Souvaine
Computational Geometry and Statistical Depth Measures 283

D.W. Scott
Partial Mixture Estimation and Outlier Detection in Data and Regression 297

D. Suter and H. Wang
Robust Fitting Using Mean Shift:
Applications in Computer Vision 307

S.S. Syed Yahaya, A.R. Othman and H.J. Keselman
Testing the Equality of Location Parameters for Skewed Distributions
Using S_1 with High Breakdown Robust Scale Estimators 319

S. Taskinen, A. Kankainen and H. Oja
Rank Scores Tests of Multivariate Independence 329

L. Thomassen and M. Van Wouwe
The Influence of a Stochastic Interest Rate on
the n-fold Compound Option 343

A. Toma
Robust Estimations for Multivariate Sinh^{-1}-Normal Distribution 355

B. Vandewalle, J. Beirlant and M. Hubert
A Robust Estimator of the Tail Index Based on an
Exponential Regression Model 367

S. Vanlanduit and P. Guillaume
Robust Processing of Mechanical Vibration Measurements 377

G. Zioutas
Quadratic Mixed Integer Programming Models in Minimax
Robust Regression Estimators 387

Preface

After two successful meetings in Vorau (2001) and Vancouver (2002), the International Conference on Robust Statistics 2003, ICORS 2003, took place in Antwerp, Belgium, from July 13–18. The conference was intended to be a forum where all aspects of robust statistics could be discussed. As such the scientific program offered a wide range of talks on new developments and practice of robust statistics with applications in finance, chemistry, engineering and others. Of equal interest were interactions between robustness and other fields of statistics, and science in general.

The conference was attended by 134 participants of 28 different countries from all over the world. The program contained 78 oral and 14 poster presentations. This volume presents a wide range of papers that were presented at the conference. As most of the contributions contain both theoretical and empirical results, we prefer to present them in alphabetical order of the first author. A rough classification could be made as follows: *new methods and theoretical results* are presented by J.G. Adrover and V.J. Yohai, R.B. Dimova and N.M. Neykov, P.J. Hingley, A. Kankainen et al., A. Kharin, A. Marazzi and V.J. Yohai, L. Mašíček, D.J. Olive, M.R. Oliveira et al., J.F. Ortega, C. Ortelli and F. Trojani, D.W. Scott, S.S. Syed Yahaya et al., S. Taskinen et al., A. Toma, and B. Vandewalle et al. *Empirical properties* of robust methods are studied in A. Christmann, P. Čížek, A. Durio and E.D. Isaia, S. Engelen et al., K. Joossens and C. Croux, and J. Jurečková and J. Picek. *Computational aspects* are treated in C. Chen, S. Copt and M.-P. Victoria-Feser, S. Morgenthaler et al., D.M. Mount et al., and G. Zioutas. *Applications* in finance, engineering and computer vision are emphasized in L. Aucremanne et al., G. Bassett et al., E. Ollila and V. Koivunen, D. Suter and H. Wang, L. Thomassen and M. Van Wouwe, and S. Vanlanduit and P. Guillaume. A *survey* of the connection between computational geometry and statistical depth is presented in E. Rafalin and D.L. Souvaine.

We are very grateful to all those who where involved in the organization of the conference and this proceedings volume. Our sincere thanks go to the local organizers from the University of Antwerp: Guy Brys, Ellen Vandervieren, Katrien Van Driessen, Martine Van Wouwe, Sabine Verboven and Gert Willems. We highly appreciate their enthusiasm and accuracy. We thank Peter Rousseeuw for his indispensable help and financial support. Michiel Debruyne, Sanne Engelen and Karlien Vanden Branden helped us with the type-setting of this volume. Our thanks go to the members of the scientific committee and many anonymous referees who carefully reviewed the papers in this volume. Their names are listed behind the table of contents. We also express our thanks to the University of Antwerp, the Minerva Foundation at Princeton – New Jersey, the Fund for Scientific Research Flanders, the Belgian Statistical Society, KBC Banking and Insurance and SAS Institute for their generous financial support. Finally, we thank all participants for making ICORS 2003 a successful meeting.

We hope that this book will provide a state-of-the-art overview of robust statistics, and that it will encourage both researchers and practitioners to study and to apply robust statistical methods in the future.

Mia Hubert
Greet Pison
Anja Struyf
Stefan Van Aelst

List of Reviewers

Scientific Committee

Christophe Croux
Rudolf Dutter
Luisa Fernholz
Ursula Gather
Xuming He
Mia Hubert
Jana Jurečková
Ricardo Maronna
Hannu Oja
Ana Pires
Greet Pison
Elvezio Ronchetti
Peter Rousseeuw
Anja Struyf
Stefan Van Aelst
Victor Yohai
Ruben Zamar

Anonymous referees

Jorge Adrover
Claudio Agostinelli
Fatemah Alqallaf
Fraga Alves
Gilbert Bassett
Claudia Becker
José Ramón Berrendero
Ana Bianco
Graciela Boente
Oscar Bustos
Eva Cantoni
Andreas Christmann
Pavel Čížek
Frank Critchley
Robert Fernholz
Chris Field
Joanna Flemming
Ricardo Fraiman
Marc Genton
Gentiane Haesbroeck
Marc Hallin
Christian Hennig
Ping-Hung Hsieh
Georges Hübner
Chris Jones
Kristel Joossens
Annaliisa Kankainen
Alfio Marazzi
Marianthi Markatou
Libor Mašíček
Stephan Morgenthaler
David Olive
Davy Paindaveine
Luke Prendergast
Jan Picek
Matías Salibián
Charles Stewart
Arnold Stromberg
Johan Suykens
Sabine Verboven
Maria-Pia Victoria-Feser
Yijun Zuo

Statistics for Industry and Technology, 1–12

Bias Behavior of the Minimum Volume Ellipsoid Estimate

J.G. Adrover and V.J. Yohai

Abstract. Rousseeuw introduced the Minimum Volume Ellipsoid (MVE) estimates of covariance matrix and multivariate location. These estimates, which are broadly used, are affine equivariant and have high breakdown point. Croux et al. (2002) derived the maximum bias curve for point mass contaminations of a non equivariant version of the MVE. In this paper we obtain a similar result for the equivariant version of the MVE estimates.

Mathematics Subject Classification (2000). Primary 62F35; Secondary 62H12.
Keywords. Minimum volume ellipsoid, maximum bias, robust estimates, multivariate location.

1. Introduction

The first equivariant estimate of multivariate location and covariance matrix with high breakdown point for all p was proposed independently by Stahel (1981) and Donoho (1982). Rousseeuw (1985) proposed the Minimum Volume Ellipsoid (MVE) estimates of multivariate location and scatter. These broadly used robust estimates are defined as the center and scatter of the ellipsoid of minimum volume covering half of the data. These estimates are affine equivariant and have asymptotic breakdown points equal to 0.5.

Let $\mathbf{Z} = \{\mathbf{z}_1, \ldots, \mathbf{z}_n\}$ be a dataset in R^p. For a vector $\mathbf{t}$ and a matrix $V \in \mathcal{P}_p$, where $\mathcal{P}_p$ denotes the set of positive definite $p \times p$ matrices, the squared Mahalanobis distances are defined as

$$d_i^2 = d_i^2(\mathbf{t}, V) = (\mathbf{z}_i - \mathbf{t})' V^{-1} (\mathbf{z}_i - \mathbf{t}). \tag{1.1}$$

Actually, the MVE belongs to a broader family of estimates based on a scale estimate, which are defined as follows. Consider a scale estimate $S_n : R^n \to R_{\geq 0}$,

Partially supported by grant PICT–99 03–06277 from the Agencia Nacional de Promoción de la Ciencia y la Tecnología, grant X611 from the Universidad de Buenos Aires, Argentina and a grant from Fundación Antorchas, Argentina.

($R_{\geq 0}$ is the set of non negative real numbers) satisfying $S_n(\lambda \mathbf{x}) = |\lambda| S_n(\mathbf{x})$. Then the estimates $(\widehat{\mathbf{T}}, \widehat{V}^*)$ are defined by

$$(\widehat{\mathbf{T}}, \widehat{V}^*) = \underset{\mathbf{t}\in R^p,\ V\in\mathcal{P}_p,\ \det(V)=1}{\operatorname{argmin}} S_n(d_1(\mathbf{t}, V), \ldots, d_n(\mathbf{t}, V)). \tag{1.2}$$

The vector $\widehat{\mathbf{T}}$ is the estimate of location and the matrix $\widehat{V}^*$ gives the shape of the covariance matrix. To account also for the size, the estimate of the covariance matrix is defined by

$$\widehat{V} = C_p S_n^2(d_1(\widehat{\mathbf{T}}, \widehat{V}^*), \ldots, d_n(\widehat{\mathbf{T}}, \widehat{V}^*))\widehat{V}^*,$$

where C_p is a constant. Generally C_p is chosen to make the estimate consistent for the multivariate normal family. If $S_n(x_1, \ldots, x_n) = ((x_1^2 + \cdots + x_n^2)/n)^{1/2}$, then $\widehat{\mathbf{T}}$ and $\widehat{V}$ are respectively the sample mean and the sample covariance matrix. If $S_n(x_1, \ldots, x_n) = \text{median}\{|x_1|, \ldots, |x_n|\}$, we have the MVE. Davies (1987) studies the case where S_n is an M-estimate of scale.

If $\mathbf{z}$ is a random vector, we denote by $Q(g(\mathbf{z}), F, \alpha)$ the quantile α of $g(\mathbf{z})$ when $\mathbf{z}$ has distribution F. Then the location and scatter shape MVE estimating functionals $\mathbf{T}(F), V^*(F)$ are defined by

$$(\mathbf{T}(F), V^*(F)) = \underset{\mathbf{t}\in R^p, V\in\mathcal{P}_p, \det(V)=1}{\operatorname{argmin}} Q((\mathbf{z}-\mathbf{t})'V^{-1}(\mathbf{z}-\mathbf{t}), F, 0.5) \tag{1.3}$$

When there are more than one value attaining this minimum, $(\mathbf{T}(F), V^*(F))$ is taken to be the subset of all these values. The objective function $Q(\cdot, F, 0.5)$ evaluated at the solutions given in (1.3) is the appropriate factor to stretch the shape matrix and yield an ellipsoid covering half of the data.

A good measure of the robustness of an estimate is the maximum bias (maxbias) curve, which gives the maximum asymptotic bias of the estimate caused by a given fraction of contamination. Yohai and Maronna (1990) obtained the maxbias curve for the MVE covariance estimate in case the location is known. Croux et al. (2002) considered an orthogonal equivariant version of the MVE, the Minimum Volume Ball (MVB) which is defined as

$$\mathbf{T}(F) = \underset{\mathbf{t}\in R^p,}{\operatorname{argmin}}\, Q((\mathbf{z}-\mathbf{t})'(\mathbf{z}-\mathbf{t}), F, 0.5). \tag{1.4}$$

The authors calculated the maxbias curve of the MVB for point mass contaminations at the normal central model assuming that the covariance matrix is known. Croux and Haesbroeck (2002) derived the maxbias curve of the MVE location estimator with unknown scale parameter in the univariate setting under all types of contamination.

In this paper we derive the point mass maxbias curves for the MVE simultaneous estimates of location and covariance matrix shape assuming an elliptical unimodal and continuous density. These maxbias curves are tabulated for the case of the multivariate normal model.

2. Maxbias Curve for the MVE

Let F_0 be a distribution on R^p and $0 < \varepsilon < 1$, then the point mass contamination neighborhood of size ε is defined by

$$\mathcal{N}_{F_0,\varepsilon} = \{F : \ F = (1-\varepsilon)F_0 + \varepsilon\delta_{\mathbf{z}}, \ \mathbf{z} \in R^p\}.$$

Consider an elliptical family of distributions $F_{\boldsymbol{\mu},\Sigma}$ on R^p, where $\boldsymbol{\mu} \in R^p$ and $\Sigma \in \mathcal{P}_p$. Then $F_{\boldsymbol{\mu},\Sigma}$ has density of the form

$$f(\mathbf{z},\boldsymbol{\mu},\Sigma) = \frac{1}{\det(\Sigma)^{1/2}} g((\mathbf{z}-\boldsymbol{\mu})'\Sigma^{-1}(\mathbf{z}-\boldsymbol{\mu})). \tag{2.1}$$

Let $(\mathbf{T}(F), V^*(F))$ be affine equivariant and Fisher consistent estimating functionals of location and of the shape of the covariance matrix respectively. Then $\mathbf{T}(F_{\boldsymbol{\mu},\Sigma}) = \boldsymbol{\mu}$. The maxbias curves of $\mathbf{T}(F)$ and $V^*(F)$ at $F_{\boldsymbol{\mu},\Sigma}$ for point mass contaminations of size ε are defined by

$$B(\mathbf{T}, F_{\boldsymbol{\mu},\Sigma}, \varepsilon) = \sup\{(\mathbf{T}(F)-\boldsymbol{\mu})'\Sigma^{-1}(\mathbf{T}(F)-\boldsymbol{\mu}) : \ F \in \mathcal{N}_{F_{\boldsymbol{\mu},\Sigma},\varepsilon}\}$$

and

$$B(V^*, F_{\boldsymbol{\mu},\Sigma}, \varepsilon) = \sup\left\{\frac{\lambda_p\left(\Sigma^{-1/2}V^*(F)\Sigma^{-1/2}\right)}{\lambda_1\left(\Sigma^{-1/2}V^*(F)^*\Sigma^{-1/2}\right)} : \ F \in \mathcal{N}_{F_{\boldsymbol{\mu},\Sigma,\varepsilon}}\right\},$$

respectively, where given a positive definite $p \times p$ matrix A, $\lambda_1(A) \leq \lambda_2(A) \leq \cdots \leq \lambda_p(A)$ are its eigenvalues.

Due to the affine equivariance property of $\mathbf{T}$, it is immediate that the maxbias curves of $\mathbf{T}$ and V^* do not depend on $\boldsymbol{\mu}$ and Σ, and therefore we can write

$$B(\mathbf{T}, F_{\mathbf{0},I}, \varepsilon) = \sup\{||\mathbf{T}(F)|| : \ F \in \mathcal{N}_{F_{\mathbf{0},I},\varepsilon}\}$$

and

$$B(V^*, F_{\mathbf{0},I}, \varepsilon) = \sup\left\{\frac{\lambda_p\left(V^*(F)\right)}{\lambda_1\left(V^*(F)\right)} : \ F \in \mathcal{N}_{F_{\mathbf{0},I},\varepsilon\cdot}\right\}.$$

According to this definition, when $(\mathbf{T},V^*)$ is the MVE estimating functional, we have

$$B(\mathbf{T}, F_{\mathbf{0},I}, \varepsilon) = \sup\{||\mathbf{t}_0|| : (\mathbf{t}_0, V_0, \mathbf{z}_0) \in \mathcal{M}\},$$

and

$$B(V^*, F_{\mathbf{0},I}, \varepsilon) = \left\{\sup \frac{\lambda_p\left(V_0\right)}{\lambda_1\left(V_0\right)} : (\mathbf{t}_0, V_0, \mathbf{z}_0) \in \mathcal{M}\right\},$$

where

$$\begin{aligned} \mathcal{M} = & \ \{(\mathbf{t}_0, V_0, \mathbf{z}_0) : \\ (\mathbf{t}_0, V_0) = & \ \underset{\mathbf{t}, V \in \mathcal{P}_p, \det(V)=1}{\operatorname{argmin}} \quad Q((\mathbf{z}-\mathbf{t})\ 'V^{-1}(\mathbf{z}-\mathbf{t}), (1-\varepsilon)F_{\mathbf{0},I} + \varepsilon\delta_{\mathbf{z}_0}, 0.5)\}. \end{aligned}$$

Because of the equivariance of the MVE, it is easy to show that without loss of generality we can restrict these minima to points $\mathbf{z}_0$ of the form $(z_0, 0, \ldots, 0)$. Then it is possible to prove that we can also restrict $\mathbf{t}$ to points of the form $(\mu, 0, \ldots, 0)$ and V^* to matrices of the form $\text{diag}(v, v^{-1/(p-1)}, \ldots, v^{-1/(p-1)})$, where $v > 0$ (see

Adrover and Yohai, 2003). In this case we can write $(\mathbf{z}-\mathbf{t})'V^{-1}(\mathbf{z}-\mathbf{t}) = h(\mathbf{z}, v, \mu)$, where

$$h(\mathbf{z}, v, \mu) = \frac{(z_1 - \mu)^2}{v} + v^{1/(p-1)} \sum_{i=2}^{p} z_i^2.$$

Then we can also write

$$B(\mathbf{T}, F_{\mathbf{0},I}, \varepsilon) = \max\{\mu_0 : (\mu_0, v_0, z_0) \in M\}$$

and

$$B(V^*, F_{\mathbf{0},I}, \varepsilon) = \max\{v_0^{p/(p-1)} : (\mu_0, v_0, z_0) \in M\},$$

where

$$\begin{aligned} M &= \{(\mu_0, v_0, z_0) : \mu_0 \geq 0,\ v_0 > 0,\ z_0 \geq 0, \\ (\mu_0, v_0) &= \operatorname*{argmin}_{\mu, v} Q(h(\mathbf{z}, v, \mu), (1-\varepsilon)F_{\mathbf{0},I} + \varepsilon\delta_{(z_0,0,\ldots,0)}, 0.5)\}. \end{aligned}$$

From now on, ε is going to be fixed, then we are not going to make explicit the dependence on ε of some of the functions we define.

Put $\varepsilon_1 = (0.5-\varepsilon)/(1-\varepsilon)$, $\varepsilon_2 = 0.5/(1-\varepsilon)$ and $c_i = Q(||\mathbf{z}||^2, F_{\mathbf{0},I}, \varepsilon_i)$, $i = 1, 2$. Let

$$m(v, \mu) = Q(h(\mathbf{z}, v, \mu), F_{\mathbf{0},I}, \varepsilon_1).$$

Then we have the following lemma.

Lemma 2.1. *Assume that the function g in (2.1) is strictly decreasing and continuous. Then*

1. *The function $m(v, \mu)$ is continuous.*
2. *The unique minimum of $m(v, \mu)$ is attained at $v = 1$ and $\mu = 0$.*
3. *The function $m(v, \mu)$ is strictly increasing on μ for fixed v and $\mu \geq 0$. We also have $\lim_{\mu\to\infty} m(v, \mu) = \infty$.*
4. *The function $m(v, 0)$ is strictly increasing on v for $v \geq 1$. We also have $\lim_{v\to\infty} m(v, 0) = \infty$.*
5. *For any $c_1 \leq c \leq c_2$ there exists a unique $\overline{v}(c) \geq 1$ such that $m(\overline{v}(c), 0) = c$. The function $\overline{v}(c)$ is strictly increasing and continuous and $\overline{v}(c_1) = 1$.*
6. *For any $c_1 \leq c \leq c_2$ and $1 \leq v \leq \overline{v}(c)$, there exists a unique value $l(c, v) \geq 0$ such that $m(v, l(c, v)) = c$. The function $l(c, v)$ is continuous and strictly increasing on c. Put $v_2 = \overline{v}(c_2)$, then $l(c_2, v_2) = 0 = l(c_1, 1)$.*
7. *Define for any c, $c_1 \leq c \leq c_2$, the values $r^*(c)$, $v^*(c)$ and $\mu^*(c)$ by*

$$\begin{aligned} v^*(c) &= \operatorname*{argmax}_{1\leq v\leq \overline{v}(c)} l(c, v) + (cv)^{1/2}, \\ r^*(c) &= l(c, v^*(c)) + (cv^*(c))^{1/2}, \\ \mu^*(c) &= l(c, v^*(c)). \end{aligned}$$

Then, the function $r^*(c)$ is strictly increasing and continuous.

Proof (1) We can write

$$P(h(\mathbf{z},v,\mu) \leq \delta) = \int_{C_\delta} g\left(v(y_1+\mu)^2 + \frac{1}{v^{1/(p-1)}}\sum_{i=2}^{p} y_i^2\right) d\mathbf{y},$$

where $C_\delta = \{\mathbf{y} \in R^p: ||\mathbf{y}|| \leq \delta\}$. Then part (1) of the Lemma follows from the Dominated Convergence Theorem.

(2) It is enough to show that for any $(v,\mu) \neq (1,0)$ and $\delta > 0$

$$P(h(\mathbf{z},v,\mu) \leq \delta) < P(h(\mathbf{z},1,0) \leq \delta). \tag{2.2}$$

Let $D_0 = \{\mathbf{z} \in R^p: ||\mathbf{z}||^2 \leq \delta\}$ and

$$D_1 = \{\mathbf{z} \in R^p: (z_1-\mu)^2/v + v^{1/(p-1)}\sum_{i=2}^{p} z_i^2 \leq \delta\}.$$

Put $E_0 = D_0 - D_0 \cap D_1$ and $E_1 = D_1 - D_0 \cap D_1$. Then, since

$$\int_{D_0} d\mathbf{z} = \int_{D_1} d\mathbf{z},$$

we also have

$$\int_{E_0} d\mathbf{z} = \int_{E_1} d\mathbf{z}. \tag{2.3}$$

Since g is strictly decreasing, we have

$$\begin{aligned} P(h(\mathbf{z},1,0) \leq \delta) &= \int_{D_0\cap D_1} g(||\mathbf{z}||^2)d\mathbf{z} + \int_{E_0} g(||\mathbf{z}||^2)d\mathbf{z} \\ &> \int_{D_0\cap D_1} g(||\mathbf{z}||^2)d\mathbf{z} + g(\delta^2)\int_{E_0} d\mathbf{z} \end{aligned} \tag{2.4}$$

and

$$\begin{aligned} P(h(\mathbf{z},v,\mu) \leq \delta) &= \int_{D_0\cap D_1} g(||\mathbf{z}||^2)d\mathbf{z} + \int_{E_1} g(||\mathbf{z}||^2)d\mathbf{z} \\ &< \int_{D_0\cap D_1} g(||\mathbf{z}||^2)d\mathbf{z} + g(\delta^2)\int_{E_1} d\mathbf{z}. \end{aligned} \tag{2.5}$$

From (2.3), (2.4) and (2.5) we get (2.2), and part (2) of the Lemma follows.

(3) Take $\nu \geq 1$ and $\mu_0 < \mu_1$. Put $D_i = \{\mathbf{z} \in R^p: h(\mathbf{z},v,\mu_i) \leq \delta\}$, $i = 0,1$ and $E_i = D_i - D_0 \cap D_1$, $i = 0,1$. Then to prove that $m(v,\mu)$ is strictly increasing on μ, it will be enough to show that

$$\int_{E_0} g(||\mathbf{z}||^2)d\mathbf{z} > \int_{E_1} g(||\mathbf{z}||^2)d\mathbf{z}. \tag{2.6}$$

Let $\bar{\mu} = (\mu_0 + \mu_1)/2$, then

$$\mathbf{z} = (z_1, \ldots, z_p) \in E_0 \Rightarrow z_1 < \bar{\mu} \tag{2.7}$$

and

$$\mathbf{z} = (z_1, \ldots, z_p) \in E_1 \Rightarrow z_1 > \bar{\mu}. \tag{2.8}$$

Consider the transformation $\kappa(z_1,\ldots,z_p)=(2\bar{\mu}-z_1,z_2,\ldots,z_p)$. Then it is easy to show that $\kappa(E_0)=E_1$. Moreover by (2.7), (2.8) and the fact that g is strictly decreasing we obtain

$$g(\|k(\mathbf{z})\|^2)<g(\|\mathbf{z}\|^2)\ \forall\mathbf{z}\in E_0. \tag{2.9}$$

Since the Jacobian of κ has absolute value one, using the change of variable formula in multiple integrals and (2.9), we get

$$\int_{E_1} g(\|\mathbf{z}\|^2)d\mathbf{z}=\int_{E_0} g(\|\kappa(\mathbf{z})\|^2)d\mathbf{z}<\int_{E_0} g(\|\mathbf{z}\|^2)d\mathbf{z},$$

and therefore (2.6) holds. The proof of $\lim_{\mu\to\infty} m(v,\mu)=\infty$ is immediate from the fact that $H(\mathbf{z},v,\mu)\to\infty$ in probability when $\mu\to\infty$.

(4) The fact that $m(v,0)$ is strictly increasing in v for $v>1$ follows from the Lemma in Section 3 of Yohai and Maronna (1990). The proof of $\lim_{v\to\infty} m(v,0)=\infty$ is immediate from the fact that $H(\mathbf{z},v,0)\to\infty$ in probability when $v\to\infty$.

(5) From the definition of $\overline{v}(c)$, it follows that $\overline{v}(c_1)=1$. Then, since $m(1,0)=c_1<c$ and $\lim_{v\to\infty} m(v,\mu)=\infty$, the continuity of m implies that there exists $\overline{v}(c)$ satisfying $m(\overline{v}(c),0)=c$. Part (4) of the Lemma implies that $\overline{v}(c)$ is strictly increasing. The continuity of $\overline{v}(c)$ follows easily from the continuity of m.

(6) The existence and uniqueness of $l(c,v)$ follows from the continuity of m, the facts that $m(v,0)<c$ and $\lim_{\mu\to\infty} m(v,0)=\infty$. Part (3) of the Lemma implies that $l(c,v)$ is strictly increasing on c. To show the continuity of l, consider a sequence $(c_n,v_n)\to(c,v)$. Suppose that $l(c_n,v_n)\to l(c,v)$ does not hold, then without loss of generality we can assume that

$$l(c_n,v_n)\to l_0\neq l(c,v). \tag{2.10}$$

However by the continuity of m we have

$$\lim_{n\to\infty} m(v_n,l(c_n,v_n))=c=m(v,l_0).$$

Then $l_0=l(c,v)$, contradicting (2.10).

(7) Take $c<c^*$. Then, since $v^*(c)\le\bar{v}(c)\le\bar{v}(c^*)$ by part (6) of the Lemma we have

$$\begin{aligned} r^*(c^*)&\ge l(c^*,v^*(c))+(c^*v^*(c))^{1/2}\\ &>l(c,v^*(c))+(cv^*(c))^{1/2}=r^*(c), \end{aligned}$$

and therefore $r^*(c)$ is strictly increasing.

We now show that r^* is continuous at any point c. Suppose this is not true, then there exists a sequence c_n such that $c_n\to c$ and $v^*(c_n)\to v_0\neq v^*(c)$. Then

$$l(c,v_0)+(cv_0)^{1/2}<l(c,v^*(c))+(cv^*(c))^{1/2}. \tag{2.11}$$

On the other hand, by the continuity of $l(c,v)$, we have

$$l(c,v_0)+(cv_0)^{1/2}=\lim_{n\to\infty} l(c_n,v^*(c_n))+(c_nv^*(c_n))^{1/2} \tag{2.12}$$

and

$$l(c,v^*(c))+(cv^*(c))^{1/2}=\lim_{n\to\infty} l(c_n,v^*(c))+(c_nv^*(c))^{1/2}. \tag{2.13}$$

From (2.11), (2.12) and (2.13) we obtain that there exists n_0 such that

$$l(c_{n_0}, v^*(c_{n_0})) + (c_{n_0} v^*(c_{n_0}))^{1/2} < l(c_{n_0}, v^*(c)) + (c_{n_0} v^*(c))^{1/2},$$

and this contradicts the definition of $v^*(c_{n_0})$.

The next lemma gives the median of any statistics under a point mass contaminated distribution.

Lemma 2.2. *Given a distribution G on R^p, and a Borel measurable function $t : R^p \to R$, we have*

$$Q(t(\mathbf{z}), F_{\mathbf{0},I}, \varepsilon_1) \leq Q(t(\mathbf{z}), (1-\varepsilon)F_{\mathbf{0},I} + \varepsilon G, 0.5) \leq Q(t(\mathbf{z}), F_{\mathbf{0},I}, \varepsilon_2).$$

Moreover

$$Q(t(\mathbf{z}), (1-\varepsilon)F_{\mathbf{0},I} + \varepsilon\delta_{\mathbf{z}_0}, 0.5) = \begin{cases} Q(t(\mathbf{z}), F_{\mathbf{0},I}, \varepsilon_1) & \text{if } t(\mathbf{z}_0) < Q(t(\mathbf{z}), F_{\mathbf{0},I}, \varepsilon_1) \\ t(\mathbf{z}_0) & \text{if } Q(t(\mathbf{z}), F_{\mathbf{0},I}, \varepsilon_1) \leq t(\mathbf{z}_0) \leq Q(t(\mathbf{z}), F_{\mathbf{0},I}, \varepsilon_2) \\ Q(t(\mathbf{z}), F_{\mathbf{0},I}, \varepsilon_2) & \text{if } t(\mathbf{z}_0) > Q(t(\mathbf{z}), F_{\mathbf{0},I}, \varepsilon_2). \end{cases}$$

In the following theorem we see that the location MVE estimate turns out to be the origin for point mass sufficiently large or small. In any other case, the location estimate under contamination coincides with the center of the ellipsoid whose contour contains the point mass.

Theorem 2.3. *Assume that the function g in (2.1) is strictly decreasing and continuous, and let $(\mathbf{T}, V^*)$ be the MVE estimates of location and covariance matrix shape. Then,*

(i) *Let $k_i = r^*(c_i)$, $i = 1, 2$, then, for any $k_1 < k < k_2$ we have*

$$\begin{array}{l} \mathbf{T}((1-\varepsilon)F_{\mathbf{0},I} + \varepsilon\delta_{(k,0,\ldots,0)}) = (\mu^*(r^{*-1}(k)), 0, \ldots, 0) \\ V^*((1-\varepsilon)F_{\mathbf{0},I} + \varepsilon\delta_{(k,0,\ldots,0)}) = \\ \operatorname{diag}(v^*(r^{*-1}(k)), v^*(r^{*-1}(k))^{-1/(p-1)}, \ldots, v^*(r^{*-1}(k))^{-1/(p-1)}). \end{array} \tag{2.14}$$

(ii) *If $k > k_2$ or $k \leq k_1$*

$$\begin{array}{ll} \mathbf{T}((1-\varepsilon)F_{\mathbf{0},I} + \varepsilon\delta_{(k,0,\ldots,0)}) & = (0, 0, \ldots, 0) \\ V^*((1-\varepsilon)F_{\mathbf{0},I} + \varepsilon\delta_{(k,0,\ldots,0)}) & = I. \end{array} \tag{2.15}$$

(iii) *For $k = k_2$ there are two solutions for $(\mathbf{T}((1-\varepsilon)F_{\mathbf{0},I} + \varepsilon\delta_{(k,0,\ldots,0)}), V^*((1-\varepsilon)F_{\mathbf{0},I} + \varepsilon\delta_{(k,0,\ldots,0)}))$, the ones given in (2.14) and (2.15).*

Proof. We start proving (i). Put $c = r^{*-1}(k)$. Then

$$Q(h(\mathbf{z}, v^*(c), \mu^*(c)), F_{\mathbf{0},I}, \varepsilon_1) = m(v^*(c), \mu^*(c)) = c$$

and

$$h((k, 0, \ldots, 0), v^*(c), \mu^*(c)) = c.$$

Then, according to Lemma 2.2 we have

$$Q(h(\mathbf{z}, v^*(c), \mu^*(c)), (1-\varepsilon)F_{\mathbf{0},I} + \varepsilon\delta_{(k,0,\ldots,0)}, 0.5) = c. \tag{2.16}$$

Now we prove that if $(\mu, v) \neq (\mu^*(c), v^*(c))$, then

$$Q(h(\mathbf{z}, v, \mu), (1-\varepsilon)F_{\mathbf{0},I} + \varepsilon\delta_{(k,0,\ldots,0)}, 0.5) > c. \tag{2.17}$$

Suppose first that $\mu > l(c,v)$. Then

$$Q(h(\mathbf{z},v,\mu),F_{\mathbf{0},I},\varepsilon_1) = m(v,\mu) > m(v,l(c,v)) = c,$$

and by Lemma 2.2, (2.17) holds. Suppose now that $\mu \leq l(c,v)$. In this case

$$k = l(c,v^*(c)) + (cv^*(c))^{1/2} > l(c,v) + (cv)^{1/2} \geq \mu + (cv)^{1/2},$$

and therefore

$$h((k,0,\ldots,0),v,\mu) = \frac{(k-\mu)^2}{v} > c. \tag{2.18}$$

We also have

$$Q(h(\mathbf{z},v,\mu),F_{\mathbf{0},I},\varepsilon_2) \geq Q(h(\mathbf{z},1,0),F_{\mathbf{0},I},\varepsilon_2) = c_2 > c. \tag{2.19}$$

(2.18) and (2.19) imply (2.16). Then (i) is proved.

Take $k > k_2$. Observe first that $Q(h(\mathbf{z},1,0),F_{\mathbf{0},I},\varepsilon_2) = c_2$ and therefore, because of Lemma 2.2, for any k

$$Q(h(\mathbf{z},1,0),(1-\varepsilon)F_{\mathbf{0},I} + \varepsilon\delta_{(k,0,\ldots,0)},0.5) \leq c_2. \tag{2.20}$$

Given any $(\mu,v) \neq (0,1)$, put $c = m(v,\mu)$. There are three possible cases (a) $v > v_2$, (b) $v \leq v_2$, $\mu > l(c_2,v)$ and (c) $v \leq v_2$, $\mu \leq l(c_2,v)$. In cases (a) and (b) we have $m(v,\mu) > c_2$ and therefore by Lemma 2.2

$$Q(h(\mathbf{z},v,\mu),(1-\varepsilon)F_{\mathbf{0},I} + \varepsilon\delta_{(k,0,\ldots,0)},0.5) > c_2. \tag{2.21}$$

In case (c),

$$k > l(c_2,v^*(c_2)) + (c_2v^*(c_2))^{1/2} \geq l(c_2,v) + (c_2v)^{1/2} \geq \mu + (c_2v)^{1/2},$$

and therefore

$$h((k,0,\ldots,0),v,\mu) = \frac{(k-\mu)^2}{v} > c_2. \tag{2.22}$$

Since $Q(h(\mathbf{z},v,\mu),F_{\mathbf{0},I},\varepsilon_2) > c_2$, using Lemma 2.1 and (2.22) we get once again (2.21). This shows (ii) when $k > k_2$.

Finally consider the case $k \leq k_1$. Since $Q(h(\mathbf{z},1,0),F_{\mathbf{0},I},\varepsilon_1) = c_1$, $k_1 = c_1^{1/2}$ and

$$h((k,0,\ldots,0),1,0) = k^2 \leq k_1^2 = c_1, \tag{2.23}$$

by Lemma 2.2 we have

$$Q(h(\mathbf{z},1,0),(1-\varepsilon)F_{\mathbf{0},I} + \varepsilon\delta_{(k,0,\ldots,0)},0.5) \leq c_1. \tag{2.24}$$

Let now $(\mu,v) \neq (0,1)$, then we have $Q(h(\mathbf{z},v,\mu),F_{\mathbf{0},I},\varepsilon_1) > c_1$, and by Lemma 2.2

$$Q(h(\mathbf{z},v,\mu),(1-\varepsilon)F_{\mathbf{0},I} + \varepsilon\delta_{(k,0,\ldots,0)},0.5) > c_1. \tag{2.25}$$

Then, by (2.24) and (2.25) we conclude that (ii) holds for $k \leq k_1$. (iii) is proved combining the proofs of (i) and (ii).

Corollary 2.4. *Assume that the function g in (2.1) is strictly decreasing and continuous and let* $(\mathbf{T}, V^*)$ *be the MVE estimates of location and covariance matrix shape. Then*

$$B(\mathbf{T}, F_{\mathbf{0},I}, \varepsilon) = \sup_{c_1 \leq c \leq c_2} \mu^*(c).$$

and

$$B(V^*, F_{\mathbf{0},I}, \varepsilon) = \sup_{c_1 \leq c \leq c_2} v^*(c)^{p/p-1}.$$

Remark 2.5. In the normal case there is numerical evidence that μ^* and v^* are strictly monotone. Therefore

$$B(\mathbf{T}, F_{\mathbf{0},I}, \varepsilon) = \mu^*(c_2) \tag{2.26}$$

and

$$B(V^*, F_{\mathbf{0},I}, \varepsilon) = v^*(c_2)^{p/(p-1)}.$$

Remark 2.6. The derivation of Lemma 2.1 and Theorem 2.3 relies on the fact of having a unique maximum for $l(c, v) + (cv)^{1/2}$ for each $c \in [c_1, c_2]$. By these means, $v^*(c)$ is well defined. We have not dealt with the case of multiple values achieving the maximum to avoid a relentlessly technical presentation of the problem.

Remark 2.7. Croux and Haesbroeck (2002) studied the maxbias curve for the MVE estimate of location in the univariate setting. If $F = F_{0,1}$ and $H_\varepsilon^+(x)$ and $H_\varepsilon^-(x)$ stand for the solutions s of the equations

$$\begin{aligned} F(x+s) - F(x-s) &= \frac{1}{2(1-\varepsilon)} \\ F(x+s) - F(x-s) &= \frac{1-2\epsilon}{2(1-\varepsilon)} \end{aligned}$$

respectively, then Proposition 3 of Croux and Haesbroeck (2002) states that the maximum bias of the univariate MVE estimate of location is the solution b of the equation

$$H_\varepsilon^-(b) = H_\varepsilon^+(0). \tag{2.27}$$

The authors emphasize that an interesting feature of the maxbias of the MVE location estimator is that the most unfavorable contaminating distribution is a point mass distribution δ_z with z not at infinity but much closer to the center of the distribution. Actually, Corollary 2.4 and (2.26) amounts to (2.27). To see that, let us observe that the formulation for R^1 is much more simple because $v = 1$. Then, the dependence on v of the functions $m(v, \mu)$, $l(c, v)$ and $h(z, v, \mu)$ can be omitted. Consequently, $\mu^*(c_2) = l(c_2)$ and

$$\begin{aligned} &[H_\varepsilon^+(0)]^2 = Q(z^2, F, \varepsilon_2) = c_2 = m(l(c_2)) \\ &= Q(h(z, l(c_2)), F, \varepsilon_1) = [H_\varepsilon^-(l(c_2))]^2. \end{aligned}$$

In Table 1 and Figure 1 we show the maxbias curves of the MVE estimate of multivariate location and in Table 2 and Figure 2 those of the covariance shape. In Table 3 we show the values of z_0 such that $F = (1-\varepsilon)F_{\mathbf{0},I} + \varepsilon\delta_{(z_0,0,\ldots,0)}$ is the least favorable point mass contaminated distribution. For comparison with other estimates, see Table 2 of Adrover and Yohai (2002). Adrover and Yohai (2002) observed that the maximum biases of the MVE and the Stahel-Donoho estimate (1982) increase with the dimension in the location case. The bias performance of MVE compares favorably with respect to the Minimum Covariance Determinant (Rousseeuw, 1985) although its performance is rather poor compared to the projection estimate (Tyler, 1994).

TABLE 1. Maximum biases of the MVE location estimate.

p	ε				
	0.01	0.05	0.10	0.15	0.20
2	0.17	0.40	0.63	0.91	1.29
3	0.17	0.41	0.69	1.03	1.48
4	0.17	0.42	0.72	1.08	1.57
5	0.16	0.42	0.73	1.12	1.65
6	0.16	0.42	0.74	1.14	1.71
7	0.16	0.42	0.74	1.16	1.78
8	0.15	0.41	0.75	1.18	1.84
9	0.15	0.42	0.75	1.20	1.90
10	0.14	0.41	0.75	1.22	1.97
15	0.14	0.40	0.76	1.31	2.30
20	0.13	0.39	0.77	1.41	2.65

TABLE 2. Maximum biases of the MVE covariance shape estimate.

p	ε				
	0.01	0.05	0.10	0.15	0.20
2	1.36	4.45	9.49	18.66	37.33
3	1.43	3.58	6.86	12.62	24.09
4	1.47	3.45	6.65	12.56	24.94
5	1.51	3.48	6.90	13.54	28.12
6	1.55	3.57	7.31	14.90	32.37
7	1.57	3.67	7.84	16.47	37.35
8	1.60	3.81	8.30	18.21	43.00
9	1.63	3.92	8.84	20.07	49.32
10	1.66	4.86	9.41	22.06	56.31
15	1.76	4.71	12.45	33.83	102.63
20	1.85	5.37	15.04	48.90	172.45

TABLE 3. Least favorable z_0.

p	ε				
	0.01	.05	0.10	0.15	0.20
2	1.55	2.18	2.86	3.68	4.75
3	2.01	2.91	3.80	4.98	6.37
4	2.40	3.41	4.65	6.22	8.45
5	2.73	3.93	5.46	7.48	10.43
6	3.04	4.43	6.26	8.75	11.95
7	3.33	4.90	6.75	10.05	14.74
8	3.59	5.36	7.84	11.39	17.06
9	3.85	5.81	8.62	12.75	19.52
10	4.09	6.24	9.40	14.15	22.11
15	5.17	8.30	13.36	21.71	37.14
20	6.12	10.27	17.47	30.31	56.04

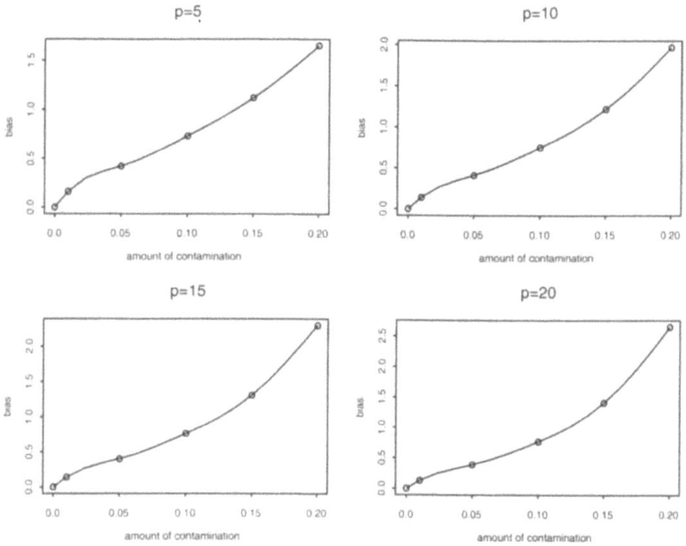

FIGURE 1. Maxbias curves for the MVE location estimate.

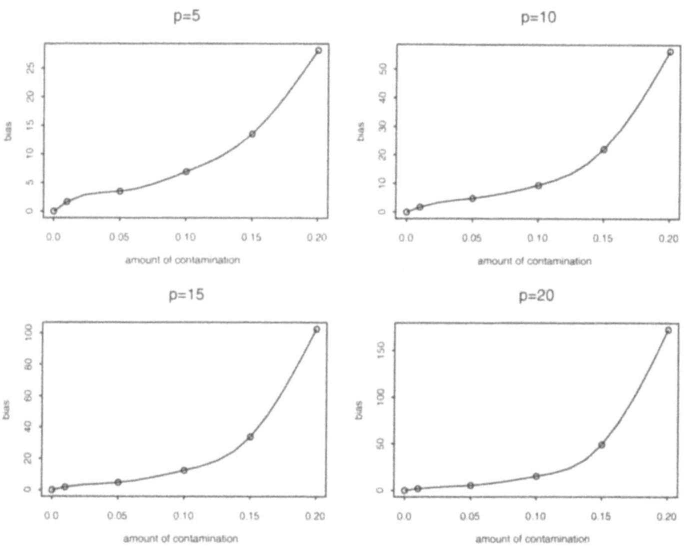

FIGURE 2. Maxbias curves for the MVE covariance shape estimate.

References

[1] J.G. Adrover and V.J. Yohai, *Projection Estimates of Multivariate Location.* Ann. Statist. **30** (2002), 1760–1781.

[2] J.G. Adrover and V.J. Yohai, *Bias Behavior of the Minimum Volume Ellipsoid Estimate.* Technical report, available at *http://mate.dm.uba.ar/~vyohai/mvebias.pdf.*

[3] C. Croux and G. Haesbroeck, *Maxbias Curves of Location Estimators based on Subranges.* J. Nonparametr. Stat. **14** (2002), 295–306.

[4] C. Croux, G. Haesbroeck, and P.J. Rousseeuw, *Location Adjustment for the Minimum Volume Ellipsoid Estimator.* Stat. Comput. **12** (2002), 191–200.

[5] P.L. Davies, *Asymptotic Behavior of S-Estimates of Multivariate Location Parameters and Dispersion matrices.* Ann. Statist. **15** (1987), 1269–1292.

[6] D.L. Donoho, *Breakdown Properties of Multivariate Location Estimators.* Ph.D. Thesis, Dept. of Statistics, Harvard University, 1982.

[7] P.J. Rousseeuw, *Multivariate Estimation With High Breakdown Point.* In W. Grossmann, G. Pflug, I. Vincze, and W. Wertz, editors, Mathematical Statistics and Applications, pages 283–297. Reidel Publishing Company, Dordrecht, 1985.

[8] W.A. Stahel, *Breakdown of Covariance Estimators.* Research Report 31, Fachgruppe für Statistik, ETH, Zürich, 1981.

[9] D.E. Tyler, *Finite Sample Breakdown Points of Projection Based Multivariate Location and Scatter Statistics.* Ann. Statist. **22** (1994), 1024–1044.

[10] V.J. Yohai and R.A. Maronna, *The Maximum Bias of Robust Covariances.* Comm. Statist. Theory and Methods **19** (1990), 3925–3933.

J.G. Adrover
Facultad de Matemática, Astronomía y Física and CIEM
Universidad Nacional de Córdoba
Avda. Medina Allende s/n
CIUDAD UNIVERSITARIA
(5000) Córdoba
Argentina
e-mail: `adrover@mate.uncor.edu`

V.J. Yohai
Departamento de Matemática
Facultad de Ciencias Exactas y Naturales
Ciudad Universitaria, Pabellon 1
1428 Buenos Aires
Argentina
e-mail: `vyohai@uolsinectis.com.ar`

Statistics for Industry and Technology, 13–25

A Study of Belgian Inflation, Relative Prices and Nominal Rigidities using New Robust Measures of Skewness and Tail Weight

L. Aucremanne, G. Brys, M. Hubert, P.J. Rousseeuw and A. Struyf

Abstract. This paper studies the distribution of Belgian consumer price changes and its interaction with aggregate inflation over the period June 1976–September 2000. Given the fat-tailed nature of this distribution, both classical and robust measures of location, scale, skewness and tail weight are presented. The chronic right skewness of the distribution, revealed by the robust measures, is cointegrated with aggregate inflation, suggesting that it is largely dependent on the inflationary process itself and would disappear at zero inflation.

Mathematics Subject Classification (2000). Primary 91B84; Secondary 62G32.
Keywords. Belgian inflation data, robustness, skewness, tail weight.

1. Introduction

The aim of this paper is to study the cross-sectional properties of Belgian inflation data. The international literature on this issue highlights, broadly speaking, three not mutually exclusive aspects. First, many papers discuss the positive relationship between inflation on the one hand and the dispersion and/or the asymmetry of relative prices on the other. Often these relationships have been interpreted as being symptomatic of nominal rigidities of one form or another. Ball and Mankiw's menu cost model (Ball and Mankiw, 1995) has recently been the focal point of this strand in the literature. Second, in line with the findings of Bryan et al. (1997), it is often documented that inflation data are fat-tailed and this motivates the stochastic approach to core inflation in which robust estimators of location are proposed. Third, in quite a few countries researchers found not only a considerable degree of frequently switching left and right skewness, but, on average, also a tendency towards right skewness.

The last author is Postdoctoral Fellow of the Fund for Scientific Research – Flanders (Belgium).

This type of analysis often uses the classical characteristics of location, scale, skewness and kurtosis. These measures are, however, very sensitive to outlying values (Ruppert, 1987). In this paper we will compare them with robust alternatives, not only for location, as is typically done in the core inflation part of this literature, but also for scale, skewness and tail weight. We will address the question whether robust measures are applicable in this context, because inflation is often strongly influenced by outliers which are both correct and important. Overly strong robustness may downweight outliers too much and thus yield little sensitivity. Next, we study the relation between skewness and inflation, and interpret the results in the framework of menu cost models (Ball and Mankiw, 1994).

2. Description of the Data

This study is based on Belgian monthly consumer price index (CPI) data for the period June 1976 up to September 2000, yielding a total of 292 months. It was not possible to start at an earlier point in time, for instance during the low inflation regime of the sixties, for data availability reasons. For each month, we have aggregated price indices of 60 different product categories such as meat, clothing, tobacco, electricity, recreational and cultural services, etc. The index of product category i $(i = 1, \ldots, 60)$ at month t $(t = 0, \ldots, 291)$ is denoted by $I_{i,t}$.

We started by transforming these data into percentage 1-monthly price changes $\Pi_{i,t}$ $(t = 1, \ldots, 291)$, defined by $\Pi_{i,t} = (I_{i,t}/I_{i,t-1}) - 1$. For a motivation and more information about selecting $\Pi_{i,t}$ instead of any other transformation of the raw data, we refer to Aucremanne et al. (2002).

Summarizing, we can represent our final dataset with a matrix that consists of 291 rows (indicating the different periods) and 60 columns (for the product categories). A cross-section contains the price changes of one particular month, and thus corresponds with one row of the data matrix. Additionally, we have taken account of the fact that each product category has a time-varying weight $w_{i,t}$ which is obtained by multiplying for each period t its fixed Laspeyres-type weight w_i by the change in its relative price between the base period and period t. The Laspeyres-type weights w_i of the CPI reflect the importance of each category in total household consumption expenditure in the base period. The weights satisfy $\sum_{i=1}^{60} w_{i,t} = 1$ for all t. In so doing, the weighted mean of the 60 product-specific price changes $\Pi_{i,t}$ corresponds to aggregate inflation.

3. Detection of Outliers

In this section we will mainly show that the Belgian price changes contain a substantial number of outlying values. For this, we will consider the cross-sections and compute robust measures of tail weight for each of them. To show the importance of the tails of univariate data (as the cross-sections are), we can use the classical measure of kurtosis. In general, the kurtosis is said to characterize the fatness of the tails, or equivalently the tail weight, but it also reflects the shape of the density in the center. Moreover, the classical measure of kurtosis is based on moments of the dataset, and thus it is strongly influenced by outliers. Despite the fact that outliers determine the tail weight, we can construct robust measures to compute it. For this purpose, we first need a robust estimator of location and scale for univariate data. In general, we denote the sample by x_i $(i = 1, \ldots, n)$ and the corresponding weights by w_i. As usual, the weights sum to 1. Time subscripts are omitted to simplify notations.

3.1. Robust Measure of Location

A measure of location should estimate a value that characterizes the central position of the data. The best known measure of location is the weighted mean or average, defined as $\overline{x} = \sum_{i=1}^{n} w_i x_i$. Note that with the dataset we analyze, the weighted mean corresponds to observed inflation.

A typical robust measure of location is the median. Here, we need the weighted variant of the median. In general, all measures based on percentiles can be modified easily to their weighted version by filling in the weighted percentiles. That is the main reason why all robust measures presented in this paper are based on percentiles. Initial work on other robust measures based on couples and triples of observations was abandoned precisely because it was not straightforward to construct them in weighted terms. Indeed, Brys et al. (2003) and Brys et al. (2004) discuss some unweighted robust skewness measures based on couples and triples. For the $p\%$ weighted percentile, denoted by Q_p, we first sort the observations from smallest to largest, and then take the first value with cumulative weight higher than $p\%$. The weighted median equals the 50% weighted percentile, or $Q_{0.50}$.

3.2. Robust Measure of Scale

Scale characteristics measure how "spread out" the data values are. The classical measure of scale is the standard deviation, which is given by:

$$s = \sqrt{\frac{\sum_{i=1}^{n} w_i (x_i - \overline{x})^2}{1 - \sum_{i=1}^{n} w_i^2}} \tag{3.1}$$

where $\overline{x}$ stands for the mean. This equation is based on the unbiased sample moments of unequally-weighted frequency distributions presented in Roger (2000).

The interquartile range or IQR is a robust measure of scale which, like the median, is based on percentiles and is easy to calculate. The (weighted) IQR is the distance between the 75% percentile $Q_{0.75}$ and the 25% percentile $Q_{0.25}$, or

$$IQR = Q_{0.75} - Q_{0.25} \tag{3.2}$$

3.3. Robust Measures of Tail Weight

We propose several tailmass measures as alternative to the classical kurtosis. They simply count the outliers in a data set. To distinguish the outlying values from the regular ones we use the following outlier rules:

1. The *general boxplot rejection rule* defines outliers as points outside the interval:

$$[Q_{0.50} - \frac{3}{2}IQR, Q_{0.50} + \frac{3}{2}IQR] \tag{3.3}$$

This criterion is based on the definition of the whiskers in the univariate boxplot (Tukey, 1977). Note that this interval corresponds approximately to $[\overline{x} - 2s, \overline{x} + 2s]$ in case of the normal distribution.

2. The *asymmetric boxplot rule.* This rule is a special case of the bivariate bagplot (Rousseeuw et al., 1999), which is a bivariate generalization of the boxplot. Points are considered outlying when they lie outside the interval:

$$[Q_{0.50} - 3(Q_{0.50} - Q_{0.25}), Q_{0.50} + 3(Q_{0.75} - Q_{0.50})] \tag{3.4}$$

For each outlier rule, we define left outliers as those data points which are smaller than the lower bound of the interval defined in (3.3) or (3.4). Analogously the right outliers are larger than the upper bound of this interval. We then obtain the left-tailmass measures as the sum of the weights of the left outliers, i.e.,

$$\text{left-tailmass} = \sum_{i=1}^{n} w_i \theta_i \tag{3.5}$$

where θ_i equals 1 if x_i is a left outlier and 0 if not. This leads to left-tailmass(box) and left-tailmass(abp). Considering the right outliers, we obtain right-tailmass (box) and right-tailmass(abp).

3.4. Results for Belgian Inflation Data

All measures defined above were applied to the cross-sections of the Belgian inflation data. In Figure 1 the resulting time series are plotted, by considering all 291 consecutive cross-sections. On the plots, the scatter points of the different measures are depicted, together with a smoother (solid line). We have chosen to use a lowess smoother, which is a robust scatter plot smoother (Cleveland, 1979). Lowess uses robust locally linear fits, by placing a window about each point and weighting the points that are inside the window so that nearby points get the most weight.

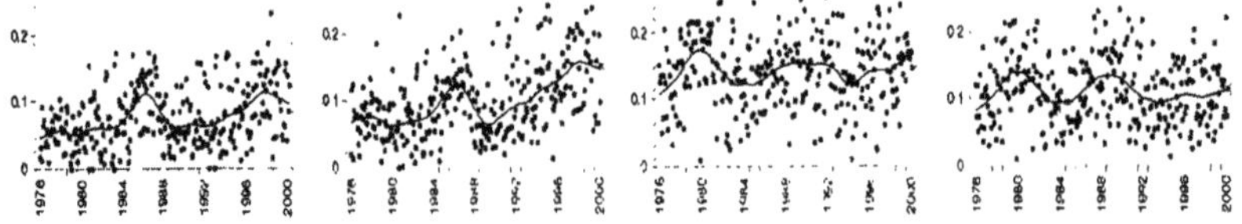

FIGURE 1. Time series of the tail weight measures.

All the tailmass measures show substantial proportions of outliers and these proportions are, moreover, very volatile. This is why we will use robust measures of location, scale and skewness to analyze the data further on. Before doing so, we have a closer look at the tails. On the basis of the smoothed curves, left-tailmass(box) amounted to roughly 5% at the beginning of the sample and increased to approximately 10% at the end of the sample. Right-tailmass(box), in contrast, tends to oscillate around 15% during the whole sample. These observed tail weights are substantially in excess of those for the normal distribution, for which the corresponding tailmass(box) measures amount to 2.15% for each tail. This clearly illustrates the fat-tailed nature of Belgian inflation data, a finding which is in line with the results in the international literature on this issue. It should, however, be stressed that most of the international evidence is based on the classical kurtosis, which may for several reasons underestimate the importance of outliers (see for instance Aucremanne, 2000, on this so-called masking issue).

The fact that right-tailmass(box) tends to have larger values than the corresponding left alternative is a first indication of the existence of chronic right skewness, albeit apparently decreasing over time. A possible disadvantage of the general boxplot rejection rule is that it takes the left part of the data into account when defining a right outlier and vice versa. Using the asymmetric boxplot rule overcomes this problem. In so doing, we find that left-tailmass(abp) increases and right-tailmass(abp) decreases relative to their symmetrical alternatives, which can be seen as another indication of chronic right skewness. These asymmetrically constructed tailmass measures confirm the tendency for the left-hand tail to increase and the more stationary behavior of the right-hand tail.

4. Classical and Robust Measures of Location and Scale

The classical and robust measures of location and scale are plotted in Figure 2. The robust alternatives were already used in the previous section to construct robust measures of tail weight. In this section we will compare them to their classical counterparts.

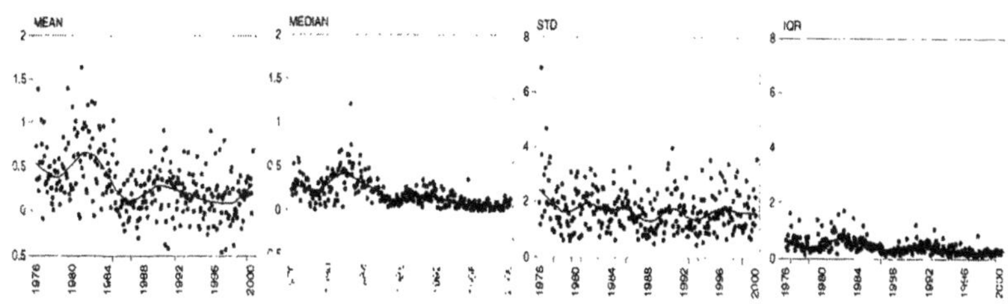

FIGURE 2. Time series of the location and scale measures.

4.1. Location

As the use of robust measures of location for inflation data is well-developed in the core inflation literature (see for instance Aucremanne, 2000, for references), we only give some brief comments when comparing the mean to the median. In Table 1 the unit root test for the classical and robust measures of location, scale and skewness is given. Given the monthly frequency of the data, for each estimator 11 lags were necessary to produce residuals without autocorrelation. The first column gives the value of the augmented Dickey Fuller test statistic, while the second column reports the significance of the constant in the test equation only if the null hypothesis of a unit root is rejected. The 95% critical value for rejection of the null hypothesis of a unit root is −2.87. Inclusion of a trend in the test equations for those variables for which the null of a unit root was not rejected did not change the results. In other words, they were not trend stationary either.

TABLE 1. Unit root tests for the classical and robust measures of location, scale and skewness.

	ADF statistic	signif		ADF statistic	signif
mean	−1.74		skew(class)	−3.82	0.12
median	−1.58		dskew(class)	−5.80	0.02
			dskew(125)	−2.55	
standard deviation	−3.97	0.00	dskew(250)	−3.44	0.00
IQR	−1.74		meme	−2.59	

As the ADF-tests reported in Table 1 fail to reject the null hypothesis of a unit root for the classical and the robust measures of location, both time series contain a stochastic trend. The smoothed curves in Figure 2 seem to confirm this conclusion. The median is substantially less volatile around its trend than the mean, thus confirming that using a robust estimator of location yields a substantial gain in efficiency in the case of a fat-tailed distribution. This could motivate the use of the median as a measure of core inflation statistically. The first robust core inflation measure was indeed the median (Bryan and Pike, 1991). Subsequently, a wide range of alternative robust estimators of core inflation have been proposed.

A cointegration equation of two non-stationary weighted time series X and Y is a linear combination of both time series which is itself stationary (Verbeek, 2000). As mean and median both are non-stationary, we can report the normalized cointegration equation (CE) in Table 2. The 95% critical value is 19.96 for rejection of the null hypothesis of no cointegration equation and 9.24 for rejection of the null hypothesis of at most 1 cointegration equation. The cointegration rank and the specification of the deterministic components of the model were determined jointly on the basis of the general test procedure discussed in Johansen (1992). However, models having trends in the levels of the variables (constants in the VAR) were not considered. As can be seen from Table 2, the median is cointegrated with the mean, but with a coefficient (0.5934) that is substantially less than one. In Figure 2 it can indeed be verified that the median tends to be lower than the mean. This is another indication of chronic right skewness in the Belgian inflation data. Note that the constant in the cointegration relation is very close to zero, suggesting that the chronic asymmetry would disappear completely at zero inflation.

TABLE 2. Cointegration of the robust measure of location (median), of the robust measures of skewness (dskew(125) and dskew(250)) and of the classical skewness measures (skew(class) and dskew(class)) with actual inflation (mean). Between brackets the standard errors are given. Lag number is 11.

	median	dskew(125)	dskew(250)	skew(class)	dskew(class)
CE:					
Constant	0.0002	0.0412	−0.0533	−0.4711	−11.0449
	(0.0200)	(0.0566)	(0.0209)	(0.3879)	(5.5618)
Mean	−0.5934	−1.5037	−0.2579	−0.2853	6.3588
	(0.0563)	(0.1597)	(0.0589)	(1.0913)	(15.7954)
Trace statistic:					
No CE	35.09	30.86	24.88	28.82	40.62
At most 1 CE	4.70	3.41	3.39	4.37	3.79

4.2. Scale

Comparing the classical standard deviation with IQR indicates that relying on a robust alternative dramatically alters the characteristics of the measure of scale. According to the ADF-tests in Table 1, IQR contains a unit root whereas the classical measure appears as a stationary variable. Moreover, Figure 2 shows that the variability of IQR around its smoothed curve is substantially lower than the variability of the classical standard deviation. We owe this to the fact that the classical measure is dominated by the impact of outliers or, in economic terms, by large relative price shocks. In the absence of outliers, IQR would in contrast tend to be larger than the standard deviation (in the case of the normal distribution IQR $\approx 1.35s$). Evidently, the impact of outliers on the classical measure of scale is magnified by squaring their deviations from the mean (see formula (3.1)).

5. Classical and Robust Measures of Skewness

Skewness reflects the shape of the distribution. A symmetric distribution has zero skewness, a distribution which is asymmetric with the largest tail to the right has a positive skewness, and a distribution with a longer left tail has a negative skewness.

5.1. Classical Measure of Skewness

Normally, the standardized, unitless measures of asymmetry (skewness) are used. The best known measure of skewness is the classical skewness (Roger, 2000)

$$\text{skew(class)} = \frac{\sum_{i=1}^{n} w_i (x_i - \overline{x})^3}{\left(1 - 3\sum_{i=1}^{n} w_i^2 + 2\sum_{i=1}^{n} w_i^3\right) s^3} \tag{5.1}$$

However, we will consider unstandardized measures of asymmetry as well. In this way skewness becomes dependent on scale, but this correlation does not bother us. Indeed, the economic models we rely on suggest that scale and skewness interact. In Ball and Mankiw (1995) it is shown that scale magnifies the effect of skewness on inflation, while the asymmetric range of inaction in Ball and Mankiw (1994), which is a possible source of right skewness in the data, is also unscaled. All unstandardized measures of asymmetry are expressed in some units, and the robust measures have the same units as the data and the measures of location and scale. The unstandardized classical skewness is defined as dskew(class) = s^3skew(class).

5.2. Robust Measures of Skewness

A disadvantage of the classical measures of skewness is that they change dramatically when we introduce an outlier. This is clearly shown by the empirical study of Brys et al. (2003). On the other hand, skewness is supposed to measure the asymmetry of the observations, so we may not downweight outliers too heavily. The first robust alternative we propose is dskew(125), which is given by:

$$\text{dskew}(125) = (Q_{0.875} - Q_{0.50}) - (Q_{0.50} - Q_{0.125}) \tag{5.2}$$

with Q_p again the p% (weighted) percentile. The following measure is very similar, but it is based on other percentiles. We compute dskew(250) as:

$$\text{dskew}(250) = (Q_{0.75} - Q_{0.50}) - (Q_{0.50} - Q_{0.25}) \tag{5.3}$$

Note that (5.2) and (5.3) are unstandardized versions of the class of skewness measures introduced by Hinkley (1975).

Another measure of skewness can be obtained by standardizing the data as $z_i = x_i - Q_{0.50}$ and then taking the first moment of these standardized observations which corresponds to the mean-median difference, yielding meme = $\overline{z}$. This measure is not robust, because it uses the mean. However, we consider it here as an alternative measure of skewness, as it has, compared to the classical measures, the advantage that the influence of what happens in the tails is far less accentuated.

5.3. Chronic Right Skewness

We have applied both types of measures of skewness on the cross-sections of the Belgian inflation data. The resulting time series are plotted in Figure 3. The classical measures of skewness show a substantial degree of frequently switching left and right skewness, which is a typical result in the literature in regard to the analysis of inflation data. Indeed, this property has been documented for several other countries (see Aucremanne et al., 2002, for references). However, they do not show a clear tendency towards positive values. As for scale, the time series properties of the robust measures of skewness are fundamentally different from those of the classical ones. As the ADF tests in Table 1 indicate, the null hypothesis of a unit root in the classical measures was rejected at the 99% significance level. Moreover, the constant term in the test equation was not different from zero at the conventional significance levels for skew(class). For dskew(class) the constant was significant at the 95% level, but not at the 99% level. Also, in this case the estimated value of the constant is small relative to the observed variability. Hence, the classical measures hardly reveal any form of chronic right skewness in a significant way, notwithstanding the fact that we provided evidence of a tendency towards right skewness.

In contrast, the robust measures dskew(125) and dskew(250) show a more pronounced tendency towards positive values. Hence, they better reveal the chronic right skewness of the data. The mean-median difference meme also shows such a tendency, although it takes negative values more frequently than the two other robust measures. We owe the latter to the fact that meme is in fact not robust. The null hypothesis of a unit root is rejected for dskew(250) at the 95% significance level, suggesting mean reversion around a significant positive constant. In contrast, for dskew(125) and meme, the null of a unit root is not rejected, suggesting a time-varying degree of chronic right skewness. Particularly in the second half of the sample the tendency towards right skewness seems less pronounced, thus confirming the findings of Sections 3 and 4. A chronic tendency towards right skewness is also found in several other countries and some of these studies also mention that this tendency is less pronounced in the most recent period, when inflation is lower (see Aucremanne et al., 2002, for references).

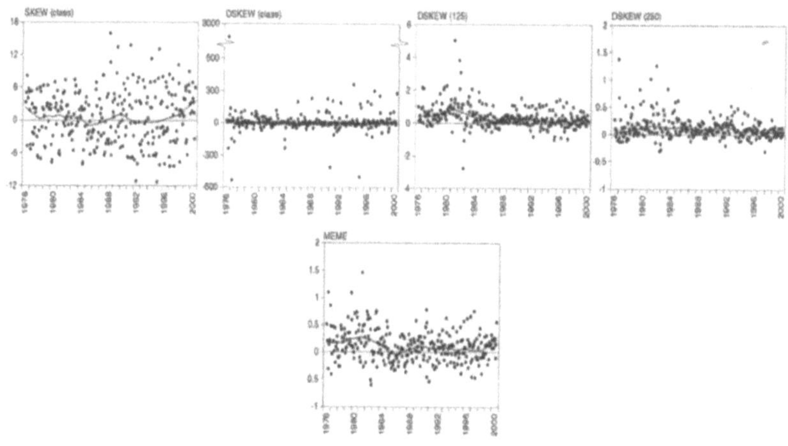

FIGURE 3. Time series of the skewness measures.

6. Inflation Affecting Relative Price Changes

In view of the chronic right skewness of the distribution of Belgian consumer price changes revealed in the previous section by the robust measures, it is important to stress that two alternative economic models can generate this phenomenon. The two models have however different, empirically testable implications for the relation between skewness and aggregate inflation.

The first model relies on an exogenous form of downward nominal rigidity of prices of the Tobin-type (Tobin, 1972). This model predicts a negative relation between inflation and skewness: higher aggregate inflation leads to less right skewness (see Yates, 1998, on this issue). The second model relies on menu costs, i.e. it assumes that changing prices involves a (menu) cost, which is fixed in the sense that it does not depend on the magnitude of the price change. Examples of this type of costs are changes to price lists, catalogues, advertising, etc. This menu cost produces a range of inaction: firms react only to large price shocks, for which the advantage of changing prices outweighs the menu cost. In Ball and Mankiw (1994) it is shown that this range of inaction becomes asymmetric in the presence of trend inflation. In other words, (positive) trend inflation reduces the likelihood of observing a downward adjustment of prices relative to the likelihood of an upward adjustment. Therefore, this model implies a positive relation between inflation and skewness: higher aggregate inflation leads to more right skewness.

In view of this, the relation between aggregate inflation and skewness is studied by cointegrating the robust measures of skewness with actual inflation. As is shown in Table 2, using the robust measure dskew(125) reveals a clear positive cointegration relationship between asymmetry and inflation. Each percentage point of additional (monthly) inflation tends to increase dskew(125) by 1.5 percentage points. Moreover, the constant term in the cointegration relation is very small both in statistical and in economic terms, suggesting that the chronic right skewness would disappear completely at zero inflation. As such, the long term behavior of the chronic right skewness measured by dskew(125) seems to be an integral part of the inflationary process itself. A similar conclusion was obtained on the basis of the cointegration relation between the median and the mean.

Also for dskew(250), the Johansen cointegration test reveals the existence of a positive cointegration relation with inflation, notwithstanding the fact that the ADF-test reported in Table 1 suggests that dskew(250) is stationary. In such conflicting case the Johansen test dominates the ADF-test. This measure of asymmetry tends to increase by 0.26 percentage points for each percentage of (monthly) inflation. The constant in the cointegration relation for dskew(250) suggests the existence of some form of chronic right skewness which is independent of the inflationary process. The impact of this constant is, however, relatively small compared to the impact of inflation. Indeed, each 0.2% of additional monthly inflation (approximately 2.4% on a yearly basis) increases dskew(250) by the same amount as that resulting from the constant. This suggests that a substantial part of the chronic asymmetry measured by dskew(250) disappears at zero inflation.

These findings thus tend to provide evidence in favor of menu cost models, while they strongly contradict the exogenously assumed downward nominal rigidity of Tobin (1972). Distinguishing between these two models is important from a monetary policy perspective, as they have different implications for the optimal rate of inflation. The menu cost favors a zero inflation rate in order to minimize the disturbing impact of aggregate inflation on the distribution of relative price changes, whereas the Tobin-model model argues in favor of a positive inflation rate, that is sufficiently high to overcome the inefficiencies resulting from the downward rigidity.

A similar analysis which uses the classical measures of skewness did not allow to describe this long-run behavior of the chronic right skewness. Although with skew(class) the Johansen test suggests the existence of a cointegration relation, we do not interpret this as meaningful because of its coefficients which are estimated very imprecisely (large standard errors for both the coefficient of the mean and the constant). As a matter of fact, in this case the Johansen test exactly confirms the findings of the ADF-test as skew(class) is cointegrated with the constant zero (or, it is stationary around a non-significant constant). Hence this measure does not reveal any form of chronic right skewness in a significant way, nor a systematic long-run impact of the inflation rate on the asymmetry. A similar conclusion can be made with dskew(class), although the constant term is estimated somewhat more precisely. Both tests come to a similar conclusion, namely a stationary behavior around a positive constant which is (according to the ADF-test) statistically significant at the 95% level, but not at the 99% level. Nevertheless, this constant is small compared to the variability observed in the series of dskew(class), such that we may conclude that dskew(class) hardly reveals any chronic right skewness.

7. Conclusions

In this paper we have studied the properties of the distribution of Belgian consumer price changes, by using both classical and robust measures of location, scale, skewness and tail weight. The robust measures of tail weight showed clearly (i) the fat-tailed nature of the distribution, (ii) the short run volatility of the tail weights, (iii) the tendency for the right-hand tail to dominate the left-hand tail, which points towards chronic right skewness and (iv) the increase over time of the left-hand tail and the more stationary behavior of the right-hand tail.

It was found that the chronic right skewness was cointegrated with aggregate inflation. Moreover, we found little evidence of chronic right skewness not dependent on the inflationary process, suggesting that the asymmetry would disappear at zero inflation. Overall, these results are in line with the predictions of menu cost models and they are symptomatic of nominal rigidities, in the sense that prices are adjusted infrequently. However, they do not point in the direction of specific downward rigidities, other than those endogenously generated by the prevailing inflation rate. These findings provide arguments in favor of a price stability-oriented monetary policy.

References

[1] L. Aucremanne, *The Use of Robust Estimators as Measures of Core Inflation.* National Bank of Belgium Working Papers – Research Series **2**, 2000, available at *http://www.nbb.be/Sg/En/Produits/publication/4410e.htm.*

[2] L. Aucremanne, G. Brys, M. Hubert, P.J. Rousseeuw, and A. Struyf, *Inflation, Relative Prices and Nominal Rigidities.* National Bank of Belgium Working Papers – Research Series **20**, 2002, available at *http://www.nbb.be/Sg/En/Produits/publication/4410e.htm.*

[3] L. Ball and N.G. Mankiw, *Asymmetric Price Adjustment and Economic Fluctuations.* Econom. J. **104** (1994), 247–262.

[4] L. Ball and N.G. Mankiw, *Relative-price Changes as Aggregate Supply Shocks.* Quart. J. Econom. **1** (1995), 161–193.

[5] M.F. Bryan, S.G. Cecchetti, and R.L. Wiggins, *Efficient Inflation Estimation.* NBER Working Paper Series **5793**, 1997, available at *http://www.nber.org/papers/.*

[6] M.F. Bryan and C.J. Pike, *Median Price Changes: An Alternative Approach to Measuring Current Monetary Inflation.* Federal Reserve Bank of Cleveland Economic Commentary, pages 1–4, 1991.

[7] G. Brys, M. Hubert, and A. Struyf, *A Comparison of Some New Measures of Skewness.* In R. Dutter, P. Filzmoser, U. Gather, and P.J. Rousseeuw, editors, Developments in Robust Statistics, International Conference on Robust Statistics 2001, pages 98–113. Springer-Verlag, Heidelberg, 2003.

[8] G. Brys, M. Hubert, and A. Struyf, *A Robust Measure of Skewness.* J. Comput. Graph. Statist. (2004), to appear.

[9] W.S. Cleveland, *Robust Locally Weighted Regression and Smoothing Scatterplots.* J. Amer. Statist. Assoc. **74** (1979), 829–836.

[10] D.V. Hinkley, *On Power Transformations to Symmetry.* Biometrica **62** (1975), 101–111.

[11] S. Johansen, *Determination of the Cointegration Rank in the Preference of a Linear Trend.* J. Econometrics **52** (1992), 389–402.

[12] S. Roger, *Relative Prices, Inflation and Core Inflation.* IMF Working Paper **58**, 2000, available at *http://www.imf.org/external/pubs/cat/.*

[13] P.J. Rousseeuw, I. Ruts, and J.W. Tukey, *The Bagplot: a Bivariate Boxplot.* Amer. Statist. **53** (1999), 382–387.

[14] D. Ruppert, *What is Kurtosis? An Influence Function Approach.* Amer. Statist. **41** (1987), 1–5.

[15] J. Tobin, *Inflation and Unemployment.* Amer. Econom. Rev. **62** (1972), 1–18.

[16] J.W. Tukey, *Exploratory Data Analysis.* Addison-Wesley, Reading, M.A., 1977.

[17] M. Verbeek, *A Guide to Modern Econometrics.* John Wiley and Sons, Chichester, England, 2000.

[18] A. Yates, *Downward Nominal Rigidity and Monetary Policy.* Bank of England Working Paper Series **82**, 1998, available at
http://www.bankofengland.co.uk/workingpapers/.

L. Aucremanne
Research Department
National Bank of Belgium (NBB)
Boulevard de Berlaimont 14
B-1000 Brussel
Belgium
e-mail: luc.aucremanne@nbb.be

G. Brys
Faculty of Applied Economics
University of Antwerp – Stadscampus
Prinsstraat 13
B-2000 Antwerpen
Belgium
e-mail: guy.brys@ua.ac.be

M. Hubert
Department of Mathematics
Katholieke Universiteit Leuven (KULeuven)
W. de Croylaan 54
B-3001 Leuven
Belgium
e-mail: mia.hubert@wis.kuleuven.ac.be

P.J. Rousseeuw and A. Struyf
Department of Mathematics and Computer Science
University of Antwerp – Middelheimcampus
Middelheimlaan 1
B-2020 Antwerpen
Belgium
e-mail: {peter.rousseeuw, anja.struyf}@ua.ac.be

Statistics for Industry and Technology, 27–37

Robust Strategies for Quantitative Investment Management

G. Bassett, G. Gerber and P. Rocco

Abstract. We show how quantile estimation combined with robust methods can be used in quantitative investment management. A portfolio manager uses a quantitative model to select securities. The objective is to outperform a benchmark portfolio, subject to risk constraints. Traditional stock selection models express expected returns as a function of factors where all parts of the return distribution are affected similarly. This is subsumed by the quantile approach in which a stock's entire return distribution is a conditional function of factors. Robust methods insure that our estimates do not depend on a small subset of the data. Regression quantile estimates then detect the potentially different impact of factors at the center and tails of the return distribution. This is illustrated in assessing a model's forecasting accuracy while controlling for return differences between economic sectors. We are thereby able to detect forecasting properties that would have been missed by a conventional analysis of the data.

Mathematics Subject Classification (2000). Primary 91B28; Secondary 62G35.

Keywords. Quantitative portfolio management, robust models, quantile regression.

1. Introduction

We illustrate how the combination of robust and quantile-based methods can improve quantitative investment management strategies. A portfolio manager uses a quantitative model to select securities in the context of a systematic and disciplined approach to asset valuation and investing. The objective is to build and then periodically rebalance a portfolio so that, subject to risk constraints, it outperforms a benchmark portfolio. The risk constraints limit expected deviations between the portfolio and benchmark returns. Asset selection is based on a multivariate model that exploits "anomalies" in which stocks' risk-adjusted returns above a benchmark (residual returns) are associated with various characteristics of particular stocks. These characteristics (e.g., P/E ratios and earnings growth rates) are termed "factors". Among the best-known anomalies are those associated with value and small capitalization stocks, and market reaction to revised earnings estimates. (The literature in finance on such anomalies is quite vast. Extended discussions are provided in Fama and French, 1992, DeBondt and Thaler, 1985, and Lakonishok et al., 1994.)

Quantitative methods in investment management utilize very large data sets. Historical databases with daily returns and firm characteristics for thousands of companies going back many years are used to develop and backtest models. Updated data is used for diagnostics, factor attribution, and periodic revision of portfolio weights. The large amount of data means it is difficult and costly to eliminate mistakes or detect correct, but influential, observations. Mistakes arise because of transcription errors, changed ticker symbols, unrecorded stock splits, and so on. Extreme observations occur for the stocks of the formerly exuberant New Economy as well as more recently for companies using the latest accounting "innovations".

The consequences of data outliers are well known: only a few discrepant observations have a huge impact on standard estimation methods. The methods are tricked because they are extremely sensitive to only a few deviant observations – they are not robust. Robust data analysis produces estimates that are determined by the bulk of the data without undue influence by a few observations; see Rousseeuw (1984) and Koenker (1982). Knowing that estimates are based on the bulk of the data is advantageous for model development. It signals important determinants of returns that otherwise would be obscured by a few discrepant observations. Conversely, factors deemed important because of only a few influential observations are revealed by robust estimates as lacking explanatory power for the bulk of the data. Hence, robustness is important for investment management because of the existence of outliers in very large databases together with the lack of robustness of standard methods. The importance of robustness is magnified by the fact that quantitative models are often used to select a relatively limited number of stocks; mistakes can be costly.

Investment valuation models are also characterized by very low signal-to-noise ratios. Finding factors that correctly signal even slight departures, on average, from market returns can be quite profitable even when departures from the average are large. The signal provides the basis for a portfolio manager's ability to earn excess returns while the noise means there can be large deviations between actual and expected returns. Some of the deviation can be diversified away by exploiting independence and negative correlations. The sheer magnitude of the noise however means that portfolios with a large number of stocks still have high model risk. Further, signal strength varies over time thus requiring frequent rebalancing and hence large transaction costs for large portfolios. As a result of high noise, portfolio returns are subject to potentially large departures from what is predicted by the signal. This leads to consideration of how factors affect the tails of the return distribution.

The traditional framework for the robust approach has been a data generating process with a few gross errors from a "true" primary model. The objective is to estimate the central tendency of the main model, without undue influence by a few outlying, discrepant, observations. In our case the primary model specifies the relation between expected returns and other variables. Features of the distribution at places other than the expected value, however, are of critical importance. Rather than "mistakes" the deviations are (or, at least can be) functions of the same factors that determine central tendency. Hence, we want to allow factors that determine central tendency to also affect the tails of the return distribution. Factors that may have no influence on expected returns cannot be ruled out as important in forecasting other parts of the return distribution. Alternatively, factors that do influence the center can have different effects in the tails.

This leads to our data analysis that combines robust and quantile methods. For an overview of quantile modelling, see Koenker and Hallock (2001). The combination is useful for quantitative investment because there are influential observations in our large data sets, and because there are quantile effects in which models work differently in the tails and in the central region of the return distribution. The quantile approach therefore provides information about how factors affect the entire return distribution, not just central tendency. The standard model in which quantile effects are constant is nested in the quantile model. While the traditional approach has factors affecting the mean and other parts of the return distribution equally, the more comprehensive quantile approach allows differences to be detected.

We have used this approach in the development, assessment, and calibration of investment strategies. In this paper we focus on how the quantile-based, robust approach can be used to assess potential modifications of a stock selection model. Section 2 explains how the stock selection model fits into the overall investment process. The selection model itself is estimated using robust methods. Section 3 presents an assessment of predictive ability based on expected values, the usual criterion in the investment management field. It also indicates our enhancement in which the entire conditional return distribution is considered. Section 4 presents regression quantile estimates of the relation between actual and predicted returns where we control for sector effects. Section 5 discusses results and indicates further enhancements using the robust-quantile approach.

2. Overview of the Investment Process

The investment process can be divided into four interrelated stages. In the first stage, a portfolio manager identifies an investment universe from which individual stocks will be ranked and evaluated for inclusion in a portfolio. The investment universe is typically designed to client specifications; it might consist of constituents in a broad-market index or stocks in a specific range of market values. The universe for our illustration consists of the largest 1000 stocks (by market value) in the U.S.

In the next stage, the manager identifies a benchmark portfolio of stocks (often an index). In the investment management industry the benchmark plays two roles. It sets the target rate of return that a portfolio seeks to exceed. For our illustration the benchmark is the S&P 500 so that the manager seeks to maximize returns relative to the S&P 500 index.

The benchmark also serves as the reference point for defining the extent to which a portfolio can differ from the benchmark. A client will typically specify a constraint that limits deviations between portfolio (P_t) and benchmark (B_t) returns during period t. This constraint, known as "tracking error", is defined as (an estimate of) the standard deviation of $P_t - B_t$. It is denoted by $TE(P_t, B_t)$ and is required to stay below a prespecified value. Tracking error is typically estimated using industry-standard risk models supplied by vendors. The benchmark therefore sets the target rate against which performance is judged, and constrains the expected difference between P_t and B_t.

Let P_t denote portfolio returns for period t where $P_t = \sum_{i \in U} w_{i,t} R_{i,t}$ with U denoting the investment universe, $w_{i,t}$ the weight on stock i at time t, and $R_{i,t}$ the return to stock i during period t. The manager's objective is to select weights $w_{i,t}$ to maximize to the difference between P_t and B_t subject to $TE(P_t, B_t) < x$.

The third stage in the process makes use of a selection model to determine the relative weight of stocks in the managed portfolio. Our analysis is based on a multi-factor model that assigns a so-called "alpha"-score, $\alpha_{i,t+1}$, to each stock i which serves as a forecast of a stock's relative residual return in month $t+1$. A linear-model for the score $\alpha_{i,t}$ may be represented as,

$$\alpha_{i,t+1} = b_{1,t+1} F_{1,i,t} + b_{2,t+1} F_{2,i,t} + \cdots + b_{k,t+1} F_{k,i,t} \tag{2.1}$$

where $F_{j,i,t}$ is the exposure of factor j for stock i at the end of period t, and $b_{j,t+1}$ is the "weight"on factor j during period $t+1$. Typical factors might include well-known financial ratios (e.g., E/P), measures of earnings growth and dispersion, and past price movements. The model "weights", $b_{j,t+1}$, are computed using historical data on factors and returns. Robust and quantile methods are used at this stage of the investment process to control for outliers, and assess quantile effects in the way factors influence returns; estimators include least median of squares, least absolute deviations, and regression quantiles; see Rousseeuw (1984), Bassett and Koenker (1978), and Koenker and Bassett (1978).

The weights, $b_{j,t+1}$, are estimated based on data known at time t and then updated to reflect the success of the associated factors in forecasting returns. The α-values are generally in the range -2 to 2. Note that the range of α-values is limited because extreme factor values are trimmed or Winsorized, see Grinold and Kahn (2000, p. 382). For our assessment, we have a sequence of $\alpha_{i,t}$ for i=1 to 1000 and t each month from January 1991 to September 2002. The goal is to assess the ability of the model to discriminate between the stocks in the investment universe with respect to future returns.

The final stage of the process consists of constructing a portfolio by maximizing the "alpha" of the portfolio subject to the risk constraint associated with the

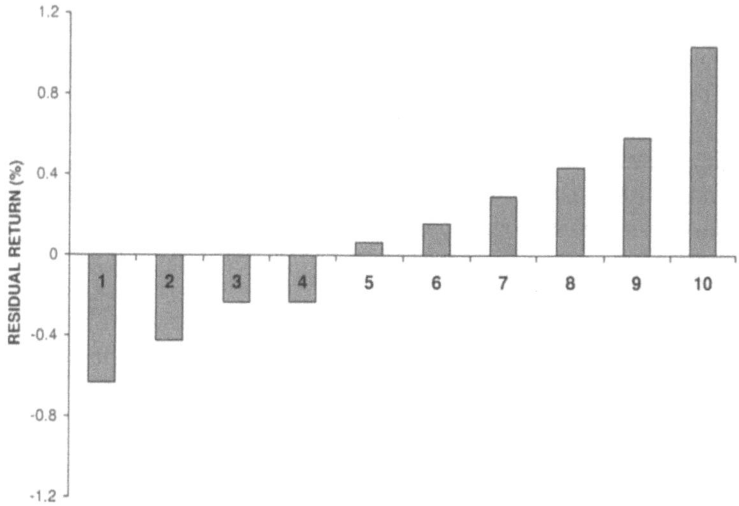

FIGURE 1. Average monthly residual returns grouped by α decile (Decile 10 is best).

benchmark portfolio and turnover considerations. The process of buying and selling of stocks to adjust the holdings in the managed portfolio entails costs. As such, there is a trade-off between capturing the signals from the valuation model and paying the turnover costs. Note that this process reflects a setting in which only purchases of stocks is permitted. The issues involved when purchases and sales of borrowed stocks (shorts) are permitted is more complex (see Gerber, 1996).

3. Assessing Model Performance

3.1. Conditional Expectation

As indicated, our analysis is based on α's generated for each of approximately 1000 stocks for each month from January 1991 to September 2002.

The weights on each factor, b_j, vary over time depending on the recent performance of the factor. The method used to generate factor weights is similar to that employed in other stock selection models we have investigated. Backtests of models are typically based on the ability of α to forecast one-month ahead returns. For example, Figure 1 shows *average* residual monthly return associated with portfolios created by sorting stocks into deciles based on α-score. The graph shows stocks in the top α decile had an average monthly residual return of 1.1% while the stocks in the lowest decile had returns of -0.7% over the entire period.

Gross returns (exclusive of transactions cost) from a strategy of buying the stocks in decile 10 and shorting those in decile 1 would have produced an average monthly gain of 1.8%. The nice "step-ladder" pattern in the graph indicates the valuation model is able to discriminate among stocks based on future expected returns. While the α's are suggestive of a profitable portfolio strategy, further analysis would need to consider transaction costs. Further, the returns pictured in Figure 1 do not account for the tracking error. The positive returns associated with α, for example, might be concentrated in only a few sectors and hence entail large tracking error. If the highest α's tended to concentrate in, say, Technology, then a portfolio tilted toward large α would be heavily exposed to Technology and subject to large deviations from the overall market.

3.2. Conditional Distribution

A more complete picture of model performance that is motivated by our concern about the entire return distribution is depicted in Figure 2. It shows residual returns at each α decile. The average return in the previous figure is indicated by E(). The picture shows returns tend to increase with α decile, but it also shows a very wide distribution around the central tendency. Further, the conditional distributions at each α decile are not just location shifts of one another. For example, we see that the Q(.9) part of the distribution tends to increase with α, but at low α deciles the function is actually decreasing. This means that the top end of the return distribution decreases in a move from α decile 1 to 2, contrary to what occurs in the center of the distribution. In a similar fashion, Q(.1) decreases at high α. Shifting a portfolio into higher α stocks increases average returns, but decreases returns in the lower tail.

4. Estimating Conditional Quantile Effects

In this section we present regression quantile estimates of the relation between α and one-month ahead actual excess returns. Our model controls for sector effects with sector dummy variables corresponding to the 15 economic sectors in the 1000 stock universe. The model for returns is,

$$Q(\theta \mid s, \alpha) = a_s(\theta)S + \beta(\theta)\alpha \tag{4.1}$$

where $Q(\theta \mid s, \alpha)$ is θ^{th} quantile of the return distribution conditional on the stock's sector and α score, S is an indicator for the stock's sector, a_s is the sector coefficient, α is the stock's "alpha", and $\beta(\theta)$ is the regression quantile coefficient.

The model is estimated separately for $\alpha > 0$ and $\alpha < 0$. This provides separate $\beta(\theta)$ estimates for stocks predicted to outperform the market ($\alpha > 0$) and those predicted to underperform the market ($\alpha < 0$).

Graphical versions of the estimates are presented in Figures 3 and 4. Quantile regression coefficients as a function of the θth quantile for $\alpha > 0$ (Figure 3) and $\alpha < 0$ (Figure 4) are shown. The horizontal lines in Figure 3 depict the least squares estimate and confidence bounds. It shows a positive and constant relation between predicted and actual excess returns. The estimates indicate that for $\alpha > 0$ expected excess returns are an increasing function of α, after controlling for sector

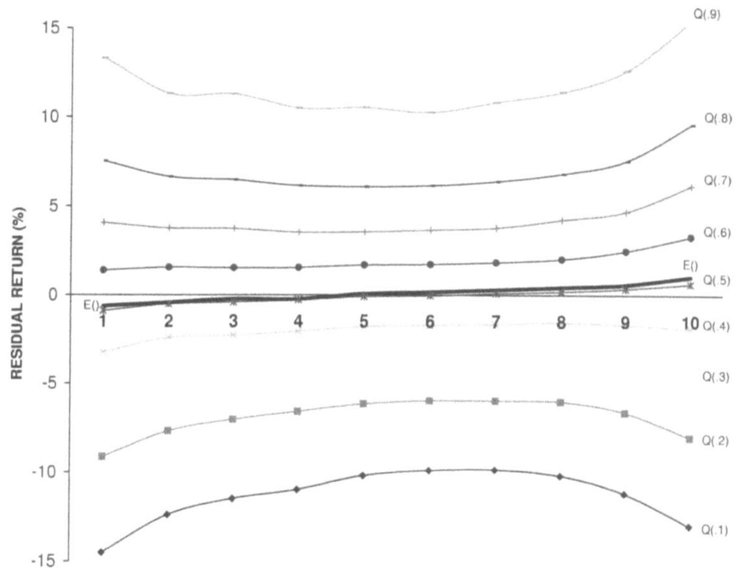

FIGURE 2. Quantile residual returns by α decile.

effects. The figure also shows, however, there are quantile effects: after controlling for sector affects the impact of α on excess returns is not the same throughout the return distribution. Compared to the effect at the mean the impact of α on returns in the upper tail is even more powerful. At the lower quantiles the effect diminishes and at the lowest part of the distribution actually becomes negative. A negative coefficient means that as α increases the associated quantile of the return distribution decreases, contrary to how we would want the model to perform.

The situation for $\alpha < 0$ is shown in Figure 4. The coefficient at the expected value is positive but smaller than for $\alpha > 0$. The expected average performance of the model is good for all α, but better for the stocks that are predicted to outperform the market ($\alpha > 0$). But there are again quantile effects; high quantile parts of the distribution show a negative association between alpha and returns.

Figure 5 presents the same results, but in "data space". It shows returns as a function of α controlling for the separate sector intercepts included in the model. There is a positive relation between expected and forecast returns. There is still a low signal-to-noise ratio, which is evidenced by the wide range of conditional quantiles. Since we are controlling for sectors the low signal-to-noise evident in Figure 2 is not due to return differences between sectors. The Figure 5 also shows the poor performance of α in the lower tails for $\alpha > 0$, and in the upper tails for $\alpha < 0$. The

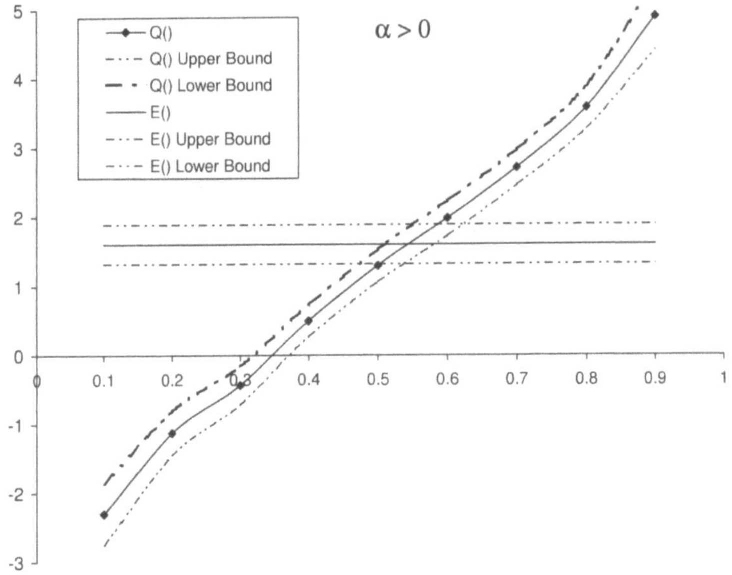

FIGURE 3. Regression quantile and least squares, E(), estimates for residual returns for $\alpha > 0$, controlling for sector effects.

results raise the question of whether this poor performance might be improved by including additional explanatory variables in our quantile specification, a topic to be explored in further research.

5. Discussion

In spite of their advantages, robust methods have not been widely used in quantitative finance. Like other areas of applied statistics, the absence of robust methods has been due in part to the failure to recognize the extreme sensitivity of standard methods, or the belief that standard methods were already "optimal" or "best". The optimality label however is based on assumptions that are only more or less appropriate for financial data. Indeed, some of the earliest questioning of standard assumptions came from analysis of financial data that showed too many outlying realizations relative to the standard normal distribution; see Mandelbrot (1965) and Fama (1963). The fact that data tended to have fatter tails relative to the center of the distribution meant not only that standard methods were suboptimal, but that they could be highly misleading relative to robust methods.

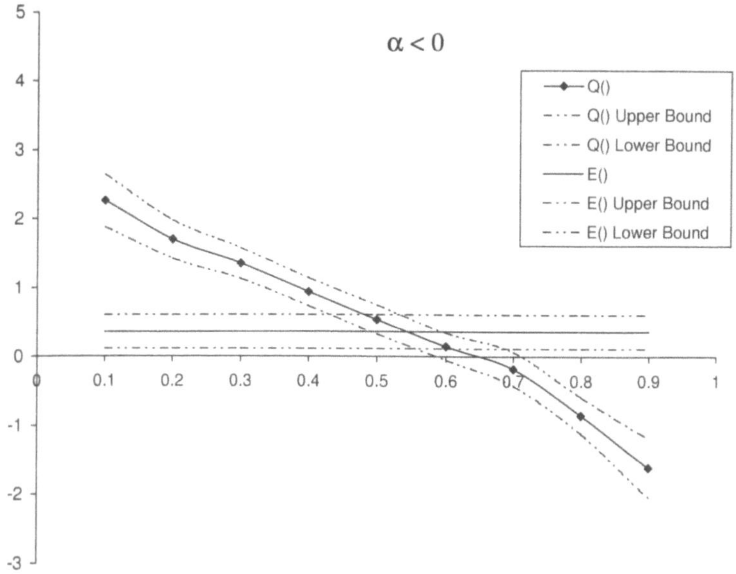

FIGURE 4. Regression quantile and least squares, E(), estimates for residual returns for $\alpha < 0$, controlling for sector effects.

Another reason for the absence of robust methods is a perception that preprocessing data to remove univariate outliers results in robust estimates. Univariate screening and trimming of data helps identify influential values of dependent and explanatory variables, but it does not assure robust estimates. In the multivariate setting with no outliers for any single variable, there can be large differences between standard and robust estimates. Trimming extreme observations is motivated by robustness considerations, but it does not assure estimates that are determined by the bulk of the data.

Statistical applications of quantile estimation in Finance have also not been used widely (but for a recent application, see Bassett and Chen, 1991). Data analysis has tended to focus on estimating central tendency with only secondary attention on determinants of tail events. Or models were restricted to simple location shifts as a function of explanatory variables. The low signal-to-noise inherent in financial models, together with increasing concern about risk, makes quantile methods particularly appealing in quantitative portfolio management. The quantile model allows the relation between a set of factors and tail events to be estimated directly. This allows the relation between predicted and actual returns to be assessed comprehensively. The application presented here has shown that the

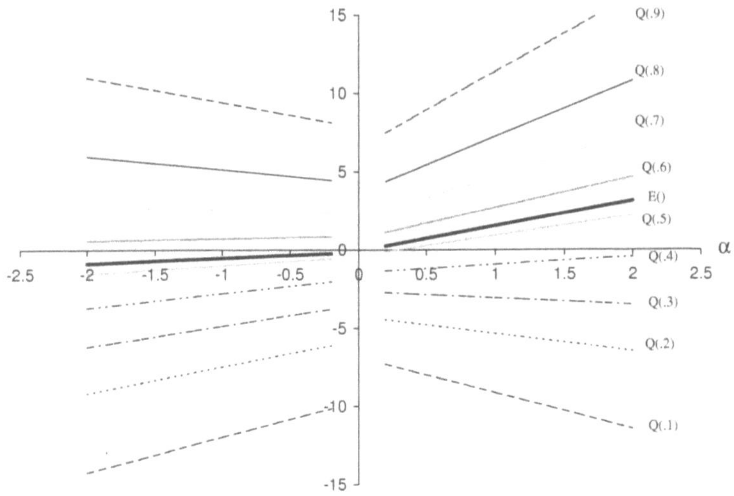

FIGURE 5. Regression quantile predictions as a function of α distribution, controlling for sector effects.

α-scores work well for expected returns, but result in increased dispersion at extreme α values, a feature that would have been missed in a conventional analysis focusing only on conditional expectations.

References

[1] G. Bassett and H. Chen, *Portfolio Style: Return-based Attribution Using Quantile Regression.* Empirical Economics **26** (1991), 293–305.

[2] G. Bassett and R. Koenker, *The Asymptotic Theory of Least Absolute Error Regression.* J. Amer. Statist. Assoc. **73** (1978), 618–622.

[3] W. DeBondt and R. Thaler, *Does the Stock Market Overreact?* J. Finance **40** (1985), 793–806.

[4] E. Fama, *Mandelbrot and the Stable Paretian Hypothesis.* J. Business **36** (1963), 420–429.

[5] E. Fama and K. French, *The Cross-Section of Expected Stock Returns.* J. Finance **47** (1992), 427–466.

[6] G. Gerber, *Using a Nonparametric Approach to Market-Neutral Investing.* In J. Lederman and R. Klein, editors, Market Neutral: State-of-the-Art Strategies for Every Market Environment, pages 93–103. Irwin, 1996.

[7] R. Grinold and R. Kahn, *Active Portfolio Management.* McGraw-Hill, 2000.

[8] R. Koenker, *Robust Methods in Econometrics.* Econometric Rev. **1** (1982), 213–225.

[9] R. Koenker and G. Bassett, *Regression Quantiles.* Econometrica **46** (1978), 33–50.

[10] R. Koenker and K. Hallock, *Quantile Regression.* J. Econom. Perspectives **15** (2001), 143–156.

[11] J. Lakonishok, A. Schleifer, and R. Vishny, *Contrarian Investment, Extrapolation and Risk.* J. Finance **49** (1994), 1541–1578.

[12] B. Mandelbrot, *The Variation of Certain Speculative Prices.* J. Business **36** (1963), 394–419.

[13] P.J. Rousseeuw, *Least Median of Squares Regression.* J. Amer. Statist. Assoc. **79** (1984), 871–880.

Acknowledgment

Thanks to two reviewers for their helpful comments on an earlier version. Research support from Twin Capital Management, Inc. is gratefully acknowledged.

G. Bassett
Department of Finance
University of Illinois at Chicago
601 South Morgan (MC 168)
Chicago, Illinois 60607-7121, USA
e-mail: gib@uic.edu

G. Gerber and P. Rocco
Twin Capital Management, Inc.
3244 Washington RD STE 202 McMurray, PA 15317-3153, USA
e-mail: geoff@twincapital.com, research@twincapital.com

Statistics for Industry and Technology, 39–48

An Adaptive Algorithm for Quantile Regression

C. Chen

Abstract. In this article, we introduce an algorithm to compute the regression quantile functions. This algorithm combines three algorithms – the simplex, interior point, and smoothing algorithm. The simplex and interior point algorithms come from the background of linear programming (Portnoy and Koenker, 1997). While the simplex method can handle small to middle sized data sets, the interior point method can handle large to huge data sets efficiently. The smoothing algorithm is specially designed for the L_1 or quantile regression type of problems (Chen, 2002), and it outperforms the other two algorithms for *fat* data sets. Combining these three algorithms produces an algorithm, which is adaptive in the sense that it can intelligently detect the input data sets and select one of the three algorithms to efficiently compute the regression quantile functions.

Mathematics Subject Classification (2000). Primary 62F35; Secondary 62J99.

Keywords. Quantile regression, heteroscedasticity, finite smoothing algorithm, simplex, interior point, median regression, linear programming.

1. Introduction

Quantile regression, which was introduced by Koenker and Bassett (1978), estimates and conducts inference about conditional quantile functions. A real valued random variable may be characterized by its distribution function,

$$F(y) = \text{ Prob } (Y \leq y), \tag{1.1}$$

while for any $0 < \tau < 1$,

$$Q(\tau) = \text{ inf } \{y : F(y) \geq \tau\} \tag{1.2}$$

is called the τth quantile of Y. The median, $Q(1/2)$, plays the central role. Like the distribution function, the quantile function provides a complete characterization of the random variable Y.

For a random sample $\{y_1, \ldots, y_n\}$ of Y, it is well known that the sample median is the solution of the optimization problem

$$\min_{\xi \in \boldsymbol{R}} \sum_{i=1}^{n} |y_i - \xi|. \tag{1.3}$$

The general τth sample quantile $\xi(\tau)$, which is an analogue of $Q(\tau)$, may be formulated as the solution of the optimization problem

$$\min_{\xi \in \boldsymbol{R}} \sum_{i=1}^{n} \rho_\tau(y_i - \xi), \tag{1.4}$$

where $\rho_\tau(z) = z(\tau - I(z < 0))$, $0 < \tau < 1$.

As estimating the unconditional mean, viewed as the minimizer,

$$\hat{\mu} = \operatorname{argmin}_{\mu \in \boldsymbol{R}} \sum_{i=1}^{n} (y_i - \mu)^2 \tag{1.5}$$

can be extended to estimation of the linear conditional mean function $E(Y|\mathbf{x} = x) = x'\beta$ by solving

$$\hat{\beta} = \operatorname{argmin}_{\beta \in \boldsymbol{R}^p} \sum_{i=1}^{n} (y_i - x_i'\beta)^2, \tag{1.6}$$

the linear conditional quantile function, $Q_\tau(\tau|\mathbf{x} = x) = x'\beta(\tau)$, can be estimated by solving

$$\hat{\beta}(\tau) = \operatorname{argmin}_{\beta \in \boldsymbol{R}^p} \sum_{i=1}^{n} \rho_\tau(y_i - x_i'\beta). \tag{1.7}$$

The median case, $\tau = 1/2$, which is equivalent to minimizing the sum of absolute values, is usually known as the L_1 regression.

An LP Problem

Let $\mu = [y - X\beta]_+$, $\nu = [X\beta - y]_+$, $\phi = [\beta]_+$, and $\varphi = [-\beta]_+$, where $y = (y_1, \ldots, y_n)'$, $X = (x_1, \ldots, x_n)'$, and $[z]_+$ is the nonnegative part of z.

Let $D_{LAR}(\beta) = \sum_{i=1}^{n} |y_i - x_i'\beta|$ and $D_{\rho_\tau}(\beta) = \sum_{i=1}^{n} \rho_\tau(y_i - x_i'\beta)$. For the L_1 problem, the simplex approach solves $\min_\theta D_{LAR}(\theta)$ by the reformulation

$$\min_{\beta}\{e'\mu + e'\nu | y = X\beta + \mu - \nu, \{\mu, \nu\} \in \boldsymbol{R}_+^n\}, \tag{1.8}$$

where e denotes an n-vector of ones.

Let $B = [X \;\; -X \;\; I \;\; -I]$, $\theta = (\phi' \; \varphi' \; \mu' \; \nu')'$, and $d = (\mathbf{0}' \; \mathbf{0}' \; e' \; e')'$ where $\mathbf{0}' = (0 \; 0 \; \ldots \; 0)_p$. The reformulation presents a standard LP problem:

$$(P) \qquad \begin{aligned} &\min_\theta d'\theta \\ &B\theta = y \\ &\theta \geq 0. \end{aligned}$$

This problem has the dual formulation

$$(D) \qquad \max_{d} y'z$$
$$B'z \leq d.$$

It can be simplified as

$$\max_{z}\{y'z|X'z = 0, z \in [-1,1]^n\}.$$

By setting $\eta = \frac{1}{2}z + \frac{1}{2}e$, $b = \frac{1}{2}X'e$, it is

$$\max_{\eta}\{y'\eta|X'\eta = b, \eta \in [0,1]^n\}. \tag{1.9}$$

For quantile regression $\min_\beta \sum \rho_\tau(y_i - x_i'\beta)$, a similar processing presents the dual formulation:

$$\max_{z}\{y'z|X'z = (1-\tau)X'e, z \in [0,1]^n\}. \tag{1.10}$$

2. Simplex

Since the early 1950's it has been recognized that median regression (L_1 regression) can be formulated as linear programming problems and efficiently solved with some form of the simplex algorithm. An efficient version of the simplex algorithm was developed by Barrodale and Roberts (1974). This algorithm solves the primary LP problem (P) by two stages, which were developed according to the special structure of the coefficient matrix B. The first stage only picks the columns in X or $-X$ as pivotal columns. The second stage only interchanges the columns in I or $-I$ as basic or nonbasic columns. The algorithm gets an optimal solution by executions of these two stages interactively. Also because of the special structure of B, only the main data matrix X is stored in the current memory.

This special version of the simplex algorithm for median regression can be naturally implemented to compute quantile regression with any given quantile, even the whole quantile process (Koenker and d'Orey, 1993).

Although the worst case for this simplex algorithm is regarded as computationally highly demanding for large data sets, careful and proper coding of the algorithm still makes it suitable for data sets with less than 5000 observations and 50 variables with the present popular hardware (e.g., 1GHz CPU and 256MB of RAM).

3. Interior Point

To solve large to huge LP problems, alternative algorithms have been developed. Rather than moving from vertex to vertex around the outer surface of the constraint set as dictated by the simplex, the interior point approach of Karmarkar (1984) solves a sequence of quadratic problems in which the relevant interior of the constraint set is approximated by an ellipsoid. The worst-case performance of the interior point algorithm is demonstrated better than that of the simplex algorithm.

The excellent paper by Portnoy and Koenker (1997) revives the enthusiasm of applying L_1 regression or quantile regression to large or huge data sets by introducing the faster interior point algorithm into this area.

There are many variations of interior point algorithms. The most popularly used algorithm for L_1 regression or quantile regression is the Primal-Dual with Predictor-Corrector algorithm.

Let $c = y$, $b = (1-\tau)X'e$, and $A = X'$, the dual problem (1.10) with a general upper bound u is

$$\begin{aligned} & \max\{c'z\} \\ \text{subject to} \quad & Az = b \\ & 0 \le z \le u. \end{aligned}$$

To solve this LP problem, $0 \le z \le u$ is split into $z \ge 0$ and $z \le u$. Let v be primal slack so that $z + v = u$, and associate dual variables w with these constraints.

The interior point solves the system of equations to satisfy the Karush-Kuhn-Tucker (KKT) conditions for optimality:

$$\begin{aligned} \text{(KKT)} \quad & Az = b \\ & z + v = u \\ & A't + s - w = c \\ & ZSe = 0 \\ & VWe = 0 \\ & z, s, v, w \ge 0, \end{aligned}$$

where $Z = \text{diag}(z)$, (that is, $Z_{i,j} = z_i$ if $i = j$, $Z_{i,j} = 0$ otherwise), $S = \text{diag}(s)$, $W = \text{diag}(w)$, $V = \text{diag}(v)$.

These are the conditions for feasibility, with the *complementarity* conditions $ZSe = 0$ and $VWe = 0$ added. $c'z = b't - u'w$ must occur at the optimum. Complementarity forces the optimal objectives of the primal and dual to be equal, $c'z_{\text{opt}} = b't_{\text{opt}} - u'w_{\text{opt}}$.

The interior point algorithm works by iteratively using Newton's method to find a direction $(\Delta z^k, \Delta t^k, \Delta s^k, \Delta v^k, \Delta w^k)$ to move from the current solution $(z^k, t^k, s^k, v^k, w^k)$ toward a better solution.

To do this, two steps are used. The first step is called an *affine* step, which solves a linear system using Newton's method to find a direction (Δz^k_{aff}, Δt^k_{aff}, Δs^k_{aff}, Δv^k_{aff}, Δw^k_{aff}) to reduce the complementarity toward zero. The second step is called a *centering* step, which solves another linear system to determine a centering vector (Δz^k_c, Δt^k_c, Δs^k_c, Δv^k_c, Δw^k_c) to further reduce the complementarity. The centering step may not reduce too much of the complementarity; however, it builds up the *central path* and makes substantial progress toward the optimum in the next iteration. With these two steps, then

$$\begin{aligned}
&(\Delta z^k, \Delta t^k, \Delta s^k, \Delta v^k, \Delta w^k) \\
&= (\Delta z_{aff}, \Delta t_{aff}, \Delta s_{aff}, \Delta v_{aff}, \Delta w_{aff}) \\
&\quad + (\Delta z_c, \Delta t_c, \Delta s_c, \Delta v_c, \Delta w_c)
\end{aligned}$$

$$\begin{aligned}
&(z^{k+1}, t^{k+1}, s^{k+1}, v^{k+1}, w^{k+1}) \\
&= (z^k, t^k, s^k, v^k, w^k) \\
&\quad + \alpha(\Delta z^k, \Delta t^k, \Delta s^k, \Delta v^k, \Delta w^k)
\end{aligned}$$

where, α is the *step length* assigned a value as large as possible but not so large that a z_i^{k+1}, s_i^{k+1}, v_i^{k+1}, or w_i^{k+1} is "too close" to zero.

Although the Predictor-Corrector variant entails solving two linear systems instead of one, fewer iterations are usually required to reach the optimum. The additional overhead of calculating the second linear system is small, as the factorization of the $(A\Theta^{-1}A')$ matrix has already been performed to solve the first linear system. Refer to Wright (1996) for more details about this algorithm.

4. Smoothing

Algorithms developed for a general LP problem may not fully deploy the properties of the original L_1 or quantile regression and have their own shortcomings. For example, the interior point algorithm can only give the approximate solutions of the original problem and rounding has to been done if one requires the same accuracy as that of the simplex algorithm. For some problems, this rounding step requires some significant extra computing time. In this case, some heuristic approaches demonstrate advantages on both speed and accuracy. One of them is the finite smoothing algorithm.

The finite smoothing algorithm was used by Clark and Osborne (1986), Madsen and Nielsen (1993) for the L_1 regression. It can be naturally extended to compute regression quantiles. In the following we briefly describe this algorithm, refer to Chen (2002) for details.

The non-differentiable

$$D_{\rho_\tau}(\beta) = \sum_{i=1}^{n} \rho_\tau(y_i - x_i'\beta)$$

can be approximated by the smooth function

$$D_{\gamma,\tau}(\beta) = \sum_{i=1}^{n} H_{\gamma,\tau}(r_i(\beta))$$

where

$$H_{\gamma,\tau}(t) = \begin{cases} t(\tau-1) - \frac{1}{2}(\tau-1)^2\gamma & \text{if } t \le (\tau-1)\gamma \\ \frac{t^2}{2}\gamma & \text{if } (\tau-1)\gamma \le t \le \tau\gamma \\ t\tau - \frac{1}{2}\tau^2\gamma & \text{if } t \ge \tau\gamma \end{cases}$$

See Figure 1 for a plot of $H_{\gamma,\tau}$ and ρ_τ.

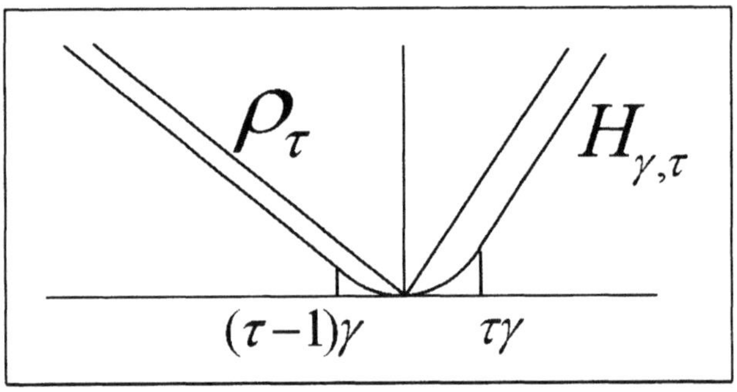

FIGURE 1. Objective Functions $H_{\gamma,\tau}$ and ρ_τ.

The function $H_{\gamma,\tau}$ is determined by whether $r_i(\beta) \le (\tau-1)\gamma$, $r_i(\beta) \ge \tau\gamma$, or $(\tau-1)\gamma \le r_i(\beta) \le \tau\gamma$. These inequalities divide $\boldsymbol{R}^p$ into subregions separated by the parallel hyperplanes $r_i(\beta) = (\tau-1)\gamma$ and $r_i(\beta) = \tau\gamma$. The set of all such hyperplanes is denoted by B_γ:

$$B_\gamma = \{\beta \in \boldsymbol{R}^p | \exists i : r_i(\beta) = (\tau-1)\gamma \text{ or } r_i(\beta) = \tau\gamma\}$$

Define the sign vector $s_\gamma(\beta) = (s_1(\beta), \ldots, s_n(\beta))'$ by

$$s_i = s_i(\beta) = \begin{cases} -1 & \text{if } r_i(\beta) \le (\tau-1)\gamma \\ 0 & \text{if } (\tau-1)\gamma \le r_i(\beta) \le \tau\gamma \\ 1 & \text{if } r_i(\beta) \ge \tau\gamma \end{cases}$$

and introducing

$$w_i = w_i(\beta) = 1 - s_i^2(\beta)$$

thus,

$$H_{\gamma,\tau}(r_i(\beta)) = \frac{1}{2}\gamma w_i r_i^2(\beta)$$
$$+s_i[\frac{1}{2}r_i(\beta) + \frac{1}{4}(1-2\tau)\gamma + s_i(r_i(\beta)(\tau-\frac{1}{2}) - \frac{1}{4}(1-2\tau+2\tau^2)\gamma)]$$

yielding

$$D_{\gamma,\tau}(\beta) = \frac{1}{2}\gamma r' W_\gamma r + v'(s)r + c(s)$$

where W_γ is the diagonal n by n matrix with diagonal elements $w_i(\beta)$, $v'(s) = (s_1((2\tau-1)s_1+1)/2, \ldots, s_n((2\tau-1)s_n+1)/2)$, $c(s) = \sum[\frac{1}{4}(1-2\tau)\gamma s_i - \frac{1}{4}s_i^2(1-2\tau+2\tau^2)\gamma]$, and $r(\beta) = (r_1(\beta), \ldots, r_n(\beta))'$.

The gradient of $D_{\gamma,\tau}$ is given by

$$D^{(1)}_{\gamma,\tau}(\beta) = -X'[\frac{1}{\gamma}W_\gamma(\beta)r(\beta) + v(s)]$$

and for $\beta \in \boldsymbol{R}^p \backslash B_\gamma$ the Hessian exists and is given by

$$D^{(2)}_{\gamma,\tau}(\beta) = \frac{1}{\gamma}X'W_\gamma(\beta)X$$

The gradient is a continuous function in $\boldsymbol{R}^p$, whereas the Hessian is piecewise constant.

s is called a *γ-feasible sign vector* if there exists $\beta \in \boldsymbol{R}^p \backslash B_\gamma$ with $s_\gamma(\beta) = s$. If s is *γ-feasible* then Q_s is defined as the quadratic, which is deduced from $D_{\gamma,\tau}(\beta)$ by inserting s instead of s_γ. Thus, for any β with $s_\gamma = s$,

$$Q_s(\alpha) = \frac{1}{2}(\alpha - \beta)' D^{(2)}_{\gamma,\tau}(\beta)(\alpha - \beta) + D^{(1)}_{\gamma,\tau}(\beta)(\alpha - \beta) + D_{\gamma,\tau}(\beta)$$

Clearly $D_{\gamma,\tau}(\alpha) = Q_s(\alpha)$ in the domain

$$C_s = \{\alpha | s_\gamma(\alpha) = s\}$$

For each $\gamma > 0$ and $\beta \in \boldsymbol{R}^p$, there is one or several corresponding quadratics Q_s. If $\beta \notin B_\gamma$ then Q_s is characterized by β and γ, but for $\beta \in B_\gamma$ the quadratic is not unique. Therefore, a *reference*

$$(\gamma, \beta, s)$$

determines the quadratic. Define
(γ, β, s) be a *feasible reference* if s is a *γ-feasible sign vector* with $\beta \in C_s$, and
(γ, β, s) be a *solution reference* if it is feasible and β minimizes $D_{\gamma,\tau}$.

The smoothing algorithm for minimizing D_{ρ_τ} is based on minimizing $D_{\gamma,\tau}$ for a set of decreasing γ. For every new value of γ, information from the previous solution is utilized. Finally, when γ is small enough, a solution can be found by the modified Newton-Ralphon algorithm. The algorithm is described as:

> find an initial solution reference $(\gamma, \beta_\gamma, s)$
> **repeat**
> decrease γ
> find a solution reference $(\gamma, \beta_\gamma, s)$
> **until** $\gamma = 0$
> β_0 is the solution.

The initial solution reference is found by letting β_γ be the least squares solution and then choosing γ and s appropriately. β_0 can be found after finite iterations when γ becomes small enough.

A similar strategy by Madsen and Nielsen (1993) can be used to decrease γ. The computation involved is not significant comparing with the Newton-Ralphon step.

For the L_1 and quantile regression, it turns out that the smoothing algorithm is very efficient and competitive, especially for fat data sets (e.g., $\frac{p}{n} \geq .05$).

5. Combining the Three Algorithms

Each of the previous three algorithms we described has its own advantages. None of them can dominate others. Based on this fact we construct an adaptive algorithm by combining these three algorithms. In the adaptive algorithm, we select one of the three algorithms according to our experience through a large number of simulations for various data sets.

If $n \le 5000$ and $p \le 50$, use **simplex**
else if $\frac{p}{n} \ge .05$, use **smoothing**
else use **interior point**

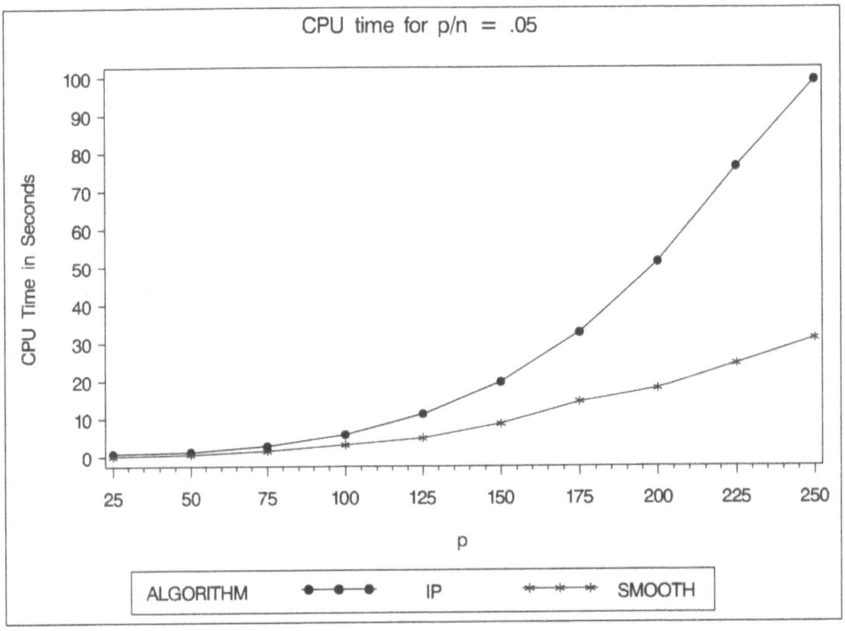

FIGURE 2. CPU Time for $\frac{p}{n} = .05$.

Comparisons between the simplex algorithm and the interior point algorithm have been thoroughly done in Portnoy and Koenker (1997). Here we do some simple comparisons between the interior point algorithm and the smoothing algorithm. More details can be found in Chen (2002).

A data set is called *fat* if the ratio between the number of variables p and the number of observations n exceeds certain percentage. Here our empirical value for this percentage is 5%. With these kind of data sets, the smoothing algorithm out-performances the interior point algorithm in average. The main reason is the

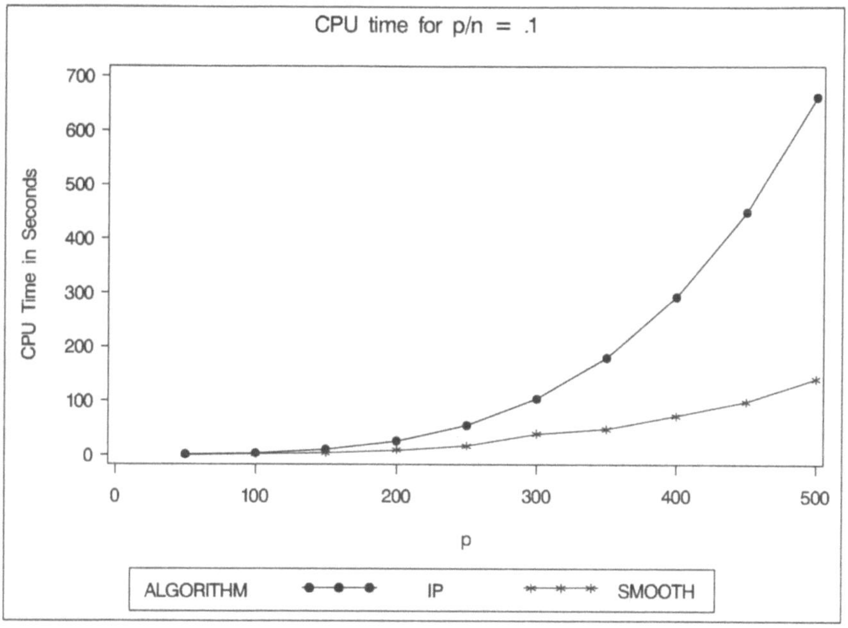

FIGURE 3. CPU Time for $\frac{p}{n} = .1$.

smoothing algorithm finds the optimum solution by fewer iterations and that contributes fewer QR decompositions. This saves computing time, especially for the *fat* data sets.

To support this argument, we carried two experiments with the L_1 regression. The first one has a fixed ratio $\frac{p}{n} = .05$ and the second one has $\frac{p}{n} = .1$. The sizes of the data sets run from 500 to 5000 by 500. The variables are randomly generated normal variables. The computing was carried out by a Dell PC with 1.33GHz CPU and 512MB RAM. The CPU time (in seconds) is displayed on the two plots.

For data set with large n and small p, like $\frac{p}{n} < .001$, smoothing algorithm can be faster by using a fast threshold decreasing procedure. The difference in computing time of the two algorithms is not as significant as for the *fat* data set. For data set with $.001 < \frac{p}{n} < .01$, our test cases show that these two algorithms are competitive.

It should be pointed out that the interior point algorithm here does not include the preprocessing described in Portnoy and Koenker (1997). The smoothing algorithm can also implement the preprocessing. The effect of the preprocessing to the smoothing algorithm is under investigation.

References

[1] I. Barrodale and F.D.K. Roberts, *An Improved Algorithm for Discrete l_1 Linear Approximation.* SIAM J. Numer. Anal. **10** (1973), 839–848.

[2] D.I. Clark and M.R. Osborne, *Finite Algorithms for Huber's M-estimator.* SIAM J. Sci. Statist. Comput. **6** (1986), 72–85.

[3] C. Chen, *A Finite Smoothing Algorithm for Quantile Regression.* Submitted, 2002.

[4] P.J. Huber, *Robust Statistics.* John Wiley & Sons, New York, 1981.

[5] N. Karmarkar, *A New Polynomial-time Algorithm for Linear Programming.* Combinatorica **4** (1984), 373–395.

[6] R. Koenker and G. Bassett, Jr, *Regression Quantiles.* Econometrica **46** (1978), 33–50.

[7] R. Koenker and V. d'Orey, *Computing Regression Quantiles.* Appl. Statist. **43** (1993), 410–414.

[8] K. Madsen and H.B. Nielsen, *A Finite Smoothing Algorithm for Linear L_1 Estimation.* SIAM J. Optim. **3** (1993), 223–235.

[9] S. Portnoy and R. Koenker, *The Gaussian Hare and the Laplacian Tortoise: Computation of Squared-error vs. Absolute-error Estimators.* Statist. Sci. **12** (1997), 279–300.

[10] S.J. Wright, *Primal-Dual Interior Point Algorithms.* SIAM, Philadelphia, 1996.

Acknowledgment

Many thanks to the organizers of the ICORS 2003. The author would also like to thank Xuming He and Roger Koenker for their discussions.

C. Chen
R4117 R&D
SAS Institute Inc.
Cary, NC 27513
USA
e-mail: lin.chen@sas.com

Statistics for Industry and Technology, 49–58

On Properties of Support Vector Machines for Pattern Recognition in Finite Samples

A. Christmann

Abstract. The support vector machine proposed by Vapnik belongs to a class of modern statistical learning methods based on convex risk minimization. Other special cases are AdaBoost, kernel logistic regression and least squares. The support vector machine has the advantage that it usually leads to a reduction of complexity, because only the support vectors and not all observations contribute to the prediction of a new response. This paper addresses robustness properties of the support vector machine for pattern recognition in finite samples. Sensitivity curves in the sense of J.W. Tukey are used to investigate the possible impact of a single data point.

Mathematics Subject Classification (2000). Primary 62G35, 62H30; Secondary 62G05, 47N30.

Keywords. Convex risk minimization, pattern recognition, reduction of complexity, robustness, sensitivity function, statistical learning methods, support vector machine.

1. Introduction

In statistical machine learning for pattern recognition two major goals are the estimation of a functional relationship $y \approx f(x)$ between an outcome y and predictors $x = (x_1, \ldots, x_k)' \in \mathbb{R}^d$ and the prediction of an unobserved outcome y_{new} based on an observed value x_{new} of predictors. The function f is unknown. One needs the implicit assumption that the relationship between x_{new} and y_{new} is – at least almost – the same as in the training data set (x_i, y_i), $i = 1, \ldots, n$. Otherwise, it is useless to extract knowledge on f from the training data set. The classical assumption in machine learning is, that the training data (x, y) are independent and identically generated from an underlying unknown distribution $\mathbb{P}$ for random

The financial support of the Deutsche Forschungsgemeinschaft (SFB 475, "Reduction of complexity in multivariate data structures") and of the Forschungsband DoMuS ("Modelling and Simulation") are gratefully acknowledged.

variables (X, Y). In practical applications the training data set is often quite large, high dimensional and complex. The quality of the predictor $f(x)$ is measured by some loss function $L(y, f(x))$. The goal is to find a predictor $f(x)$ which minimizes the expected loss of f:

$$R_{L,\mathbb{P}}(f, b) = \mathrm{E}_{\mathbb{P}} L(Y, f(X) + b). \tag{1.1}$$

In this paper we are interested in binary classification, where $y \in \{-1, +1\}$. The classification error is given by $I(y, f(x)) = \mathbb{I}(yf(x) < 0) + \mathbb{I}(f(x) = 0)\mathbb{I}(y = -1)$, where $\mathbb{I}$ denotes the indicator function. Based on the law of large numbers one might estimate f by the minimizer of the empirical classification error

$$\frac{1}{n}\sum_{i=1}^{n} I(y_i, f(x_i) + b). \tag{1.2}$$

Unfortunately, the classification function I is not convex and the minimization of (1.2) is NP-hard (Höffgen et al., 1995). To circumvent this problem, one minimizes a convex upper bound of the classification error function $I(y, f)$ (Schölkopf and Smola, 2002; Vapnik, 1998). If $L : X \to \mathbb{R}$ is an appropriate convex function, one considers the (approximate) minimization of the empirical risk (Steinwart, 2002):

$$(\hat{f}_{n,\lambda}, \hat{b}_{n,\lambda}) = \arg \min_{f \in H,\, b \in \mathbb{R}} \lambda \|f\|_H^2 + \frac{1}{n}\sum_{i=1}^{n} L(y_i, f(x_i) + b), \tag{1.3}$$

where λ is a small regularization parameter and H is a reproducing kernel Hilbert space (RKHS) of a kernel k. Often λ is expressed via a positive penalizing constant C by $\lambda = (2Cn)^{-1}$. Let $\Phi : X \to H$ be the feature map of H, i.e. $\Phi(x) := k(x, .)$. Hence, $f(x) = \Phi(x)'\theta + b$. Classical kernels k are linear kernels and the (exponential) radial basis function kernel $k(x, x') = \exp(-\gamma\|x - x'\|^2)$, $\gamma > 0$. The prediction of f for the label (x, y) is $\mathrm{sign}(f(x) + b)$. This problem can be interpreted as a stochastic approximation of the minimization of the theoretical regularized risk given in (1.4) (Vapnik,1998; Zhang, 2001; Steinwart, 2002):

$$(f_{\mathbb{P},\lambda}, b_{\mathbb{P},\lambda}) = \arg \min_{f \in H,\, b \in \mathbb{R}} \lambda \|f\|_H^2 + E_{\mathbb{P}} L(Y, f(X) + b). \tag{1.4}$$

The objective function in (1.4) is denoted by $R^{reg}_{L,\mathbb{P},\lambda,b}$. Note that the solution of (1.4) is unique for convex loss functions. This is one reason why convexity plays a prominent role in classical SVM theory. This is in contrast to robust statistics, where often non-convex loss functions are used, of course with two important exceptions: L1 and Huber's loss function. Suykens et al. (2002) investigate weighted least squares support vector machines for regression problems in order to deal with non-convex loss functions as well.

In this paper we will consider the support vector machine (SVM) which uses the loss function $L(y, f(x) + b) = \max\{1 - y(f(x) + b), 0\}$. The support vector machine has the inherent property that it leads to a reduction of complexity, as usually not all observations but only the support vectors contribute to the predictions. Other special cases are AdaBoost, kernel logistic regression, least squares and modified least squares. Zhang (2001) shows for many convex loss functions that the classifiers based on (1.3) are universally consistent if $\lambda_n \to 0$ and $\lambda_n n \to \infty$, i.e. the classification error of $\hat{f}_{n,\lambda}(.)$ converges to the optimal Bayes error in probability. Steinwart (2002) characterizes the loss functions which lead to universally consistent classifiers and establishes universal consistency for classifiers based on (1.3). Furthermore, he shows that there exist solutions of the minimization problems of the theoretical and of the empirical problems. Steinwart (2002) also gives lower asymptotical bounds on the number of support vectors, i.e., on the data points with non-vanishing coefficients, and investigates the asymptotic behavior of $\hat{f}_{n,\lambda}(.)$ in terms of the loss function L. The RKHS part of the solutions of (1.4) are unique. Schölkopf and Smola (2002) describe other support vector machines and give an overview on algorithms to solve the minimization problems corresponding to SVMs.

Obviously, the proof that many classifiers based on convex loss functions are universally consistent under weak conditions is a strong argument in favor of these statistical learning methods. Nevertheless, it is important to investigate robustness properties for such statistical learning methods for the following reasons. In practice one has to apply the methods to data sets with a finite sample size which contain a positive percentage of gross errors. Hampel et al. (1986, p. 27) note that 1% to 10% of gross errors in routine data seem to be more the rule than the exception. The data quality is sometimes far from being optimal, especially in large data mining problems (Hipp et al., 2001). From our point of view it is therefore important to investigate the impact a small amount of contamination of the 'true' probability measure $\mathbb{P}$ can have on the support vector machine.

The rest of the paper is organized as follows. Section 2 gives the definition of the sensitivity curve, which is the robustness concept we are dealing with. Section 3 contains the results and Section 4 contains the conclusion.

2. Sensitivity Curve

The sensitivity curve proposed by J.W. Tukey has the interpretation, that it measures the impact of just one additional data point z on the empirical quantity of interest.

Definition 2.1. *Sensitivity curve.* The sensitivity curve of an estimator T_n at a point z given a data set $z_1, \ldots, z_{n-1}$ is defined by

$$SC_n(z; T_n) = n\big(T_n(z_1, \ldots, z_{n-1}, z) - T_{n-1}(z_1, \ldots, z_{n-1})\big)\,. \tag{2.1}$$

The Dirac distribution at the point z is denoted by Δ_z. If the estimator T_n is defined via $T(\mathbb{P}_n)$, where $\mathbb{P}_n$ denotes the empirical distribution of the data points $z_1, \ldots, z_n$, then it holds for $\varepsilon_n = 1/n$:

$$SC_n(z; T_n) = \frac{T\big((1-\varepsilon_n)\mathbb{P}_{n-1} + \varepsilon_n \Delta_z\big) - T(\mathbb{P}_{n-1})}{\varepsilon_n}\,. \tag{2.2}$$

For many estimators it holds, that the sensitivity curve converges to the influence function, as n tends to infinity. Counterexamples are given, e.g., in Davies (1993). The sensitivity curve can be interpreted as a finite sample version of the influence function, see Hampel et al. (1986, p. 93).

Here, we like to investigate robustness properties of the SVM for *finite* samples. In the next section, the sensitivity curve is used to study the impact of a single data point on the results of the SVM. For an investigation of the influence function of certain statistical learning methods based on convex risk minimization methods, e.g. the support vector machine, kernel logistic regression, AdaBoost, and least squares, see Christmann and Steinwart (2003).

3. Results

We study the impact an additional data point can have on the SVM for pattern recognition on

- the estimation of $\hat{f}$,
- the fitted value of y,
- the empirical misclassification error,
- the estimated parameter θ, and
- on the prediction areas of y.

First, we consider the well-known mixture data having $n = 200$ data points given in Hastie et al. (2001), see Figure 1. The computations were done using the software SVM^{light} developed by Joachims (1999). We consider an exponential radial basis function (RBF) kernel $\Phi(x, x') = \exp(-\gamma\|x-x'\|^2)$. Appropriate values for γ and for the penalizing constant C are important for the SVM and are often determined by cross validation (Schölkopf and Smola, 2002). A cross validation based on the leave-one-out error for the training data set was carried out by a two-dimensional grid search on $\gamma \in \{0.1, 0.25, 0.5, 0.75, 1, 1.5, 2, 3, 4, 5, 10, 20\}$ and $C \in \{0.5, 0.75, 1, 1.25, 1.5, 1.75, 2, 5, 10, 20\}$. As a result of the cross validation, the tuning parameters for the SVM with RBF kernel were set to $\gamma = 3$ and $C = 1$. For $n = 200$ this results in $\lambda = (2Cn)^{-1} = 0.0025$.

Figure 2 shows the sensitivity curve of $\hat{f}$, if we add a single point $z = (x, y)$ to the original data set, where $x_1 = 2$, $x_2 = -2$, and $y = +1$. The additional data point has only a local and smooth impact on $\hat{f}$, if one uses the RBF kernel. As was to be expected, the local impact near by z on $\hat{f}$ is relatively large (but finite) because λ is very small. The outlying point would have an impact in a broader region for higher values of λ. From Figure 3 where plots of $\hat{f}$ are given for the original data set and for the modified data set, which contains the additional data point z, we see that the RBF kernel yields $\hat{f}$ approximately equal to zero outside a central region, as almost all data points are lying inside the central region. Comparing the plot of $\hat{f}$ for the modified data set with the corresponding plot for the original data set, it is obvious that the additional smooth peak is due to the new data point located at $x = (2, -2)$ with $y = +1$.

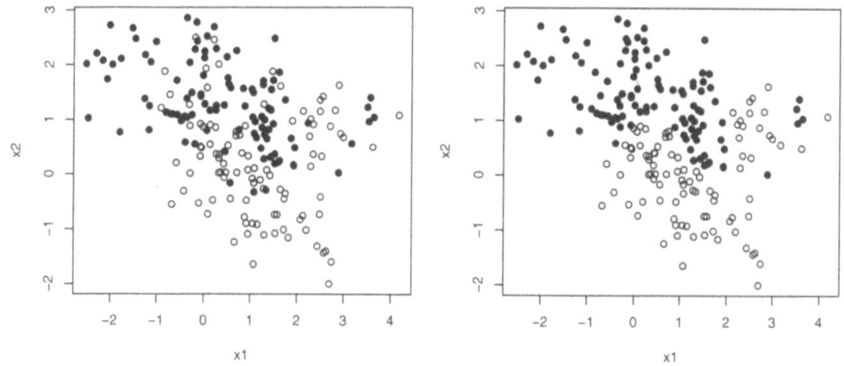

FIGURE 1. The mixture data set (Hastie et al., 2001). Left: data points. Right: fitted values. Legend: ∘ for $y = -1$, • for $y = +1$.

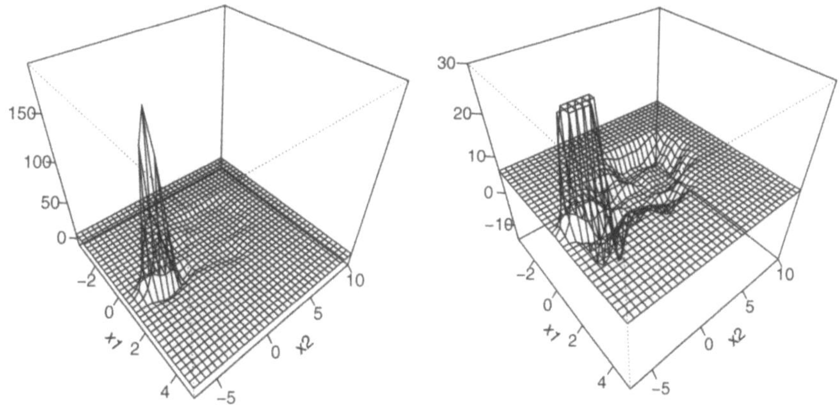

FIGURE 2. Left: Sensitivity function of $\hat{f}$, if the additional data point z is located at $z = (x, y)$, where $x = (2, -2)$ and $y = +1$. Right: identical figure, but the $z-$axis has been truncated by 30 to increase the visibility.

Now, we study the impact of an additional data point $z = (x, y)$, where $y = +1$, on the percent of classification errors and on the fitted $y-$value for z. We vary z over a grid in the $x-$coordinates. Figure 4 shows that the percent of misclassification errors is approximately constant outside the central region that contains almost all data points. The response of the additional data point was correctly estimated by $\hat{y} = +1$ outside the central region.

Next, we study the impact of an additional data point located at $z = (x, y)$ with $y = +1$ on the estimated parameters $\hat{\theta}$ and $\hat{b}$, see Figure 5. We vary z

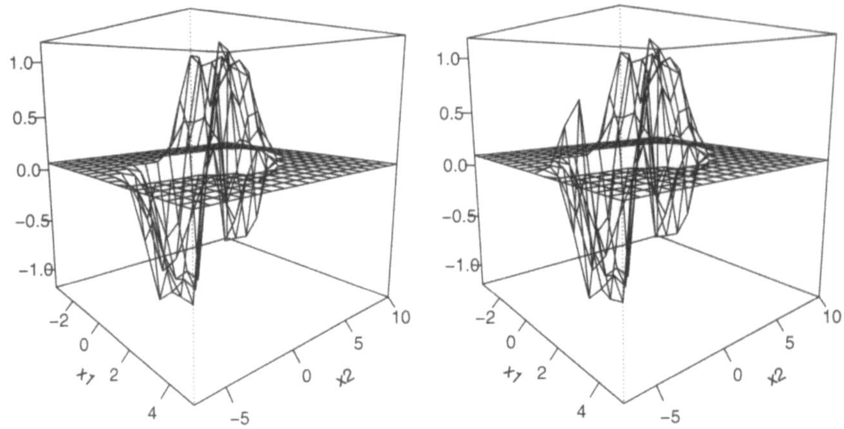

FIGURE 3. Plot of $\hat{f}$. Left: RBF kernel, original data set. Right: RFB kernel, modified data set. The modified data set contains the additional data point $z = (x, y)$, where $x = (2, -2)$ and $y = +1$.

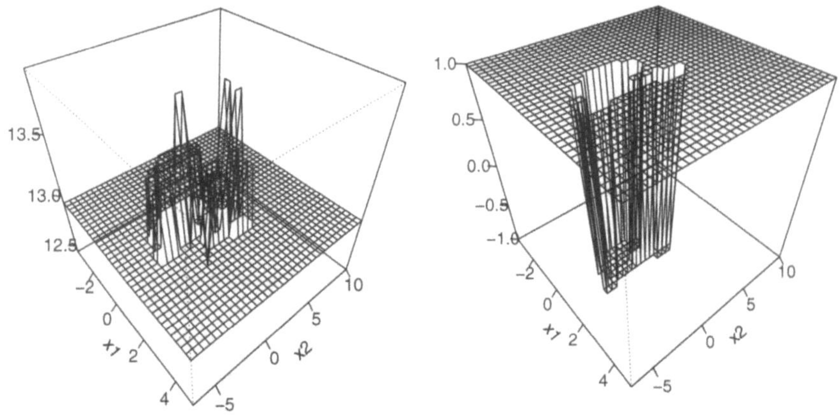

FIGURE 4. Left: Percent of classification errors if one data point $z = (x, y = 1)$ is added to the original data set, where x varies over the grid. Right: Fitted y-value for new observation if one data point $z = (x, y = 1)$ is added to the original data set, where x varies over the grid.

over a grid in the x-coordinates in the same manner as before. The sensitivity curves for the slopes estimated by the SVM with a RBF kernel are similar to an affine hyperplane outside the central region, which contains almost all data points. However, this does *not* mean, that the predictions of outlying points are bad, see

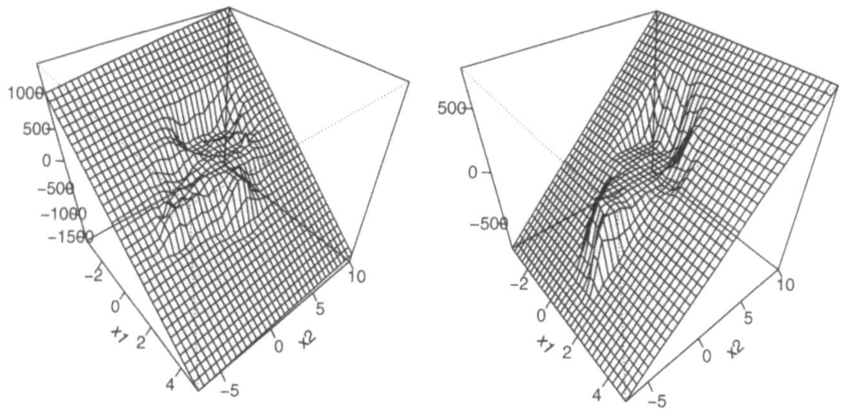

FIGURE 5. Left: Sensitivity curve for the intercept $\hat{b}$. Right: Sensitivity curve for $\hat{\theta}_2$. The sensitivity curve for $\hat{\theta}_1$ is not shown here, as it looks very similar.

Figure 4. In the central region, there is a smooth transition between regions with higher sensitivity values and regions with lower sensitivity values.

For studying the effect a single data point can have on the prediction areas of the response y depending on the penalizing constant C, which affects λ, we generated a data set with $n = 500$ data points x_i from a bivariate Student's t_3 distribution with location parameter $\mu = (0, 0)$ and scatter matrix Σ. The diagonal elements of Σ were set to 1, whereas the off-diagonal elements were set to 0.25. The responses y_i were generated from a classical logistic regression model with intercept for the parameter vector $\theta = (-1, 1)$ and $b = 1$, such that $P(Y_i = +1) = [1 + \exp(-[b + x_i'\theta])]^{-1}$ and $P(Y_i = -1) = 1 - P(Y_i = +1)$.

The upper part of Figure 6 shows that the choice of the penalizing constant C can be quite important for making predictions based on a support vector machine with an RBF kernel. This holds true especially, if predictions are made for a response with an x-value outside the bulk of the x-values of the training data set. In this sense, the SVM can produce unstable predictions and should be used with some care, if the value of C is not appropriate for the present data set. In the lower part of Figure 6, corresponding prediction areas are given if two of the data points are moderate outliers. The SVM with an RBF kernel is able to accommodate outliers due to the consistency property.

Similar sensitivity analyzes were done for some other situations: increased sample size, situations where a complete separation of both response groups is possible, and for a multivariate normal distribution instead of a multivariate Student distribution to generate the x-values. The results are not given here, because the results were qualitatively similar to those described in this section.

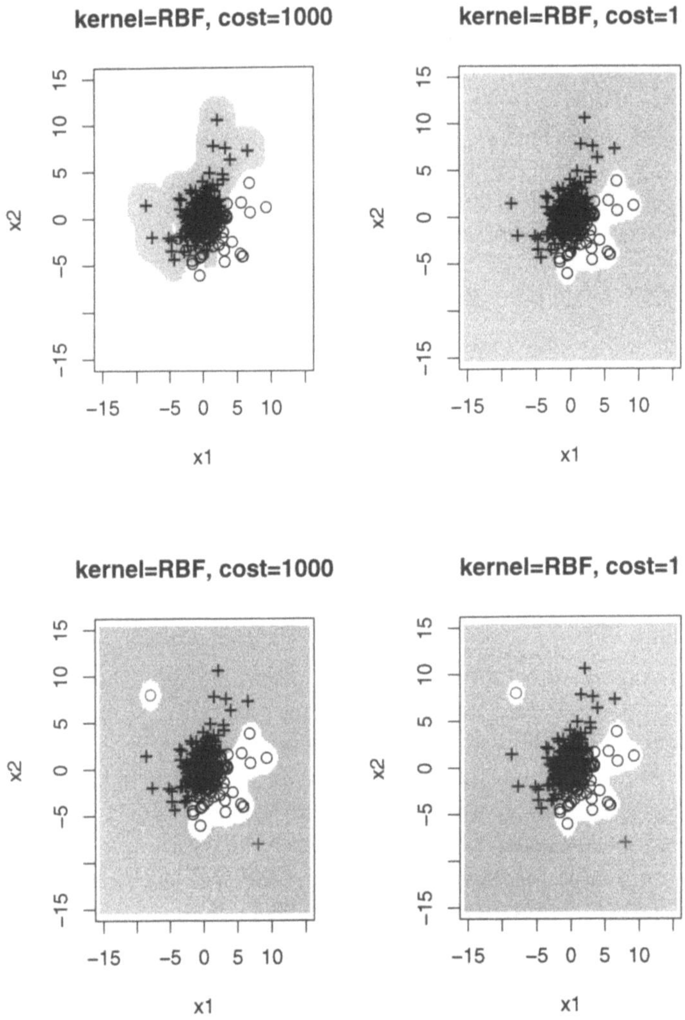

FIGURE 6. Prediction areas from a classical logistic regression model. Upper: simulated data set. Lower: simulated data set with two moderate outliers located at $x_A = (8, -8)$ with $y_A = 1$ and $x_B = (-8, 8)$ with $y_B = -1$. Legend: $+$ for $y = 1$, $\circ$ for $y = -1$. The prediction area for $\hat{y} = 1$ is shaded.

4. Conclusion

In this paper, we used J.W. Tukey's sensitivity curve to study the impact of a data point on the support vector machine which is a popular statistical learning method based on convex risk minimization methods. Our results show that the use of an exponential radial basis function kernel leads to a smooth sensitivity curve for $\hat{f}$ which may have a peak around the additional data point if the response is unprobable. Varying the position of one additional data point had a smooth and local impact on $\hat{f}$ and on the estimated parameters $\hat{\theta}$ and $\hat{b}$, if one uses an RBF kernel. Of course, the height of the peak depends on the regularizing parameter λ and on the tuning parameter γ used by the kernel. Sensitivity curves for $\hat{f}$ for a support vector machine with a linear kernel were not shown here, but indicate that the impact of varying one additional data point seems to be more globally than locally.

An investigation concerning existence and uniform bounds with respect to $\mathbb{P}$ and z of the influence function of certain statistical learning methods based on convex risk minimization methods is given in Christmann and Steinwart (2003). Support vector machine and kernel logistic regression are treated as special cases. Suykens et al. (2002) investigate robustness properties of weighted least squares support vector machines for regression problems.

For a numerical comparison between the support vector machine with a linear kernel and the regression depth method recently proposed by Rousseeuw and Hubert (1999) see Christmann and Rousseeuw (2001) and Christmann et al. (2002).

References

[1] A. Christmann, P. Fischer, and T. Joachims, *Comparison Between Various Regression Depth Methods and the Support Vector Machine to Approximate the Minimum Number of Misclassifications.* Comput. Statist. **17** (2002), 273–287.

[2] A. Christmann and P.J. Rousseeuw, *Measuring Overlap in Logistic Regression.* Comput. Statist. Data Anal. **37** (2001), 65–75.

[3] A. Christmann and I. Steinwart, *On Robustness Properties of Convex Risk Minimization Methods for Pattern Recognition.* Submitted, 2003, available at *http://www.statistik.uni-dortmund.de/download/ mitarbeiter/christmann/christmann03a.pdf.*

[4] P.L. Davies, *Aspects of Robust Linear Regression.* Ann. Statist. **21** (1993), 1843–1899.

[5] F.R. Hampel, E.M. Ronchetti, P.J. Rousseeuw, and W.A. Stahel, *Robust Statistics: The Approach Based on Influence Functions.* Wiley, New York, 1986.

[6] T. Hastie, R. Tibshirani, and J. Friedman, *The Elements of Statistical Learning. Data Mining, Inference and Prediction.* Springer, New York, 2001.

[7] J. Hipp, U. Güntzer, and U. Grimmer, *Data Quality Mining - Making a Virtue of Necessity.* Workshop on Research Issues in Data Mining and Knowledge Discovery DMKD. Santa Barbara, CA, 2001, available at *http://www.cs.cornell.edu/ johannes/papers/dmkd2001-papers/p5_hipp.pdf.*

[8] K.U. Höffgen, H.-U. Simon, and K.S. van Horn, *Robust Trainability of Single Neurons.* J. Comput. System Sci. **50** (1995), 114–125.

[9] T. Joachims, *Making Large-Scale SVM Learning Practical.* In B. Schölkopf, C. Burges, and A. Smola, editors, Advances in Kernel Methods – Support Vector Learning. MIT-Press, 1999.

[10] P.J. Rousseeuw and M. Hubert, *Regression Depth.* J. Amer. Statist. Assoc. **94** (1999), 388–433.

[11] B. Schölkopf and A.J. Smola, *Learning with Kernels. Support Vector Machines, Regularization, Optimization, and Beyond.* MIT Press, 2002.

[12] I. Steinwart, *Support Vector Machines are Universally Consistent.* J. Complexity **18** (2002), 768–791.

[13] I. Steinwart, *Consistency of Support Vector Machines and Other Regularized Kernel Machines.* Preprint, 2002, available at *http://www.minet.uni-jena.de/~ingo/publications/info-02.ps.*

[14] I. Steinwart, *Sparseness of Support Vector Machines.* Preprint, 2003, available at *http://www.minet.uni-jena.de/~ingo/publications/jmlr-03.ps.*

[15] J.A.K. Suykens, T. Van Gestel, J. De Brabanter, B. De Moor, and J. Vandewalle, *Support Vector Machines.* World Scientific, Singapore, 2002.

[16] J.A.K. Suykens, J. De Brabanter, L. Lukas, and J. Vandewalle, *Weighted Least Squares Support Vector Machines: Robustness and Sparse Approximation.* Neurocomputing **48** (2002), 85–105.

[17] V. Vapnik, *Statistical Learning Theory.* Wiley, New York, 1998.

[18] T. Zhang, *Statistical Behaviour and Consistency of Classification Methods Based on Convex Risk Minimization.* Ann. Statist. (2001), to appear.

A. Christmann
University of Dortmund
Department of Statistics
D-44221 Dortmund
Germany
e-mail: `christmann@statistik.uni-dortmund.de`

Statistics for Industry and Technology, 59–70

Smoothed Local L-Estimation With an Application

P. Čížek

Abstract. The Nadaraya-Watson regression estimator is known to be highly sensitive to the presence of outliers in the sample. A possible robustification consists in using local L-estimates of regression. Whereas the local L-estimation is traditionally done using the empirical conditional distribution function, Tamine et al. (2003) have recently proposed to use a smoothed conditional distribution function instead. This work studies computational aspects and small-sample properties of the smoothed L-estimation approach. The smoothed nonparametric L-estimator is applied to the estimation of the so-called implied volatilities, which describe the conditional variance of high-frequency financial time series (such as exchange rates or stock prices) inferred from the prices of related financial derivatives.

Mathematics Subject Classification (2000). Primary 62G08; Secondary 62G35.

Keywords. Nonparametric regression, L-estimation, smoothed cumulative distribution function.

1. Introduction

One of the most widely used nonparametric regression estimators is the Nadaraya-Watson estimator by Nadaraya (1964) and Watson (1964). This estimator, being a local average of the response variable, is highly sensitive to the presence of outliers in the data: outliers do not only increase the variance of the estimator, but can also create fictitious peaks and structures in the estimated function. There have been many attempts to make the Nadaraya-Watson estimator and its extensions more robust. Many of them rely on local M-estimation, see for example Härdle and Tsybakov (1988), Fan and Jiang (1999), and Beran et al. (2001). On the other hand, Boente and Fraiman (1994) proposed to use a local L-estimate such as the local α-trimmed mean instead of a locally weighted average. Their procedure

The research was supported by grant GA ČR number 402/03/0084.

consists in using an empirical conditional distribution function to estimate the conditional quantiles and to determine data to trim.

The median smoothing and local L_1-estimation are special cases of local L-estimation. Since they are easy to implement and their performance and robustness in nonparametric regression are quite good (see Čížek and Härdle, 2003, for instance), the use of local L-estimates may be a quite appealing alternative. Unfortunately, the use of the empirical distribution function, which is a step function by definition, increases the variance of the estimates, and additionally, the computational demands grow significantly with the sample size. To remedy this, Tamine et al. (2003) proposed to use a smoothed cumulative distribution function instead of the empirical one (see Fernholz, 1997, for theoretical arguments in the case of non-conditional L-estimates).

In Section 2, after introducing some basic concepts and results, we propose an algorithm for the smoothed local L-estimation. Section 3 covers its finite-sample properties, whereas Section 4 presents an application to the estimation of implied volatility surfaces.

2. Local L-Estimation

Boente and Fraiman (1994) defined the conditional regression L-estimate of Y on X at some $x \in R^p$ by

$$m_L(x) = \int yJ\{F_x(y)\}\, dF_x(y), \tag{2.1}$$

where the L-score function J is continuously differentiable a.e. on a compact support $[a, b] \subset (0, 1)$ and $F_x(y)$ is the conditional distribution function of Y conditional on $X = x$. This definition covers for example the α-trimmed mean and median: $J(z) = (1 - 2\alpha)^{-1} I_{\langle\alpha;1-\alpha\rangle}(z)$, where $\alpha \in \langle 0, 1/2\rangle$ represents the amount of trimming (median corresponds to $\alpha = 1/2$).

Given a sample $(X_i, Y_i)_{i=1}^n$, Boente and Fraiman (1994) proposed to estimate the unknown conditional distribution function $F_x(y)$ by the empirical conditional distribution function

$$\tilde{F}_x(y) = \sum_{i=1}^{n} \frac{K\left(\frac{x-X_i}{h_x}\right)}{\sum\limits_{j=1}^{n} K\left(\frac{x-X_j}{h_x}\right)} I(Y_i \leq y), \tag{2.2}$$

where K is a p-dimensional kernel function, h_x is a bandwidth, and I represents an indicator function. Plugging it in the definition (2.1) of the conditional L-estimator, one obtains the local empirical L-estimator

$$\tilde{m}_L(x) = \sum_{i=1}^{n} \frac{K_x\left(\frac{x-X_i}{h_x}\right) Y_i}{\sum\limits_{j=1}^{n} K\left(\frac{x-X_j}{h_x}\right)} J\left\{\tilde{F}_x(Y_i)\right\}. \tag{2.3}$$

Apart from $\tilde{F}_x(y)$ being discontinuous, using $\tilde{m}_L(x)$ is also computationally demanding for large samples, since the number of operations needed to evaluate $\tilde{m}_L(x)$ at one point x grows as $O(n^2)$ (every element of the sum requires estimating $\tilde{F}_x(Y_i)$).

Therefore, motivated by studies on smoothed conditional quantile estimators by Mehra et al. (1991) and smoothed unconditional L-estimators by Fernholz (1997), Tamine et al. (2003) proposed to estimate $F_x(y)$ by a smoothed conditional distribution function

$$\hat{F}_x(y) = \sum_{i=1}^{n} \frac{K\left(\frac{x-X_i}{h_x}\right)}{\sum_{j=1}^{n} K\left(\frac{x-X_j}{h_x}\right)} K_I\left(\frac{y-Y_i}{h_y}\right), \tag{2.4}$$

where K_I is a univariate cumulative distribution function (the integral of a kernel) and h_y is another bandwidth, and they proved its asymptotic normality under mild β-mixing conditions. This estimator inherits the regularity properties of the univariate kernel K_I and is typically differentiable. Moreover, the function $m_L(x)$ can then be estimated by the local smoothed L-estimator

$$\hat{m}_L(x) = \int yJ\left\{\hat{F}_x(y)\right\} d\hat{F}_x(y), \tag{2.5}$$

whereby the integral is approximated using classical numerical integration routines. Given a conditional distribution function, we typically need for numerical integration with a prescribed precision a grid with a fixed number k of points, and hence, $\hat{F}_x(y)$ has to be evaluated only at a fixed and finite number of points. Tamine et al. (2003) presented a detailed analysis of computational demands for both the empirical ($O(n^2)$ operations) and smoothed local L-estimation ($O(kn)$ operations). Consequently, the smoothed local L-estimation is faster to compute and should exhibit better small-sample properties because of the continuity of $\hat{F}_x(y)$. On the other hand, the asymptotic distribution of the empirical and smoothed L-estimates are equivalent (Tamine et al., 2003).

In the following Section 3, we will compare in more details the finite-sample performance of the empirical and smoothed local L-estimates and other existing nonparametric smoothers. Before doing so, note that the integration procedure used here differs from the one used by Tamine et al. (2003), who used an equidistant integration grid defined on $(\max_i Y_i + |\max_i Y_i|, \min_i Y_i - |\min_i Y_i|)$. Since this requires a relatively large number of grid points to achieve a good approximation of (2.5), we propose to use such an equidistant grid with a step c only between $s_l = \max_{|X_i - x|<h_x} Y_i$ and $i_l = \min_{|X_i - x|<h_x} Y_i$, that is only between conditional minima and maxima of Y_i. Outside of this range, the grid points are defined by $s_l + 1/(kc)$ and $i_l - 1/(kc)$ for $k = 1, \ldots, \lfloor 1/c^2 \rfloor$. Thus, the whole real line is covered and the grid is most dense in the regions, where $F_x(y)$ changes most.

3. Finite-Sample Comparison

In this section, we attempt to compare the finite sample properties of the empirical and smoothed local L-estimators with an α-trimming L-score function using Monte Carlo simulations. Relying on arguments established by Fernholz (1997) in a non-conditional setting, the smoothed local L-estimator should have better finite-sample properties than the empirical one. This is probably surprising since an additional smoothing is used to estimate $\hat{F}_x(\cdot)$, which may cause an additional bias. Nevertheless, it is accompanied by a decrease in the variance of the estimator so that the mean square error of smoothed estimates actually decreases compared to the empirical ones. Besides the local L-estimators with trimming varying from 10% to 50%, we include for comparison the classical Nadaraya-Watson estimator.

TABLE 1. Comparison of the Nadaraya-Watson, the empirical (L-empir) and smoothed (L-smooth) local L-estimators by the mean square error under normal errors. All mean square errors are multiplied by 10^3.

Method Grid	Nadaraya Watson	L-empir	L-smooth ($h_y = 0.15$)	L-smooth ($h_y = 0.25$)
0.05	2.52	2.15	2.65	2.61
0.10	0.73	1.00	0.77	0.74
0.15	0.68	0.94	0.66	0.66
0.20	0.58	0.82	0.56	0.56
0.25	0.51	0.71	0.50	0.50
0.30	0.48	0.69	0.48	0.48
0.35	0.45	0.64	0.46	0.45
0.40	0.38	0.55	0.38	0.38
0.45	0.35	0.52	0.35	0.35
0.50	0.38	0.55	0.38	0.37
0.55	0.42	0.62	0.42	0.41
0.60	0.46	0.69	0.46	0.46
0.65	0.50	0.76	0.52	0.50
0.70	0.50	0.75	0.50	0.49
0.75	0.46	0.71	0.49	0.47
0.80	0.46	0.75	0.49	0.47
0.85	0.51	0.86	0.58	0.54
0.90	0.83	0.96	0.89	0.86
0.95	2.59	2.01	2.60	2.63

Simulations were performed for a univariate data generating process $Y = m(x) + \varepsilon$, where the regression function is given by $m_L(x) = -1 + \sqrt{x} - x^2$ and the regressor x is uniformly distributed on interval $(0, 1)$. The error term ε comes from

Gaussian, heavy tailed, and contaminated distributions. The results presented here are based on 500 repeated simulations with samples of 100 observations, although we also examined smaller and larger sample sizes, getting the same qualitative results. All simulations were performed using XploRe.

3.1. Clean Data

In this subsection, we concentrate on the behavior of estimators in models where the error ε comes from the same distribution for all observations. The performance of all estimators is compared for the following error distributions: normal $N(0, 0.1)$, Student $\sigma_t t_3$, and exponential $Exp(\sigma_e)$, where $\sigma_t \doteq 0.57$ and $\sigma_e \doteq 0.71$ are chosen so that the variance of errors is always equal to 0.1. For the sake of brevity, we report results only for sample size $n = 100$, Gaussian kernel and for bandwidth $h_x = 0.075$, which is the MISE-minimizing bandwidth for the Nadaraya-Watson estimator on the interval $(0.07, 0.93)$ used for graphical comparison of estimators by MSE (see Figures 1–3). The simulations using larger sample sizes and other bandwidth choices produced qualitatively similar results. The methods for choosing h_x and h_y in practice are discussed in Subsection 3.3.

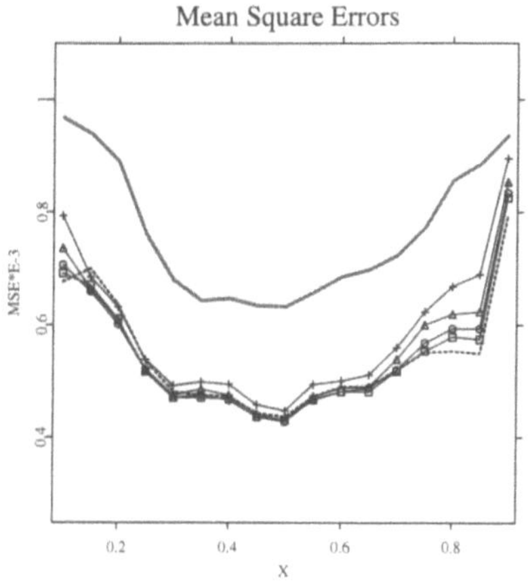

$(\varepsilon \sim N, \alpha = 0.25, c = 2\sigma)$

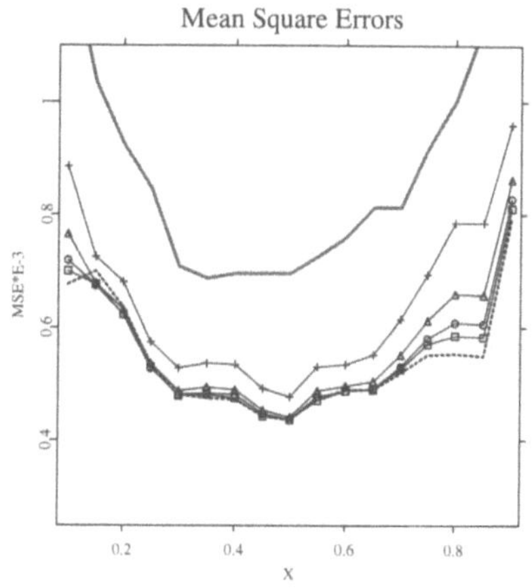

$(\varepsilon \sim N, \alpha = 0.5, c = \sigma)$

FIGURE 1. Mean square errors of the Nadaraya-Watson (dotted line), empirical (thick solid line) and smoothed (solid lines with crosses for $h_y = 0.05$, circles for $h_y = 0.10$, triangles for $h_y = 0.15$, and squares for $h_y = 0.20$) local L-estimators. The trimming level of L-estimator is denoted α.

A typical set of results is presented in Table 1, where simulations for $\varepsilon \sim N(0, 0.1)$ are summarized. There are several interesting points to observe even in this simple setting. First, notice that the smoothed L-estimator performs better than the empirical one at all, but boundary points. This is very typical indeed:

in the inner points of the interval, the bias introduced by additional smoothing of a distribution function is much smaller than the decrease in variance caused by the same smoothing and thus the smoothed L-estimator performs better; at the boundaries, where less observations are available, it is vice versa. Second, the smoothed L-estimator performs almost as well as the Nadaraya-Watson estimator. The difference between these two estimators is much smaller than in the same simulations done by Tamine et al. (2003), which can be attributed to the improved integration procedure introduced in Section 2.

To enable a more extensive comparison, the rest of the results are compared graphically using plots of the mean square errors of various estimators at the grid used in Table 1. For every simulation, there is a pair of graphs—one containing the empirical and smoothed L-estimates with 25% trimming, the other containing the L-estimates with 50% trimming; both cases always include the Nadaraya-Watson estimates for reference and in both cases, the smoothing bandwidth h_y ranges from 0.05 to 0.20. Figure 1 shows once again the mean square errors of all estimators under normally distributed errors. Notice that, even for 50% trimming, the smoothed L-estimator is still as good as the Nadaraya-Watson estimator, at least for $h_y > 0.10$ (for values h_y close to 0, the smoothed L-estimator converges to the empirical one).

Figure 2 shows the same simulations under the Student and exponential distributions. Under the Student distribution, the performance of the Nadaraya-Watson estimator is clearly the worst one. Further, the smoothed local L-estimator is for all choices of the bandwidth h_y slightly better than its empirical counterpart. Under the exponential distribution, the Nadaraya-Watson estimator is inferior to all smoothed L-estimators. The smoothed local L-estimator outperforms the empirical one even in this case, although just for bandwidths $h_y \leq 0.10$ on the whole interval $(0.1, 0.9)$. This together with the fact that, contrary to the Gaussian case, the smallest bandwidth h_y is the best choice in the case of the double exponential distribution indicates the importance of a bandwidth-choice procedure (see Subsection 3.3).

3.2. Contaminated Data

To study robust properties of local L-estimators, we used the same model as in Subsection 3.1 and add 3 observations at $x = 0.25, 0.50, 0.75$ with y-values coming from the uniform distribution on $(-1.5, 1.5)$. The simulation results for all the estimators discussed in this paper are summarized in Figure 3. Difference to the previous simulations in Subsection 3.1 is that a small level of trimming $\alpha = 0.1$ is used instead of $\alpha = 0.25$ to highlight the behavior under small and large levels of trimming. First of all, one can notice the sharp peaks in the mean square errors of the Nadaraya-Watson estimator around the outliers (they actually exceed the range used on the MSE axis). Second, the smoothed L-estimator is apparently better than its empirical version, especially for 50% trimming. The difference between MSE of the smoothed and empirical L-estimators in the case of contaminated data sets seems to be roughly the same as in the case of clean Gaussian data, see Figure 1.

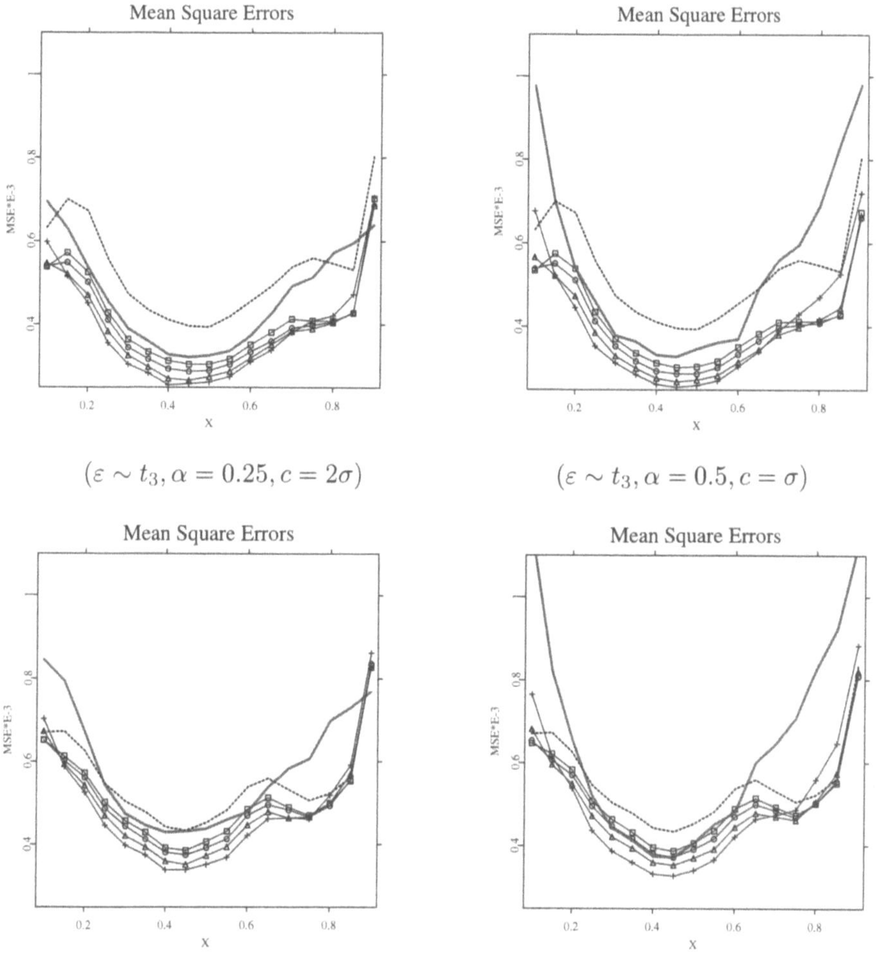

$(\varepsilon \sim t_3, \alpha = 0.25, c = 2\sigma)$ $(\varepsilon \sim t_3, \alpha = 0.5, c = \sigma)$

$(\varepsilon \sim Exp, \alpha = 0.25, c = 2\sigma)$ $(\varepsilon \sim Exp, \alpha = 0.5, c = \sigma)$

FIGURE 2. Mean square errors of the Nadaraya-Watson (dotted line), empirical (thick solid line) and smoothed (solid lines with crosses for $h_y = 0.05$, circles for $h_y = 0.10$, triangles for $h_y = 0.15$, and squares for $h_y = 0.20$) local L-estimators. The trimming level of L-estimator is denoted α.

On the other hand, it is interesting to see how the smoothed L-estimator may react to outliers, although only to a limited extent, for some choices of bandwidth h_y. Especially for $\alpha = 0.10$ trimming, the effect of outliers on the regression estimates increases with the cdf smoothing bandwidth h_y. This is caused

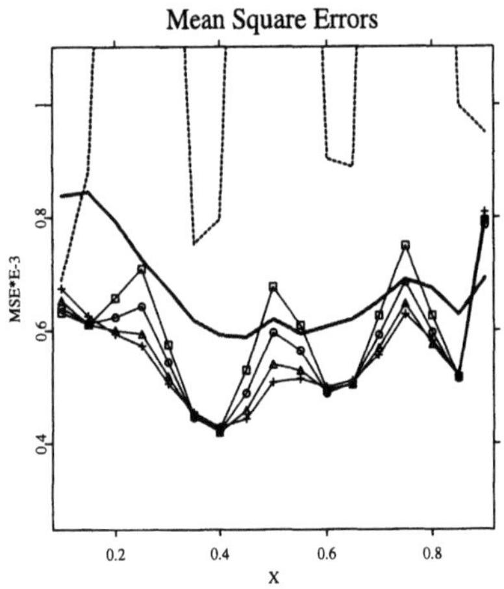

(3 outliers; $\varepsilon \sim N, \alpha = 0.1, c = 2\sigma$)

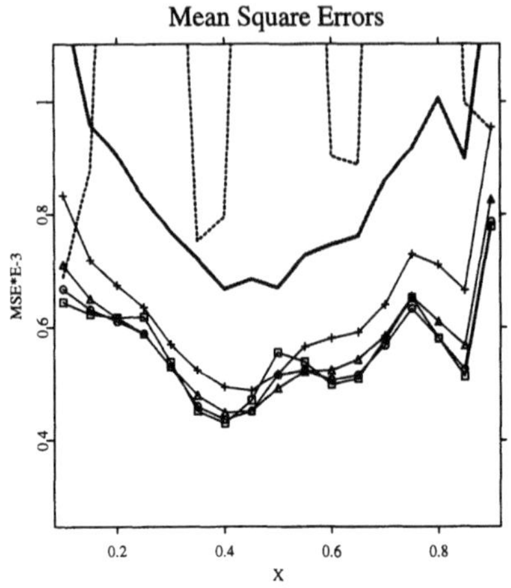

(3 outliers; $\varepsilon \sim N, \alpha = 0.5, c = \sigma$)

FIGURE 3. Mean square errors of the Nadaraya-Watson (dotted line), empirical (thick solid line) and smoothed (solid lines with crosses for $h_y = 0.05$, circles for $h_y = 0.10$, triangles for $h_y = 0.15$, and squares for $h_y = 0.20$) local L-estimators. The trimming level of L-estimator is denoted α.

by the smoothing term $K_I[(y - Y_i)/h_y]$ in equation (2.4): (i) an outlier $Y_i^o > y$ has a small positive impact on any value of the smoothed cdf $\hat{F}_x(y)$ as long as $K_I[(y - Y_i^o)/h_y] > 0$ (this is always true for the Gaussian kernel used here); (ii) this effect increases with h_y because K_I is an increasing function (it is the integral of a kernel).

3.3. Bandwidth Choice

As gradually became clear in Subsections 3.1 and 3.2, the use of smoothed cdf improves the performance of local L-estimator, but the extent of the improvement depends significantly on the choice of the bandwidth h_y (the other bandwidth h_x is common to all nonparametric regression estimators). To determine the optimal bandwidth pair (h_x, h_y), one should use a robust bandwidth-selection procedure because the standard methods for the bandwidth selection, such as least-squares cross validation, can exhibit downward bias in presence of outliers. Although we did not explore the bandwidth selection to a greater extent, established robust bandwidth-selection methods such as, for example, the L_1 cross validation by Wang and Scott (1994) and robust plug-in methods by Boente et al. (1997) can generally be recommended.

4. Application: Implied Volatility Surface

The analysis of volatility, that is, of the conditional variance of financial time series (such as exchange rates or stock prices), is nowadays one of the most important issues in modern finance. Moreover, instead of volatilities estimated from historical data, the so-called implied volatility approach is gaining importance. The implied volatility of a stock or an index is derived from the currently observed prices of related financial instruments (options) on the market and thus reflects the current market expectations. Having data on a sufficient number of transactions, one can estimate a mapping that gives implied volatilities for various combinations of maturities and prices, which is called implied volatility surface (see Härdle et al., 2002, Chapter 6, for more details). The series of these surfaces are then usually studied in time to reveal the dynamics of implied volatilities and to predict future development. Thus, estimating these surfaces is of great importance.

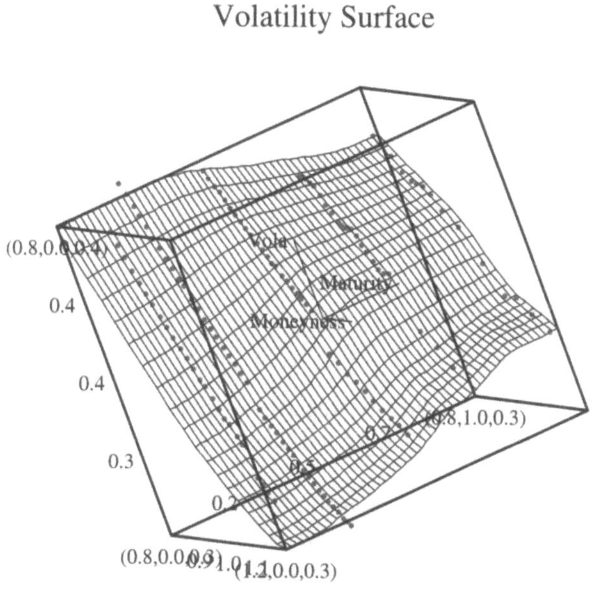

FIGURE 4. Implied volatility surface for DAX (German stock index) on January 4, 1999.

The implied volatility surface on a specified grid is typically estimated by a two-dimensional nonparametric smoothing procedure; mostly, the Nadaraya-Watson estimator with a quartic kernel is employed, see Aït-Sahalia and Lo (1998,

2000) or Fengler et al. (2002), for instance. An example of an estimated surface is on Figure 4. On the other hand, the data on implied volatilities are often scarce or of low quality for some combinations of maturities and strike prices. Since errors in estimation are transferred to any subsequent analysis, it would be wise to use a more robust local smoothing procedure than the Nadaraya-Watson estimator. Given the simulation results and practical aspects, the smoothed local L-estimator seems to be a very good candidate. It can be applied to this problem, since both in the nonparametric regression and the implied volatility estimation one estimates an expectation of a random variable conditional on values of exogenous variables.

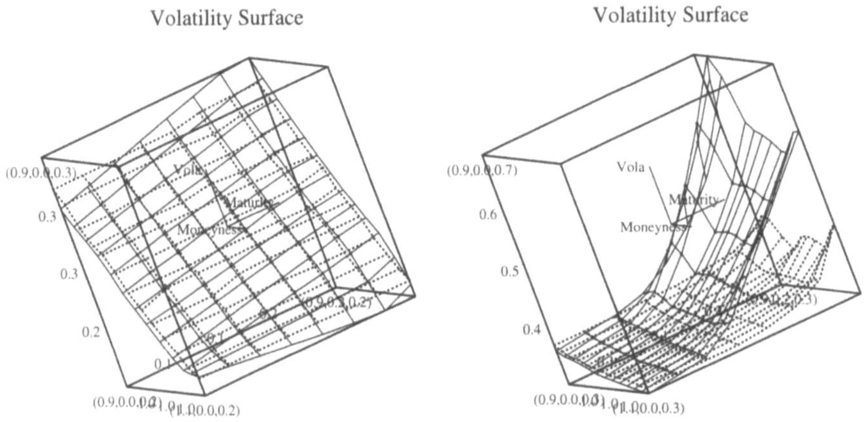

(April 16, 1999) (February 26, 1999)

FIGURE 5. Implied volatility surfaces estimated by the Nadaraya-Watson (solid line) and smoothed local L-estimator (dotted line).

To demonstrate the impact of using a robust smoothing procedure instead of the standard Nadaraya-Watson smoother, we estimated the implied volatility surfaces for DAX (a German stock index) on two selected days. The options on DAX are most actively traded contracts at the derivative exchange EUREX and should therefore provide observations on different maturities and strike prices in sufficient quantity and quality. Moreover, data are in practice cleared from unusual contracts and big errors. Nevertheless, as we demonstrate, the use of simple Nadaraya-Watson procedure may still introduce large errors that are then inherited by any subsequently analysis.

Let us compare the implied-volatility surface estimates on two days: April 16, 1999, and February 26, 1999. The estimates produced by the Nadaraya-Watson and smoothed local L-estimate for maturities ranging from 10 days to 3 months are on Figure 5 (25% trimming is used; the estimated surfaces are almost the same also for trimming levels from 10% to 40%). In the first case, the estimated surfaces are almost identical and this is probably the most usual case. In the second case, there is quite a big difference between the estimated surfaces. The most likely reason is that there is a large difference in volatility between shorter maturities (one to three months) and longer ones: since the discrete nature of observed maturities often enforces oversmoothing, the different volatilities at longer maturities can adversely affect the Nadaraya-Watson estimates. This adverse effect is much less pronounced in the case of the smoothed local L-estimation as can be verified by looking at the average mean square error for observations within the surface area. One can also notice that the volatility surfaces estimated by the smoothed L-estimator have roughly the same magnitude on both days, while the volatility magnitudes of Nadaraya-Watson estimates differ greatly. Hence, a more robust procedure can "survive" a strange behavior on the market and can actually detect it by comparing with standard nonparametric estimators.

Concluding Remarks

Following the Tamine et al. (2003) proposal of the smoothed local L-estimator, we improved the computation procedure and studied its small sample behavior. The smoothed local L-estimation exhibits better finite-sample properties than the empirical one and is faster to compute. As demonstrated, the use of such a robust technique may be highly desirable, for example, in modeling volatilities of financial instruments.

References

[1] Y. Aït-Sahalia and A.W. Lo, *Nonparametric Estimation of State-Price Densities Implicit in Financial Assets*. J. Finance **LIII** (1998), 499–547.

[2] Y. Aït-Sahalia and A.W. Lo, *Nonparametric Risk Management and Implied Risk Aversion*. J. Econometrics **94** (2000), 9–51.

[3] J. Beran, Y. Feng, S. Ghosh, and P. Sibbertsen, *On Robust Local Polynomial Estimation With Long-Memory Errors*. Internat. J. Forecasting (2003), to appear.

[4] G. Boente and R. Fraiman, *Local L-Estimators For Nonparametric Regression Under Dependence*. J. Nonparametr. Stat. **4** (1994), 91–101.

[5] G. Boente, R. Fraiman, and J. Meloche, *Robust Plug-In Bandwidth Estimators in Nonparametric Regression*. J. Statist. Plann. Inference **57** (1997), 109–142.

[6] P. Čížek and W. Härdle, *Robust Adaptive Estimation of Dimension Reduction Space*. Submitted, 2003, available at *http://sfb.wiwi.hu-berlin.de/papers/index.html*.

[7] J. Fan and J. Jiang, *Variable Bandwidth and One-step Local M-Estimator*. Sci. China Ser. A **29** (1999), 1–15.

[8] M. Fengler, W. Härdle, and C. Villa, *The Dynamics of Implied Volatilities: A Common Principal Components Approach*. Discussion paper 38, SFB 373, Humboldt-Universität zu Berlin, 2001, available at *http://sfb.wiwi.hu-berlin.de/papers/2001.html*.

[9] L.T. Fernholz, *Reducing the Variance By Smoothing.* J. Statist. Plann. Inference **57** (1997), 29–38.

[10] W. Härdle, T. Kleinow, and G. Stahl, *Applied Quantitative Finance.* Springer, Germany, 2002.

[11] W. Härdle and A. Tsybakov, *Robust Nonparametric Regression With Simultaneous Scale Curve Estimation.* Ann. Statist. **16** (1988), 120–135.

[12] K.L. Mehra, S.M. Rao, and S.P. Upadastra, *A Smoothed Conditional Quantile Estimator and Related Applications of Conditional Empirical Processes.* J. Multivariate Anal. **37** (1991), 151–179.

[13] E.A. Nadaraya, *On Estimating Regression.* Theory. Probab. Appl. **10** (1964), 186–190.

[14] J. Tamine, P. Čížek, and W. Härdle, *Smoothed L-Estimation of Regression Function.* Submitted, 2003, available at *http://sfb.wiwi.hu-berlin.de/papers/2002.html.*

[15] F.T. Wang and D.W. Scott, *The L_1 Method For Robust Nonparametric Regression.* J. Amer. Statist. Assoc. **89** (1994), 65–76.

[16] G.S. Watson, *Smooth Regression Analysis.* Sankhyā, Ser. A **26** (1964), 359–372.

Acknowledgment

I am grateful to Matthias Fengler, who provided me the data used in this work. I would also like to thank two anonymous referees for their comments, which allowed me to greatly improve the paper.

P. Čížek
Tilburg University
Dept. of Econometrics & OR
Room B 516
P.O. Box 90153
5000 LE Tilburg
The Netherlands
e-mail: P.Cizek@uvt.nl

Statistics for Industry and Technology, 71–82

Fast Algorithms for Computing High Breakdown Covariance Matrices with Missing Data

S. Copt and M.-P. Victoria-Feser

Abstract. Robust estimation of covariance matrices when some of the data at hand are missing is an important problem. It has been studied by Little and Smith (1987) and more recently by Cheng and Victoria-Feser (2002). The latter propose the use of high breakdown estimators and so-called hybrid algorithms (see, e.g., Woodruff and Rocke, 1994). In particular, the minimum volume ellipsoid of Rousseeuw (1984) is adapted to the case of missing data. To compute it, they use (a modified version of) the forward search algorithm (see e.g. Atkinson, 1994). In this paper, we propose to use instead a modification of the C-step algorithm proposed by Rousseeuw and Van Driessen (1999) which is actually a lot faster. We also adapt the orthogonalized Gnanadesikan-Kettenring (OGK) estimator proposed by Maronna and Zamar (2002) to the case of missing data and use it as a starting point for an adapted S-estimator. Moreover, we conduct a simulation study to compare different robust estimators in terms of their efficiency and breakdown.

Mathematics Subject Classification (2000). Primary 62G35; Secondary 62H12.

Keywords. C-step algorithm, minimum covariance determinant, outliers, robust statistics, S-estimators, orthogonalized Gnanadesikan-Kettering robust estimator.

1. Introduction

The statistical literature contains several proposals for high breakdown estimators of the mean and covariance in multivariate data when it is suspected that the data contain outliers or extreme observations. A well known one is the minimum covariance determinant (MCD) of Rousseeuw (1984). However, little has been done

Both authors acknowledge the financial support of the Swiss National Science Foundation (grant 610-057883.99).

to consider also the case of missing data which is in practice very common. Only Little and Smith (1987) and Cheng and Victoria-Feser (2002) propose different solutions. In this paper we actually concentrate on robust estimators with missing data, in particular we propose the use of faster algorithms for their computation and compare them through extensive simulations in terms of their robustness properties when data are contaminated and also in terms of the speed of two different algorithms used to compute the robust estimators. In particular, we adapt the orthogonalized Gnanadesikan-Kettenring (OGK) estimator proposed by Maronna and Zamar (2002) to the case of missing data and use it as a starting point for an adapted S-estimator. All our programs are readily available (upon request) in the form of an Splus library which has been used to produce the results and graphics presented in this paper.

2. A General Class of Estimators with Missing Data

The aim is to estimate the parameters μ and $\mathbf{\Sigma}$, i.e., the mean and covariance of an underlying multivariate variable $Y = (Y_1, \ldots, Y_p)$ that has supposedly generated the sample $\mathbf{y}_i, i = 1, \ldots, n$ at hand. As it often happens in practice, we suppose that some of the observations might be missing in that some of the y_{ij} are observed for some $j \in \{1, \ldots, p\}$ and the others are not observed or missing for the other j's. In other terms, $\mathbf{y}_i = [\mathbf{y}_{[oi]}^T, \mathbf{y}_{[mi]}^T]^T$ so that a distinction is made between the observed (oi) and the missing (mi) data. We suppose that the data are missing at random (see Rubin, 1976), a sufficient condition for correct likelihood-based inferences. Most known estimators of mean and covariance with missing data fall in the class proposed by Cheng and Victoria-Feser (2002), i.e.,

$$\frac{1}{n}\sum_{i=1}^{n} w_i^{\mu}(\mu - \widehat{\mathbf{y}}_i) = 0 \tag{2.1}$$

$$\frac{1}{n}\sum_{i=1}^{n} \left[w_i^{\delta}\mathbf{\Sigma} - w_i^{\eta}((\widehat{\mathbf{y}}_i - \mu)(\widehat{\mathbf{y}}_i - \mu)^T - \mathbf{C}_i)\right] = 0 \tag{2.2}$$

where

$$\begin{aligned} \widehat{\mathbf{y}}_i &= \left[\mathbf{y}_{[oi]}^T, E\left[\mathbf{y}_{[mi]} \middle| \mathbf{y}_{[oi]}, \mu, \mathbf{\Sigma}\right]^T\right]^T \\ &= \left[\mathbf{y}_{[oi]}^T, \mu_{[mi]}^T + (\mathbf{y}_{[oi]} - \mu_{[oi]})^T \mathbf{\Sigma}_{[ooi]}^{-1} \mathbf{\Sigma}_{[omi]}\right]^T, \end{aligned} \tag{2.3}$$

and

$$\begin{aligned} \mathbf{C}_i &= \begin{bmatrix} 0 & 0 \\ 0 & \operatorname{cov}\left[\mathbf{y}_{[mi]}, \mathbf{y}_{[mi]}^T \middle| \mathbf{y}_{[oi]}, \mu, \mathbf{\Sigma}\right] \end{bmatrix} \\ &= \begin{bmatrix} 0 & 0 \\ 0 & \mathbf{\Sigma}_{[mmi]} - \mathbf{\Sigma}_{[moi]} \mathbf{\Sigma}_{[ooi]}^{-1} \mathbf{\Sigma}_{[omi]} \end{bmatrix} \end{aligned} \tag{2.4}$$

where for example $\mathbf{\Sigma}_{[ooi]}$ denotes the partition of $\mathbf{\Sigma}$ corresponding to the observed part of $\mathbf{y}_i$, etc. The different estimators are actually defined through the data weighting system given by w_i^{μ}, w_i^{δ} and w_i^{η} in (2.1) and (2.2) which in turn also depend on the parameters μ and $\mathbf{\Sigma}$ (see below). To compute the estimators, one can use an iterative procedure in which given current estimates $\widehat{\mu}^{(t)}$ and $\widehat{\mathbf{\Sigma}}^{(t)}$, the $\widehat{\mathbf{y}}_i$, $\mathbf{C}_i$ and the weights are first computed, and the values are then updated by

$$\widehat{\mu}^{(t+1)} = \frac{1}{n}\sum_{i=1}^{n} \widehat{w}_i^{\mu}\widehat{\mathbf{y}}_i \Big/ \frac{1}{n}\sum_{i=1}^{n} \widehat{w}_i^{\mu} \tag{2.5}$$

$$\widehat{\mathbf{\Sigma}}^{(t+1)} = \left[\frac{1}{n}\sum_{i=1}^{n} \widehat{w}_i^{\eta}((\widehat{\mathbf{y}}_i - \mu^{(t+1)})(\widehat{\mathbf{y}}_i - \mu^{(t+1)})^T - \mathbf{C}_i)\right] \Big/ \left[\frac{1}{n}\sum_{i=1}^{n} \widehat{w}_i^{\delta}\right]. \tag{2.6}$$

We have not worked out the conditions for (2.1) and (2.2) in particular to have a unique solution and for (2.5) and (2.6) to converge to that unique solution. However, the reader is referred to Davies (1987) for general conditions for *S*-estimators.

The classical *MLE* is obtained when $w_i^{\mu} = w_i^{\eta} = w_i^{\delta} = 1 \; \forall i$, and (2.5) and (2.6) define the *EM* algorithm (Dempster et al., 1977). However, with complete data it is well known that the *MLE* of mean and covariance are not robust. When there are missing data, the situation does not change; see Cheng and Victoria-Feser (2002). Little and Rubin (1987) propose to base the *M*-step on a robust estimator belonging to the general class of *M*-estimator (Huber, 1981). They call the resulting procedure the *ER* algorithm. Their estimator is defined by (2.5) and (2.6) with

$$(w_i^{\mu})^2 = w_i^{\eta} = w_i^{\delta} = w_i = \omega(d_{oi})/d_{oi}^2 \tag{2.7}$$

where

$$d_{oi}^2 = d_{oi}^2(\mu, \mathbf{\Sigma}) = \left(\mathbf{y}_{[oi]} - \mu_{[oi]}\right)^T \mathbf{\Sigma}_{[ooi]}^{-1} \left(\mathbf{y}_{[oi]} - \mu_{[oi]}\right) \tag{2.8}$$

is the squared Mahalanobis distance corresponding to the observed part of $\mathbf{y}_i$. See Little and Smith (1987) for the choice of the weight function ω. The iteration step for the covariance matrix (2.6) does not exactly correspond to the same step in the *ER* algorithm in that the weights w_i^{η} are not applied to the correction matrix $\mathbf{C}_i$. We will however, in what follows consider this slight modification of the ER algorithm (Cheng and Victoria-Feser, 2002). If the case i is uncontaminated, the data are normal and missing values are missing at random, then (2.8) is asymptotically $\chi^2_{p_i}$ where $p_i = \dim\left(\mathbf{y}_{[oi]}\right)$. The Wilson-Hilferty transformation of the chi-squared distribution yields $(d_i^2/p_i)^{1/3} \sim N(1-2/(9p_i), 2/(9p_i))$. Following Little and Smith (1987), we also propose a probability plot of

$$Z_i = \frac{(d_i^2/p_i)^{1/3} - 1 + 2/(9p_i)}{\sqrt{2/(9p_i)}} \tag{2.9}$$

versus standard normal order statistics, that should reveal atypical observations.

Little and Smith (1987) proposed as starting point of the *ER* algorithm, the *MLE* on the data where the missing ones have been replaced by the median of the corresponding observations. Although the ER algorithm is relatively simple to implement, it suffers from an important drawback : its breakdown point is at most $1/(p+1)$ because it is based on a weighting scheme that is not redescending. This drawback will be highlighted by the simulation results. This means that if the proportion of outliers exceeds this value (or even is near it) the robust estimator is not robust anymore.

To construct a high breakdown estimator of mean and covariance matrix in multivariate data when some observations are missing, Cheng and Victoria-Feser (2002) propose two strategies. The first one is to provide a high breakdown estimator such as the MCD estimator as starting point for the ER algorithm and the second is to also adapt a high breakdown estimator such as an S-estimator (Rousseeuw and Yohai, 1984) to incomplete data. The resulting estimator which is called the $ERTBS$ is then defined through (2.5) and (2.6) with

$$w_i^{\mu} = w_i^{\eta}/p = w_i^{\delta} \big/ \tilde{d}_i^2 = \psi(\tilde{d}_i) \Big/ \tilde{d}_i \ , \tag{2.10}$$

$\tilde{d}_i = d_{oi}\left(\widehat{\mu}, \widehat{\mathbf{\Sigma}}\right)/k$ with $\widehat{\mu}$ and $\widehat{\mathbf{\Sigma}}$ being the current values of the high breakdown point estimator, and

$$k = \frac{d_{[q]}}{\sqrt{(\chi_p^2)^{-1}(q/(n+1))}} \ , \tag{2.11}$$

where $d_{[q]}$ denotes the qth ordered distance (based on the $d_{oi}\left(\widehat{\mu}, \widehat{\mathbf{\Sigma}}\right)$), $q = \lfloor (n+p+1)/2 \rfloor$ with $\lfloor x \rfloor$ denoting the integer part of x, and

$$\psi(d; c, M) = \begin{cases} d & 0 \le d < M \\ d\left(1 - \left(\frac{d-M}{c}\right)^2\right)^2 & M \le d \le M + c \\ 0 & d > M + c. \end{cases}$$

This ψ function defines the translated biweight S-estimator proposed by Rocke (1996). The parameters M and c control the breakdown point ε^* and the asymptotic rejection probability ARP α of the $ERTBS$. The ARP can be interpreted as the probability for an estimator, in large samples under a reference distribution, to give a null (or nearly null) weight. M and c are found implicitly by

$$\begin{aligned} \varepsilon^* \max_d \rho(d; c, M) &= E_{\chi_p^2}[\rho(d; c, M)] \ , \\ M + c &= \sqrt{(\chi_p^2)^{-1}(1-\alpha)} \ ; \end{aligned}$$

where ρ is the primitive of ψ (see Rocke, 1996). The choices for ε^* and α are to be made by the analyst. The former is the suspected maximal amount of contaminated data and for the latter Cheng and Victoria-Feser (2002) propose choices between 0.1% and 1%.

As Rocke (1996) noted, it is very important to choose a good starting point for any algorithm defining a high breakdown point estimator, otherwise the latter can loose its high breakdown properties. For the $ERTBS$, Cheng and Victoria-Feser (2002) therefore propose an adaptation of the MCD estimator as a starting point as well as an algorithm to compute it. However, to compute the MCD one needs algorithms that are based on random starting subsamples. This can lead to situations in which the MCD is very slow to compute, if not impossible. Therefore, in the following section, we propose a fast algorithm to compute the MCD by adapting the *FAST-MCD* of Rousseeuw and Van Driessen (1999) and as an even faster alternative, we propose a modified version of the OGK estimator adapted to the case of missing data to be used as a starting point for the $ERTBS$.

3. Starting Point Robust Estimators with Missing Data

3.1. The Modified MCD

The objective of the MCD estimator is to find h observations (out of n) whose covariance matrix has the lowest determinant. The MCD mean estimator is then the sample mean of those h points, and the MCD covariance estimator is their sample covariance matrix. To compute the MCD, one needs an algorithm for finding the best subset of h points, which usually involves the repeated computation of the sample mean and covariance as well as Mahalanobis distances. When some observations are missing, Cheng and Victoria-Feser (2002) propose to use the EM algorithm to compute the sample means and covariances at all steps of the algorithm and to base the Mahalanobis distances on the observed part of the observation as in (2.8). The latter are standardized by means of the Wilson-Hilferty transformation given in (2.9), so that one takes into account the unequal number of missing values for each observation.

A choice needs to be made on h and one way is to choose it such that the MCD has the highest breakdown. In this case, the minimal value of h is (Rousseeuw and Leroy, 1987) $h := \lfloor \frac{n+p+1}{2} \rfloor$. But this is also the choice that gives the largest efficiency loss. So when we suspect that the sample is not heavily contaminated we can reasonably choose a larger value for the proportion of points of say 75% or 80% so we can take $h := \lfloor 0.75n \rfloor$ or $h := \lfloor 0.80n \rfloor$.

To run the MCD, Cheng and Victoria-Feser (2002) adapt the forward search algorithm proposed by Atkinson (1993, 1994). However, more recently Rousseeuw and Van Driessen (1999) have proposed a new algorithm called *FAST-MCD* supposed to be even faster than the forward search algorithm and able to deal with very large data sets. In this paper, we propose to adapt it to compute the MCD when there are missing data.

A key idea of the *FAST-MCD* algorithm is the fact that starting from any approximation to the MCD, it is possible to find an approximation with a lower determinant. Indeed Rousseeuw and Van Driessen (1999) observed that from a subset H_k of size h in which μ, Σ and the Mahalanobis distances are computed, one can create a subset H_{k+1} by taking among the n observations the h ones with the smallest Mahalanobis distances with the property that the determinant of Σ

based on H_{k+1} is smaller. Each step is called a C-step. The initial subset is created by choosing randomly $p+1$ observations on which the Mahalanobis distances are computed to order the n observations. The first h ones define the initial subset H_1. If the determinant of Σ based on the randomly chosen $p+1$ observations is null, one adds one randomly chosen observation at the time until the determinant becomes positive. If for any subset H_k there are missing values, we compute μ_k and Σ_k with the EM algorithm. The Mahalanobis distances are also changed as in (2.8) and standardized using the Wilson-Hilferty transformation. The absolute value of the latter is used to order the observations. The initial subset is created choosing randomly $p+1$ observations among the fully observed ones. For more details, see Copt and Victoria-Feser (2003).

Through extensive simulations we compare, in Section 4, the forward search algorithm and the *FAST-MCD* algorithm for the computation of the *MCD* with missing data.

3.2. The Modified OGK

Maronna and Zamar (2002) base their *OGK* on the robust estimator for covariances σ_{jk} proposed by Gnanadesikan and Kettenring (1972) which is very simple to compute. Indeed the latter is defined for a pair of random variables (i.e., $p=2$) as

$$\frac{1}{4}\left(\sigma(Y_j+Y_k)^2-\sigma(Y_j-Y_k)^2\right)$$

where $\sigma()$ is a standard deviation function applied on its argument. A robust estimator for σ_{jk} is obtained when $\sigma()$ is a robust function. When $p>2$, the covariance matrix $\mathbf{\Sigma}$ is estimated by replacing all its elements by all pairwise estimates. It is well known that such an estimator may produce non positive definite matrices and the estimator is not affine equivariant. To overcome the lack of positive definiteness, Maronna and Zamar (2002) propose an estimator defined by the following four steps:

1. Let $\mathbf{D}=\operatorname{diag}\left(\sigma\left(Y_j\right)\right)|_{j=1,\ldots,p}$ and define $\mathbf{x}_i=\mathbf{D}^{-1}\mathbf{y}_i, i=1,\ldots,n$, i.e., realizations from $X=(X_1,\ldots,X_p)$.
2. Compute the matrix $\mathbf{U}=(u_{jk})$ with
$$u_{jk}=\begin{cases}\frac{1}{4}\left(\sigma(X_j+X_k)^2-\sigma(X_j-X_k)^2\right) & j\neq k\\ 1 & j=k.\end{cases}\tag{3.1}$$
3. Decompose $\mathbf{U}$ as $\mathbf{U}=\mathbf{E}\Lambda\mathbf{E}^T$ with $\Lambda=\operatorname{diag}(\lambda_1,\ldots,\lambda_p)$.
4. Define $\mathbf{z}_i=\mathbf{E}^T\mathbf{x}_i$, i.e., realizations from $Z=(Z_1,\ldots,Z_p)$ and $\mathbf{A}=\mathbf{DE}$. The estimator of $\mathbf{\Sigma}$ is $\mathbf{A}\Gamma\mathbf{A}^T$ with $\Gamma=\operatorname{diag}\left(\sigma\left(Z_j\right)^2\right)\Big|_{j=1,\ldots,p}$.

A location estimator for μ is given by $\mathbf{A}\nu$ with $\nu=\left(m\left(Z_j\right)\right)|_{j=1,\ldots,p}$, $m()$ being a (robust) mean function. The procedure can be iterated by replacing $\mathbf{U}$ in step 2 by $\mathbf{E}\Gamma\mathbf{E}^T$ until convergence. For the choice of $\sigma()$ and $m()$, see Maronna and Zamar (2002). The latter argue that to improve the efficiency of the *OGK*, one could use it as a hard rejection tool in that a reweighted estimator as in (2.1) is used in which $\widehat{\mathbf{y}}_i = \mathbf{y}_i \forall i$ and $w_i^{\mu} = w_i^{\eta} = w_i^{\delta} = w_i$ with

$$w_i = \begin{cases} 1 & (\mathbf{y}_i - \widehat{\mu}_{\mathrm{OGK}})^T \widehat{\mathbf{\Sigma}}_{\mathrm{OGK}}^{-1} (\mathbf{y}_i - \widehat{\mu}_{\mathrm{OGK}}) \le \chi_p^2(.9) \\ 0 & \text{otherwise.} \end{cases}$$

The resulting estimator will be called the reweighted *OGK* (*rOGK*). Note that this strategy is also used most of the times with the *MCD* but with the quantile 0.975 (instead of 0.9) of the χ_p^2. We will call the resulting estimator the *rMCD*. We have not derived the formal conditions for convergence to a unique solution of the algorithm used to compute the (*rOGK*), but in our simulations we found that it converged in all our samples.

To extend the *OGK* or *rOGK* to the case of missing data, we propose to impute the missing values by means of the $\widehat{\mathbf{y}}_i$ in (2.3) obtained by the EM algorithm, i.e., with μ and $\mathbf{\Sigma}$ estimated by (2.1) where all weights are equal to 1. The reason is that the EM algorithm is very fast, and although it leads to biased estimates of μ and $\mathbf{\Sigma}$ and therefore of the imputed values $\widehat{\mathbf{y}}_i$, we found through extensive simulation studies, that in practice the *rOGK* is not or very little affected (see Section 4). In future research, we will seek a better adaptation of the *OGK* to the case of missing data.

4. Simulation Study

4.1. The Design

The model is the multivariate normal distribution $N(\mu, \mathbf{\Sigma})$. Because the *OGK* is not affine equivariant, following Maronna and Zamar (2002), we choose to simulate correlated data, i.e., $\mathbf{\Sigma} = \mathbf{R}(\rho)^2$ where $\mathbf{R}(\rho) = \rho$ for the elements $j \neq k$ and 1 for the others, with $\rho = 0.2$. We also chose $\mu = \mathbf{0}$. The data were contaminated by so-called shift-outliers (Woodruff and Rocke, 1993), i.e., an ϵ-proportion was generated using $N((\sqrt{p} + \beta/\sqrt{2})\, \mathbf{e}_p, \mathbf{\Sigma})$ where $\mathbf{e}_p$ is a p-dimensional vector of ones. We set $\beta = 1.6$ and $\epsilon = 0, 0.02, 0.05$ and 0.1. The missing data, if any, were chosen randomly among the mixture distribution between the good and the bad data. We chose proportions $miss = 0.1, 0.2$ and 0.3. Table 1 shows the different values for n and p used in the simulations. Each robust estimator requires a decision on

TABLE 1. Values for n and p.

	$p = 10$	$p = 20$	$p = 50$
$n =$	50	100	200
$n =$	100	200	400
$n =$	500	500	600

its initialization parameters. For the *MCD* estimator, $h = [0.6n]$ was chosen. For the *OGK*, $c_1 = 4.5$ and $c_2 = 3$ were chosen (see Maronna and Zamar, 2002). For the *ERTBS* estimator we chose for our simulations the breakdown point $\varepsilon^* = 0.3$

and the *ARP* $\alpha = 0.001$. All computational experiments were done on a Athlon 1900Mhz with 512 MB of memory. The core of the program was written in Fortran 77 and Splus was used as a front-end (to produce the various graphics). For all combinations of parameters, 1000 samples were generated.

4.2. Computational Times

The computational time to compute the ER and the $ERTBS$ depends essentially on the choice of the starting point. Therefore, we compute here the time (as function of the sample size n) needed to compute the *rOGK* or the $rMCD$, when the latter is computed using the adapted *FAST-MCD* algorithm ($rMCD/FAST$) or the forward search algorithm ($rMCD/FWD$). We chose the reweighted version of the two starting point estimators, because as is shown in Copt and Victoria-Feser (2003), the non-reweighted versions can lead to biased estimates. For each value of $miss$ and ϵ and the values of Table 1, a time in seconds has been computed. Figure 1 shows the results (in a log-scale) for the datasets with $\epsilon = 10\%$ and $miss = 30\%$ (for other combinations the results are comparatively similar). We notice the following features. The speed for the $rMCD/FWD$ as expected is

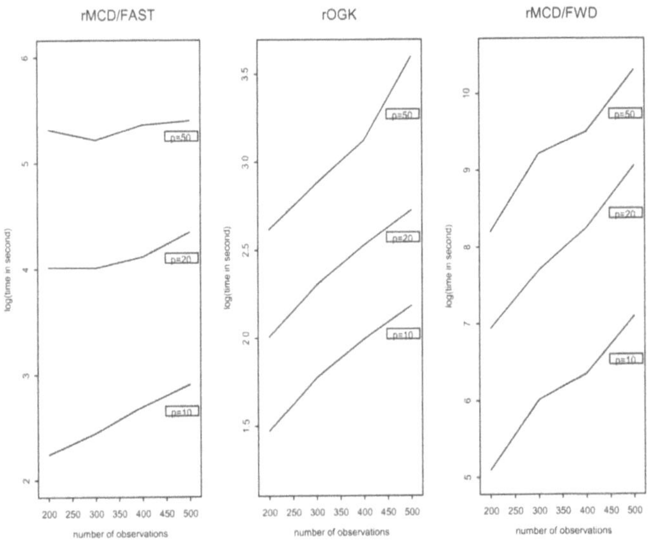

FIGURE 1. Log of mean time in seconds needed to compute the *rOGK* and the $rMCD$ by means of the forward search (FWD) algorithm and the FAST-MCD algorithm, as a function of the sample size and the data dimension p.

slower than the speed of the $rMCD/FAST$, with an increasing difference as the sample size increases. The $rMCD/FAST$ can be up to 150 times faster than the $rMCD/FWD$. However, when the *rOGK* is used as a starting point, the computational times decrease drastically, with sometimes a ratio of 18 compared with the $rMCD/FAST$. However, the speed of the $rMCD/FAST$ does not depend very much on the sample size n, whereas the $rOGK$ does quite substantially.

4.3. Comparing Estimators

The aim of this subsection is to study the robustness properties (bias versus efficiency) of the different estimators proposed with incomplete data by means of simulations. For the MCD, all calculations were made using the modified *FAST-MCD* for missing data. It should be stressed that this exercise has not been done in Cheng and Victoria-Feser (2002). Copt and Victoria-Feser (2003) compare the behavior of the MCD, $rMCD$, OGK and $rOGK$. They conclude that both the $rMCD$ and $rOGK$ behave nicely (in terms of bias) even with correlated data ($\mathbf{\Sigma} = \mathbf{R}(.2)^2$), whereas the OGK for the variances and covariances can be biased with contaminated data.

The estimators we consider here are the final estimators namely the MLE computed via the EM algorithm (which is taken as a benchmark), the ER algorithm with the MLE as starting point (ER/MLE), the ER algorithm with the $rMCD$, and $rOGK$ as starting point ($ER/rMCD$ and $ER/rOGK$), the $ERTBS$ algorithm with the $rMCD$ and $rOGK$ as starting point ($ERTBS/rMCD$ and $ERTBS/rOGK$). The data were generated using the designs presented in Section 4.1. The percentage of missing observations and the sizes n and p do not seem to have an influence on the behavior of the different estimators. The influential factor is the percentage of contamination and potentially, the covariance structure when the $rOGK$ is chosen as starting point. We therefore chose the $N(\mathbf{0}, \mathbf{R}(.2)^2)$ as data generating model. We use boxplots to compare the estimators. They are built on the estimated biases of one of the elements of the mean vector, one of the diagonal elements of the covariance matrix and one of the off-diagonal elements of the covariance matrix. Only the results for μ_1, σ_{11}, and σ_{12} are represented, since for other parameters, the same pattern is found.

Figures 2 and 3 show the boxplots of the sampling distributions of the final estimators using different amounts of data contamination. The MLE clearly fails even if the contamination is small. However it is the most efficient with no contamination but the efficiency loss for the robust estimators seems to be quite small. The ER/MLE breaks down at 5% of data contamination. Finally, the $ER/rMCD$, $ER/rOGK$, $ERTBS/rMCD$ or $ERTBS/rOGK$ are very robust and can withstand at least 10% of data contamination.

If we want to see a difference between the ER and $ERTBS$ with the same high breakdown starting point, we have to push the percentage of contamination up to 30%. We have not done a full coverage of such situation since it is very unlikely that someone will want to study data sets with such a percentage of contamination. However, Copt and Victoria-Feser (2003) show a simulated example with 30% of contaminated data in which the $ER/rOGK$ (and also the $ER/rMCD$) clearly fails to detect the outliers, whereas the $ERTBS/rOGK$ is not affected by their presence in the sample.

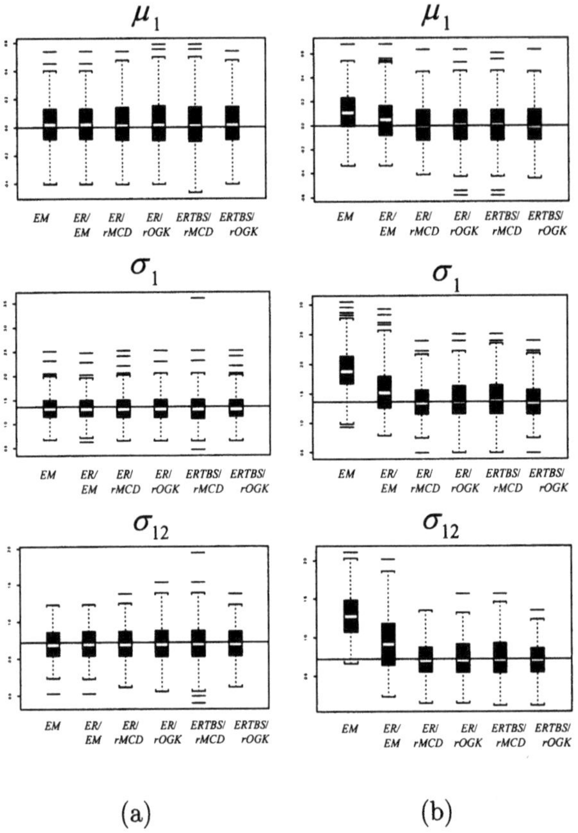

FIGURE 2. Sampling distribution of the final high breakdown estimators with missing data for (a) 0% and (b) 2% of data contamination.

5. Conclusion

In this paper, we have considered (efficient) high breakdown estimators of the mean and covariance of a multivariate normal distribution with missing data. The computational speed of the estimators depends on the computational speed of their starting point. The fastest is the modified OGK, although its speed depends on the sample size, which is not so strongly the case for the modified MCD using the $FAST$-MCD algorithm to compute it. For their performance in terms of bias, efficiency and breakdown point, the conclusion is that a high breakdown estimator is crucial as starting point, among which the (modified) OGK can be biased when the data are correlated but the (modified) $rOGK$ is not, and that the $ERTBS$ has a larger breakdown point as the ER. The EM is not robust and the ER/EM has a very small breakdown point. Finally, under no contamination, the EM is the most efficient estimator but the $ERTBS$ has a very small efficiency loss.

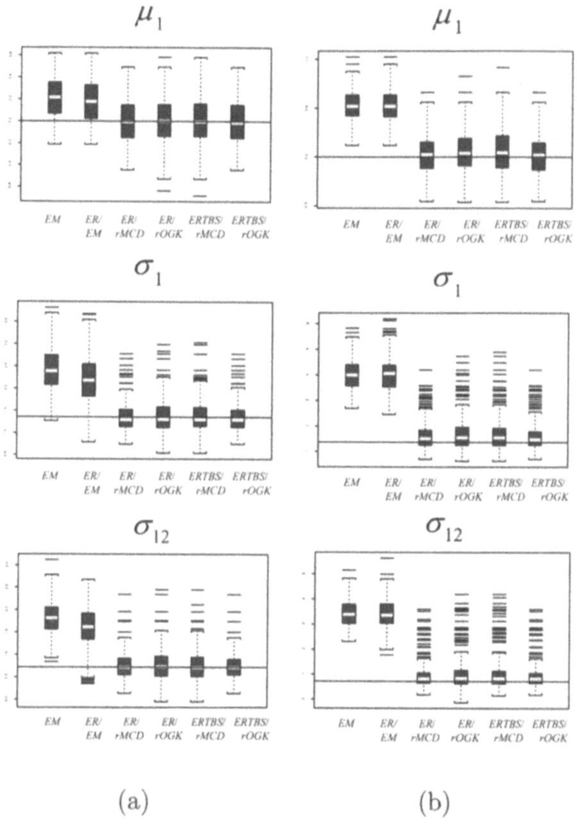

FIGURE 3. Sampling distribution of the final high breakdown estimators with missing data for (a) 5% and (b) 10% of data contamination.

References

[1] A.C. Atkinson, *Stalactite Plots and Robust Estimation for the Detection of Multivariate Outliers.* In S. Morgenthaler, E. Ronchetti, and W.A. Stahel, editors, New Directions in Statistical Data Analysis and Robustness. Birkhäuser, Basel, 1993.

[2] A.C. Atkinson, *Fast Very Robust Methods for the Detection of Multiple Outliers.* J. Amer. Statist. Assoc. **89** (1994), 1329–1339.

[3] T.-C. Cheng and M.-P. Victoria-Feser, *High Breakdown Estimation of Multivariate Location and Scale with Missing Observations.* British J. Math. Statist. Psych. **55** (2002), 317–335.

[4] S. Copt and M.-P. Victoria-Feser, *Fast Algorithms for Computing High Breakdown Covariance Matrices with Missing Data.* Cahiers du département d'Econométrie no 2003.04, University of Geneva, CH-1211 Geneva, 2003, available at *http://www.unige.ch/ses/metri.*

[5] P.L. Davies, *Asymptotic Behaviour of S-estimators of Multivariate Location Parameters and Dispersion Matrices.* Ann. Statist. **15** (1987), 1269–1292.

[6] A.P. Dempster, M.N. Laird, and D.B. Rubin, *Maximum Likelihood from Incomplete Data via the EM Algorithm.* J. R. Stat. Soc. Ser. B **39** (1977), 1–22.

[7] R. Gnanadesikan and J.R. Kettenring, *Robust Estimates, Residuals, and Outlier Detection with Multiresponse Data.* Biometrics **28** (1972), 81–124.

[8] P.J. Huber, *Robust Statistics.* Wiley, New York, 1981.

[9] R.J.A. Little and D.B. Rubin, *Statistical Analysis with Missing Data.* Wiley, New York, 1987.

[10] R.J.A. Little and P.J. Smith, *Editing and Imputing for Quantitative Survey Data.* J. Amer. Statist. Assoc. **82** (1987), 58–68.

[11] R.A. Maronna and R.H. Zamar, *Robust Multivariate Estimates for High-Dimensional Datasets.* Technometrics **44** (2002), 307–317.

[12] D.M. Rocke, *Robustness Properties of S-estimators of Multivariate Location and Shape in High Dimension.* Ann. Statist. **24** (1996), 1327–1345.

[13] P.J. Rousseeuw, *Least Median of Squares Regression.* J. Amer. Statist. Assoc. **79** (1984), 871–880.

[14] P.J. Rousseeuw and A.M. Leroy, *Robust Regression and Outlier Detection.* Wiley, New York, 1987.

[15] P.J. Rousseeuw and K. Van Driessen, *A Fast Algorithm for the Minimum Covariance Determinant Estimator.* Technometrics **41** (1999), 212–223.

[16] P.J. Rousseeuw and V.J. Yohai, *Robust Regression by means of S-estimators.* In J. Franke, W. Härdle, and D. Martin, editors, Robust and Nonlinear Time Series Analysis, pages 256–272. Springer-Verlag, New York, 1984.

[17] D.B. Rubin, *Inference and Missing Data.* Biometrika **63** (1976), 581–592.

[18] D.L. Woodruff and D.M. Rocke, *Heuristic Search Algorithm for the Minimum Volume Ellipsoid.* J. Comput. Graph. Statist. **2** (1993), 69–95.

[19] D.L. Woodruff and D.M. Rocke, *Computable Robust Estimation of Multivariate Location and Shape in High Dimension using Compound Estimators.* J. Amer. Statist. Assoc. **89** (1994), 888–896.

Acknowledgements

Both authors are grateful to two anonymous referees for their constructive comments.

S. Copt and M.-P. Victoria-Feser
Department of Econometrics and HEC
University of Geneva
40, Bd. du Pont d'Arve
CH 1211 Geneva
Switzerland
e-mail: {maria-pia.victoriafeser, SamuelCopt}@hec.unige.ch

Statistics for Industry and Technology, 83–91

Generalized d-fullness Technique for Breakdown Point Study of the Trimmed Likelihood Estimator with Application

R.B. Dimova and N.M. Neykov

Abstract. The d-fullness technique of Vandev (1993) for the finite sample breakdown point study of the Weighted Trimmed Likelihood Estimator is extended. The proposed generalized d-fullness technique is illustrated over the generalized logistic regression model.

Mathematics Subject Classification (2000). Primary 62J12; Secondary 62F35.

Keywords. Breakdown point, d-fullness, trimmed likelihood estimator, generalized logistic regression.

1. Introduction

The classical Maximum Likelihood Estimator (MLE) can be very sensitive to outliers in the data. In fact, even a single outlier can ruin completely the MLE. To overcome this problem many robust alternatives of the MLE have been developed (see Atkinson and Riani, 2000; Bednarski and Clarke, 1993; Beran, 1982; Christmann, 1994; Field and Smith, 1994; Hampel et al., 1986; Huber, 1981; Hubert, 1997; Marazzi and Yohai, 2003; Markatou et al., 1997; Neykov and Neytchev, 1990; Shane and Simonoff, 2001; Vandev and Neykov, 1993; Windham, 1995; Choi et al., 2000).

Hadi and Luceño (1997) and Vandev and Neykov (1998) introduced a robust parametric modification of the MLE called the Weighted Trimmed Likelihood (WTL) estimator. The basic idea behind the trimming in the proposed estimator is in the removal of those observations whose values would be highly unlikely to occur if the fitted model was true. These authors showed that under appropriate choices of the trimming parameter and the weights, the WTL estimator reduces to the MLE, to the LMS and LTS estimators of Rousseeuw (1984) in the case of normal regression, and to the MVE and MCD estimators of the multivariate location and scatter introduced by Rousseeuw (1985) in the multivariate normal case.

The $\sqrt{n}$ consistency of the WTL estimator is derived by Cizek (2002).

Algorithms for TL estimation in the univariate case were developed by Hadi and Luceño (1997), whereas Neykov and Müller (2002) proposed a FAST-TLE algorithm in the framework of the generalized linear models.

The breakdown point (BP) properties of the WTL estimator were studied by Vandev and Neykov (1998) and Müller and Neykov (2003) using the d-fullness technique of Vandev (1993). According to Vandev and Neykov (1998), a set $F = \{f_1, \ldots, f_n\}$ of arbitrary functions $f_i : \Theta \to \mathbb{R}^+$, $\Theta \subseteq \mathbb{R}^q$, is called d-full if for every subset $J \subset \{1, \ldots, n\}$ of cardinality d ($|J| = d$) the function $g_J(\theta) = \max_{j\in J} f_j(\theta)$, $\theta \in \Theta$, is subcompact. A function $g : \Theta \to \mathbb{R}$, $\Theta \subseteq \mathbb{R}^q$ is called subcompact if its Lebesgue set $L_g(C) = \{\theta \in \Theta : g(\theta) \leq C\}$ is contained in a compact set for every real constant C, as defined by Müller and Neykov (2003).

The requirement for d-fullness of the set F is restrictive, more precisely, the condition *for every real constant* C in the definition of a subcompact function is not always satisfied. For instance the corresponding set F of log likelihoods for the mixtures of univariate/multivariate normal or binomial distributions, just to name but a few, are not d-full in the above sense.

In this paper a generalized d-fullness technique is proposed to study the BP of the WTL estimator for a wider class of functions containing the class of subcompact functions. Section 2 defines the concept of breakdown point and generalized d-fullness. This technique is illustrated for the generalized logistic regression model in Section 3. The proofs of the lemmas and propositions are given in the Appendix.

2. Definitions and Generalized d-fullness Technique

To aid the presentation we remind the replacement variant of the finite sample BP given in Hampel et al. (1986), which is closely related to that introduced by Donoho and Huber (1983). Let $X = \{x_i \in \mathcal{X} \subseteq \mathbb{R}^p, \text{ for } i = 1, \ldots, n\}$ be a sample of size n.

Definition 2.1. The BP of an estimator T at X is given by

$$\varepsilon_n^*(T) = \frac{1}{n} \max\{m : \sup_{\tilde{X}_m} \|T(\tilde{X}_m)\| < \infty\},$$

where $\tilde{X}_m$ is a sample obtained from X by replacing any m of the points in X by arbitrary values from $\mathcal{X}$, and $\|.\|$ is the Euklidean norm.

We now recall the definition of the Weighted Trimmed estimator given in Vandev and Neykov (1998). Let $f : X \times \Theta \to \mathbb{R}^+$, where $\Theta \subseteq \mathbb{R}^q$ be an open set, and $F = \{f_i(\theta) = f(x_i, \theta), \text{ for } i = 1, \ldots, n\}$.

Definition 2.2. The Weighted Trimmed estimator is defined as

$$\mathrm{W}_k := \arg\min_{\theta\in\Theta} \sum_{i=1}^{k} w_{\nu(i)} f_{\nu(i)}(\theta), \tag{2.1}$$

where $f_{\nu(1)}(\theta) \leq f_{\nu(2)}(\theta) \leq \ldots \leq f_{\nu(n)}(\theta)$ are the ordered values of f_i at θ, $\nu = (\nu(1), \ldots, \nu(n))$ is the corresponding permutation of the indices, which depends on θ, k is the trimming parameter, the weights $w_i \geq 0$ for $i = 1, \ldots, n$ are associated with the functions $f_i(\theta)$ and are such that $w_{\nu(k)} > 0$.

The W_k estimator is too general to be of practical use. However, many high breakdown statistical estimators can be derived from it. For instance, let $f_i(\theta) = g(|r_i(\theta)|)$ for $i = 1, \ldots, n$, where g be a continuous monotonic function such that $g(0) = 0$ and $r_i(\theta) = y_i - x_i^T\theta$ be the ith linear regression residual, generated by the observation $(y_i, x_i^T) \in \mathbb{R}^{p+1}$ and the unknown parameter $\theta \in \Theta \subseteq \mathbb{R}^p$. Then the LMS, LTS and LQS regression estimators of Rousseeuw (1984), the h-trimmed weighted L_q estimator of Müller (1995), and the Hössjer (1994) rank-based estimator can be obtained as special cases of the W_k estimator. If $x_i \in \mathcal{X}$ for $i = 1, \ldots, n$ be i.i.d. observations with probability density function $\phi(x, \theta)$, which depends on an unknown parameter $\theta \in \Theta \subseteq \mathbb{R}^q$ and $f(x_i, \theta) = -\log \phi(x_i, \theta)$, then the W_k estimator coincides with the WTL_k estimator proposed by Hadi and Luceño (1997) and Vandev and Neykov (1998). In particular, when $\phi(x, \theta)$ is the multivariate normal density function, the W_k reduces to the MVE and MCD estimators of the multivariate location and scatter considered by Rousseeuw (1985) (see Vandev and Neykov, 1998). Vandev and Neykov (1998) prove that the BP of the W_k is not less than $(n - k)/n$ if the set F is d-full, $n \geq 3d$ and $(n + d)/2 \leq k \leq n - d$. Müller and Neykov (2003) extend their result finding the lower bound of the BP without additional assumptions on k and n. Moreover, for the generalized linear models it was shown that the fullness parameter d is related with the quantity $\mathcal{N}(X)$ of Müller (1995), where X is the matrix of the explanatory variables which may not be in general position, as is the case with the designed experiment, or are generated by qualitative factors.

We need the following notations in the presentation of the generalized d-fullness technique.

Let g be a function such that $g : \Theta \to \mathbb{R}$, $\partial\Theta$ be the set of the boundary points of Θ, and $\Theta_\infty = \{\{\theta_k\}_{k=1}^{\infty} : \theta_k \in \Theta, \|\theta_k\| \to \infty\}$ be the set of all sequences whose norm tends to infinity. Then $\underline{g}$ is defined as

$$\underline{g} = \begin{cases} \inf\limits_{\theta^* \in \partial\Theta} \liminf\limits_{\theta_k \to \theta^*} g(\theta_k), & \text{if } \Theta \text{ is bounded, or} \\ \\ \inf\limits_{\theta^* \in \partial\Theta} \liminf\limits_{\substack{\theta_k \to \theta^* \\ \{\theta_k\} \in \Theta_\infty}} g(\theta_k), & \text{if } \Theta \text{ is unbounded.} \end{cases} \tag{2.2}$$

Let us introduce the following conditions:

A1. $F = \{f_i(\theta) \geq 0,\ i = 1, \ldots, n,\ \text{for } \theta \in \Theta\}$ be a set of continuous functions;

A2. There exists $\theta_0 \in \Theta$, such that for every subset $J \subset \{1, \ldots, n\}$ of cardinality d, $c^{**} g_J(\theta_0) < C$, where $g_J(\theta) = \max\limits_{j \in J} f_j(\theta)$ and $C = \inf\limits_J \underline{g_J}$, $c^{**} = \frac{c^*}{w^*}$, $c^* = \sum\limits_{i=1}^{k} w_{\nu(i)}$, and $w^* = \min\{w_j > 0,\ j = 1, \ldots, n\}$.

Remark 2.3. The d-full sets class of functions is a special case of the class of sets of functions satisfying A1 and A2 conditions, because if $\underline{g} = \infty$, then g is a subcompact function. This follows from Lemma 4.1 given in the Appendix.

The following proposition gives the necessary conditions under which there exists a solution of the optimization problem (2.1).

Proposition 2.4. *If $k \geq d$, A1 and A2 hold, then* W_k *is a non-empty compact set.*

The next proposition gives a lower bound for the BP of W_k for a set of functions F satisfying A1 and A2 conditions. It is a generalization of the corresponding result of Vandev and Neykov (1998) who required d-fullness of F.

Proposition 2.5. *The BP of* W_k *is not less than* $\frac{1}{n}\min(n-k, k-d)$ *if A1 and A2 hold.*

Remark 2.6. The above propositions hold for the WTL_k estimator. Thus, if one wishes to study the BP of the WTL_k estimators for a particular distribution, one has to establish the validity of the conditions A1 and A2 for the corresponding set of functions $f_i(\theta) = -\log\phi(x_i, \theta)$ for $i = 1, \ldots, n$. Then the BP can be exemplified by the range of values of k by Proposition 2.5.

3. Application on a Generalized Logistic Regression Model

As an illustration of the above propositions we consider the grouped binary linear regression model with generalized logistic link. The type of the data under consideration is of the form (y_i, x_i^T) for $i = 1, \ldots, N$. It is assumed that, y_i is binomially distributed, $b(y_i \mid n_i, \pi_i)$, where the group size is n_i, the probability of success is π_i, and x_i is a p-dimensional vector of covariates (explanatory variables). The total number of observations is $n = n_1 + n_2 + \cdots + n_N$. We will assume that $0 < y_i < n_i$ for each i, and π_i follows the Prentice (1976) generalized logistic distribution

$$\pi_i = (1 + \exp(-\eta_i))^{-a},$$

where $a > 0$, $\eta_i = x_i^T\beta$ is the linear predictor and β is a p-dimensional vector of unknown parameters.

The particular case, when a=1, is considered by Müller and Neykov (2003) who proved that the BP of the WTL_k estimator is equal to $\min(N-k+1, k-\mathcal{N}(X))/N$, where $\mathcal{N}(X) = \max_{0 \neq \beta \in \mathbb{R}^p} \mathrm{card}\{i \in \{1, \ldots, N\}; x_i^T\beta = 0\}$.

We will show that the set $F = \{f(y_i, \eta_i, a), i = 1, \ldots, N\}$, where $f(y_i, \eta_i, a) = -\log\binom{n_i}{y_i} + y_i a \log(1 + e^{-\eta_i}) - (n_i - y_i)\log(1 - (1 + e^{-\eta_i})^{-a})$, satisfies Conditions A1 and A2.

It is obvious that $\lim_{a \to 0} f(y_i, \eta_i, a) = +\infty$, $\lim_{a \to +\infty} f(y_i, \eta_i, a) = +\infty$, and $\lim_{\eta_i \to \pm\infty} f(y_i, \eta_i, a) = +\infty$. Therefore $f(y_i, \eta_i, a)$ is a subcompact function because $\underline{f} = +\infty$.

Proposition 3.1. *The set* $\{f(y_i, x_i, \beta, a), i = 1, \ldots, N\}$ *is* $\mathcal{N}(X) + 1$*-full.*

As a consequence of this proposition, the following corollary is obtained.

Corollary 3.2. *The set WTL_k for the grouped binary linear regression model with generalized logistic link is a non-empty compact set if $k \geq \mathcal{N}(X) + 1$.*

Applying Theorem 2 of Müller and Neykov (2003) we get the following

Corollary 3.3. *If $\lfloor (N + \mathcal{N}(X) + 1)/2 \rfloor \leq k \leq \lfloor (N + \mathcal{N}(X) + 2)/2 \rfloor$, then the BP of the WTL_k estimator for the grouped binary linear regression model with generalized logistic link is*

$$\varepsilon_N^*(WTL_k) \geq \frac{1}{N} \left\lfloor \frac{N - \mathcal{N}(X) + 1}{2} \right\rfloor .$$

We remind that $\lfloor z \rfloor := \max\{n : n \leq z\}$.

4. Appendix

The proofs in this section are based on the following Lemma.

Lemma 4.1. *Let $g : \Theta \to \mathbb{R}$ be continuous function, $\Theta \subseteq \mathbb{R}^q$ be an open set. If there exists $\theta_0 \in \Theta$ and a real constant $a \geq 1$, such that $ag(\theta_0) < C$, where $C \leq \underline{g}$, then the set $S = \{\theta : g(\theta) < C\}$ is bounded and non-empty.*

Proof. We shall note that the set S is non-empty since $\theta_0 \in S$. Let us assume that S is unbounded, $\{\theta_j\}_{j=1}^{\infty}$ be a sequence from S such that $\|\theta_j\| \underset{j\to\infty}{\longrightarrow} \infty$ and $r_j = \max\{\|\theta_1\|, \ldots, \|\theta_j\|\}$. Then we have that $\inf\limits_{\|\theta_j\| \geq r_j} g(\theta) \leq g(\theta_j) < C$. Taking a limit we get $\underline{g} < C$, which is a contradiction with $C \leq \underline{g}$.

We will use the representation $f_{(k)}(\theta) = \min\limits_{I \in I_k} \max\limits_{i \in I} f_i(\theta)$ which holds at any fixed θ and I_k is the set of all subsets of $\{1, \ldots, n\}$ consisting of k elements in the propositions proof (see Krivulin, 1992).

Proof of Proposition 2.4. The following inclusions hold

$$\begin{aligned}
\mathrm{W}_k &= \left\{\theta : \sum_{i=1}^{k} w_{\nu(i)} f_{\nu(i)}(\theta) \leq \sum_{i=1}^{k} w_{\nu(i)} f_{\nu(i)}(\vartheta) \ \forall\, \vartheta \in \Theta \right\} \\
&= \left\{\theta : \sum_{i=1}^{k} w_{\nu(i)} f_{\nu(i)}(\theta) \leq \inf_{\vartheta \in \Theta} \sum_{i=1}^{k} w_{\nu(i)} f_{\nu(i)}(\vartheta) \right\} \\
&\subseteq \left\{\theta : \sum_{i=1}^{k} w_{\nu(i)} f_{\nu(i)}(\theta) \leq \inf_{\vartheta \in \Theta} \sum_{i=1}^{k} w_{\nu(i)} f_{\nu(k)}(\vartheta) \right\} \\
&\subseteq \left\{\theta : \min_{I \in I_k} \sum_{i \in I} w_i f_i(\theta) \leq c^* \inf_{\vartheta \in \Theta} f_{\nu(k)}(\vartheta) \right\}
\end{aligned}$$

$$
\begin{aligned}
&\subseteq \bigcup_{I \in I_k} \left\{ \theta : \sum_{i \in I} w_i f_i(\theta) \leq c^* \inf_{\vartheta \in \Theta} f_{\nu(k)}(\vartheta) \right\} \\
&= \bigcup_{I \in I_k} \left\{ \theta : \sum_{i \in I} w_i f_i(\theta) \leq c^* \inf_{\vartheta \in \Theta} \min_{I \in I_k} \max_{i \in I} f_i(\vartheta) \right\} \\
&\subseteq \bigcup_{I \in I_k} \left\{ \theta : \sum_{i \in I} w_i f_i(\theta) \leq c^* \inf_{\vartheta \in \Theta} \max_{i \in I} f_i(\vartheta) \right\} \\
&\subseteq \bigcup_{I \in I_k} \left\{ \theta : w^* \max_{i \in I} f_i(\theta) \leq c^* \inf_{\vartheta \in \Theta} \max_{i \in I} f_i(\vartheta) \right\} \\
&\subseteq \bigcup_{I \in I_k} \left\{ \theta : \max_{i \in I} f_i(\theta) \leq c^{**} \max_{i \in I} f_i(\theta_0) \right\} \\
&\subseteq \bigcup_{I \in I_k} \bigcup_{J \subset I} \left\{ \theta : \max_{j \in J} f_j(\theta) \leq c^{**} \max_{j \in J} f_j(\theta_0) \right\} \\
&= \bigcup_{I \in I_k} \bigcup_{J \subset I} \{ \theta : g_J(\theta) \leq c^{**} g_J(\theta_0) \} \\
&\subset \bigcup_{I \in I_k} \bigcup_{J \subset I} \{ \theta : g_J(\theta) < C \}
\end{aligned}
$$

The latter set is a non-empty bounded set according to Lemma 4.1, since A2 condition is satisfied. Therefore, W_k is a compact set, since the functions from F are continuous.

Proof of Proposition 2.5. Let $\tilde{F} = \{\tilde{f}_i(\theta) = f(\tilde{x}_i, \theta) \text{ for } i = 1, \ldots, n\}$ be obtained from F upon replacement of $m = \min\{n-k, k-d\}$ observations from the sample X with arbitrary ones from $\mathcal{X}$, the weights w_i correspond to the functions that belong to F and $\tilde{\mathrm{W}}_k$ is the corresponding analog of W_k, defined over $\tilde{F}$. The number of the original functions in $\tilde{F}$ is $n - m \geq k$. Hence there exists $I^* \in I_k$ such that $\tilde{f}_i(\theta) \equiv f_i(\theta)$ for $i \in I^*$. The following inequalities hold $\tilde{f}_{\nu(k)}(\theta) = \min_{I \in I_k} \max_{i \in I} \tilde{f}_i(\theta) \leq \max_{i \in I^*} f_i(\theta)$ and $f_{\nu(d)}(\theta) \leq \tilde{f}_{\nu(k)}(\theta)$ (see Müller and Neykov, 2003). Then we have the inclusions

$$
\begin{aligned}
\tilde{\mathrm{W}}_k &= \left\{ \theta : \sum_{i=1}^{k} w_{\nu(i)} \tilde{f}_{\nu(i)}(\theta) \leq \inf_{\vartheta \in \Theta} \sum_{i=1}^{k} w_{\nu(i)} \tilde{f}_{\nu(i)}(\vartheta) \right\} \\
&\subseteq \left\{ \theta : w^* \tilde{f}_{\nu(k)}(\theta) \leq c^* \inf_{\vartheta \in \Theta} \tilde{f}_{\nu(k)}(\vartheta) \right\} \\
&\subseteq \left\{ \theta : w^* \tilde{f}_{\nu(k)}(\theta) \leq c^* \inf_{\vartheta \in \Theta} \max_{i \in I^*} f_i(\vartheta) \right\}
\end{aligned}
$$

$$
\begin{aligned}
&\subseteq \left\{\theta : w^* \tilde{f}_{\nu(k)}(\theta) \le c^* \max_{i\in I^*} f_i(\theta_0)\right\} \\
&\subseteq \bigcup_{\substack{J\subset I^* \\ |J|=d}} \left\{\theta : w^* \tilde{f}_{\nu(k)}(\theta) \le c^* \max_{j\in J} f_j(\theta_0)\right\} \\
&\subseteq \bigcup_{\substack{J\subset I^* \\ |J|=d}} \left\{\theta : w^* f_{\nu(d)}(\theta) \le c^* \max_{j\in J} f_j(\theta_0)\right\} \\
&\subset \bigcup_{\substack{J\subset I^* \\ |J|=d}} \{\theta : f_{\nu(d)}(\theta) < C\}
\end{aligned}
$$

According to Lemma 4.1 the final set is a non-empty bounded set. Therefore, $\tilde{\mathrm{W}}_k$ is a compact set, since the functions from $\tilde{F}$ are continuous.

Proof of Proposition 3.1. Let $C \in \mathbb{R}$ is arbitrary. Since $f(y_i, \eta_i, a)$ is a subcompact function of η_i and a, there exist constants B_i and A_i for $i \in I \subset \{1, \dots, N\}$, $\mathrm{card}(I) = \mathcal{N}(X) + 1$, such that the set

$$
\begin{aligned}
&\{\beta \in \mathbb{R}^p, a > 0 : \max_{i\in I} f(y_i, x_i, \beta, a) \le C\} \\
= &\bigcap_{i\in I} \{\beta \in \mathbb{R}^p, a > 0 : f(y_i, x_i, \beta, a) \le C\} \\
= &\bigcap_{i\in I} \{\beta \in \mathbb{R}^p, a > 0 : f(y_i, \eta_i = x_i^T\beta, a) \le C\} \\
\subset &\bigcap_{i\in I} \{\{\beta \in \mathbb{R}^p : |x_i^T\beta| \le B_i\} \times \{a : 0 < a \le A_i\}\}
\end{aligned}
$$

is contained in a bounded set. (The set $\{\beta \in \mathbb{R}^p | x_i^T\beta| \le B_i\}$ is bounded for all B_i according to Lemma 3 of Müller and Neykov, 2003.)

References

[1] A.C. Atkinson and M. Riani, *Robust Diagnostic Regression Analysis.* Springer, New York, 2000.

[2] D.V. Atanasov and N.M. Neykov, *On the Finite Sample Breakdown Point of the Weighted Trimmed Likelihood Estimators and the d-fullness of a Set of Continuous Functions.* In S. Aivazian, Yu. Kharin, and H. Reider, editors, Proceedings of the VI Intern. Conf. "Computer Data Analysis and Modelling", Vol 1, pages 52–57. Minsk, Belarus, 2001.

[3] T. Bednarski and B.R. Clarke, *Trimmed Likelihood Estimation of Location and Scale of the Normal Distribution.* Austral. J. Statist. **35** (1993), 141–153.

[4] R. Beran, *Robust Estimation in Models for Independent Nonindentically Distributed Data.* Ann. Statist. **10** (1982), 415–428.

[5] E.P. Choi, P. Hall, and B. Presnell, *Rendering Parametric Procedures More Robust by Empirically Tilting the Model.* Biometrika **87** (2000), 453–465.

[6] P. Cizek, *Robust Estimation in Nonlinear Regression and Limited Dependent Variable Models.* Submitted, 2002, available at *http://econpapers.hhs.se/paper/wpawuwpem/0203003.htm.*

[7] A. Christmann, *Least Median of Weighted Squares in Logistic Regression with Large Strata.* Biometrika **81** (1994), 413–417.

[8] E.Z. Demidenko, *Optimization and Regression.*(in Russian) Nauka, Moscow, 1989.

[9] D.L. Donoho and P.J. Huber, *The Notion of Breakdown Point.* In P.J. Bickel, K.A. Doksum, and J.L. Hodges, editors, A Festschrift for Erich L. Lehmann, pages 157–184. Wadsworth, Belmont, CA, 1983.

[10] C. Field and B. Smith, *Robust Estimation – a Weighted Maximum Likelihood Approach.* Int. Statist. Rev. **62** (1994), 405–424.

[11] A.S. Hadi and A. Luceño, *Maximum Trimmed Likelihood Estimators: a Unified Approach, Examples, and Algorithms.* Comput. Statist. Data Anal. **25** (1997), 251–272.

[12] F.R. Hampel, E.M. Ronchetti, P.J. Rousseeuw, and W.A. Stahel, *Robust Statistics. The Approach Based on Influence Functions.* John Wiley & Sons, New York, 1986.

[13] O. Hössjer, *Rank-based Estimates in the Linear Model with High Breakdown Point.* J. Amer. Statist. Assoc. **89** (1994), 149–158.

[14] P. Huber, *Robust Statistics.* John Wiley & Sons, New York, 1981.

[15] M. Hubert, *The Breakdown Value of the L_1 Estimator in Contingency Tables.* Statist. Probab. Lett. **33** (1997), 419–425.

[16] N. Krivulin, *An Analysis of the Least Median of Squares Regression Problem.* In Y. Dodge and J. Whittaker, editors, Proceedings in Computational Statistics, pages 471–476. Physica-Verlag, Heidelberg, 1992.

[17] A. Marazzi and V. Yohai, *Adaptively Truncated Maximum Likelihood Regression with Asymmetric Errors.* J. Statist. Plann. Inference (2003), in press.

[18] M. Markatou, A. Basu, and B.C. Lindsay, *Weighted Likelihood Estimating Equations: the Discrete Case with Applications to Logistic Regression.* J. Statist. Plann. Inference **57** (1997), 215–232.

[19] Ch.H. Müller, *Breakdown Points for Designed Experiments.* J. Statist. Plann. Inference **45** (1995), 413–427.

[20] Ch.H. Müller and N.M. Neykov, *Breakdown Points of the Trimmed Likelihood and Related Estimators in Generalized Linear Models.* J. Statist. Plann. Inference **116** (2003), 503–519.

[21] N.M. Neykov and Ch.H. Müller, *Breakdown Point and Computation of Trimmed Likelihood Estimators in Generalized Linear Models.* In R. Dutter, P. Filzmoser, U. Gather, and P. Rousseeuw, editors, Developments in Robust Statistics, pages 277–286. Physica-Verlag, Heidelberg, 2002.

[22] N.M. Neykov and P.N. Neytchev, *A Robust Alternative of the Maximum Likelihood Estimator.* In Short communications of COMPSTAT'90, pages 99–100. Dubrovnik, 1990.

[23] R.L. Prentice, *A Generalization of the Probit and Logit Methods for Dose Response Curves.* Biometrics **32** (1976), 761–768.

[24] P.J. Rousseeuw, *Least Median of Squares Regression.* J. Amer. Statist. Assoc. **79** (1984), 851–857.

[25] P.J. Rousseeuw, *Multivariate Estimation with High Breakdown Point.* In W. Grossmann, G. Pflug, I. Vincze, and W. Wertz, editors, Mathematical Statistics and Applications, Vol B, pages 283–297. Reidel Publishing Company, Dordrecht, 1985.

[26] K.V. Shane and J.S. Simonoff, *A Robust Approach to Categorical Data Analysis.* J. Comput. Graph. Statist. **10** (2001), 135–157.

[27] D.L. Vandev, *A Note on Breakdown Point of the Least Median Squares and Least Trimmed Squares.* Statist. Probab. Lett. **16** (1993), 117–119.

[28] D.L. Vandev and N.M. Neykov, *Robust Maximum Likelihood in the Gaussian Case.* In S. Morgenthaler, E. Ronchetti, and W.A. Stahel, editors, New Directions in Data Analysis and Robustness, pages 259–264. Birkhäuser Verlag, Basel, 1993.

[29] D.L. Vandev and N.M. Neykov, *About Regression Estimators with High Breakdown Point.* Statistics **32** (1998), 111–129.

[30] M.P. Windham, *Robustifying Model Fitting.* J. Roy. Statist. Soc. Ser. B **57** (1995), 599–609.

Acknowledgment

The authors wish to thank the editors and the anonymous referees for helpful comments on a previous draft of the paper. We are grateful both Prof. D. Vandev and Prof. D. Atanasov of Faculty of Mathematics and Informatics, "St. Kliment Ohridski" Sofia University for discussions on the generalized d-fullness technique. N. Neykov would like to express his deepest and heartfelt thanks to Heidi and Peter Filzmoser who made his ICORS'03 participation possible as well as to the ICORS'03 organizers for the financial support.

R.B. Dimova
Faculty of Mathematics and Informatics, "St. Kliment Ohridski" Sofia University
J. Baucher str.
Sofia, Bulgaria
e-mail: `rdimova@fmi.uni-sofia.bg`

N.M. Neykov
National Institute of Meteorology and Hydrology
Bulgarian Academy of Sciences
66 Tsarigradsko chaussee
1784 Sofia, Bulgaria
e-mail: `neyko.neykov@meteo.bg`

Statistics for Industry and Technology, 93–104

On Robustness to Outliers of Parametric L_2 Estimate Criterion in the Case of Bivariate Normal Mixtures: a Simulation Study

A. Durio and E.D. Isaia

Abstract. The purpose of this work is to investigate on the use of the Integrated Square Error, or L_2 distance, as a practical tool for parameters estimation of mixtures of two normal bivariates in presence of outliers, situations in which maximum likelihood estimators are usually unstable. Theory is outlined, closed expressions for the L_2 minimizing estimate criterion for bivariate gaussian mixtures are given. In order to evaluate robustness of maximum likelihood and L_2 minimizing estimate criteria we compare results arising from a Monte Carlo simulation for some mixtures of gaussian bivariates in occurrence of different outliers positioning and consistency, matching some typical situations that frequently arise in industrial and chemical fields.

Mathematics Subject Classification (2000). Primary 62F35; Secondary 62H12.
Keywords. Bivariate normal mixtures, integrated squared error, minimum distance estimation, outliers detection, robust multivariate estimation.

1. Introduction

In this paper we investigate on the use of the Integrated Square Error, or L_2 distance, as a practical tool for parameters estimation of mixtures of two normal bivariates in presence of outliers, situations in which Maximum Likelihood Estimators (MLE) are usually unstable.

Following Scott (2001), theory is outlined and a closed expression for the L_2 minimizing estimate criterion (L_2E) for a mixture of two bivariate normal densities is presented.

In order to evaluate robustness of ML and L_2 estimates, the latter being by nature less influenced by outliers (e.g., Basu et al., 1998; Durio and Isaia, 2003) we compare the results, in terms of average and standard error, arising from a Monte

This research was supported partly by the MURST 60% funds, grant EDI.es.fin.02 and AD.es.fin.02.

Carlo simulation for some specific cases of mixtures of two normal bivariates in occurrence of different outliers positioning and consistency. There are some real situations in which few outliers lie in specific positions, for instance in the field of quality control process the outliers are scarce and come from a sudden state out of registration of the production process or the measurement system. In this paper we consider the outliers as realizations of rare events rather than observations belonging to clusters not contemplated by the model. Therefore, in order to match some typical situations that frequently arise in scientific areas, outliers have been positioned in a deterministic way and their contamination percentage has been kept quite low.

2. The L_2 Estimator

It is well known (e.g., Terrel, 1990) that the L_2E criterion originates in the derivation of the nonparametric least squares cross-validation algorithm to choose the bandwidth for the kernel estimate of a density, e.g., Wand and Jones (1995). In the parametric case a multivariate r.v. $\boldsymbol{X}$ is given of dimension $p \geq 2$, with probability density $g(\boldsymbol{x}|\boldsymbol{\theta}_0)$, depending on the unknown parameters vector $\boldsymbol{\theta}_0$ for which we introduce the model $f(\boldsymbol{x}|\boldsymbol{\theta})$. In order to have an estimate of $\boldsymbol{\theta}$ we minimize the L_2 distance with respect to $\boldsymbol{\theta}$

$$\int_{\mathbb{R}^p} [f(\boldsymbol{x}|\boldsymbol{\theta}) - g(\boldsymbol{x}|\boldsymbol{\theta}_0)]^2\, d\boldsymbol{x}$$
$$= \left[\int_{\mathbb{R}^p} f(\boldsymbol{x}|\boldsymbol{\theta})^2 d\boldsymbol{x} - 2\mathbb{E}\left[f(\boldsymbol{x}|\boldsymbol{\theta})\right] + \int_{\mathbb{R}^p} g(\boldsymbol{x}|\boldsymbol{\theta}_0)^2 d\boldsymbol{x}\right]. \quad (2.1)$$

In (2.1) the integral $\int_{\mathbb{R}^p} g(\boldsymbol{x}|\boldsymbol{\theta}_0)^2\, d\boldsymbol{x}$ does not depend on $\boldsymbol{\theta}$. So if we estimate the expected height of the density $\mathbb{E}\left[f(\boldsymbol{x}|\boldsymbol{\theta})\right] = \int_{\mathbb{R}^p} f(\boldsymbol{x}|\boldsymbol{\theta})\, g(\boldsymbol{x}|\boldsymbol{\theta}_0) d\boldsymbol{x}$ with $\widehat{\mathbb{E}}\left[f(\boldsymbol{x}|\boldsymbol{\theta})\right] = n^{-1}\sum_{i=1}^{n} f(\boldsymbol{x}_i|\boldsymbol{\theta})$, the proposed estimator for $\boldsymbol{\theta}_0$ will be

$$\hat{\boldsymbol{\theta}}_{L_2E} = \arg\min_{\boldsymbol{\theta}} \left[\int_{\mathbb{R}^p} f(\boldsymbol{x}|\boldsymbol{\theta})^2\, d\boldsymbol{x} - \frac{2}{n}\sum_{i=1}^{n} f(\boldsymbol{X}_i|\boldsymbol{\theta})\right]. \quad (2.2)$$

Thus, for instance, in the case of a multivariate gaussian density $\phi(\boldsymbol{x}|\boldsymbol{\mu}, \boldsymbol{\Sigma})$, equation (2.2) becomes

$$\hat{\boldsymbol{\theta}}_{L_2E} = \arg\min_{\boldsymbol{\mu}, \boldsymbol{\Sigma}} \left[\frac{1}{2^p\, \pi^{p/2} |\boldsymbol{\Sigma}|^{1/2}} - \frac{2}{n}\sum_{i=1}^{n} \phi(\boldsymbol{X}_i|\boldsymbol{\mu}, \boldsymbol{\Sigma})\right]$$

since $\int_{\mathbb{R}^p} \phi(\boldsymbol{x}|\boldsymbol{\mu}, \boldsymbol{\Sigma})^2\, d\boldsymbol{x} = \phi(\mathbf{0}|\mathbf{0}, 2\,\boldsymbol{\Sigma})$, e.g., Scott (2001).

Let us now consider a mixture of two bivariate gaussian densities (e.g., Titterington et al., 1985)

$$f(\boldsymbol{x}|\boldsymbol{\mu}_1, \boldsymbol{\mu}_2, \boldsymbol{\Sigma}_1, \boldsymbol{\Sigma}_2) = w\,\phi(\boldsymbol{x}|\boldsymbol{\mu}_1, \boldsymbol{\Sigma}_1) + (1-w)\,\phi(\boldsymbol{x}|\boldsymbol{\mu}_2, \boldsymbol{\Sigma}_2) \quad (2.3)$$

where the mixing probability w satisfies $0 \leq w \leq 1$.

In this situation the integral that appears in (2.2) becomes

$$\int_{I\!R^2} f(\boldsymbol{x}|\boldsymbol{\mu}_1, \boldsymbol{\mu}_2, \boldsymbol{\Sigma}_1, \boldsymbol{\Sigma}_2)^2 d\boldsymbol{x} = w^2 \int_{I\!R^2} \phi(\boldsymbol{x}|\boldsymbol{\mu}_1, \boldsymbol{\Sigma}_1)^2\, d\boldsymbol{x} + (1-w)^2 \int_{I\!R^2} \phi(\boldsymbol{x}|\boldsymbol{\mu}_2, \boldsymbol{\Sigma}_2)^2\, d\boldsymbol{x} + 2\,w\,(1-w) \int_{I\!R^2} \phi(\boldsymbol{x}|\boldsymbol{\mu}_1, \boldsymbol{\Sigma}_1)\, \phi(\boldsymbol{x}|\boldsymbol{\mu}_2, \boldsymbol{\Sigma}_2)^2\, d\boldsymbol{x}.$$

If we now observe that

$$\int_{I\!R^2} \phi(\boldsymbol{x}|\boldsymbol{\mu}_1, \boldsymbol{\Sigma}_1)\, \phi(\boldsymbol{x}|\boldsymbol{\mu}_2, \boldsymbol{\Sigma}_2)\, d\boldsymbol{x} = \phi(\mathbf{0}|\boldsymbol{\mu}_1 - \boldsymbol{\mu}_2, \boldsymbol{\Sigma}_1 + \boldsymbol{\Sigma}_2)$$

it follows that

$$\int_{I\!R^2} f(\boldsymbol{x}|\boldsymbol{\mu}_1, \boldsymbol{\mu}_2, \boldsymbol{\Sigma}_1, \boldsymbol{\Sigma}_2)^2\, d\boldsymbol{x} = w^2\, \phi(\mathbf{0}|\mathbf{0}, 2\,\boldsymbol{\Sigma}_1) + (1-w)^2\, \phi(\mathbf{0}|\mathbf{0}, 2\,\boldsymbol{\Sigma}_2) + 2\,w\,(1-w)\, \phi(\mathbf{0}|\boldsymbol{\mu}_1 - \boldsymbol{\mu}_2, \boldsymbol{\Sigma}_1 + \boldsymbol{\Sigma}_2)$$

and hence equation (2.2) reduces to

$$\hat{\boldsymbol{\theta}}_{L_2E} = \underset{w, \boldsymbol{\mu}_1, \boldsymbol{\mu}_2, \boldsymbol{\Sigma}_1, \boldsymbol{\Sigma}_2}{\arg\min} \Bigg[w^2\, \phi(\mathbf{0}|\mathbf{0}, 2\,\boldsymbol{\Sigma}_1) + (1-w)^2\, \phi(\mathbf{0}|\mathbf{0}, 2\,\boldsymbol{\Sigma}_2) + 2\,w\,(1-w)\, \phi(\mathbf{0}|\boldsymbol{\mu}_1 - \boldsymbol{\mu}_2, \boldsymbol{\Sigma}_1 + \boldsymbol{\Sigma}_2) - \frac{2}{n} \sum_{i=1}^{n} f(\boldsymbol{X}_i|\boldsymbol{\mu}_1, \boldsymbol{\mu}_2, \boldsymbol{\Sigma}_1, \boldsymbol{\Sigma}_2) \Bigg]. \tag{2.4}$$

Since closed forms for equations (2.4) are unlikely to be found, we have to resort to numerical minimization algorithms, such as those based on quasi-Newton optimization. It is important to recall that the convergence of any algorithm to optimal solutions strongly depends on the initialization of the parameters.

3. The Simulation

For our simulation study we decided to draw $k = 1,000$ samples of size $n = 200$ each from different mixtures of two bivariate gaussian densities, infected by m points that, according to specific rules, may be considered as outliers. Since our goal was to study the behavior of both ML and L_2 estimators in situations that frequently arise in industrial and chemical fields, we chose not to position outliers at random and to keep their contamination percentage quite low; in fact we set $m = 8\,(16, 32)$.

We performed the simulation varying the distance between each component of the mixture, assuming either independence of the components of both the bivariate gaussian densities or dependence of the components of one of the bivariate normal densities. Furthermore, in order to examine the influence of the mixing probability on the estimates, we repeated the simulation with mixing probability $w = .5$ and $w = .25$.

More precisely, referring to (2.3) and letting for $i = 1, 2$

$$\boldsymbol{\mu}_i = \begin{pmatrix}\mu_{i1} & \mu_{i2}\end{pmatrix}^T \qquad \boldsymbol{\Sigma}_i = \begin{pmatrix}\sigma_{i1}^2 & \rho_i\sigma_{i1}\sigma_{i2} \\ \rho_i\sigma_{i1}\sigma_{i2} & \sigma_{i2}^2\end{pmatrix}$$

we considered the following eight experimental conditions

case	$w\,\phi(\mu_{11},\mu_{12},\sigma_{11},\sigma_{12},\rho_1) + (1-w)\,\phi(\mu_{21},\mu_{22},\sigma_{21},\sigma_{22},\rho_2)$
a	$.5\,\phi(-1,-1,1/3,1/3,0) + .5\,\phi(1,1,1/3,1/3,0)$
b	$.5\,\phi(-1,-1,1/3,1/3,-.7) + .5\,\phi(1,1,1/3,1/3,0)$
c	$.25\,\phi(-1,-1,1/3,1/3,0) + .75\,\phi(1,1,1/3,1/3,0)$
d	$.25\,\phi(-1,-1,1/3,1/3,-.7) + .75\,\phi(1,1,1/3,1/3,0)$
e	$.5\,\phi(-1,-1,2/3,2/3,0) + .5\,\phi(1,1,2/3,2/3,0)$
f	$.5\,\phi(-1,-1,2/3,2/3,-.7) + .5\,\phi(.5,.5,2/3,2/3,0)$
g	$.25\,\phi(-1,-1,2/3,2/3,0) + .75\,\phi(1,1,2/3,2/3,0)$
h	$.25\,\phi(-1,-1,2/3,2/3,-.7) + .75\,\phi(.5,.5,2/3,2/3,0)$

infected by outliers according to the following rules

• cases (a) and (c), *well separated components and independence*: we introduced (e.g., Figure 1) $m\,w$ outliers equally spaced on the first quadrant of a circle centered on $\boldsymbol{\mu}_1$ and $m\,(1-w)$ outliers equally spaced on the first quadrant of a circle centered on $\boldsymbol{\mu}_2$, both the circles having radius $r = r_1 = r_2 = 1.00\,(1.17, 1.34)$. The circle of radius $1.00\,(1.17, 1, 34)$ gives the region wherein the two gaussian bivariates fall with common probability $p = 0.9889\,(0.9978, 0.9997)$.

• cases (b) and (d), *well-separated components and dependence*: we introduced (e.g., Figure 1) $m\,(1-w)$ outliers equally spaced on the first quadrant of a circle centered on $\boldsymbol{\mu}_2$ and, as we did in cases (a) and (c), with $r_2 = 1.00\,(1.17, 1.34)$ and $m\,w$ outliers equally spaced on the first quadrant of an ellipse centered on $\boldsymbol{\mu}_1$, having length of the major axis $r_1 = 1.30\,(1.52, 1.75)$, so that the ellipse gives the region wherein the first component of the mixture falls with the same probability as the second component, i.e., $p = 0.9889\,(0.9978, 0.9997)$.

• cases (e) and (g), *close components and independence*: we introduced (e.g., Figure 2) $m\,w$ outliers equally spaced on the third quadrant of a circle centered on $\boldsymbol{\mu}_1$ and $m\,(1-w)$ outliers equally spaced on the first quadrant of a circle centered on $\boldsymbol{\mu}_2$. The circles have radius $r = r_1 = r_2 = 2.00\,(2.33, 2.69)$ so they meet the same probability regions of cases (a) and (b).

• cases (f) and (h), *close components and dependence*: we introduced (e.g., Figure 2) $m\,(1-w)$ outliers equally spaced on the first quadrant of a circle centered on $\boldsymbol{\mu}_2$

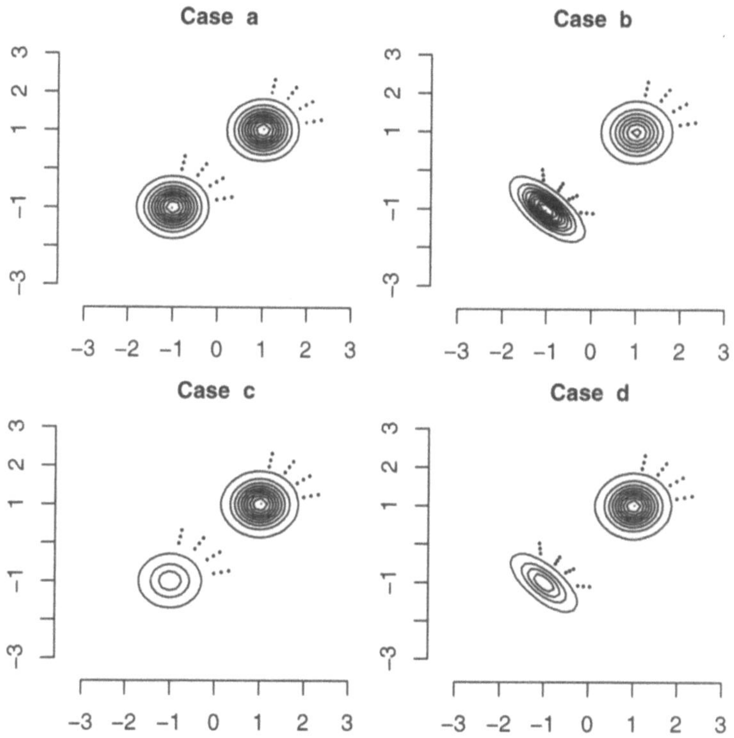

FIGURE 1. Experimental conditions of simulation with $m = 4$ outliers for any position considered.

with $r_2 = 2.00\,(2.33, 2.69)$ and $m\,w$ outliers equally spaced on the third quadrant of an ellipse centered on $\boldsymbol{\mu}_1$ with major axis $r_1 = 2.61\,(3.04, 3.50)$, so they meet the same probability regions of previous cases.

We chose cases (a), (c), (e) and (g) in order to study the behavior of L_2E versus MLE under the assumption of independence of the components of both the bivariate gaussian densities; in these situations the variance-covariance matrix of equation (2.4) becomes, for $i = 1, 2$, $\boldsymbol{\Sigma_i} = \boldsymbol{\sigma_i}\,\boldsymbol{I}$ where $\boldsymbol{\sigma_i} = (\sigma_{i1}^2, \sigma_{i2}^2)^T$, $\boldsymbol{I}$ is a 2×2 identity matrix and $\hat{\boldsymbol{\theta}}_{L_2E}$ is a random vector of dimension 9. In case (a), where the components of the mixture are well separated, we were interested in robustness of MLE and L_2E with respect to location and scale parameters, while in case (e), where the two components of the mixture are close to each other, we focused our attention to scale parameters estimation.

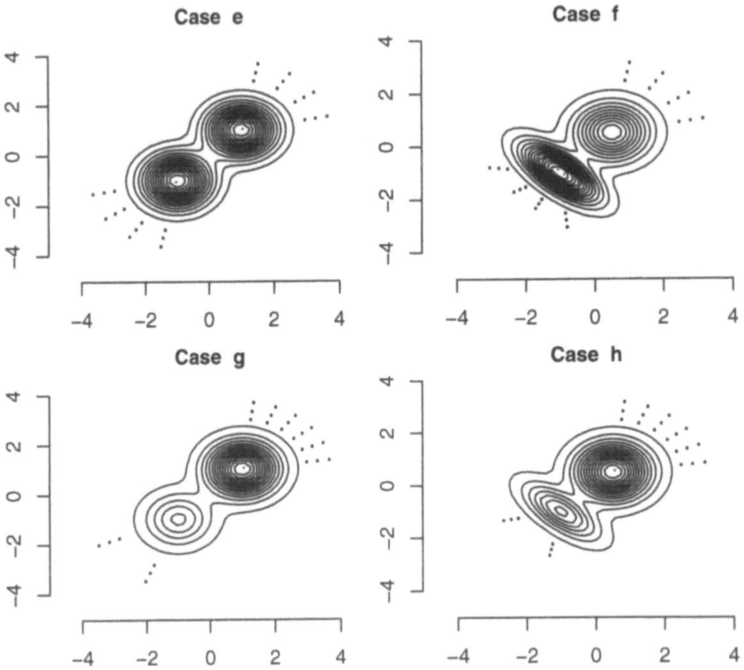

FIGURE 2. Experimental conditions of simulation with $m = 4$ outliers for any position considered.

In cases (b), (d), (f) and (h), where one of the bivariate gaussian densities of the mixture has dependent components ($\rho \neq 0$), we were mainly interested in robustness of MLE and L_2E with respect to the correlation parameter. We recall that in these cases the random vector $\hat{\boldsymbol{\theta}}_{L_2E}$ of equation (2.4) has dimension 11.

Comparing the results arising from cases (c), (d), (g) and (h) with the ones from cases (a), (b), (e) and (f), we observed the influence of the mixing probability on the estimates.

Before going on, it is worthwhile to underline that in order to obtain the MLE of the parameters of the mixture (2.3) we maximized the likelihood function resorting to numerical algorithms; the random vector of ML estimates looks like

$$\hat{\boldsymbol{\theta}}_{MLE} = \underset{w,\boldsymbol{\mu}_1,\boldsymbol{\mu}_2,\boldsymbol{\Sigma}_1,\boldsymbol{\Sigma}_2}{\arg\max} \prod_{i=1}^{n} w\,\phi(\boldsymbol{X}_i|\boldsymbol{\mu}_1,\boldsymbol{\Sigma}_1) + (1-w)\,\phi(\boldsymbol{X}_i|\boldsymbol{\mu}_2,\boldsymbol{\Sigma}_2). \tag{3.1}$$

We resorted also to the E-M algorithm (e.g., Flury, 1997; Gray, 1994) obtaining results that are close to the ones of the direct maximization of (3.1). We chose however the latter since the E-M algorithm is more sensitive to the initial guesses.

All MLE and L_2E were performed using the `nlim` routine implemented in R computing environment. The convergence of the algorithm to local optimal solutions was (almost always) ensured setting the vector of initial guesses equals to the vector of true parameters.

4. Some Results

For each of the $k = 1,000$ samples we estimated simultaneously location, scale, correlation parameters and mixing probability according to both ML and L_2 criteria. Comparison between the estimators has been done on the basis of average and standard error of samples results.

At a first glance we may state that MLE and L_2E of the mixing probability is in average accurate and this for all the experimental situations.

In all the cases we considered, the variability of the two estimators is of the same order for location parameters, while for scale parameters and mixing probability L_2E shows systematically a somewhat greater variability than MLE.

Comparing the results arising from cases (a), (b), (e) and (f) with those from cases (c), (d), (g) and (h), we observed that the mixing probability influences the variability of the estimates of all parameters. Obviously for the first component of the mixture (for which $w = .25$) the variability of the estimates is greater than the one we observed in the cases for which $w = .5$.

In the following we report the results of the simulation, in terms of average of the two estimators, for cases (a), (b), (e) and (f), since the same considerations could be made for cases (c), (d), (g) and (h).

• **case (a)**: the L_2 criterion shows a good performance in terms of location and scale parameters estimation and seems to be less sensitive to outliers than the ML criterion, which tends to overestimate the parameters as m increases. Furthermore, for a fixed m, the ML criterion tends to overestimate the parameters and especially the variances as r increases, while L_2 estimates don't seem to be inflated by the position of the outliers. Obviously, in this situation, the two estimators behave in the same way with respect to the two components of the mixture. Table 1 reports the means and the standard errors (in brackets) of the estimates arising from simulation and this for every m and every r.

• **case (e)**: the means of the ML estimates always overestimate the true parameters and as in case (a) this is due to the presence and positioning of outliers and also to the fact that in this case the two components of the mixture are close to each other. In average L_2 criterion overestimates too the variances, but in a mild way. In fact the vector of L_2E for $m = 32$ and $r = 2.69$ (the "worst" situation we

TABLE 1. Case (a): average of the estimates of the parameters of the mixture (standard errors in brackets).

$r=1.00$		$\hat{\mu}_{11}$	$\hat{\mu}_{12}$	$\hat{\sigma}_{11}$	$\hat{\sigma}_{12}$	$\hat{\mu}_{21}$	$\hat{\mu}_{22}$	$\hat{\sigma}_{21}$	$\hat{\sigma}_{22}$	w
$m=8$	L_2E	−0.998	−1.001	0.335	0.335	0.999	1.003	0.337	0.339	0.499
		(0.031)	(0.033)	(0.029)	(0.027)	(0.035)	(0.035)	(0.030)	(0.030)	(0.015)
	MLE	−0.976	−0.977	0.353	0.353	1.023	1.026	0.352	0.354	0.500
		(0.031)	(0.033)	(0.021)	(0.021)	(0.033)	(0.032)	(0.022)	(0.022)	(0.007)
$m=16$	L_2E	−0.999	−0.994	0.342	0.346	1.002	1.005	0.346	0.345	0.500
		(0.034)	(0.033)	(0.029)	(0.030)	(0.033)	(0.031)	(0.029)	(0.027)	(0.015)
	MLE	−0.953	−0.952	0.368	0.369	1.045	1.049	0.369	0.370	0.500
		(0.030)	(0.031)	(0.019)	(0.021)	(0.030)	(0.030)	(0.019)	(0.019)	(0.005)
$m=32$	L_2E	−0.993	−0.993	0.363	0.363	1.009	1.012	0.364	0.363	0.499
		(0.033)	(0.033)	(0.034)	(0.036)	(0.034)	(0.034)	(0.037)	(0.032)	(0.012)
	MLE	−0.914	−0.914	0.396	0.396	1.087	1.088	0.395	0.395	0.500
		(0.029)	(0.028)	(0.019)	(0.019)	(0.029)	(0.030)	(0.019)	(0.017)	(0.004)
$r=1.17$										
$m=8$	L_2E	−1.001	−0.999	0.334	0.338	1.004	0.999	0.335	0.336	0.501
		(0.032)	(0.032)	(0.027)	(0.029)	(0.033)	(0.034)	(0.028)	(0.030)	(0.014)
	MLE	−0.972	−0.971	0.360	0.361	1.032	1.028	0.362	0.362	0.500
		(0.033)	(0.032)	(0.021)	(0.021)	(0.031)	(0.032)	(0.021)	(0.022)	(0.006)
$m=16$	L_2E	−0.998	−1.000	0.342	0.340	0.999	1.000	0.339	0.344	0.500
		(0.033)	(0.032)	(0.028)	(0.029)	(0.033)	(0.033)	(0.031)	(0.031)	(0.015)
	MLE	−0.945	−0.946	0.386	0.385	1.055	1.055	0.386	0.386	0.500
		(0.030)	(0.030)	(0.019)	(0.019)	(0.029)	(0.030)	(0.020)	(0.020)	(0.008)
$m=32$	L_2E	−0.998	−0.996	0.357	0.356	1.002	1.002	0.358	0.357	0.500
		(0.031)	(0.033)	(0.032)	(0.031)	(0.033)	(0.032)	(0.027)	(0.030)	(0.013)
	MLE	−0.898	−0.897	0.422	0.420	1.102	1.103	0.422	0.422	0.500
		(0.027)	(0.028)	(0.018)	(0.019)	(0.029)	(0.030)	(0.016)	(0.017)	(0.007)
$r=1.34$										
$m=8$	L_2E	−1.001	−1.001	0.333	0.334	0.998	1.004	0.336	0.337	0.499
		(0.034)	(0.034)	(0.027)	(0.027)	(0.033)	(0.035)	(0.030)	(0.028)	(0.013)
	MLE	−0.968	−0.968	0.371	0.372	1.031	1.034	0.375	0.376	0.499
		(0.035)	(0.033)	(0.020)	(0.021)	(0.033)	(0.035)	(0.023)	(0.022)	(0.009)
$m=16$	L_2E	−1.002	−1.000	0.342	0.340	1.002	1.002	0.341	0.343	0.500
		(0.034)	(0.033)	(0.033)	(0.031)	(0.034)	(0.035)	(0.032)	(0.032)	(0.012)
	MLE	−0.939	−0.937	0.403	0.402	1.061	1.062	0.409	0.408	0.499
		(0.029)	(0.028)	(0.018)	(0.018)	(0.029)	(0.030)	(0.019)	(0.018)	(0.010)
$m=32$	L_2E	−1.001	−0.998	0.359	0.356	1.003	0.997	0.353	0.354	0.502
		(0.034)	(0.035)	(0.035)	(0.031)	(0.027)	(0.026)	(0.023)	(0.035)	(0.011)
	MLE	−0.884	−0.884	0.452	0.451	1.116	1.114	0.455	0.456	0.499
		(0.037)	(0.036)	(0.022)	(0.023)	(0.025)	(0.025)	(0.021)	(0.021)	(0.015)

considered) is similar to the vector of MLE for $m = 8$ and $r = 2$ (the "best" situation we considered). Figure 3 shows the contour plot of the mixture density with parameters the means of the estimates of our simulation according to L_2 and ML criteria, for $m = 32$ and $r = 2.69$. We finally observe that in this case too L_2 estimates don't seem to be inflated by the position of the outliers; in fact, for a fixed m, we obtain the same estimates as r increases. Conversely, for a fixed m, ML estimates increase as r increases.

• **case (b)**: for the second component of the mixture (for which $\rho_2 = 0$) both MLE and L_2E for location and scale parameters behave in the same way as outlined for case (a). The estimates of ρ_2 are in average more accurate if we resort to the L_2 criterion. The MLE for ρ_2 show a dependence between the components of the bivariate gaussian density that increases as both m and r increase (they range between .107 and .320).

For the first component of the mixture (for which $\rho_1 = -.70$) the ML criterion tends to overestimate both location and scale parameters. In particular, see

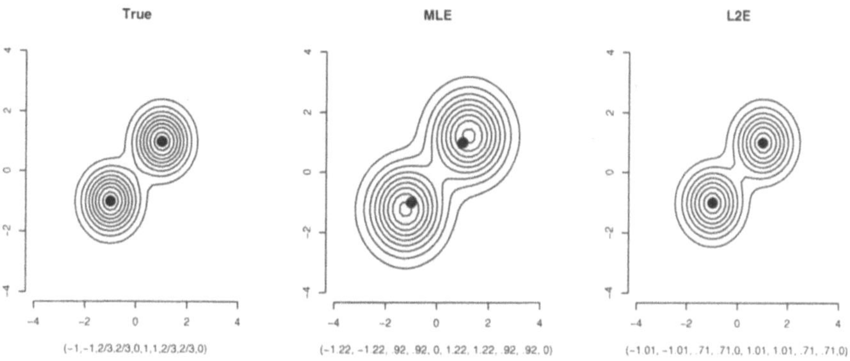

FIGURE 3. Contour plots of case (e) with $m = 32$ and $r = 0.69$.

Table 2, it seems that μ_{11} and σ_{11} are more overestimated than μ_{12} and σ_{12}, but this is probably due to the position of the outliers on the ellipse.

On the other hand, L_2 estimates of ρ_1 are in average very accurate, while MLE shows an increasing independence between the components of the bivariate normal density as both m and r increase.

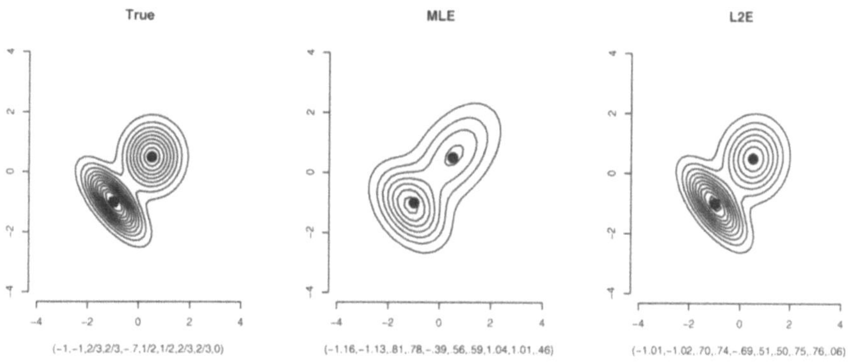

FIGURE 4. Contour plots of case (f) with $m = 32$, $r_1 = 3.50$ and $r_2 = 2.69$.

• **case (f)**: in this situation some problems arise in estimating the vectors of parameters, since the two components of the mixture are very close to each other and for the first of them we set $\rho_1 = -.7$. The main problem in this case is finding the right values of the initial guesses of the numerical algorithm for solving equations (2.4) and (3.1). For all previous cases the true parameters vector was a good

TABLE 2. Case (b): average of the estimates of the parameters of the first component of the mixture (standard errors in brackets).

$r_1 = 1.30$		$\hat{\mu}_{11}$	$\hat{\mu}_{12}$	$\hat{\sigma}_{11}$	$\hat{\sigma}_{12}$	$\hat{\rho}_1$
$m = 8$	L_2E	−1.000	−0.996	0.333	0.331	−0.688
		(0.032)	(0.031)	(0.028)	(0.028)	(0.064)
	MLE	−0.982	−0.988	0.338	0.332	−0.627
		(0.032)	(0.032)	(0.023)	(0.022)	(0.055)
$m = 16$	L_2E	−0.996	−1.001	0.341	0.337	−0.675
		(0.033)	(0.031)	(0.029)	(0.028)	(0.074)
	MLE	−0.965	−0.980	0.349	0.335	−0.575
		(0.032)	(0.030)	(0.021)	(0.021)	(0.056)
$m = 32$	L_2E	−0.990	−0.995	0.352	0.346	−0.633
		(0.033)	(0.033)	(0.029)	(0.029)	(0.065)
	MLE	−0.935	−0.963	0.359	0.338	−0.499
		(0.030)	(0.030)	(0.020)	(0.020)	(0.061)
$r_1 = 1.52$						
$m = 8$	L_2E	−0.997	−1.003	0.336	0.335	−0.701
		(0.033)	(0.034)	(0.029)	(0.031)	(0.068)
	MLE	−0.978	−0.988	0.346	0.336	−0.606
		(0.031)	(0.032)	(0.023)	(0.022)	(0.061)
$m = 16$	L_2E	−0.999	−1.001	0.339	0.341	−0.693
		(0.032)	(0.033)	(0.029)	(0.030)	(0.068)
	MLE	−0.959	−0.977	0.358	0.341	−0.541
		(0.031)	(0.032)	(0.023)	(0.022)	(0.061)
$m = 32$	L_2E	−0.999	−0.998	0.350	0.349	−0.679
		(0.033)	(0.034)	(0.029)	(0.031)	(0.067)
	MLE	−0.925	−0.955	0.376	0.348	−0.450
		(0.028)	(0.029)	(0.020)	(0.020)	(0.062)
$r_1 = 1.75$						
$m = 8$	L_2E	−1.001	−0.998	0.336	0.335	−0.698
		(0.031)	(0.032)	(0.029)	(0.029)	(0.063)
	MLE	−0.977	−0.985	0.351	0.340	−0.583
		(0.031)	(0.032)	(0.022)	(0.023)	(0.060)
$m = 16$	L_2E	−1.001	−0.998	0.341	0.339	−0.693
		(0.033)	(0.034)	(0.030)	(0.028)	(0.066)
	MLE	−0.954	−0.973	0.369	0.347	−0.495
		(0.030)	(0.031)	(0.021)	(0.021)	(0.062)
$m = 32$	L_2E	−1.002	−0.998	0.350	0.351	−0.692
		(0.031)	(0.030)	(0.029)	(0.029)	(0.065)
	MLE	−0.915	−0.950	0.396	0.362	−0.392
		(0.028)	(0.028)	(0.020)	(0.020)	(0.063)

choice for the initial guesses, but in this case it gives less performing results. In our simulation study we chose the sample moments as initial guesses.

Both the ML and L_2 estimates of the location and scale parameter of the two components of the mixture behave in general as in case (a). The estimates of ρ_1 and ρ_2 are more accurate if we resort to L_2E, and this for any outliers consistency and positioning. This behavior, which does not differ from the one obtained in case (b), is outlined in Figure 4.

5. Conclusions and Future Work

In this paper we outlined an approach in estimating the parameters of a mixture of two bivariate gaussian densities based on the L_2 criterion. From a simulation study, where a particularly care has been taken in outliers positioning and consistency, to match some typical situations that frequently arise in scientific areas (e.g., engineering, chemical, pharmaceutical), it follows that the L_2 approach generally seems to be robust over a wide case of data contamination with respect to ML estimates. Since our results are limited to the specific experimental conditions we chose for the simulation, it would be interesting to study, in a future work, the behavior of the L_2 estimator by increasing the fraction of outliers and changing the type of contamination.

An amazing subject would be exploring the possibility of constructing parameters confidence intervals using asymptotic theory or bootstrap techniques.

Further features on the topic would be, following the framework proposed by Durio and Isaia (2003) in the case of regression modelling by L_2, the improvement of a quick rule to identify and simultaneously quantify clusters in multivariate data.

Since the L_2 estimate criterion is helpful to point out the presence of outliers if evident discrepancies arise between L_2E and MLE, we may exploit this consideration in order to implement a useful algorithm to determine the appropriate number of components of a mixture.

References

[1] A. Basu, I.R. Harris, N.L. Hjort, and M.C. Jones, *Robust and Efficient Estimation by Minimizing a Density Power Divergence.* Biometrika **85** (1998), 549–559.

[2] A. Durio and E.D. Isaia, *A Parametric Regression Model by Minimum L_2 Criterion: a Study on Hydrocarbon Pollution of Electrical Transformers.* Developments in Applied Statistics, Metodološki Zvezki **19** (2003), 69–83.

[3] A. Durio and E.D. Isaia, *Parametric Multivariate Density Estimation Using L_2 Distance: a Simulation Study on Robustness.* In Proceedings of the S.I.S. Intermediate Scientific Meeting, RCE Edizioni, Napoli, 2003.

[4] B. Flury, *A First Course in Multivariate Statistics.* Springer-Verlag, New York, 1997.

[5] G. Gray, *Bias in Misspecified Mixtures.* Biometrics **50** (1994), 457–470.

[6] D.W. Scott, *Parametric Statistical Modeling by Minimum Integrated Square Error.* Technometrics **43** (2001), 274–285.

[7] G.R. Terrel, *Linear Density Estimates.* In Proceedings of the Statistical Computing Section of the American Statistical Association, pages 297–302, 1990.

[8] D.M. Titterington, A.F.M. Smith, and U.E. Markov, *Statistical Analysis of Finite Mixture Distributions.* John Wiley, Chichester, 1985.

[9] M.P. Wand and M.C. Jones, *Kernel Smoothing.* Chapman & Hall, London, 1995.

Acknowledgment
The authors are indebted to the coordinating editor and to the anonymous referees for carefully reading the manuscript and for their many important remarks and suggestions.

A. Durio and E.D. Isaia
Dep. of Statistics & Mathematics
Piazza Arbarello, 8
10128 – Turin
Italy
e-mail: {durio, isaia}@econ.unito.it

Statistics for Industry and Technology, 105–117

Robust PCR and Robust PLSR: a Comparative Study

S. Engelen, M. Hubert, K. Vanden Branden and S. Verboven

Abstract. Principal Component Regression (PCR) and Partial Least Squares Regression (PLSR) are the two most popular regression techniques in chemometrics. They both fit a linear relationship between two sets of variables. The responses are usually low-dimensional whereas the regressors are very numerous compared to the number of observations. In this paper we compare two recent robust PCR and PLSR methods and their classical versions in terms of efficiency, goodness-of-fit, predictive power and robustness.

Mathematics Subject Classification (2000). 62F35; 62H20; 92E99.

Keywords. Principal component analysis, principal component regression, partial least squares, robustness.

1. Introduction

Principal Component Regression and Partial Least Squares Regression are the two most popular regression techniques in chemometrics (Martens and Naes, 1998). Both methods can handle multicollinearity in the data and can be applied if there are more regressors than objects. Hence, it are the standard tools for multivariate calibration where the concentrations of certain constituents in samples are modelled and predicted from their spectra.

PCR and PLSR are based on a bilinear model that explains the existence of a relation between a set of p-dimensional regressors and a set of q-dimensional response variables through k-dimensional scores $\boldsymbol{t}_i$ with $k \ll p$. More precisely, for $i = 1, \ldots, n$ with n the number of observations $(\boldsymbol{x}_i, \boldsymbol{y}_i)$, we assume that

$$\boldsymbol{x}_i = \bar{\boldsymbol{x}} + P_{p,k}\boldsymbol{t}_i + \boldsymbol{f}_i \quad (1.1)$$

$$\boldsymbol{y}_i = \bar{\boldsymbol{y}} + \mathcal{A}'_{q,k}\boldsymbol{t}_i + \boldsymbol{g}_i. \quad (1.2)$$

Here $\bar{\boldsymbol{x}}$ is the mean of the x-variables, $\bar{\boldsymbol{y}}$ the mean of the y-variables, $P_{p,k}$ the matrix of x-loadings and $\mathcal{A}_{k,q}$ represents the slope matrix in the regression of $\boldsymbol{y}_i$ on $\boldsymbol{t}_i$. The superscript $'$ is used for the transpose of a vector or matrix, and subscripts as in $P_{p,k}$ indicate that the matrix P has p rows and k columns. The error terms are denoted by $\boldsymbol{f}_i$ and $\boldsymbol{g}_i$. In terms of the original predictor variables, this bilinear model can be written as

$$\boldsymbol{y}_i = \boldsymbol{\beta}_0 + \mathcal{B}'_{q,p}\boldsymbol{x}_i + \boldsymbol{e}_i$$

with

$$\begin{aligned} \mathcal{B}_{p,q} &= P_{p,k}\mathcal{A}_{k,q} && (1.3)\\ \boldsymbol{\beta}_0 &= \bar{\boldsymbol{y}} - \mathcal{B}'_{q,p}\bar{\boldsymbol{x}}. && (1.4) \end{aligned}$$

According to the bilinear model, both PCR and PLSR proceed in two major stages. In the first step, following (1.1), they summarize the high-dimensional observations $\boldsymbol{x}_i$ in scores $\boldsymbol{t}_i$ of dimension $k \ll p$. The selection of the number of components k can be done by means of various different criteria, which will be discussed in Section 3. These k latent variables then become the regressors in the second step of the algorithm (1.2). Finally, estimates for $\mathcal{B}$ and $\boldsymbol{\beta}_0$ are obtained via (1.3) and (1.4).

The main difference between PCR and PLSR lies in the construction of the scores $\boldsymbol{t}_i$. In PCR the scores are obtained by extracting the most relevant information present in the x-variables by performing a principal component analysis on the predictor variables and thus using a variance criterion. No information concerning the response variables is yet taken into account. In contrast, the PLSR scores are calculated by maximizing a covariance criterion between the x- and y-variables. Hence also information present in the responses is used in the first stage of the algorithm. By construction we thus expect that PLSR requires less components than PCR. This has been confirmed by several studies (De Jong, 1993b), Frank and Friedman (1993), Martens and Naes (1998).

In this paper we want to investigate how recent robust methods of PCR and SIMPLS behave with respect to each other in terms of efficiency, goodness-of-fit, predictive ability and robustness. In Section 2 we shortly describe the four methods involved in this paper: classical PCR (CPCR), robust PCR (RPCR), SIMPLS and robust SIMPLS (RSIMPLS). In Section 3 a comparison between the four methods is given on the basis of a simulation study. All four methods will be illustrated on a real example in Section 4 whereas Section 5 summarizes our conclusions.

2. Robust Calibration Methods

2.1. Principal Component Regression

As explained in the introduction, CPCR starts by performing a Principal Component Analysis (PCA) on the x-variables. The PCA loading matrix $P_{p,k}$ then contains the first k dominant eigenvectors of the empirical covariance matrix of the $\boldsymbol{x}_i$, and the scores satisfy $\boldsymbol{t}_i = P'_{k,p}(\boldsymbol{x}_i - \bar{\boldsymbol{x}})$. In the second step of CPCR multiple linear least squares regression (MLR) is applied on the $(\boldsymbol{t}_i, \boldsymbol{y}_i)$ to obtain an estimate of the slope matrix $\mathcal{A}_{k,q}$ in (1.2).

In Hubert and Verboven (2003) a robust PCR method is proposed by robustifying both steps of CPCR. First a robust PCA method is applied on the regressors. For low-dimensional data ($p < n/2$), the MCD estimator (Rousseeuw, 1984) is used as a robust estimator of the covariance matrix of the $\boldsymbol{x}_i$, and for high-dimensional data the ROBPCA method (Hubert et al., 2004). This estimator combines projection pursuit techniques with robust covariance estimation in low dimensions. Next a robust regression method is applied. If there is only one response variable the reweighted LTS regression (Rousseeuw, 1984) is preferred, else the MCD regression (Rousseeuw et al., 2004) is performed.

2.2. Partial Least Squares Regression

We consider the SIMPLS algorithm (De Jong, 1993a) being the leading PLSR method because of its speed and efficiency. Let $\tilde{X} = \{(\boldsymbol{x}_i - \bar{\boldsymbol{x}})'\}_{i=1}^n$ and $\tilde{Y} = \{(\boldsymbol{y}_i - \bar{\boldsymbol{y}})'\}_{i=1}^n$ be the centered data matrices. The first normalized PLSR weight vectors $\boldsymbol{r}_1$ and $\boldsymbol{q}_1$ are obtained as linear combinations of $\tilde{X}$ and $\tilde{Y}$ that maximize

$$\mathrm{cov}(\tilde{X}\boldsymbol{r}_1, \tilde{Y}\boldsymbol{q}_1).$$

The solution of this maximization problem is found by taking $\boldsymbol{r}_1$ and $\boldsymbol{q}_1$ as the first left and right singular eigenvectors of $S_{xy} = \tilde{X}'\tilde{Y}/(n-1)$, the cross-covariance matrix of the x- and y-variables. For each observation the first coordinate of the score $\boldsymbol{t}_i$ is computed as $t_{i1} = \tilde{\boldsymbol{x}}_i'\boldsymbol{r}_1$.

The other PLSR weight vectors $\boldsymbol{r}_a$ and $\boldsymbol{q}_a$ for $a = 2, \ldots, k$ are obtained by imposing an orthogonality constraint to the elements of the scores. If we require that $\sum_{i=1}^n t_{ia}t_{ib} = 0$ for $a \neq b$, a deflation of the cross-covariance matrix S_{xy} provides the solutions for the other PLSR weight vectors. This deflation is carried out by first calculating the x-loading $\boldsymbol{p}_a = S_x\boldsymbol{r}_a/(\boldsymbol{r}_a'S_x\boldsymbol{r}_a)$ with S_x the empirical variance-covariance matrix of the x-variables. Next an orthonormal base $\{\boldsymbol{v}_1, \ldots, \boldsymbol{v}_a\}$ of $\{\boldsymbol{p}_1, \ldots, \boldsymbol{p}_a\}$ is constructed and S_{xy} is deflated as

$$S_{xy}^a = S_{xy}^{a-1} - \boldsymbol{v}_a(\boldsymbol{v}_a'S_{xy}^{a-1})$$

with $S_{xy}^1 = S_{xy}$. In general the PLSR weight vectors $\boldsymbol{r}_a$ and $\boldsymbol{q}_a$ are obtained as the left and right singular vector of S_{xy}^a. Finally, when the scores are k-dimensional, MLR is performed of the responses $\boldsymbol{y}_i$ on these scores $\boldsymbol{t}_i$.

A robust method RSIMPLS has recently been developed in Hubert and Vanden Branden (2003). It starts by applying ROBPCA (Hubert et al., 2004) on the x- and y-variables in order to replace S_{xy} and S_x by robust estimates and then proceeds analogously to the SIMPLS algorithm. Similar to RPCR a robust regression method (ROBPCA regression) is performed in the second stage. In Vanden Branden and Hubert (2004) it is proved that for low-dimensional data the RSIMPLS approach yields bounded influence functions for the weight vectors $\boldsymbol{r}_a$ and $\boldsymbol{q}_a$, and for the regression estimates. Also the breakdown value is inherited from the MCD estimator.

The computational complexity of ROBPCA and RSIMPLS is discussed in detail in Hubert et al. (2004) and Hubert and Vanden Branden (2003). The computation time remains feasible due to the FAST-MCD algorithm (Rousseeuw and Van Driessen, 1999). To give an example, on a Pentium IV with 1.60 GHz, it requires in a full Matlab implementation approximately 7 seconds to perform RSIMPLS on a data set with $n = 100, p = 500$ and $k = 5$.

3. Experimental Study

We will compare the efficiency, the goodness-of-fit (GOF), the predictive power and the robustness of CPCR, RPCR, SIMPLS and RSIMPLS by performing a simulation study on uncontaminated and contaminated data. Note that the robustness of RPCR and RSIMPLS has also been shown through simulations in Hubert and Verboven (2003) and Hubert and Vanden Branden (2003), but there the emphasis was put only on the parameter estimation and not on the predictive performance of the methods. The experiments described in this section consider univariate responses ($q = 1$) which are mostly used in practice. We thus consider the regression model

$$y_i = \beta_0 + \mathcal{B}'_{1,p}\boldsymbol{x}_i + e_i$$

with $\mathcal{B}_{p,1} = (\beta_1, \ldots, \beta_p)'$. The regression vector including the intercept is denoted as $\boldsymbol{\beta} = (\beta_0, \beta_1, \ldots, \beta_p)'$.

3.1. Simulation Settings

We compare the algorithms on high-dimensional data sets $X_{50,100}$ of size $n = 50$ and $p = 100$ and low-dimensional data sets $X_{100,6}$ of size $n = 100$ and $p = 6$. They were constructed according to the bilinear model (1.1) and (1.2):

$$\begin{aligned} T &\sim N_2(\mathbf{0}_2, \Sigma_t) \\ X &= TI_{2,p} + N_p(\mathbf{0}_p, 0.1I_p) \\ Y &= T\mathcal{A}_{2,1} + N(0,1) \end{aligned}$$

with $\mathbf{0}_2 = (0,0)'$, $I_{k,p} = \delta_{ij}$ and $\mathcal{A}_{2,1} = (1,1)'$. Furthermore we set $\Sigma_t = \begin{pmatrix} 8 & 0 \\ 0 & 2 \end{pmatrix}$ when $p = 100$ and $\Sigma_t = \begin{pmatrix} 4 & 0 \\ 0 & 2 \end{pmatrix}$ for $p = 6$. Hence the optimal number of components $k_{\text{opt}} = 2$.

Next, contamination is added by replacing 10% of the observations by different types of outliers. Denote T_ϵ, X_ϵ, and Y_ϵ as the contaminated parts of the data.

1. *Bad leverage points* were constructed by substituting $T_\epsilon \sim N_2((15,15)', \Sigma_t)$ in equation (1.1):

$$X_\epsilon = T_\epsilon I_{2,p} + N_p(\mathbf{0}_p, 0.1I_p).$$

   Note that the corresponding y-values did not change.

2. *Vertical outliers* have uncontaminated x-values, but their y-values were changed by adjusting the error term in (1.2):

$$Y_\epsilon = T\mathcal{A}_{2,1} + N(15, 0.1).$$

For each situation, $m = 100$ data sets were generated and they were analyzed with $k = 1, 2$ and 3 components.

The efficiency of the considered methods is evaluated by means of the MSE of the estimated regression parameters $\hat{\boldsymbol{\beta}}$. It is defined by

$$\mathrm{MSE}_k(\hat{\boldsymbol{\beta}}) = \frac{1}{m}\sum_{l=1}^{m} \|\hat{\boldsymbol{\beta}}_k^{(l)} - \boldsymbol{\beta}\|^2$$

where $\hat{\boldsymbol{\beta}}_k^{(l)}$ denotes the estimated parameter based on k components in the lth simulation. The MSE indicates to what extent the slope and intercept are correctly estimated. So the goal is to obtain an MSE value close to zero.

Next we want to study how well the methods fit the regular data points. Because of the simulation settings, we know exactly their indices which we store in the set G_r. Then we define the goodness-of-fit criterion as

$$\mathrm{GOF}_k = 1 - \frac{\operatorname*{var}_{i \in G_r}(r_{i,k})}{\operatorname*{var}_{i \in G_r}(y_i)} \tag{3.1}$$

with $r_{i,k}$ the residual of the ith observation when k components are computed. The objective is to obtain a GOF close to 1. Note that this GOF is an adaptation of the robust R^2 proposed in Hubert and Verboven (2003) where G_r contains the non-outlying data points detected by the estimation procedure itself (RPCR or RSIMPLS). The value of k where the R^2-curve stabilizes is then selected as the optimal one.

Finally, we measure the predictive ability of the methods by means of the Root Mean Squared Error (RMSE) as in Hubert and Vanden Branden (2003). First we generate a test set G_t of uncontaminated points with size $n_t = 50$, and then we compute

$$\mathrm{RMSE}_k = \sqrt{\frac{1}{n_t}\sum_{i=1}^{n_t}(y_i - \hat{y}_{i,k})^2} \tag{3.2}$$

with $\hat{y}_{i,k}$ the predicted y-value of observation i from the test set when the regression parameter estimates are based on the training set (X, Y) of size n and k scores are retained. The optimal number of components is often selected as that k for which this RMSE value (or its cross-validated version) is minimal.

The results of the simulations are listed in Tables 1–3 for the high-dimensional data sets and in Tables 4–6 for the low-dimensional situation.

TABLE 1. $n = 50, p = 100$, no contamination.

		Method			
		CPCR	RPCR	SIMPLS	RSIMPLS
$k = 1$	MSE($\hat{\beta}$)	1.12	1.14	0.63	0.65
	GOF	0.69	0.69	0.81	0.80
	RMSE	1.81	1.82	1.50	1.51
$k = 2$	MSE($\hat{\beta}$)	0.16	0.22	0.23	0.29
	GOF	0.89	0.89	0.91	0.90
	RMSE	1.14	1.15	1.13	1.15
$k = 3$	MSE($\hat{\beta}$)	0.20	0.29	3.19	0.91
	GOF	0.89	0.89	0.97	0.91
	RMSE	1.14	1.16	1.27	1.19

TABLE 2. $n = 50, p = 100, 10\%$ bad leverage points.

		Method			
		CPCR	RPCR	SIMPLS	RSIMPLS
$k = 1$	MSE($\hat{\beta}$)	1.93	1.14	1.92	0.65
	GOF	0.17	0.71	0.31	0.82
	RMSE	3.07	1.84	2.85	1.50
$k = 2$	MSE($\hat{\beta}$)	2.49	0.22	2.80	0.31
	GOF	0.39	0.89	0.45	0.91
	RMSE	2.68	1.15	2.68	1.16
$k = 3$	MSE($\hat{\beta}$)	2.78	0.28	20.24	1.02
	GOF	0.41	0.89	0.80	0.93
	RMSE	2.68	1.15	3.01	1.19

3.2. Simulation Results

When no contamination is added (Tables 1 and 4), the classical methods perform somewhat better than their robust versions, as we would expect. At the optimal $k_{\text{opt}} = 2$, CPCR and SIMPLS can hardly be distinguished. At high-dimensional data, the MSE(CPCR) is minimal, but the GOF and RMSE values are slightly in favor of SIMPLS. At low-dimensional data, we see almost no differences. When only one component is selected ($k = 1$), SIMPLS outperforms CPCR noticeably. This confirms the findings of Frank and Friedman (1993) that PLSR constructs

TABLE 3. $n = 50, p = 100, 10\%$ vertical outliers.

		Method			
		CPCR	RPCR	SIMPLS	RSIMPLS
$k = 1$	MSE($\hat{\beta}$)	3.39	1.15	3.06	0.62
	GOF	0.65	0.69	0.76	0.81
	RMSE	2.44	1.81	2.25	1.48
$k = 2$	MSE($\hat{\beta}$)	2.68	0.22	14.97	0.29
	GOF	0.80	0.89	0.66	0.90
	RMSE	2.11	1.15	2.55	1.15
$k = 3$	MSE($\hat{\beta}$)	3.37	0.28	60.56	1.01
	GOF	0.77	0.89	0.57	0.92
	RMSE	2.13	1.16	3.26	1.20

TABLE 4. $n = 100, p = 6$, no contamination.

		Method			
		CPCR	RPCR	SIMPLS	RSIMPLS
$k = 1$	MSE($\hat{\beta}$)	1.06	1.11	0.27	0.27
	GOF	0.55	0.54	0.77	0.77
	RMSE	1.79	1.81	1.28	1.27
$k = 2$	MSE($\hat{\beta}$)	0.02	0.03	0.03	0.05
	GOF	0.83	0.83	0.83	0.83
	RMSE	1.10	1.11	1.10	1.11
$k = 3$	MSE($\hat{\beta}$)	0.10	0.13	0.51	0.60
	GOF	0.83	0.83	0.84	0.84
	RMSE	1.11	1.11	1.12	1.13

its components more efficiently than PCR. However, when we choose more components than required, here $k = 3$, we notice that SIMPLS suffers much more from overfitting than CPCR. This is reflected in the large MSE values of SIMPLS. At high-dimensional data $\text{MSE}_3(\text{SIMPLS}) = 3.19$ is even considerably larger than $\text{MSE}_3(\text{RSIMPLS}) = 0.91$.

If we add contamination, CPCR and SIMPLS clearly break down. The MSE of the regression parameter estimates increase drastically and even attain their minimum at $k = 1$ (except for CPCR in Table 3). The GOF values are very low,

TABLE 5. $n = 100, p = 6, 10\%$ bad leverage points.

		CPCR	RPCR	SIMPLS	RSIMPLS
$k = 1$	MSE($\hat{\beta}$)	1.84	1.06	1.83	0.24
	GOF	0.10	0.54	0.13	0.78
	RMSE	2.51	1.81	2.49	1.26
$k = 2$	MSE($\hat{\beta}$)	2.04	0.03	2.05	0.05
	GOF	0.18	0.83	0.18	0.83
	RMSE	2.42	1.11	2.42	1.11
$k = 3$	MSE($\hat{\beta}$)	2.56	0.17	4.52	0.65
	GOF	0.19	0.83	0.21	0.83
	RMSE	2.43	1.11	2.47	1.13

(Column header: Method)

TABLE 6. $n = 100, p = 6, 10\%$ vertical outliers.

		CPCR	RPCR	SIMPLS	RSIMPLS
$k = 1$	MSE($\hat{\beta}$)	3.31	1.10	2.55	0.24
	GOF	0.52	0.53	0.74	0.78
	RMSE	2.39	1.81	2.04	1.25
$k = 2$	MSE($\hat{\beta}$)	2.35	0.03	2.41	0.05
	GOF	0.78	0.83	0.78	0.83
	RMSE	1.97	1.10	1.96	1.11
$k = 3$	MSE($\hat{\beta}$)	4.46	0.13	10.79	0.62
	GOF	0.75	0.83	0.68	0.84
	RMSE	2.02	1.11	2.16	1.13

especially when the data contain bad leverage points. This shows that the regular data points are badly fitted. The high RMSE values indicate the low predictive ability of the classical methods. SIMPLS is more sensitive to vertical outliers than CPCR. This is probably due to the fact that the response variable is already used in the construction of the SIMPLS scores, contrary to the CPCR scores that only depend on the x-variables.

The results of the robust methods on the other hand are almost identical to the uncontaminated case. RSIMPLS is again superior to RPCR for $k = 1$ and

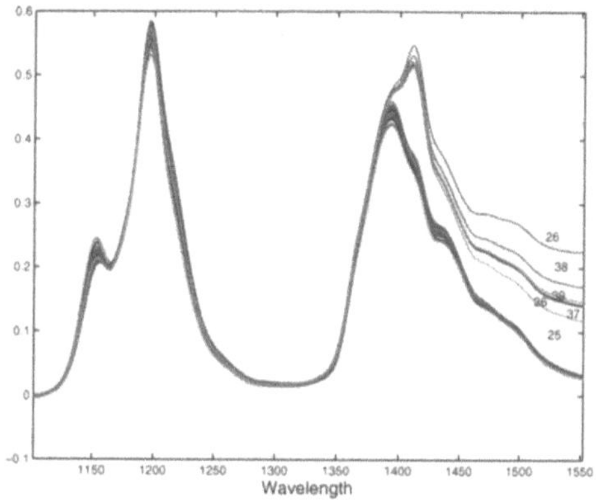

FIGURE 1. The spectra of the octane data.

$k = 2$, but shows more overfitting when $k = 3$. Note that the GOF values of RSIMPLS are always higher than those of RPCR. Both GOF and RMSE appear to be good criteria to select k. We see that the differences $\mathrm{GOF}_3 - \mathrm{GOF}_2$ are very small compared to $\mathrm{GOF}_2 - \mathrm{GOF}_1$. But we cannot conclude that we should choose k for which GOF_k is maximal. On the other hand, the minimal value of RMSE is always reached at the correct $k = 2$. This suggests to select k such that RMSE_k is minimal.

4. An Example

In this section, we compute the GOF and the RMSE values of the two classical and the two robust calibration methods on a real data set. The *octane* data (Esbensen et al., 1994) consist of NIR absorbance spectra over $p = 226$ wavelengths ranging from 1102 to 1552 nm with measurements every two nm. For each of the $n = 39$ production gasoline samples the octane number y was measured, so $q = 1$.

This data set has been studied before in Hubert et al. (2004). It is well known that the octane data set contains six outliers to which alcohol was added. In Figure 1 we see that the spectra of those six samples clearly stand out. They were also detected by applying the robust PCA method ROBPCA to the 226 regressors (Hubert et al., 2004). Therefore we put the set of regular observations G_r equal to the full data of 39 observations minus these six outliers.

For each of the four methods we computed the goodness-of-fit index as defined in (3.1). Applying (3.2) is not possible here because a test data set is beyond our reach, and given the small number of observations, we do not want to split the data set into a training and a test data set. Therefore, we use the cross-validated R-RMSECV value as proposed in Hubert and Verboven (2003) and Hubert and

Vanden Branden (2003). It is defined as

$$\text{R-RMSECV}_k = \sqrt{\frac{1}{33} \sum_{i \in G_r} (y_i - \hat{y}_{-i,k})^2}. \tag{4.1}$$

Here $\hat{y}_{-i,k}$ represents the predicted y-value for observation i based on k components when observation i was left out of the estimation of the regression parameters. Note that when the outliers are not know in advance, the set of regular observations G_r is determined by the calibration method itself. Hence different sets could be obtained for e.g. RPCR and RSIMPLS. This would make it difficult to compare R-RMSECV values of several methods.

We evaluate criteria (3.1) and (4.1) for $k = 1, \ldots, 4$ components. The results are summarized in Table 7. For $k = 1$ the classical methods perform very

TABLE 7. Analysis of the octane data.

		Method			
		CPCR	RPCR	SIMPLS	RSIMPLS
$k=1$	GOF	-0.04	0.81	0.13	0.82
	R-RMSECV	2.08	0.93	1.92	0.90
$k=2$	GOF	0.80	0.98	0.85	0.98
	R-RMSECV	0.94	0.31	0.81	0.32
$k=3$	GOF	0.98	0.98	0.98	0.99
	R-RMSECV	0.31	0.30	0.30	0.30
$k=4$	GOF	0.98	0.98	0.98	0.99
	R-RMSECV	0.31	0.30	0.29	0.29

badly. For CPCR this even results in a negative $\text{GOF}_1(\text{RPCR}) = -0.04$. Also $\text{GOF}_1(\text{SIMPLS}) = 0.13$ is very small compared to the robust values. The R-RMSECV values of CPCR and SIMPLS are approximately twice as high than those of RPCR and RSIMPLS.

When we retain $k = 2$ components, the GOF values of the classical methods increase considerably, and their R-RMSECV values have been more than halved, yielding results that are comparable with the robust ones for $k = 1$. RPCR and RSIMPLS clearly improve their fit with a very high $\text{GOF}_2 = 0.98$ and R-RMSECV values of 0.31 resp. 0.32.

For $k = 3$ the classical values correspond with the robust ones for $k \geq 2$. This clearly shows that the first component of CPCR and SIMPLS is completely determined by the outliers and it confirms the conclusion in Tenenhaus (1998) to retain the second and the third component of SIMPLS. The results obtained with

RPCR and RSIMPLS remain stable from $k = 2$ on, so these two methods suggest to retain $k_{\text{opt}} = 2$ components. Note that here the R-RMSECV value is not at its minimal value, but it does not hardly change when k is increased.

5. Conclusions

Both the simulation study and the analysis of a real data set show that CPCR and SIMPLS are very sensitive to outliers in the data, whereas RPCR and RSIMPLS can resist several types of contamination.

When the correct number of components is used in the calibration, RPCR and RSIMPLS are comparable in terms of efficiency, goodness-of-fit, predictive power and robustness. For smaller k, RSIMPLS is to be preferred, whereas RPCR is less sensitive to overfitting when a larger set of components is selected.

Finally, the proposed GOF and RMSE/R-RMSECV measures are shown to be good indicators to select the optimal number of components. To speed up the heavy computations involved in the cross-validated R-RMSECV, we are currently developing faster algorithms. This will allow to perform fast and robust model selection in multivariate calibration.

All the methods described in this paper can be downloaded from the web sites `http://www.wis.kuleuven.ac.be/stat/robust.html` and `http://www.agoras.ua.ac.be/` as part of the Matlab library for Robust Calibration (Verboven and Hubert, 2004).

Note that in this paper we have concentrated on calibration methods that are particularly useful for small data sets in high-dimensions, which are very common in chemometrics, food science and bioinformatics. Successful applications of robust PCA in bioinformatics are e.g. presented in Model et al. (2002) and Hubert and Engelen (2004). One of the referees wondered whether they also could be applied to problems in computer vision where both the number of samples and the number of variables can be very large. Currently ROBPCA is being implemented by Darren Cosker (3D Vision and Geometry, Department of Computer Science, Cardiff University, UK) to build statistical models of shape and appearance, i.e. Active Shape Models (ASM) and Active Appearance Models (AAM) (Cootes et al., 1998). In particular the method is used for building a model of mouth tracking. Other robust methods in computer vision have e.g. been proposed in Skocaj et al. (2002), Xu and Yuille (1995), De la Torre and Black (2001), and Stewart (1999). For an overview of the use of high-breakdown methods in computer vision, see e.g. Meer (2004).

References

[1] T.F. Cootes, G.J. Edwards, and C.J. Taylor, *Active Appearance Models.* In H. Burkhardt and B. Neumann, editors, Proc. European Conference on Computer Vision, Vol. 2, pages 484–498. Springer, 1998.

[2] S. de Jong, *SIMPLS: An Alternative Approach to Partial Least Squares Regression.* Chemometrics Intell. Lab. Syst. **18** (1993a), 251–263.

[3] S. de Jong, *PLS Fits Closer Than PCR.* J. Chemometrics **7** (1993b), 551–557.

[4] F. De la Torre and M.J. Black, *Robust Principal Component Analysis for Computer Vision.* In Int. Conf. on Computer Vision (ICCV-2001), IEEE Computer Society, Vancouver, B.C., Vol. 1, pages 362–369, 2001.

[5] K.H. Esbensen, S. Schönkopf, and T. Midtgaard, *Multivariate Analysis in Practice.* Camo, Trondheim, 1994.

[6] I.E. Frank and J.H. Friedman, *A Statistical View of Some Chemometrics Regression Tools.* Technometrics **35** (1993), 109–135.

[7] M. Hubert and S. Engelen, *Robust PCA and Classification in Biosciences.* Bioinformatics (2004), to appear, available at *http://www.wis.kuleuven.ac.be/stat/robust.html.*

[8] M. Hubert, P.J. Rousseeuw, and K. Vanden Branden, *ROBPCA: a New Approach to Robust Principal Component Analysis.* Technometrics (2004), to appear, available at *http://www.wis.kuleuven.ac.be/stat/robust.html.*

[9] M. Hubert and K. Vanden Branden, *Robust Methods for Partial Least Squares Regression.* J. Chemometrics **17** (2003), 537–549.

[10] M. Hubert and S. Verboven, *A Robust PCR Method for High-Dimensional Regressors.* J. Chemometrics **17** (2003), 438–452.

[11] H. Martens and T. Naes, *Multivariate Calibration.* Wiley, Chichester, 1998.

[12] P. Meer, *Robust Techniques for Computer Vision.* In G. Medioni and S.B. Kang, editors, Emerging Topics in Computer Vision. Prentice Hall, 2004.

[13] F. Model, T. König, C. Piepenbrock, and P. Ardojan, *Statistical Process Control for Large Scale Microarray Experiments.* Bioinformatics **18** Suppl 1 (2002), 55–63.

[14] P.J. Rousseeuw, *Least Median of Squares Regression.* J. Amer. Statist. Assoc. **79** (1984), 871–880.

[15] P.J. Rousseeuw, S. Van Aelst, K. Van Driessen, and J. Agulló, *Robust Multivariate Regression.* Technometrics, 2004, to appear, available at *http://www.agoras.ua.ac.be/.*

[16] P.J. Rousseeuw and K. Van Driessen, *A Fast Algorithm for the Minimum Covariance Determinant Estimator.* Technometrics **41** (1999), 212–223.

[17] D. Skocaj, H. Bischof, and A. Leonardis, *A Robust PCA Algorithm for Building Representations from Panoramic Images.* In A. Heyden, G. Sparr, M. Nielsen, and P. Johansen, editors, Computer Vision - ECCV 2002, part IV, pages 761–775. Springer-Verlag, Heidelberg, 2002.

[18] C.V. Stewart, *Robust Parametric Estimation in Computer Vision.* Siam Rev. **41** (1999), 513–537.

[19] M. Tenenhaus, *La Régression PLS.* Édition Technip, Paris, 1998.

[20] K. Vanden Branden and M. Hubert, *Robustness Properties of a Robust SIMPLS Method.* Anal. Chim. Acta (2004), to appear, available at *http://www.wis.kuleuven.ac.be/stat/robust.html.*

[21] S. Verboven and M. Hubert, *A Matlab Toolbox for Robust Calibration.* Submitted, 2004, available at *http://www.wis.kuleuven.ac.be/stat/robust.html.*

[22] L. Xu and A. Yuille, *Robust Principal Components Analysis by Self-Organizing Rules Based on Stastistical Physics Approach.* IEEE Trans. Neural Networks **6** (1995), 131–143.

S. Engelen, M. Hubert and K. Vanden Branden
Katholieke Universiteit Leuven
W. de Croylaan 54
B-3001 Leuven
Belgium
e-mail: {`sanne.engelen, mia.hubert, karlien.vandenbranden`}`@wis.kuleuven.ac.be`

S. Verboven
University of Antwerp
Middelheimlaan 1
B-2020 Antwerpen
Belgium
e-mail: `sabine.verboven@ua.ac.be`

Statistics for Industry and Technology, 119–130

Analytic Estimator Densities for Common Parameters under Misspecified Models

P.J. Hingley

Abstract. An expression is given for the exact probability density function of the parameter values that maximize the likelihood of a statistical model, where the data generating model is allowed to differ from the estimation model. The density can be used to study the robustness of estimation of alternative hypothetical models. It is described for curved exponential families, then specifically for gamma distribution models and for nonlinear regression models. An example is given in the context of alternative models for data from the biochemical ELISA test method. Finally an indication is given of how a robustness index can be calculated to assess the effects of estimation of a common parameter vector under a wrong model.

Mathematics Subject Classification (2000). Primary 62B10, 62-02; Secondary 62B15, 62E15, 62F10, 62G35, 62M10, 62P10.

Keywords. ELISA, exact estimator density, model robustness, nonlinear regression, small sample properties.

1. Introduction

How can the joint density of the maximum likelihood estimates (MLE) from an estimation model be specified, when the data generating model is different? In this case the MLE are termed quasi maximum likelihood estimates (QMLE). The QMLE still maximize the likelihood with respect to the estimation model, but also minimize the Kullbach-Leibler information criterion between the estimation model and the data generating model (White, 1994). Here an exact analytic expression will be developed for the probability density function of the QMLE, that can be applied to models that are nonlinear with respect to their parameters. It can be used for any sample size, because its derivation does not depend on asymptotic properties.

For estimation models with normal error structures, that are linear with respect to their parameters, QMLE have an exact multivariate normal distribution.

But for nonlinear models, even where estimation and data-generating models are identical, the vector of QMLE only has an asymptotic multivariate normal distribution. Experimental designs with finite numbers of sampling points often lead to markedly non-normal distributions (Bates and Watts, 1988). The proposed method is an alternative to carrying out simulation studies.

The method that will be developed here may be of particular use to scientists who need to take account of the possibility that several possible underlying mathematical models could be generating the data in their experiments.

2. Technique for Estimator Densities (TED)

Consider the n members of a sample as a $(n\times 1)$ vector w, $g_0(w)$ as the true density of w, and $g_1(w|\theta)$ as a presumed density with $(p\times 1)$ parameter vector θ to be estimated. The log likelihood corresponding to $g_1(w|\theta)$ is $l(\theta|w)$. The space of w is W, and the space of θ is Θ. A vector of independent variables z can, if necessary, be introduced to cope with the regression situation.

Regularity conditions are (with respect to θ in each case, and throughout W and Θ): $g_1(w|\theta)$ is continuous; $l(\theta|w)$ is twice differentiable, possesses a single maximum and has no other turning points. The QMLE $\hat{\theta}$ is then given by the same expression as that for the MLE in the usual situation, $l'(\theta, w)|_{\theta=\hat{\theta}} = 0$, (where $'$ indicates differentiation with respect to θ). The space of $\hat{\theta}$ is $\hat{\Theta}$, a subspace of Θ.

Consider a $(p\times 1)$ vector T.

$$T(\theta, \theta^*, w) = l'(\theta^*, w) - l'(\theta, w), \tag{2.1}$$

The parameter θ^* is fixed at an arbitrary value (it is not the asymptotic limit of $\hat{\theta}$, e.g., as represented by White, 1994, as θ_n^*).

The following Theorem 2.1 and Corollary 2.2 show that W can be divided into distinct subspaces $W_{\hat{\theta}}$, corresponding to the different values of $\hat{\theta}$, so that w determines $T(\hat{\theta}, \theta^*, w)$ within each subspace. For continuously distributed data, W is made up of an uncountably large number of such subspaces $W_{\hat{\theta}}$.

Theorem 2.1. *For fixed θ^*, each possible value of $\hat{\theta}$ within $\hat{\Theta}$ corresponds to a distinct subspace $W_{\hat{\theta}}$.*

Proof. It will be shown that the converse of the theorem implies a contradiction. For fixed θ^*, consider two unequal values of $\hat{\theta}$, denoted $\hat{\theta}_1$ and $\hat{\theta}_2$, with overlapping subspaces $W_{\hat{\theta}1}$ and $W_{\hat{\theta}2}$. Call the common shared part of these subspaces $W_{\widehat{\theta 1+2}}$. Pick any value of w from $W_{\widehat{\theta 1+2}}$, and call it w_c. Then, from (2.1), $T(\hat{\theta}_1, \theta^*, w_c) = l'(\theta^*, w_c) = T(\hat{\theta}_2, \theta^*, w_c)$. According to the regularity conditions, $l(\theta, w_c)$ possesses a single maximum over W at $\hat{\theta}$, at which point $l'(\theta, w_c)|_{\theta=\hat{\theta}} = 0$ and $T = l'(\theta^*, w_c)$. Therefore $\hat{\theta}_1$ and $\hat{\theta}_2$ are both equal to the unique $\hat{\theta}$. This contradicts inequality for $\hat{\theta}_1$ and $\hat{\theta}_2$, and so $W_{\hat{\theta}1}$ and $W_{\hat{\theta}2}$ are distinct. □

Corollary 2.2. *If θ^* and w are fixed, $T(\hat{\theta}, \theta^*, w)$ is also fixed.*

The following theorem establishes the main result (2.2) which gives the exact density for $\hat{\theta}$.

Theorem 2.3. *Assume that $W_{\hat{\theta}}$ specifies one or several distinct manifolds, that run over a single open $(n-p)$ dimensional subset of W. Then the density of $\hat{\theta}$ is given by*

$$g(\hat{\theta}) = E_w[|j(\theta, w)||\theta = \hat{\theta}] \cdot g_{[T(\hat{\theta},\hat{\theta},w)]}(0), \tag{2.2}$$

where $j(\theta, w) = -l''(\theta, w)$ is the observed information.

The left-hand term in (2.2) is a conditional expectation, that is conditional on $\theta = \hat{\theta}$ and is taken with respect to w over $g_1(w|\theta)$. The second term represents the density $g_{[T(\hat{\theta},\theta^*,w)]}(t)$, for which the known vector $\theta^* = \hat{\theta}$, and hence $t = 0$ by (2.1).

Proof. Part 1. Special case of a unique $w_{\hat{\theta}}$ for each $\hat{\theta}$:
Imagine initially that, for each $\hat{\theta}$ in $\hat{\Theta}$, $W_{\hat{\theta}}$ contains a single distinct vector, $w_{\hat{\theta}}$ say. For each possible $\hat{\theta}$, the correspondence to $w_{\hat{\theta}}$ is determined by the estimation model g_1. The conditional density of $\hat{\theta}$, for fixed θ^*, is given by a standard change of variable argument, using a Jacobian, as follows.

$$g(\hat{\theta}|\theta^*) = g_{[T(\hat{\theta},\theta^*,w)]}(t) \cdot |dT/d\theta||_{\theta=\hat{\theta},w=w_{\hat{\theta}}}. \tag{2.3}$$

The density $g_{[T(\hat{\theta},\theta^*,w)]}(t)$ is the density of a transform of $w_{\hat{\theta}}$ alone under the data generating model $g_0(w)$, even if some other value of w, $w_{\widehat{\theta+}}$ say with corresponding QMLE $\widehat{\theta+}$, gives $T(\widehat{\theta+}, \theta^*, w_{\widehat{\theta+}}) = T(\hat{\theta}, \theta^*, w_{\hat{\theta}})$. Theorem 2.1 shows that it is not possible for $\widehat{\theta+} = \hat{\theta}$.

Part 2. Usual case of one or more $w_{\hat{\theta}}$ for each $\hat{\theta}$:
For each $\hat{\theta}$ in $\hat{\Theta}$, if $W_{\hat{\theta}}$ contains one or more vectors w, then the assignment of w to $W_{\hat{\theta}}$ depends upon the estimation model g_1. For fixed θ^*, each w value within $W_{\hat{\theta}}$ corresponds to $\hat{\theta}$ (by Theorem 2.1), and specifies $T(\hat{\theta}, \theta^*, w)$ (by Corollary 2), although $T(\hat{\theta}, \theta^*, w)$ can vary between w values within $W_{\hat{\theta}}$.

The density $g(\hat{\theta}|\theta^*)$ can be found as the conditional expectation under $g_1(w|\theta = \hat{\theta})$ of (2.3) over $W_{\hat{\theta}}$. Taking a $(n-p)$ coordinate basis on the manifold $W_{\hat{\theta}}$, in coordinates $v_1, v_2, \ldots, v_{n-p} = v$, this conditional expectation can be written as a ratio of integrals with respect to variations over v.

$$g(\hat{\theta}|\theta^*) = \frac{\int_{W_{\widehat{\theta(v)}}} g_{[T(\hat{\theta},\theta^*,w(v))]}(t) \cdot |dT/d\theta||_{\theta=\hat{\theta}} \cdot g_1(w(v)|\theta = \hat{\theta}) \,.\, ||w'(v)||\, dv}{\int_{W_{\widehat{\theta(v)}}} g_1(w(v)|\theta = \hat{\theta}) \,.\, ||w'(v)||\, dv}. \tag{2.4}$$

The density $g_{[T(\hat{\theta},\theta^*,w(v))]}(t)$ is the density of a transform of $w(v)$ alone under $g_0(w)$. The term $dT/d\theta$ can be evaluated from (2.1) as $-l''(\theta, w(v))$ (the second derivative of l wrt θ), and $|l''(\theta, w(v))| = |j(\theta, w(v))|$. If the subspace $W_{\hat{\theta}}$ runs over more than one manifold, then the top and bottom of expression (2.4) are split into the sums of the corresponding integrals for each manifold before taking the ratio. The same $(n-p)$ coordinate basis is taken over all such manifolds.

Part 3: The unconditional density $g(\hat{\theta})$:
In equation (2.4) it has been assumed that θ^* is fixed and the density $g_{[T(\hat{\theta},\theta^*,w)]}(t)$ has been used to develop the conditional density $g(\hat{\theta}|\theta^*)$. A family of densities $\mathrm{Fam}[g(\hat{\theta}|\theta^*)]$ can be considered to exist for different values of θ^*.

Consider a mechanism for computing $g(\hat{\theta})$ for a given value of $\hat{\theta}$. For each value of $\hat{\theta}$ for which $g(\hat{\theta})$ is to be evaluated, select $g(\hat{\theta}|\theta^* = \hat{\theta})$ as that member of $\mathrm{Fam}[g(\hat{\theta}|\theta^*)]$ for which $\theta^* = \hat{\theta}$. From (2.1), $\theta = \hat{\theta}$ and $\theta^* = \hat{\theta}$ together imply $T = 0$. The term $g_{[T(\hat{\theta},\theta^*,w(v))]}(t)$ in (2.4) can now be written $g_{[T(\hat{\theta},\theta^*=\hat{\theta},w(v))]}(0)$. Repeating this selection process for all the required values of $\hat{\theta}$, the corresponding terms $g(\hat{\theta}|\theta^* = \hat{\theta})$ taken together constitute a new density $g(\hat{\theta})$ that is independent of θ^*.

$$g(\hat{\theta}) = \frac{\int_{W_{\widehat{\theta(v)}}} g_{[T(\hat{\theta},\theta^*=\hat{\theta},w(v))]}(0)\,.\,|j(\theta,w(v))||_{\theta=\hat{\theta}}\cdot g_1(w(v)|\theta=\hat{\theta})\,.\,||w'(v)||\,dv}{\int_{W_{\widehat{\theta(v)}}} g_1(w(v)|\theta=\hat{\theta})\,.\,||w'(v)||\,dv}. \tag{2.5}$$

Now $T(\hat{\theta},\theta^* = \hat{\theta},w(v)) = 0$ throughout $W_{\widehat{\theta(v)}}$. It follows that the factor $g_{[T(\hat{\theta},\theta^*=\hat{\theta},w(v))]}(0)$ is not a function of v and can be moved outside the integrand on the top line of (2.5).

$$g(\hat{\theta}) = \left[\frac{\int_{W_{\widehat{\theta(v)}}} |j(\theta,w(v))||_{\theta=\hat{\theta}}\cdot g_1(w(v)|\theta=\hat{\theta})\,.\,||w'(v)||\,dv}{\int_{W_{\widehat{\theta(v)}}} g_1(w(v)|\theta=\hat{\theta})\,.\,||w'(v)||\,dv}\right]\cdot g_{[T(\hat{\theta},\theta^*=\hat{\theta},w)]}(0). \tag{2.6}$$

The bracketed term in (2.6) is the conditional expectation of $(|j(\theta,w(v))||_{\theta=\hat{\theta}})$, conditional on $W_{\widehat{\theta(v)}}$ under g_1, and can be written $E_w[|j(\theta,w)||\theta=\hat{\theta}]$. □

Equation (2.2), detailed at (2.6), is the required expression for $g(\hat{\theta})$. The exact density $g(\hat{\theta})$ is invariant under linear transformation of w. It also transforms under one to one reparameterization of $\hat{\theta}$, to $\hat{\phi}$ say, as $g(\hat{\phi}) = |\delta\hat{\theta}(\hat{\phi})/\delta\hat{\phi}|\,g(\hat{\theta})$.

It should be noted that $E_w[|j(\theta,w)||\theta=\hat{\theta}]$ is the conditional expectation of the determinant of the observed information, conditional on $\hat{\theta}$. This is not generally the same as $[(E_w[|j(\theta,w)|])|\theta=\hat{\theta}]$, which is the unconditional expected information at $\hat{\theta}$. However, for some simple models there can be equivalence between these expectation terms (e.g., the examples in Section 3.2 and Section 4 below).

It is usually unnecessary to evaluate the multidimensional integrals in (2.6), because in practice terms in w can be replaced by $E_w[w|\theta=\hat{\theta}]$, terms in w^2 by $E_w[w^2|\theta=\hat{\theta}]$, etc. For example, if the model $g_1(w|\theta)$ was normal $N(h[\theta_1],\theta_2)$, terms proportional to w would be replaced by terms proportional to $h[\hat{\theta}_1]$, and terms proportional to w^2 would be replaced by terms proportional to $\hat{\theta}_2+[h(\hat{\theta}_1]^2$. Also, when W is discretised, it is generally the case that the distribution of $\hat{\theta}$ is discretised and no conditional expectation term is required in (2.2).

In the special case that the data generating model is equivalent in form to the estimation model, equation (2.2) reduces to a version of a formula given by Skovgaard (1990) for the density of minimum contrast estimators. Hillier and Armstrong (1999) give an integral formula for the density of the MLE in this special case.

3. Examples

In this section it will be shown how TED can be applied when the estimation model comes from a curved exponential family. In this case $T(\hat{\theta}, \theta^* = \hat{\theta}, w)$ turns out to be a linear transform of a functional of w, and so $g_{[T(\hat{\theta},\theta^*,w)]}(t)$ can be found relatively easily.

3.1. Curved Exponential Families as Estimation Models

Let the $(p\times 1)$ parameter vector θ appear in a $(n\times 1)$ canonical function $b(\theta, z)$ and in a $(p\times 1)$ functional $c(\theta, z)$, together with $(n\times 1)$ functionals of the data $a(w)$ and $d(w)$; all constrained to describe a valid density for w (Dobson, 1983).

$$g_1(w|\theta) = \exp[a(w)^T b(\theta, z) + 1_{(p\times 1)}{}^T c(\theta, z) + 1_{(n\times 1)}{}^T d(w)], \tag{3.1}$$

where ^{T} indicates transposition. If $f(\theta, z)$ is the unconditional expectation of $a(w)$, McCullagh and Nelder (1989) show the following.

$$b'(\theta, z)_{(p\times n)} \cdot f(\theta, z)_{(n\times 1)} = -c'(\theta, z)_{(p\times n)} \cdot 1_{(n\times 1)}. \tag{3.2}$$

Therefore

$$l'(\theta, w)_{(p\times 1)} = b'(\theta, z) . (a(w) - f(\theta, z)), \tag{3.3}$$

$$T(\hat{\theta}, \theta^*, w) = b'(\theta^*, z) . (a(w) - f(\theta^*, z)). \tag{3.4}$$

Equation (3.4) shows that $T(\hat{\theta}, \theta^*, w)$ is a linear transform of $a(w)$.

The conditional expected information term $E_w[|j(\theta, w)||\theta = \hat{\theta}]$ can be obtained after differentiating equation (3.3) again to give $-j(\theta, w)$.

$$\begin{aligned} -j(\theta, w)_{(p\times p)} &= l''(\theta, w) \\ &= [b''(\theta, z)_{(p\times p\times n)} \cdot (a(w) - f(\theta, z))] - [f'(\theta, z)_{(p\times n)} \cdot (b'(\theta, z))^T]. \end{aligned} \tag{3.5}$$

The left-hand part of (3.5) is the curvature of the estimation model (Barndorff-Nielsen and Cox, 1994, page 70), while the right-hand part is $(E_w[|j(\theta, w)|])$.

3.2. Estimation of the Rate Constant for a Gamma Distribution using the Negative Exponential Distribution

Consider i.i.d. data from a data generating model given by the gamma distribution, and an estimation model given by a negative exponential distribution.

$$g_0(w_i) = \frac{\theta_0^\alpha w_i^{\alpha-1} \exp[-\theta_0 w_i]}{\Gamma(\alpha)}, \quad w_i \geq 0, \quad i = 1, \dots, n \tag{3.6}$$

$$g_1(w_i|\theta) = \theta \exp[-\theta w_i], \quad w_i \geq 0, \quad i = 1, \dots, n.$$

The shape parameter α takes a fixed value here and is not considered to be estimated. The estimation model can be restructured in the form (3.1), where in this case the quantities T, $\hat{\theta}$ and θ^* are scalars. More precisely, $a(w) = w, b(\theta, z) = -\theta 1_{(n\times 1)}, 1_{(p\times 1)}{}^T c(\theta, z) = 1_{(p\times 1)}{}^T (log\theta) 1_{(p\times 1)}, 1_{(n\times 1)}{}^T d(w) = 0$ and

$$l'(\theta, w) = -1_{(1\times n)}(w_{(n\times 1)} - 1_{(n\times 1)} 1/\theta)$$
$$T(\hat{\theta}, \theta^*, w) = -s + (n/\theta^*), \quad s = \sum w_i.$$

The distribution of s is obtained from $g_0(w_i)$, using the reproductive property that the sum of a set of i.i.d. gamma variables itself has a gamma distribution.

$$g(s) = \frac{\theta_0^{(n\alpha)} s^{((n\alpha)-1)} \exp[-\theta_0 s]}{\Gamma(n\alpha)}, \quad s \geq 0.$$

From this, the distribution of $T(\hat{\theta}, \theta^*, w)$ is given as

$$g_{[T(\hat{\theta},\theta^*,w(v))]}(t) = \frac{\theta_0^{(n\alpha)}[(n/\theta^*) - t]^{((n\alpha)-1)}}{\Gamma(n\alpha)} \exp(-\theta_0[(n/\theta^*) - t]); \quad t \leq \frac{n}{\theta^*}.$$

In this case $E_w[|j(\theta, w)||\theta = \hat{\theta}] = [(E_w[|j(\theta, w)|])|\theta = \hat{\theta}] = n/\hat{\theta}^2$. From (2.2)

$$g(\hat{\theta}) = \frac{\theta_0^{(n\alpha)} n^{(n\alpha)}}{\hat{\theta}^{((n\alpha)+1)}\Gamma(n\alpha)} \exp[-n\theta_0/\hat{\theta}], \quad \hat{\theta} \geq 0. \tag{3.7}$$

This density can be verified using other analytic methods.

3.3. Nonlinear Regression Models with Known Homoscedastic Normal Errors

An expression will be developed for the density of a parameter vector estimated from a nonlinear regression model, where data generating model and estimation model are distinct, but both are of a multivariate normal form $MN_w(\mu(z); \Sigma)$. Here $\mu(z)$ is the $(n\times 1)$ vector of mean responses and Σ is the known $(n \times n)$ covariance matrix. In this application of a curved exponential family model, it is convenient to assume that enough previous data have been analyzed to give a good estimate of Σ, or that the whole method is to be applied conditional on $\hat{\Sigma}$.

Assume that the data generating model $g_0(w|z)$ has $\mu(z) = f_0(z)$ and the estimation model $g_1(w|\theta, z)$ has $\mu(z) = f(\theta, z)$, where $f(\theta, z)$ is nonlinear in θ. The estimation model can be restructured in the form (3.1) with $a(w) = w, b(\theta, z) = \Sigma^{-1} f(\theta, z), 1^T c(\theta, z) = -\frac{1}{2} f(\theta, z)^T \Sigma^{-1} f(\theta, z), 1^T d(w) = -\frac{1}{2}[n \log(2\pi) + \log|\Sigma| + w^T \Sigma^{-1} w]$ and

$$T(\hat{\theta}, \theta^*, w) = f'(\theta^*, z)\Sigma^{-1}[w - f(\theta^*, z)].$$

Since $T(\hat{\theta}, \theta^*, w)$ is a linear transform of w, its density is as follows (Gallant, 1987, page 122):

$$MN_T(f'(\theta^*, z)\Sigma^{-1}[f_0(z) - f(\theta^*, z)]; \; f'(\theta^*, z)\Sigma^{-1}\Sigma\Sigma^{-1}(f'(\theta^*, z))^T). \tag{3.8}$$

This applies to a general covariance matrix Σ, but attention will now be restricted to the simple i.i.d. case $\Sigma = \sigma^2 I$, where I is the $(n \times n)$ identity matrix. From (2.2), (2.6) and (3.5), the top line of $E_w[|j(\theta, w)||\theta = \hat{\theta}]$ is

$$\int_{W_{\widehat{\theta(v)}}} (|(f''(\theta, z)[w(v) - f(\theta^*, z)] - [f'(\theta, z)(f'(\theta, z))^T])\, 1/\sigma^2||_{\theta=\hat{\theta}})$$

$$\cdot\, g_1(w(v)|\theta = \hat{\theta})\,.\, ||w'(v)||\, dv.$$

In order to obtain the density $g(\hat{\theta})$ from (2.2), sample the density (3.8), where $\theta^* = \hat{\theta}$ and $T = 0$.

$$g(\hat{\theta}) = E_w[|j(\theta, w)||\theta = \hat{\theta}] \cdot |\frac{2\pi}{\sigma^2} f'(\hat{\theta}, z)(f'(\hat{\theta}, z))^T|^{-\frac{1}{2}} \tag{3.9}$$

$$\cdot \exp(-\frac{1}{2\sigma^2}[f(\hat{\theta},z) - f_0(z)]^T [f'(\hat{\theta},z)]^T [f'(\hat{\theta},z)(f'(\hat{\theta},z))^T]^{-1} [f'(\hat{\theta},z)][f(\hat{\theta},z) - f_0(z)]).$$

This reduces to a simpler form for a linear model $f(\theta, z) = X_{(n \times p)} B_{(p \times 1)}$, when $g(\hat{\theta})$ is multivariate normal $MN_{\hat{\theta}}((X^T X)^{-1} X^T f_0(z);\, (X^T X)^{-1} \sigma^2)$.

3.4. Nonlinear Regression Model for the Estimation of Antibody Affinities from the ELISA Immunoassay

In this biochemical test, the extent of reaction of various concentrations of a standard antigen mixed with a sample of blood can be measured quantitatively. The data on the extent of reaction follow a sigmoid saturation profile when plotted against log concentration. It has been suggested that the data could be modelled using a single logistic function with a homoscedastic normal error distribution (Hingley and Ouldridge, 1985; Hingley, 1991):

$$f_0(z) = \frac{10^{a_0(\gamma_0 + z)}}{1 + 10^{a_0(\gamma_0 + z)}}.$$

Here γ_0 is the mean antibody affinity; a_0 is a parameter related to the variance of the underlying distribution of antibody affinity, $0 \leq a_0 \leq 1$; and z is log concentration.

However the underlying distribution of antibodies could be assumed to be of an entirely different form, such as the following combination of two logistic functions:

$$f(\theta, z) = 0.5 \left[\frac{10^{\gamma + z - \delta}}{1 + 10^{\gamma + z - \delta}} + \frac{10^{\gamma + z + \delta}}{1 + 10^{\gamma + z + \delta}} \right].$$

Here the parameter vector $\theta = [\gamma, \delta]^T$ comprises γ, which represents the mean affinity; and δ which is related to the variance of an underlying two point affinity distribution, $\delta > 0$. A set of data were previously reported for serum from an individual animal (Hingley, 1988, Table 1). After re-expressing these data as proportions of their plateau values, the estimated parameters from a fit of the single logistic function are $\gamma_0 = -2.75$, $a_0 = 0.92$ and $\sigma^2 = 0.008787$.

Assuming the correctness of this model, a total of 5000 simulated sets of data were generated around a single logistic data generating model with these parameter values, by using normally distributed random numbers. The two-point affinity estimation model was then fitted to each set of data (Hingley, 1992). The QMLE could only be obtained from 3571 sets of the data (71 per cent of the 5000 data sets). In the other 1429 cases the log-likelihood remained increasing at the parameter boundary $\delta = 0$, and no local maximum was available in the differentiable region. However for these cases the values of γ that minimized the residual sum of squares along the boundary were distributed in similar proportions to the marginal distribution of $\hat{\gamma}$ over the remaining 3571 sets of data.

For the application of TED to this system, $g(\hat{\theta})$ was determined from equation (3.9) using a routine written in APL2. In order to obtain $E_w[|\jmath(\theta, w)||\theta = \hat{\theta}]$, the matrices $f'(\theta, z)$ and $f''(\theta, z)$ were decomposed in order to reflect the contribution of the parameters γ and δ, indexed 1 and 2 respectively.

$$f'(\theta, z)_{(2\times n)} = \begin{pmatrix} f'_{1(1\times n)} \\ f'_{2(1\times n)} \end{pmatrix}$$

$$f''(\theta, z)_{(2\times 2\times n)} = \begin{pmatrix} f''_{11(1\times n)} & f''_{12(1\times n)} \\ f''_{21(1\times n)} & f''_{22(1\times n)} \end{pmatrix}.$$

Here $f''_{12} = f''_{21}$, since the order of differentiation is immaterial.

$$\begin{aligned} E_w[|\jmath(\theta, w)||\theta = \hat{\theta}] &= (1/\sigma^2)^2([(\sigma^2 \sum[(f''_{11i})^2(f''_{22i})^2]) \\ &+ ([\sum(f'_{1i})^2][\sum(f'_{2i})^2])] - [(\sigma^2 \sum(f''_{12i})^4) - (\sum[(f'_{1i}f'_{2i})]^2)]). \end{aligned}$$

Summations in this expression are over the sample members i from 1 to n.

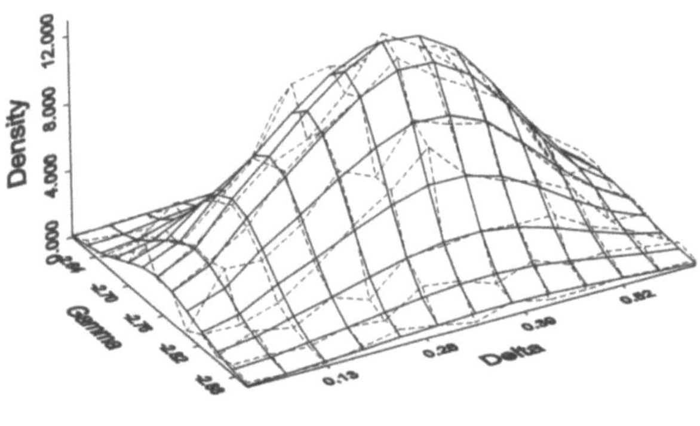

FIGURE 1. Distributions of estimates: bivariate density for ELISA data where the data generating model is logistic and the estimation model is a two point affinity distribution ($\hat{\theta} = [\hat{\gamma}, \hat{\delta}]^T$). The exact density $g(\hat{\theta})$, solid lines; Simulation results, dashed lines.

Figure 1 shows a comparison of the TED expression $g(\hat{\theta})$ to a surface fitted to a histogram of the 3571 results. This surface has been normalized so that it would be a valid two dimensional density over the whole set of 5000 simulations. The absence of 1429 results means that the volume subtended by the surface is approximately 0.71 (= 3571 / 5000), rather than 1. The distribution is fairly symmetric, with a maximum at about $\hat{\gamma} = -2.75, \hat{\delta} = 0.32$. The analytic and simulation methods give similar results, although there is some roughness in the surfaces due to the simulations. It seems that the analytic method may be modelling only the set of maxima in the differentiable region of the likelihood, accounting again for about 71 per cent of the data sets.

The technique described above for determining $E_w[|j(\theta, w)||\theta = \hat{\theta}]$ could be extended to models involving more parameters, $p > 2$, by using analytic expressions for the evaluation of determinants involving cofactors (e.g., Mardia et al., 1979, page 456).

4. Robustness

TED can be applied even if the parameter vector θ has meaning only on the estimation model. But it can also be used as a tool for robustness studies when there is a common parameter vector specified on both models. If θ is such a common parameter vector, $g_0(w)$ can be written as $g_0(w|\theta_0)$.

It is possible to calculate two densities for $\hat{\theta}$. For both densities assume that the data generating model is $g_0(w|\theta_0)$. The first density assumes that the estimation model is $g_0(w|\theta)$, and is written as $g(\hat{\theta}) = h_0(\hat{\theta}|\theta_0)$. The second density assumes that the estimation model is $g_1(w|\theta)$, and is written as $g(\hat{\theta}) = h_1(\hat{\theta}|\theta_0)$. A size α robustness index (RI) can be defined as follows.

1. Calculate a (1 -α) central interval (CI) for $\hat{\theta}$, using $h_0(\hat{\theta}|\theta_0)$.
2. Calculate RI as the probability that $\hat{\theta}$ lies within CI, using $h_1(\hat{\theta}|\theta_0)$.

It can be shown that the RI remains constant for any new metric obtained by one to one reparameterisation of $\hat{\theta}$.

The following example continues the comparison from Section 3.2 of a gamma data generating model with a negative exponential estimation model. Figure 2 shows the gamma model $g_0(w|\theta_0)$ for the case $\alpha = 2, \theta_0 = 1, n = 5$. The negative exponential estimation model $g_1(w|\theta_0)$ is also shown.

The density $h_1(\hat{\theta}|\theta_0)$ has already been given in equation (3.7). The density $h_0(\hat{\theta}|\theta_0)$ can be developed in an analogous fashion.

$$g(\hat{\theta}) = \frac{\theta_0^{(n\alpha)}(n\alpha)^{(n\alpha)}}{\hat{\theta}^{((n\alpha)+1)}\Gamma(n\alpha)} \exp[-n\alpha\theta_0/\hat{\theta}], \quad \hat{\theta} \geq 0.$$

Figure 3 shows $h_0(\hat{\theta}|\theta_0)$ and $h_1(\hat{\theta}|\theta_0)$. The 95 per cent CI is (0.585, 2.085). The size 0.05 RI is calculated as 0.352. That is, there is a 35 per cent probability that the estimate will lie in the 95 percent interval in which it would have occurred if estimation had been carried out on the "correct" model.

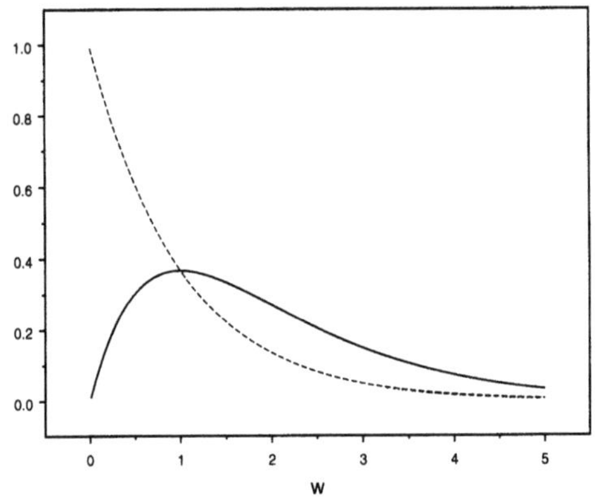

FIGURE 2. Data distributions: gamma data generating model $g_0(w|\theta_0)$, solid line; and negative exponential estimation model $g_1(w|\theta_0)$, dashed line.

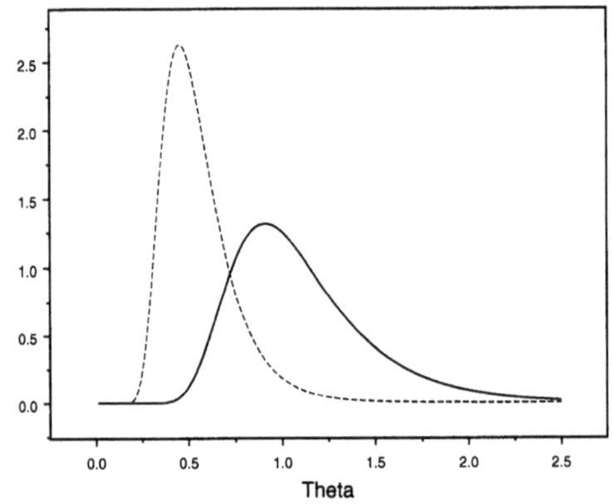

FIGURE 3. Distributions of estimates $g(\hat{\theta})$: gamma estimation model $h_0(\hat{\theta}|\theta_0)$, solid line; and negative exponential estimation model $h_1(\hat{\theta}|\theta_0)$, dashed line.

5. Discussion

The exact analytic density $g(\hat{\theta})$ for QMLE or MLE, that is given by equation (2.2), may turn out to be particularly useful for examining model robustness properties for nonlinear regression. Equation (3.9) could easily be extended to cope with multiplicative error functions, like $\sigma^2 = kf_0(z)$. Consideration of other kinds of covariance structure, such as those found in simple time series models, should in principle be straightforward. It would also be interesting to find an expression for $g(\hat{\theta})$ when the covariance matrix is an estimable parameter.

Methodological extensions could be considered. There should be little difficulty to extend TED to the case of two or more independent variables, or to situations where the error structure of the models differs instead of, or as well as, the model functions. It may also be possible to generalize the approach to cover the distributions of other types of quasi M-estimators in addition to QMLE, by studying the distributions of terms analogous to T as defined in (2.1).

The robustness calculation in Section 4 is trivial, but establishes an approach that could be applied to more interesting situations, such as nonlinear regression with a vector of estimable parameters. For a full study of robustness, it seems necessary to postulate a series of candidate data generating models against which a particular estimation model can be assessed. By comparison of the various densities produced, it can be decided whether or not the adoption of a particular estimation model actually matters. In some cases a common parameter might be better estimated on a wrong model than on the correct one.

References

[1] O.E. Barndorff-Nielsen and D.R. Cox, *Inference and Asymptotics.* Chapman and Hall, London, 1994.

[2] D.M. Bates and D.G. Watts, *Nonlinear Regression Analysis and its Applications.* Wiley, New York, 1988.

[3] A.J. Dobson, *An Introduction to Statistical Modelling.* Chapman and Hall, London, 1983.

[4] A.R. Gallant, *Nonlinear Statistical Models.* Wiley, New York, 1987.

[5] G. Hillier and M. Armstrong, *The Density of the Maximum Likelihood Estimator.* Econometrica **67** (1999), 1459–1470.

[6] P.J. Hingley, *A Clonal Selection Based Timecourse Model for Antibody Responses to Killed Vaccine, with Applications to Foot and Mouth Disease.* Biometrics **47** (1991), 1019–1047.

[7] P.J. Hingley, *Antibody Affinity Distribution-based Models for Vaccine Potency Assessment.* Santa Fe Institute studies in the science of complexity **2** Part 1 (1988), 311–325.

[8] P.J. Hingley, *Nonlinear Regression Modelling in APL.* Vector **9** (1992), 109–122.

[9] P.J. Hingley and E.O. Ouldridge, *The Use of a Logistic Model for the Quantitative Interpretation of Indirect Sandwich Enzyme Labelled Immunosorbent Assays (ELISA) for Antibodies and Antigens in Foot and Mouth Disease.* Comput. Biol. Med. **15** (1985), 137–152.

[10] K.V. Mardia, J.T. Kent, and J.M. Bibby, *Multivariate Analysis.* Academic Press, London, 1979.

[11] P. McCullagh and J.A. Nelder, *Generalized Linear Models,* 2nd Edition. Chapman and Hall, London, 1989.

[12] I.M. Skovgaard, *On the Density of Minimum Contrast Estimators.* Ann. Statist. **18** (1990), 779–789.

[13] H. White, *Estimation, Inference and Specification Analysis.* Cambridge University Press, Cambridge, 1994.

P.J. Hingley
European Patent Office
Erhardtstrasse 27
D-80331 Munich
Germany
e-mail: `phingley@epo.org`

Statistics for Industry and Technology, 131–140

Empirical Comparison of the Classification Performance of Robust Linear and Quadratic Discriminant Analysis

K. Joossens and C. Croux

Abstract. The aim of this paper is to look at the behavior of the total probability of misclassification of robust linear and quadratic discriminant analysis. The effect of outliers on the discriminant rules is studied by comparing their total probabilities of misclassification in presence of outliers.

Mathematics Subject Classification (2000). Primary 62G35; Secondary 62H30.

Keywords. Discriminant analysis, total probability of misclassification, robustness.

1. Introduction

In discriminant analysis one observes two groups of multivariate observations forming together the training sample. Using these data a discriminant rule is determined, that is used afterwards to classify new observations into one of the two groups. In this paper we restrain us to the case of two multivariate normal distributed populations. We observe p-variate observations $x_{11}, \ldots, x_{1n_1}$ coming from a first population $\wp_1 \sim H_1 = N_p(\mu_1, \Sigma_1)$ and $x_{21}, \ldots, x_{2n_2}$ coming from a second population $\wp_2 \sim H_2 = N_p(\mu_2, \Sigma_2)$.

Supposing that the observations are generated from two multivariate normal distributions, it is known that the optimal discriminant rule (i.e., the one minimizing the misclassification probability) is a quadratic function given by

$$Q(x) = -\frac{1}{2}x^t(\Sigma_1^{-1} - \Sigma_2^{-1})x + (\mu_1^t\Sigma_1^{-1} - \mu_2^t\Sigma_2^{-1})x - k(\mu_1, \mu_2, \Sigma_1, \Sigma_2), \qquad (1.1)$$

where

$$k(\mu_1, \mu_2, \Sigma_1, \Sigma_2) = \frac{1}{2}\log\left(\frac{|\Sigma_1|}{|\Sigma_2|}\right) + \frac{1}{2}(\mu_1^t\Sigma_1^{-1}\mu_1 - \mu_2^t\Sigma_2^{-1}\mu_2).$$

We assign a new p-variate observation x to $\wp_1$ if

$$Q(x) > \log\left(\frac{c_2\pi_2}{c_1\pi_1}\right) = \tau, \tag{1.2}$$

where c_1 and c_2 are the costs of misclassifying a unit of, respectively, $\wp_1$ and $\wp_2$ and π_1 and π_2 are the prior probabilities that x will belong to, respectively, $\wp_1$ and $\wp_2$. In practice these parameters are unknown and therefore we set $\tau = 0$ throughout this paper.

If we assume that the covariance matrices are equal ($\Sigma := \Sigma_1 = \Sigma_2$), we get the familiar Fisher's linear discriminant rule

$$L(x) = (\mu_1 - \mu_2)^t\Sigma^{-1}x - \frac{1}{2}(\mu_1 - \mu_2)^t\Sigma^{-1}(\mu_1 + \mu_2). \tag{1.3}$$

Since the primary goal of discriminant analysis is to classify data, we are particulary interested in the total probability of misclassification of a particular discriminant rule.

Robust linear discriminant analysis has been considered in several papers (e.g., Hawkins and McLachlan, 1997; He and Fung, 2000; Croux and Dehon, 2001). The first ones to consider robust quadratic discriminant analysis seem to be Randles et al. (1978), who used M-estimators for the means μ_1 and μ_2, and covariance matrices Σ_1 and Σ_2 in (1.1) and a rank based rule to estimate the cut-off value τ in (1.2). One of the Editors of this book also pointed out a forthcoming paper of Hubert and Van Driessen (2004), using the MCD-estimators. Note that this approach is extendable to the multiple group case. In this paper we compare the performance of robust linear and quadratic discriminant analysis using both S- and MCD-estimators. Based on simulation experiments, our main finding is that it seems to be profitable to use the quadratic over the linear discriminant rule, also in presence of outliers. Also, and not surprisingly, the robust rules outperform the classical ones when deviating from the model, while performing almost as good at the model distribution.

2. Robustification of Classical Discriminant Analysis

Outliers and atypical observations might have an influence on the results of classical discriminant analysis, since the discriminant rules are based on estimates of the population parameters. The outliers and atypical observations might shift the estimated means and they might blow up the dispersion matrices. To prevent this we make use of robust estimators of the population parameters. For our study we will look at two robust estimators, the MCD-estimator and the S-estimator.

The MCD-estimators were introduced by Rousseeuw (1985). The estimator is given by the subset of size h for which the determinant of its covariance matrix is minimal. The MCD-estimator of location is then given by the mean of these h observations and the MCD-estimator of covariance is given by their covariance matrix.

The S-estimators of location and multivariate dispersion were jointly introduced by Davies (1987) and Rousseeuw and Leroy (1987). To define these estimators, let X be a sample of p-variate observations and let n_X be the number of observations in the sample. The S-estimators of location and dispersion of this sample are defined as

$$(\hat{\mu}, \hat{\Sigma}) = \underset{\mu,\ \Sigma}{\text{argmin}}\{det\Sigma\} \quad \text{such that} \quad \frac{1}{n_X}\sum_{x\in X}\rho\left(\sqrt{(x-\mu)^t\Sigma^{-1}(x-\mu)}\right) = b,$$

where $\mu \in \mathbb{R}^p$ and $\Sigma \in \mathbb{R}^{p\times p}$ is a symmetric positive definite matrix. The function ρ needs to satisfy the following condition

(R) $\rho : \mathbb{R} \to \mathbb{R}$ is a symmetric, continuous, non decreasing function on $[0, \infty)$. Moreover, $\rho(0) = 0$ and ρ has a continuous derivative in all but finite number of points.

The constant b is set equal to $E_{F_0}[\rho(\|Z\|)]$ for F_0, the central model distribution, being $N(0, I_p)$ in our case. The most commonly used choice for ρ is the Biweight function which is defined as

$$\rho(u) = \begin{cases} \frac{u^2}{2} - \frac{u^4}{2c^2} + \frac{u^6}{6c^4} & \text{if } |u| < c \\ \frac{c^2}{6} & \text{if } |u| \geq c \end{cases}$$

where c is a tuning constant to achieve the desired value of the breakdown point.

The breakdown point of an estimator is the fraction of the data that can be completely contaminated without destroying the estimator. A breakdown point is a measure of robustness and resides between 0% and 50%. In this paper, we will use the MCD- and S-estimators with breakdown point 25%. The choice of a 25% breakdown point gives a good compromise between efficiency and robustness of the estimators (see e.g., Croux and Haesbroeck, 1999).

Another way to robustify the classical method is by detecting the influential observations and deleting them from the samples. Measures to diagnose influential observations in the context of discriminant analysis have been proposed by Fung (1995, 1996). By deleting these influential observations from the sample and using the classical discriminant analysis based on the remaining observations, the classical method becomes robust. Note that for detecting these observations, it is also advised to use robust estimates to avoid the masking effect. Indeed, it is well known that diagnostics based on non-robust estimates do not always detect all outliers.

As a measure of performance, we will look at the total probability of misclassification. Formally, the total probability of misclassification (TPM) is given by

$$\text{TPM} = \pi_1 P(Q(x) > \tau | x \sim \wp_2) + \pi_2 P(Q(x) < \tau | x \sim \wp_1).$$

The total probability of misclassification according to a rule can be estimated by classifying observations of which the source population is known and look at the fraction of misclassified observations. Under the normality assumption the probabilities of misclassification for the optimal linear and quadratic discriminant rules can be computed theoretically as function of the population parameters of location and dispersion. In the case of equal covariance matrices, the total probability of misclassification in the linear case is given by

$$\mathrm{TPM}_{\text{linear}} = \Phi\left(\frac{-\Delta}{2}\right), \tag{2.1}$$

where Φ is the cumulative standard normal distribution function and Δ is the Mahalanobis distance between the populations, namely $\sqrt{(\mu_1-\mu_2)^t\Sigma^{-1}(\mu_1-\mu_2)}$. Since the discriminant rule in the linear case is of the form $L(x) = a^t x + b$, we get in the case of populations with different covariance matrices

$$\begin{aligned}
\mathrm{TPM}_{\text{linear}} &= \pi_1 P(L(x) < 0 \mid x \in \wp_1) + \pi_2 P(L(x) > 0 \mid x \in \wp_2) \\
&= \pi_1 P(a^t x + b < 0 \mid x \sim N_p(\mu_1, \Sigma_1)) \\
&\quad + \pi_2 P(a^t x + b > 0 \mid x \sim N_p(\mu_2, \Sigma_2)) \\
&= \pi_1 P\left(z < \frac{-b - a^t\mu_1}{\sqrt{a^t\Sigma_1 a}} \mid z \sim N(0, I)\right) \\
&\quad + \pi_2 P\left(z > \frac{-b - a^t\mu_2}{\sqrt{a^t\Sigma_2 a}} \mid z \sim N(0, I)\right) \\
&= \pi_1 \Phi\left(\frac{-b - a^t\mu_1}{\sqrt{a^t\Sigma_1 a}}\right) + \pi_2\left(1 - \Phi\left(\frac{-b - a^t\mu_2}{\sqrt{a^t\Sigma_2 a}}\right)\right), \qquad (2.2)
\end{aligned}$$

where Φ is standard normal cumulative distribution function and a and b are as in (1.3), and where Σ can be estimated by the pooled covariance matrix. A theoretical expression of $\mathrm{TPM}_{\text{quadratic}}$ can also be obtained, but it needs to be evaluated numerically. For more details and expressions, we refer to Croux and Joossens (2003).

3. Simulation Experiment

We generate 1000 random normal distributed observations from two populations $\wp_1 \sim N_p(\mu_1, \Sigma_1)$ and $\wp_2 \sim N_p(\mu_2, \Sigma_2)$ of dimension $p = 3$, that construct the training sample. First, we consider the uncontaminated samples and afterwards we contaminate them by adding outliers. We will compute the classical and the robust discriminant rules for both the linear and the quadratic case.

As already mentioned for linear discriminant analysis the discriminant rule (1.3) is of the following form

$$L(x) = a^t x + b,$$

where a is a p-dimensional vector and b a scalar. In the case of quadratic discriminant analysis the discriminant rule (1.1) is of the following form

$$Q(x) = x^t A x + d^t x + e,$$

where A is a p-dimensional matrix , d is a p-dimensional vector and e is a scalar. These parameters need to be estimated, and this will be done classically and robust, using the MCD-estimators and S-estimators.

For the MCD-estimators we use the fastmcd algorithm by Rousseeuw and Van Driessen (1999) and for the S-estimators we used the algorithm developed by Ruppert (1992). For estimating the sample covariance matrix $\Sigma = \Sigma_1 = \Sigma_2$ in the linear case a pooled covariance matrix estimate is used. Programs for computing robust linear and quadratic discriminant analysis can be retrieved from the web site *http://www.econ.kuleuven.ac.be/christophe.croux*. Note that our primary interest is not in the parameter estimates of the linear and quadratic rule, but only in the probability of misclassification of the rules.

After constructing the discriminant rules we generate 5000 random normally distributed observations for each population, without contamination. These validation samples have the same distribution as the training samples (if we do not take the outliers into account). These observations are classified by all the discriminant rules. In linear discriminant analysis we assign an observation x to the first population if and only if $L(x) < 0$ and in quadratic discriminant analysis if and only if $Q(x) < 0$. Since the source populations of these observations are known, we are then able to decide whether an observation is then misclassified. The fraction of misclassified observations is then an estimate of the total probabilities of misclassification using the specific discriminant rule. By taking a number of 5000 observations in the validation sample, we aim at attaining an accurate estimate of the population misclassification rate. Indeed, for a given discriminant rule, the standard error of this estimate is always less then 0.71%.

Since the classification rules depend strongly on the generated data of the training sample, we generate 500 different training data sets, on which we apply both linear and quadratic discriminant rules based on classical and robust (MCD- and S-) estimators. Working with 500 different training sets allows us to take the estimation variability of the discriminant rules into account. Indeed, we can compute the mean and the standard deviation of the total probability of misclassification over the 500 runs for linear and quadratic, classical and robust discriminant analysis.

4. Simulation Results

We notate the total probability of misclassification in percents and put the associated standard deviations between parenthesis (also in percents). The theoretical values of the TPM, in absence of contamination, are also mentioned. Let us consider 3 cases and take rather extreme cases to illustrate the effect of outliers. For simplicity of notation, a stands for a vector $(a, a, a)^t$ and I for the 3-dimensional identity matrix. Note that similar simulation experiments, but only for the linear case, were constructed in He and Fung (2000) and Croux and Dehon (2001).

4.1. Unequal Means and Equal Covariance Matrices

The two populations have the same covariance matrix, but different means. Because of the equality of the covariance matrices, the linear method should be preferred because it is much easier then the quadratic and it should lead (asymptotically) to the same rules. Note that in the case of equal covariance matrices, the theoretical values of the total probability of misclassification for the quadratic rule are the same as for the linear rule, computed as in (2.1). The two populations consist of 1000 observations, where $\wp_1 \sim N_3(-1; I)$ and $\wp_2 \sim N_3(1; I)$. Let us now contaminate 10% of observations of each population, deviating in location from the original distributions. The outliers of the first populations follow a $N_3(9; I)$ distribution and those of the second population a $N_3(-9; I)$ distribution. As a second type of contamination, 10% of the data are switched from one group to the other.

TABLE 1. Average TPM with standard deviation between parenthesis over 500 runs, in case of equal covariance matrices. Linear and quadratic discriminant rules based on the classical, the MCD- and the S-estimators are considered.

	Theoretical	*Classic*	*MCD*	*S*
Linear	4.16	4.09 (0.02)	4.13 (0.05)	4.09 (0.02)
Quadratic	4.16	4.10 (0.02)	4.16 (0.06)	4.10 (0.02)
	in presence of 10% outliers in location of type 1			
Linear		49.95 (3.60)	4.12 (0.04)	4.09 (0.02)
Quadratic		49.88 (2.85)	4.14 (0.05)	4.10 (0.02)
	in presence of 10% outliers in location of type 2			
Linear		4.21 (0.04)	4.20 (0.03)	4.20 (0.02)
Quadratic		4.24 (0.04)	4.23 (0.04)	4.21 (0.04)

From Table 1 it seems clear that the results coming from the linear discriminant analysis are close to those of the quadratic discriminant analysis. When the populations are contaminated by extreme outliers, the robust estimators are obviously much better than the classical estimators. Note that in this case the S-estimator behaves more robust than the MCD-estimator, yielding lower values for the simulated TPM. But, taking the standard errors into account, this difference is not significant. For the second type of contamination all discriminant rules perform equally well.

4.2. Equal Means and Unequal Covariance Matrices

Let us now consider populations with the same mean, but with different covariance matrices. For the computation of the theoretical values of the total probability of misclassification in the linear case we use formula (2.2) and in the quadratic case we use the formula proposed by Croux and Joossens (2003). For the linear discriminant rule, the pooled sample covariance matrix is used.

Let $\wp_1$ and $\wp_2$ be two populations consisting each of 1000 observations following a $N_3(0; 100I)$ and a $N_3(0; I)$ distribution, for respectively the first and the second population. We contaminate again 10% of the observations, but now to get outliers in dispersion. We generate them as they would come from the wrong population. This means that 100 observations have $100I$ as covariance matrix instead of I, for the first population and vice versa for the second population. This leads to the following results.

TABLE 2. As in Table 1, but in case of equal means and unequal covariance matrices.

	Theoretical	*Classic*	*MCD*	*S*
Linear	50.00	45.73 (1.89)	43.48 (2.66)	45.56 (1.89)
Quadratic	0.82	0.83 (0.01)	0.82 (0.18)	0.83 (0.01)
	in presence of 10% outliers			
Linear		46.64 (2.21)	44.95 (2.16)	45.98 (1.70)
Quadratic		7.24 (0.46)	1.75 (0.13)	1.11 (0.03)

In case of equal means and unequal covariance matrices we notice a significant difference between the linear and the quadratic discriminant analysis. The quadratic method is much better, which is logical because the linear method assumes that the covariance matrices are equal. In case of contamination the robust method is again better than the classical and the S-estimator is better than the MCD-estimator.

4.3. Unequal Means and Unequal Covariance Matrices

Consider now two populations with different means and different covariance matrices. Since this is the more representative case, we consider three different sampling schemes with different degrees of overlap between the two populations. This results in increasing values for the total probability of misclassification. From a practical point of view, it means that Scheme 1 corresponds to "easy" classification problems, while the last scheme is a more difficult one.

Scheme 1: Let $\wp_1 \sim N_3(-1; I)$ and $\wp_2 \sim N_3(1; 0.25I)$ be two populations consisting of each 1000 observations. First we contaminate the samples, by creating 10% outliers in location such that the classical mean estimators should become close: 100 observations of $\wp_1$ follow a $N_3(9; I)$ distribution and 100 observations of $\wp_2$ follow $N_3(-9; 0.25I)$ distribution. Secondly, we change these outliers, such that they deviate in location and dispersion. This is done by changing their corresponding covariance matrices I and $0.25I$ into $0.25I$ and I, for respectively, the outliers in the first and the second population.

Scheme 2: Let $\wp_1 \sim N_3(0; 2.25I)$ and $\wp_2 \sim N_3(1; 0.25I)$. We contaminate 100 observations in each of those two populations such that they deviate in location and dispersion. The outliers of the first population follow a $N_3(3; 9I)$ distribution and those of the second population follow a $N_3(-1; I)$ distribution.

Scheme 3: Let us consider two populations, $\wp_1 \sim N_3(0;4I)$ and $\wp_2 \sim N_3(1;16I)$, consisting each of 1000 observations. The first type of outliers are generated by replacing 100 observations of $\wp_1$ by observations coming from a $N_3(4;I)$ distribution and 100 from $\wp_2$ by observations coming from a $N_3(-16;I)$ distribution. A second type of outliers is generated by replacing 100 observations from each population as if they would come from the wrong population. In other words, the first population consists of 900 observations coming from a $N_3(0;4I)$ distribution and 100 coming from a $N_3(1;16I)$ distribution and vice versa for the second population.

TABLE 3. Average TPM with standard deviation between parenthesis over 500 runs, in the most general case of unequal means and unequal covariance matrices. Linear and quadratic discriminant rules based on the classical, the MCD- and the S-estimators are considered. Three different sampling schemes are considered.

SCHEME 1	*Theoretical*	*Classic*	*MCD*	*S*
Linear	1.92	2.05 (0.08)	2.09 (0.13)	2.05 (0.08)
Quadratic	0.74	0.75 (0.08)	0.86 (0.08)	0.75 (0.09)
	in presence of 10% outliers in location			
Linear		49.89 (3.61)	2.13 (0.11)	2.12 (0.08)
Quadratic		26.44 (0.06)	0.80 (0.04)	0.87 (0.04)
	in presence of 10% outliers in location and dispersion			
Linear		49.72 (3.60)	2.05 (0.11)	2.04 (0.08)
Quadratic		26.47 (0.12)	0.78 (0.05)	0.83 (0.04)
SCHEME 2				
Linear	18.51	16.20 (0.06)	16.29 (0.14)	16.20 (0.06)
Quadratic	6.82	6.87 (0.03)	8.80 (0.08)	6.87 (0.03)
	in presence of 10% outliers in location and dispersion			
Linear		16.65 (0.45)	16.29 (0.10)	16.23 (0.01)
Quadratic		13.04 (0.26)	8.00 (0.24)	8.27 (0.15)
SCHEME 3				
Linear	38.82	37.29 (0.31)	37.40 (0.48)	37.29 (0.32)
Quadratic	20.67	20.30 (0.05)	23.40 (0.34)	20.31 (0.05)
	in presence of 10% outliers of type 1			
Linear		56.05 (0.82)	37.48 (0.44)	39.82 (0.63)
Quadratic		23.74 (0.19)	21.85 (0.23)	23.03 (0.17)
	in presence of 10% outliers of type 2			
Linear		37.86 (0.33)	38.20 (0.33)	38.12 (0.35)
Quadratic		20.73 (0.11)	20.42 (0.10)	20.36 (0.08)

In all contaminated cases (see Table 3), the quadratic rule outperforms the linear one. Classification based on robust estimates is much better than based on classical estimates in presence of outliers. (Note that for Scheme 3, with less extreme outliers as in the previous contamination schemes, the classical quadratic rule is not loosing much.) In the first scheme, for the linear case in the second scheme and for the second type of outliers in the third scheme, the robust discriminant analysis based on S-estimator is slightly better, however not significantly, than the one based on the MCD-estimator in presence of outliers (linear and quadratic). But for the first type of outliers in the third scheme and for the quadratic case in the second scheme the robust classification based on the discriminant rules using the MCD-estimator is slightly better than the one based on the S-estimator in presence of outliers.

If no outliers are present, the S-estimator yields lower misclassification rates than the MCD-estimator. From these simulations in the case of unequal means and covariance matrices, we can conclude that quadratic discriminant analysis is always preferred to linear discriminant analysis and that the robust method is better than the classical method. We notice that sometimes the robust discriminant analysis based on the MCD-estimators is better then the robust discriminant analysis based on the S-estimators and sometimes it is vice versa. Therefore we cannot say which of the robust estimators is preferred.

5. Conclusions

It is shown that quadratic discriminant analysis is needed when the covariance matrices of the populations are different. Robust estimators are needed if there are outliers in the populations. If no outliers are present in the training sample there is practical no loss of the robust procedure in classification. Therefore, the only major loss when using robust discriminant rules is computation cost. Note however that fast algorithms have been used and that computer software is available to compute the robust estimators considered in this paper. The choice for MCD- and S-estimators was motivated by this computational aspect and by their high breakdown point.

While this proceedings paper focuses on simulation aspect, a more formal and theoretical treatment is provided in Croux and Joossens (2003), where the influence of outliers on the quadratic discriminant rule and its total probability of misclassification is studied.

References

[1] C. Croux and C. Dehon, *Robust Linear Discriminant Analysis using S-estimators.* Canad. J. Statist. **29** (2001), 473–492.

[2] C. Croux and G. Haesboeck, *Influence Function and Efficiency of the Minimum Covariance Determinant Scatter Matrix Estimator.* J. Multivariate Anal. **71** (1999), 161–190.

[3] C. Croux and K. Joossens, *Influence of Observations on the Misclassification Probability in Quadratic Discriminant Analysis.* Research Report **59** (2003), Dept. of Applied Economics, K.U. Leuven, Belgium.

[4] P.L. Davies, *Asymptotic Behavior of S-estimators of Multivariate Location Parameters and Dispersion Matrices.* Ann. Statist. **15** (1987), 1269–1292.

[5] W.K. Fung, *Diagnostics in Linear Discriminant Analysis.* J. Amer. Statist. Assoc. **90** (1995), 952–956.

[6] W.K. Fung, *Diagnosing Influential Observations in Quadratic Discriminant Analysis.* Biometrics **52** (1996), 1235–1241.

[7] D.M. Hawkins and G.J. McLachlan, *High Breakdown Linear Discriminant Analysis.* J. Amer. Statist. Assoc. **92** (1997), 136–143.

[8] X. He and W.K. Fung, *High Breakdown Estimation for Multiple Populations with Applications to Discriminant Analysis.* J. Multivariate Anal. **72** (2000), 151–162.

[9] M. Hubert and K. Van Driessen, *Fast and Robust Discriminant Analysis.* Comput. Statist. Data Anal. **45** (2004), 301–320.

[10] R.H. Randles, J.D. Broffitt, J.S. Ramsberg, and R.V. Hogg, *Generalized Linear and Quadratic Discriminant Functions using Robust Estimators.* J. Amer. Statist. Assoc. **73** (1978), 564–568.

[11] P.J. Rousseeuw, *Multivariate Estimation with High Breakdown Point.* In W. Grossmann, G. Pflug, I. Vincze, and W. Wertz, editors, Mathematical Statistics and Applications, vol B, pages 283–297. Reidel Publishing Company, Dordrecht, 1985.

[12] P.J. Rousseeuw and A.M. Leroy, *Robust Regression and Outlier Detection.* Wiley, New York, 1987.

[13] P.J. Rousseeuw and K. Van Driessen, *A Fast Algorithm for the Minimum Covariance Determinant Estimator.* Technometrics **41** (1999), 212–223.

[14] D. Ruppert, *Computing S-estimators for Regression and Multivariate Location/Dispersion.* J. Comput. Graph. Statist. **1** (1992), 253–270.

Acknowledgment

We would like the thank the reviewers for their thorough revision of the paper.

K. Joossens and C. Croux
Dept. of Applied Economics
Katholieke Universiteit Leuven
Naamsestraat 69
B-3000 Leuven
Belgium
e-mail: {`kristel.joossens, christophe.croux`}`@econ.kuleuven.ac.be`

Statistics for Industry and Technology, 141–152

Estimates of the Tail Index Based on Nonparametric Tests

J. Jurečková and J. Picek

Abstract. The authors recently constructed several nonparametric tests of one-sided hypotheses on the value of the Pareto-type tail index in the family of distributions with nondegenerate right tails. Inverting the tests in the Hodges-Lehmann manner (Hodges and Lehmann, 1963), we obtain strongly consistent estimators of m. The simulation study demonstrates surprisingly good approximations of m, namely by two of the three proposed estimators.

Mathematics Subject Classification (2000). Primary 62G32; Secondary 62G30.

Keywords. Pareto-type tail index; strong approximation of empirical process.

1. Introduction

The authors recently constructed several nonparametric tests of one-sided hypotheses on the value of the Pareto-type tail index in the family of distributions with nondegenerate right tails; see Fialová et al. (2004), Jurečková (2000), Jurečková and Picek (2001) and Picek and Jurečková (2001). The proposed tests, fully nonparametric, are based on splitting the set of observations into N subsamples of sizes n and on the empirical distribution functions of N subsamples statistics of various types; namely, the subsample extremes in Jurečková and Picek (2001), the subsample means in Fialová et al. (2004), the averages of two subsubsample extremes in Picek and Jurečková (2001), and the linear combinations of extreme regression quantiles in the linear regression model (see Jurečková, 2000).

The model assumes that the distribution function F of the observations satisfies $1 - F(x) = x^{-m}L(x)$, where $m > 0$ is the parameter of interest and $L(x)$ is a function, slowly varying at infinity. As such, the problem is semiparametric in its nature, involving an unknown slowly varying function. The tests of the one-sided

Research of first author was supported by the Czech Republic Grant 201/02/0621 and by the Project CEZ: MSM 113200008 "Mathematical Methods in Stochastics". Research of second one supported by the Czech Republic Grant 205/02/0871 and by Postdoctoral Grant from the Portuguese Fundação para a Ciência e a Tecnologia (FCT) at the University of Lisbon (Centro de Estatística e Aplicações).

hypothesis $\mathbf{H}_0 : m \leq m_0$ (or $\mathbf{H}_1 : \ m \geq m_0$) are consistent against one-sided alternatives $m > m_0$ (or $m < m_0$, respectively) as $N \to \infty$ and n remains fixed; the asymptotic null distributions of the test criteria are normal. The simulation studies in the cited papers show that the tests distinguish well the distribution tails already for moderate samples.

Inverting such test in the Hodges-Lehmann manner (Hodges and Lehmann, 1963), we obtain a strongly consistent estimate M_N of m. The simulation study demonstrates a surprisingly good approximation of m by M_N, mainly by the estimators based on the tests proposed in Jurečková and Picek (2001) and Picek and Jurečková (2001).

2. Estimator of m Based on Subsample Extremes

Let us have a dataset of nN independent observations, identically distributed with distribution function F satisfying

$$1 - F(x) = x^{-m} L(x), \tag{2.1}$$

where $m, \ 0 < m < \infty$ is the parameter of interest and $L(x)$ is a function, slowly varying at infinity. To avoid an eventual effect of observations ordering, the data may be randomly permuted before starting an inference. The observations are then partitioned into N samples $\mathbf{X}_j = (X_{j1}, \ldots, X_{jn})'$ of equal fixed sizes $n, \ j = 1, \ldots, N$. Denote $X_{(n)}^{(1)}, \ldots, X_{(n)}^{(N)}$ the respective sample maxima. Then $X_{(n)}^{(1)}, \ldots, X_{(n)}^{(N)}$ is a random sample from distribution $F^*(x) = F^n(x)$. Let $\hat{F}_N^*$ be the empirical distribution function of the sample maxima, *i.e.*,

$$\hat{F}_N^*(x) = \tfrac{1}{N} \sum_{j=1}^{N} I[X_{(n)}^{(j)} \leq x]. \tag{2.2}$$

For any fixed $m > 0$, put

$$a_{N,m} = (nN^{1-\delta})^{\frac{1}{m}} \tag{2.3}$$

where $0 < \delta < \frac{1}{2}$ is a chosen constant; then $\lim_{N \to \infty} a_{N,m} = \infty$.

The test of $\mathbf{H}_{m_0} : m \leq m_0$ against $\mathbf{K}_{m_0} : m > m_0$, proposed in Jurečková and Picek (2001), rejects the hypothesis $\mathbf{H}_{m_0}$ at the asymptotic significance level $\alpha \in (0, 1)$ provided

$$\begin{aligned} &\textbf{either } 1 - \hat{F}_N^*(a_{N,m_0}) = 0, \\ &\textbf{or } 1 - \hat{F}_N^*(a_{N,m_0}) > 0 \ \text{ and} \\ &N^{\delta/2} \left[-\log(1 - \hat{F}_N^*(a_{N,m_0})) - (1-\delta) \log N \right] \geq \Phi^{-1}(1-\alpha), \end{aligned} \tag{2.4}$$

where Φ is the standard normal distribution function.

As it was proved in Jurečková and Picek (2001),

$$\lim_{N \to \infty} \mathbb{P}_{m_0} \left(0 < \hat{F}_N^*(a_{N,m_0}) < 1 \right) = 1 \tag{2.5}$$

for every F satisfying $1 - F(x) \geq x^{-m_0}$ for $\forall x > x_0$. Moreover, if F has exactly the Pareto tail corresponding to $1 - F(x) = x^{-m_0}$ for $\forall x > x_0$, then the left-hand side of (2.4) has asymptotically normal distribution, as $N \to \infty$:

$$\lim_{N\to\infty} \mathbb{P}_{m_0} \Big\{ N^{\delta/2} \Big[-\log(1 - \hat{F}_N^*(a_{N,m_0})) - (1-\delta)\log N \Big] \leq x \Big\} = \Phi(x), \ x \in \mathbb{R}.$$

If F satisfies $1 - F(x) \geq x^{-m}$ for $\forall x > x_0$ for some $m \leq m_0$, then

$$\overline{\lim}_{N\to\infty} P_{m_0} \Big\{ N^{\delta/2} \Big[-\log(1 - \hat{F}_N^*(a_{N,m_0})) - (1-\delta)\log N \Big] \geq \Phi^{-1}(1-\alpha) \Big\} \leq \alpha.$$

Analogously, we reject the hypothesis $\mathbf{H}^*_{m_0} : m \geq m_0$ against $\mathbf{K}^*_{m_0} : m < m_0$, provided

$$\begin{aligned} &\textbf{either } \hat{F}_N^*(a_{N,m_0}) = 0, \\ &\textbf{or } \hat{F}_N^*(a_{N,m_0}) > 0 \text{ and} \\ &\qquad N^{\delta/2} \Big[-\log(1 - \hat{F}_N^*(a_{N,m_0})) - (1-\delta)\log N \Big] \leq \Phi^{-1}(\alpha). \end{aligned} \tag{2.6}$$

The last inequality is equivalent to

$$N^{\delta/2} \Big[\log(1 - \hat{F}_N^*(a_{N,m_0})) + (1-\delta)\log N \Big] \geq \Phi^{-1}(1-\alpha). \tag{2.7}$$

The quantity $a_{N,m}$ defined in (2.3) is decreasing in m for any fixed N, n, hence the statistic

$$\log(1 - \hat{F}_N^*(a_{N,m})) + (1-\delta)\log N$$

is nondecreasing in m for fixed N, n. Because it is discontinuous, the equation

$$\log(1 - \hat{F}_N^*(a_{N,m})) + (1-\delta)\log N := 0 \quad [= \Phi^{-1}(\tfrac{1}{2})]$$

generally has no solution with respect to m, and we should rather look for m satisfying the approximate equation

$$1 - \hat{F}_N^*(a_{N,m})) \approx N^{-(1-\delta)}.$$

Following these ideas, we define an estimator M_N of m as

$$M_N = \tfrac{1}{2}(M_N^+ + M_N^-), \tag{2.8}$$

where

$$\begin{aligned} M_N^- &= \sup\{s :\ 1 - \hat{F}_N^*(a_{N,s})) < N^{-(1-\delta)}\}, \\ M_N^+ &= \inf\{s :\ 1 - \hat{F}_N^*(a_{N,s})) > N^{-(1-\delta)}\}. \end{aligned} \tag{2.9}$$

In this way we obtain a strongly consistent estimator of m (> 0), as it is shown in the following theorem:

Theorem 2.1. *If F has the right tail with tail exponent $m > 0$, i.e., $1 - F(x) = x^{-m}L(x)$ where $L(\cdot)$ is slowly varying at ∞, then*

$$M_N \to m \text{ a.s.} \quad \textit{as} \quad N \to \infty. \tag{2.10}$$

Proof. Because the relative increase or decrease of $L(x)$ is smaller than any power of x, it is sufficient to prove (2.10) for F with exact Pareto right tail with index m, i.e., for F satisfying $1 - F(x) = x^{-m}$ for $x > x_0$. The relations (2.8) and (2.9) imply that, for any $\varepsilon > 0$,

$$\begin{aligned} \mathbb{P}_m(|M_N - m| > \varepsilon) &\le \mathbb{P}_m\{M_N^+ > m + \varepsilon\} + \mathbb{P}_m\{M_N^- < m - \varepsilon\} \\ &\le \mathbb{P}_m\{1 - \hat{F}_N^*(a_{N,m+\varepsilon}) \le N^{-1+\delta}\} + \mathbb{P}_m\{1 - \hat{F}_N^*(a_{N,m-\varepsilon}) \ge N^{-1+\delta}\}. \end{aligned} \tag{2.11}$$

Denote by $A_{N,\varepsilon}$ the event

$$A_{N,\varepsilon} = \left[1 - \hat{F}_N^*(a_{N,m+\varepsilon}) \le N^{-1+\delta}\right]. \tag{2.12}$$

Then, for an integer $K > 0$,

$$\begin{aligned} &\mathbb{P}_m\left\{\bigcap_{k=K}^{\infty} \bigcup_{N=k}^{\infty} A_{N,\varepsilon}\right\} \\ &\le \mathbb{P}\left\{\frac{1 - \hat{F}_N^*(a_N, m + \varepsilon)}{1 - F^*(a_{N,m+\varepsilon})} \le n^{-\frac{m}{m+\varepsilon}} N^{-(1-\delta)\frac{\varepsilon}{m+\varepsilon}} \text{ i.o. for } N \ge K\right\}. \end{aligned} \tag{2.13}$$

By Komlos et al. (1975), there exists a Brownian bridge $\mathcal{B}_N$ depending on $X_{(n)}^{(1)}, \dots, X_{(n)}^{(N)}$, such that

$$\begin{aligned} N^{1/2}[\hat{F}_N^*(a_{N,m+\varepsilon}) - F^*(a_{N,m+\varepsilon})] - \mathcal{B}_N(F^*(a_{N,m+\varepsilon})) \\ = \mathcal{O}(N^{-1/2} \log N) \text{ a.s.} \end{aligned} \tag{2.14}$$

as $N \to \infty$. Hence, by (2.13) and (2.14),

$$\begin{aligned} &\mathbb{P}_m\{\textstyle\bigcap_{k=K}^{\infty} \bigcup_{N=k}^{\infty} A_{N,\varepsilon}\} \\ &\le \mathbb{P}_m\Big\{\frac{N^{-1/2}\mathcal{B}_N(F^*(a_{N,m+\varepsilon})) + \mathcal{O}_{\text{a.s.}}(N^{-1}\log N)}{1 - F^*(a_{N,m+\varepsilon})} \\ &\qquad\qquad \ge 1 - n^{-\frac{m}{m+\varepsilon}} N^{-(1-\delta)\frac{\varepsilon}{m+\varepsilon}} \text{ i.o. for } N \ge K\Big\} \\ &\le \mathbb{P}_m\left\{\frac{N^{-1/2}\mathcal{B}_N(F^*(a_{N,m+\varepsilon}))}{1 - F^*(a_{N,m+\varepsilon})}\right. \\ &\qquad\qquad \left.\ge \tfrac{1}{2}\left(1 - n^{-\frac{m}{m+\varepsilon}} N^{-(1-\delta)\frac{\varepsilon}{m+\varepsilon}}\right) \text{ i.o. for } N \ge K\right\} \\ &\quad + \mathbb{P}_m\left\{\mathcal{O}_{\text{a.s.}}\left(N^{-1+(1-\delta)\frac{m}{m+\varepsilon}} \log N\right)\right. \\ &\qquad\qquad \left.\ge \tfrac{1}{2}\left(1 - n^{-\frac{m}{m+\varepsilon}} N^{-(1-\delta)\frac{\varepsilon}{m+\varepsilon}}\right) \text{ i.o. for } N \ge K\right\}. \end{aligned} \tag{2.15}$$

If $W_N = \mathcal{O}_{\text{a.s.}}\left(N^{-1+(1-\delta)\frac{m}{m+\varepsilon}} \log N\right)$, then, for any $\eta > 0$, there exists a constant $C(\eta) > 0$ and an integer $K(\eta)$ such that

$$\mathbb{P}\left\{W_N > C(\eta) N^{-1+(1-\delta)\frac{m}{m+\varepsilon}} \log N \ \text{ i.o. for } \ N \geq K\right\} < \eta \ \text{ for } \ K > K(\eta),$$

hence,

$$\mathbb{P}\left\{W_N > \tfrac{1}{2}\left(1 - n^{-\frac{m}{m+\varepsilon}} N^{-(1-\delta)\frac{\varepsilon}{m+\varepsilon}}\right) \ \text{ i.o. for } \ N \geq K\right\} < \eta \ \text{ for } \ K > K(\eta).$$

On the other hand,

$$\begin{aligned}
\mathbb{P}_m & \left\{ \frac{N^{-1/2}\mathcal{B}_N(F^*(a_{N,m+\varepsilon}))}{1 - F^*(a_{N,m+\varepsilon})} \right. \\
& \qquad \left. \geq \tfrac{1}{2}\left(1 - n^{-\frac{m}{m+\varepsilon}} N^{-(1-\delta)\frac{\varepsilon}{m+\varepsilon}}\right) \ \text{ i.o. for } \ N \geq K \right\} \\
& = \mathbb{P}_m \left\{ Z_{N,\varepsilon} \geq \tfrac{1}{2}\left(1 - n^{-\frac{m}{m+\varepsilon}} N^{-(1-\delta)\frac{\varepsilon}{m+\varepsilon}}\right) \ \text{ i.o. for } \ N \geq K \right\} \\
& \leq \sum_{N=K}^{\infty} \mathbb{P}_m \left\{ Z_{N,\varepsilon} \geq \tfrac{1}{2}\left(1 - n^{-\frac{m}{m+\varepsilon}} N^{-(1-\delta)\frac{\varepsilon}{m+\varepsilon}}\right) \right\}
\end{aligned} \tag{2.16}$$

where $Z_{N,\varepsilon}$ is normal, with mean zero and variance $\frac{F^*(a_{N,m+\varepsilon})}{N(1-F^*(a_{N,m+\varepsilon}))}$. Hence, for sufficiently large N,

$$\begin{aligned}
\mathbb{P}_m & \left\{ Z_{N,\varepsilon} \geq \tfrac{1}{2}\left(1 - n^{-\frac{m}{m+\varepsilon}} N^{-(1-\delta)\frac{\varepsilon}{m+\varepsilon}}\right) \right\} \\
& \leq 1 - \Phi\left(\frac{N^{\frac{1}{2}\left(1-(1-\delta)\frac{m}{m+\varepsilon}\right)} n^{-\frac{m}{2(m+\varepsilon)}} \left(1 - n^{-\frac{m}{m+\varepsilon}} N^{-(1-\delta)\frac{\varepsilon}{m+\varepsilon}}\right)}{2F^*(a_{N,m+\varepsilon})} \right) \\
& \leq 1 - \Phi\left(\frac{N^{\frac{1}{2}\left(1-(1-\delta)\frac{m}{m+\varepsilon}\right)}}{4} \right) \\
& \leq \frac{4}{\sqrt{2\pi} N^{\frac{1}{2}\left(1-(1-\delta)\frac{m}{m+\varepsilon}\right)}} \exp\left\{ -\frac{1}{32} N^{1-(1-\delta)\frac{m}{m+\varepsilon}} \right\}
\end{aligned} \tag{2.17}$$

where we have used the inequality $1 - \Phi(x) \leq \frac{1}{x\sqrt{2\pi}} \mathrm{e}^{-\frac{x^2}{2}}$, $x > 0$. It follows from (2.17) that, for any $\eta > 0$, there exists an integer $K(\eta)$ such that

$$\sum_{N=K(\eta)}^{\infty} \mathbb{P}_m \left\{ Z_{N,\varepsilon} \geq \tfrac{1}{2}\left(1 - n^{-\frac{m}{m+\varepsilon}} N^{-(1-\delta)\frac{\varepsilon}{m+\varepsilon}}\right) \right\} \leq \eta. \tag{2.18}$$

It then follows from (2.13)–(2.18) that, for any $\eta > 0$, there exists an integer $K(\eta)$ such that

$$\mathbb{P}_m \left\{ \bigcap_{k=K}^{\infty} \bigcup_{N=k}^{\infty} A_{N,\varepsilon} \right\} < \eta \ \text{ for } \ K > K(\eta). \tag{2.19}$$

Similarly, denoting

$$A_{N,-\varepsilon} = \left[1 - \hat{F}_N^*(a_{N,m-\varepsilon}) \geq N^{-1+\delta}\right]$$

we shall prove that for any $\eta > 0$, there exists an integer $K(\eta)$ such that

$$\mathbb{P}_m \left\{ \bigcap_{k=K}^{\infty} \bigcup_{N=k}^{\infty} A_{N,-\varepsilon} \right\} < \eta \quad \text{for} \quad K > K(\eta) \tag{2.20}$$

and (2.11), (2.19) and (2.20) imply that $M_N \to m$ a.s. as $N \to \infty$.

3. Estimator of m Based on Subsample Means

Similarly as in Section 2, let $\mathbf{X}_j = (X_{j1}, \ldots, X_{jn})'$, $j = 1, \ldots, N$ be N independent samples from the population with distribution function F satisfying (2.1). Let $\bar{X}_n^{(1)}, \ldots, \bar{X}_n^{(N)}$ be the respective sample means. Then $\bar{X}_n^{(1)}, \ldots, \bar{X}_n^{(N)}$ is a random sample from a distribution $F^0(x)$. Denote $\hat{F}_N^0$ the empirical distribution function of the sample means, *i.e.*,

$$\hat{F}_N^0(a) = \frac{1}{N} \sum_{j=1}^{N} I[\bar{X}_n^{(j)} \leq a]. \tag{3.1}$$

Let

$$a_N = N^{(1-\delta)/m_0}, \ 0 < \delta < \tfrac{1}{2}. \tag{3.2}$$

The authors in Fialová et al. (2004) proposed a test rejecting $\mathbf{H}_{m_0} : m \leq m_0$ if

either $\hat{F}_{\bar{X}_n}^{(N)}(a_n) = 1$

or if $\hat{F}_{\bar{X}_n}^{(N)}(a_n) < 1$ and simultaneously

$$N^{\delta/2} \left[-\log \left(1 - \hat{F}_{\bar{X}_n}^{(N)}(a_n)\right) - (1-\delta) \log N \right] \geq \Phi^{-1}(1-\alpha) \tag{3.3}$$

where Φ is the standard normal distribution function. It was proved in Fialová et al. (2004) that

$$\lim_{N\to\infty} \mathbb{P}_{m_0} \left(0 < \hat{F}_N^0(a_{N,m_0}) < 1\right) = 1 \tag{3.4}$$

for every F satisfying $1 - F(x) \geq x^{-m_0}$ for $\forall x > x_0$, and if F has exactly the Pareto tail corresponding to $1 - F(x) = x^{-m_0}$ for $\forall x > x_0$, then

$$\lim_{N\to\infty} \mathbb{P}_{m_0} \left\{ N^{\delta/2} \Big[-\log(1 - \hat{F}_N^0(a_{N,m_0})) - (1-\delta) \log N \Big] \leq x \right\} = \Phi(x), \ x \in \mathbb{R}.$$

Analogously as in (2.8), we propose an estimator $\overline{M}$ of m of the following form:

$$\overline{M}_N = \tfrac{1}{2}(\overline{M}_N^+ + \overline{M}_N^-), \tag{3.5}$$

where

$$\begin{aligned} \overline{M}_N^- &= \sup\{s : \ 1 - \hat{F}_N^0(a_{N,s})) < N^{-(1-\delta)}\}, \\ \overline{M}_N^+ &= \inf\{s : \ 1 - \hat{F}_N^0(a_{N,s})) > N^{-(1-\delta)}\}. \end{aligned} \tag{3.6}$$

$\overline{M}_N$ defined in (2.8) and (2.9) is a strongly consistent estimator of m (> 0). The proof is quite analogous as the proof of Theorem 2.1; it uses the fact that, provided F satisfies (2.1), the distribution function of the sample means satisfies $1 - F^0(x) = x^{-m}L^*(x)$, $x \in \mathbb{R}^+$, with L^* being a slowly varying function (see Fialová et al., 2004).

4. Estimator of m Based on Other Subsample Characteristics

Observe N independent samples $\mathbf{X}_j = (X_{j1}, \ldots, X_{jn})'$ of the fixed size n, $j = 1, \ldots, N$. Fix ν, $1 \leq \nu \leq n-1$ and denote

$$\hat{\theta}_j^{(N)} = \tfrac{1}{4}\left(X_j^{(1)} + X_j^{(2)}\right), \tag{4.1}$$

where

$$X_j^{(1)} = \max\{X_{j1}, \ldots, X_{j\nu}\}, \quad X_j^{(2)} = \max\{X_{j,\nu+1}, \ldots, X_{jn}\}, \quad j = 1, \ldots, N.$$

Then $(\hat{\theta}_1, \ldots, \hat{\theta}_N)$ is a random sample from distribution F^{**} (say); denote $\hat{F}_N^{**}$ the corresponding empirical distribution function. Put

$$a_{N,m} = N^{(1-\delta)/m}, \ 0 < \delta < \tfrac{1}{2}, \ m > 0. \tag{4.2}$$

The authors proposed in Picek and Jurečková (2001) the test of $\mathbf{H}_0 : m \leq m_0$ with the critical region

$$\begin{aligned} \Big\{(\mathbf{X}_1, \ldots, \mathbf{X}_N) :\ & \text{either } 1 - \hat{F}_N^{**}(a_{N,m_0}) = 0 \\ & \text{or } 1 - \hat{F}_N^{**}(a_{N,m_0}) > 0 \text{ and} \\ & N^{\delta/2}\Big[-\log(1 - \hat{F}_N^{**}(a_{N,m_0})) - (1-\delta)\log N\Big] \geq \Phi^{-1}(1-\alpha)\Big\}. \end{aligned} \tag{4.3}$$

and, if F has the exact Pareto tail with index m_0,

$$\lim_{N\to\infty} P_{m_0}\Big\{N^{\delta/2}[-\log(1 - \hat{F}_N^{**}(a_{N,m_0})) - (1-\delta)\log N] \leq x\Big\} = \Phi(x), \ x \in \mathbb{R}.$$

We propose an estimator $\widetilde{M}_N$ of m in the following way:

$$\widetilde{M}_N = \tfrac{1}{2}(\widetilde{M}_N^+ + \widetilde{M}_N^-), \tag{4.4}$$

where

$$\begin{aligned} \widetilde{M}_N^- &= \sup\{s :\ 1 - \hat{F}_N^{**}(a_{N,s})) < N^{-(1-\delta)}\}, \\ \widetilde{M}_N^+ &= \inf\{s :\ 1 - \hat{F}_N^{**}(a_{N,s})) > N^{-(1-\delta)}\}. \end{aligned} \tag{4.5}$$

Then $\widetilde{M}_N$ is a strongly consistent estimator of m (> 0); the proof goes along the lines of that of Theorem 2.1 of Section 2.

5. Numerical Illustration

The performance of the proposed estimation procedures is illustrated on the simulated random samples of sizes 1000 and 200, split into $N = 200$ subsamples of sizes $n = 5$ and $N = 50$ subsamples of sizes $n = 4$ respectively; the samples were generated 1000 times from the following distributions:

Pareto $\quad F(x) = 1 - \left(\frac{1}{1+x}\right)^m, \quad x \geq 0$

Burr $\quad F(x) = 1 - \left(\frac{1}{1+x^m}\right)^\alpha, \quad x \geq 0$

Generalized Pareto

$$F(x) = \begin{cases} 1 - (1 + \frac{x}{m\beta})^{-m} & \text{if} \quad x \geq 0, \ 0 < m < \infty, \ \beta > 0, \\ 1 - (1 + \frac{x}{m\beta})^{-m} & \text{if} \quad 0 \leq x \leq -m\beta, \ m < 0, \ \beta > 0, \\ 1 - e^{-x/\beta} & \text{if} \quad m = \infty, \ \beta > 0, \\ 0 & \text{otherwise.} \end{cases}$$

For either of these distributions we proceeded as follows:

(1) Generated independent observations $\mathbf{X}_j = (X_{j1}, \ldots, X_{jn})'$, $j = 1, \ldots, N$;
(2) computed
   A) maxima $X_{(n)}^{(j)} = \max(X_{j1}, \ldots, X_{jn})$, $j = 1, \ldots, N$;
   B) sample means $\overline{X}_n^{(1)}, \ldots, \overline{X}_n^{(N)}$;
   C) means of two maxima

$$\hat{\theta}_j^{(N)} = \tfrac{1}{4}\left(\max(X_{j1}, \ldots, X_{j[n/2]}) + \max(X_{j,[n/2]+1}, \ldots, X_{jn})\right),$$

   $j = 1, \ldots, N$;
(3) found the corresponding empirical distribution function for each of the quantities A–C;
(4) computed the estimators M_N, $\overline{M}_N$, and $\widetilde{M}_N$ corresponding to A–C, respectively, for $\delta = 0.01, 0.02, \ldots, 0.49$.
(5) For a comparison, the Hill estimator

$$H(k) = \tfrac{1}{k}\sum_{i=1}^{k} \log X_{(N \cdot n - i + 1)} - \log X_{(N \cdot n - k)}$$

   was computed for $k = 1, \ldots, 999$.
(6) The steps (1)–(5) were repeated 1 000 times;
(7) selected sample statistics of pertaining estimates were computed and tabulated.

The tables show that the nonparametric estimators based on the tests can be considered as comparable with the Hill estimator, most popular among various estimators of the tail index. While the Hill estimator is better under exact Pareto tails, as it can be expected, the nonparametric estimators show a better performance when the algebraic tail is contaminated by a slowly varying function, mainly M_N and $\widetilde{M}_N$; a more extensive simulation study, as well as a study of the asymptotic bias and variance, is in a progress.

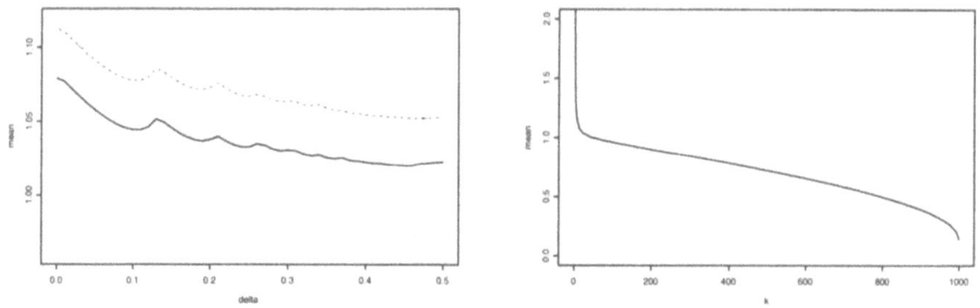

FIGURE 1. Mean values of estimators M_N, (solid) $\overline{M}_N$, (dotted) and $\widetilde{M}_N$ (dashed) plotted against δ (*left*) and of the Hill estimator plotted against k (*right*); the true distribution is Burr with $m = 1$, $\alpha = 1$; $N = 200, n = 5$.

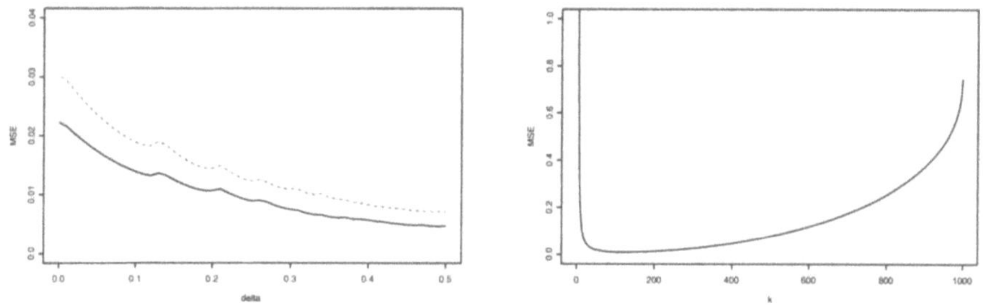

FIGURE 2. MSE values of estimators M_N, (solid) $\overline{M}_N$, (dotted) and $\widetilde{M}_N$ (dashed) plotted against δ (*left*) and of the Hill estimator plotted against k (*right*), the true distribution is Burr with $m = 1$, $\alpha = 1$; $N = 200, n = 5$.

The question is the choice of k for the Hill estimator and of δ for our procedures. To compare our estimators with the Hill estimator, we followed the standard approach of minimizing the mean squared error (MSE); the following tables (Table 1 and Table 2) give the selected sample statistics of estimators of m for various distribution shapes of errors. Figures 1–2 show the behavior of the respective means and the MSE of the estimators plotted against the choice of δ (or of k, respectively).

TABLE 1. Sample statistics of the estimates of the Pareto index under minimal MSE for various distributions; $N = 200$ and $n = 5$. The minimal MSE is denoted by $*$.

sample	method	fraction	MSE	Mean	Median	Var	MAD
Pareto	Hill	$k = 999$	*0.0003	0.5002	0.4992	0.0003	0.0160
$m = 0.5$	A	$\delta = 0.49$	0.0011	0.5093	0.5077	0.0010	0.0317
	B	$\delta = 0.01$	0.0047	0.4719	0.4719	0.0039	0.0624
	C	$\delta = 0.48$	0.0024	0.5340	0.5339	0.0012	0.0353
Pareto	Hill	$k = 999$	*0.0010	1.0003	0.9984	0.0010	0.0321
$m = 1$	A	$\delta = 0.49$	0.0043	1.0188	1.0157	0.0040	0.0634
	B	$\delta = 0.44$	0.0076	0.9642	0.9641	0.0063	0.0778
	C	$\delta = 0.49$	0.0058	1.0435	1.0421	0.0039	0.0643
Pareto	Hill	$k = 999$	*0.0091	3.0010	2.9951	0.0091	0.0962
$m = 3$	A	$\delta = 0.49$	0.0392	3.0570	3.0476	0.0360	0.1904
	B	$\delta = 0.49$	0.7612	3.8392	3.8346	0.0569	0.2347
	C	$\delta = 0.12$	0.0566	2.8928	2.8957	0.0452	0.2127
Burr	Hill	$k = 113$	0.0025	0.4758	0.4744	0.0019	0.0442
$\alpha = 1$	A	$\delta = 0.49$	*0.0012	0.5111	0.5095	0.0010	0.0324
$m = 0.5$	B	$\delta = 0.01$	0.0047	0.4721	0.4720	0.0039	0.0626
	C	$\delta = 0.43$	0.0026	0.5348	0.5349	0.0014	0.0392
Burr	Hill	$k = 113$	0.0098	0.9517	0.9489	0.0075	0.0885
$\alpha = 1$	A	$\delta = 0.49$	*0.0047	1.0226	1.0192	0.0042	0.0647
$m = 1$	B	$\delta = 0.45$	0.0077	0.9770	0.9733	0.0072	0.0827
	C	$\delta = 0.49$	0.0071	1.0528	1.0509	0.0043	0.0678
Burr	Hill	$k = 113$	0.0884	2.8551	2.8466	0.0675	0.2655
$\alpha = 1$	A	$\delta = 0.49$	*0.0423	3.0682	3.0580	0.0377	0.1942
$m = 3$	B	$\delta = 0.49$	2.6325	4.5726	4.5460	0.1597	0.3832
	C	$\delta = 0.14$	0.0555	2.9051	2.9040	0.0465	0.2177
General.	Hill	$k = 311$	*0.0010	0.4847	0.4841	0.0007	0.0261
Pareto	A	$\delta = 0.45$	0.0042	0.5519	0.5491	0.0015	0.0387
$m = 0.5$	B	$\delta = 0.26$	0.0036	0.4661	0.4639	0.0025	0.0496
β=1	C	$\delta = 0.38$	0.0080	0.5768	0.5751	0.0021	0.0455
General.	Hill	$k = 113$	0.0098	0.9517	0.9489	0.0075	0.0885
Pareto	A	$\delta = 0.49$	*0.0047	1.0226	1.0192	0.0042	0.0647
$m = 1$	B	$\delta = 0.45$	0.0077	0.9770	0.9733	0.0072	0.0827
$\beta = 1$	C	$\delta = 0.49$	0.0071	1.0528	1.0509	0.0043	0.0678
General.	Hill	$k = 24$	0.5527	2.4329	2.3598	0.2314	0.4397
Pareto	A	$\delta = 0.01$	0.6412	2.2253	2.2293	0.0411	0.2023
$m = 3$	B	$\delta = 0.11$	*0.1166	2.8500	2.8384	0.0942	0.3179
$\beta = 1$	C	$\delta = 0.01$	0.6844	2.1925	2.1974	0.0324	0.1767

TABLE 2. Sample statistics of the estimates of the Pareto index under minimal MSE for various distributions; $N = 50$ and $n = 4$. The minimal MSE is denoted by $*$.

sample	method	fraction	MSE	Mean	Median	Var	MAD
Pareto	Hill	$k = 199$	*0.0013	0.5023	0.4998	0.0013	0.0359
$m = 0.5$	A	$\delta = 0.49$	0.0036	0.5210	0.5204	0.0032	0.0589
	B	$\delta = 0.01$	0.0084	0.4772	0.4746	0.0078	0.0855
	C	$\delta = 0.49$	0.0031	0.5151	0.5148	0.0029	0.0565
Pareto	Hill	$k = 199$	*0.0050	1.0047	0.9995	0.0050	0.0718
$m = 1$	A	$\delta = 0.49$	0.0147	1.0435	1.0427	0.0128	0.1173
	B	$\delta = 0.44$	0.0204	0.9356	0.9337	0.0163	0.1307
	C	$\delta = 0.49$	0.0086	0.9851	0.9839	0.0084	0.0977
Pareto	Hill	$k = 199$	*0.0453	3.0140	2.9985	0.0452	0.2153
$m = 3$	A	$\delta = 0.49$	0.1328	3.1337	3.1302	0.1150	0.3477
	B	$\delta = 0.49$	0.3045	3.4054	3.3917	0.1403	0.3948
	C	$\delta = 0.01$	0.1430	2.7561	2.7743	0.0837	0.2857
Burr	Hill	$k = 42$	0.0069	0.4551	0.4480	0.0049	0.0682
$\alpha = 1$	A	$\delta = 0.49$	0.0045	0.5282	0.5268	0.0037	0.0627
$m = 0.5$	B	$\delta = 0.19$	0.0086	0.4460	0.4428	0.0057	0.0754
	C	$\delta = 0.49$	*0.0039	0.5230	0.5219	0.0034	0.0613
Burr	Hill	$k = 42$	0.0275	0.9102	0.8960	0.0194	0.1364
$\alpha = 1$	A	$\delta = 0.49$	0.0181	1.0581	1.0557	0.0147	0.1252
$m = 1$	B	$\delta = 0.49$	0.0233	0.9667	0.9569	0.0222	0.1566
	C	$\delta = 0.49$	*0.0105	1.0075	1.0061	0.0105	0.1089
Burr	Hill	$k = 42$	0.2471	2.7306	2.6881	0.1747	0.4093
$\alpha = 1$	A	$\delta = 0.49$	0.1641	3.1779	3.1693	0.1326	0.3745
$m = 3$	B	$\delta = 0.49$	2.6905	4.4337	4.3423	0.6357	0.7580
	C	$\delta = 0.01$	*0.1363	2.7963	2.8095	0.0949	0.3004
General.	Hill	$k = 82$	*0.0031	0.4744	0.4714	0.0025	0.0455
Pareto	A	$\delta = 0.40$	0.0116	0.5775	0.5756	0.0056	0.0747
$m = 0.5$	B	$\delta = 0.29$	0.0068	0.4781	0.4714	0.0063	0.0788
β=1	C	$\delta = 0.45$	0.0105	0.5752	0.5743	0.0048	0.0733
General.	Hill	$k = 42$	0.0275	0.9102	0.8960	0.0194	0.1364
Pareto	A	$\delta = 0.49$	0.0181	1.0581	1.0557	0.0147	0.1252
$m = 1$	B	$\delta = 0.49$	0.0233	0.9667	0.9569	0.0222	0.1566
$\beta = 1$	C	$\delta = 0.49$	*0.0105	1.0075	1.0061	0.0105	0.1089
General.	Hill	$k = 12$	0.9708	2.2516	2.1660	0.4111	0.5783
Pareto	A	$\delta = 0.01$	0.8047	2.1483	2.1436	0.0793	0.2732
$m = 3$	B	$\delta = 0.01$	*0.2903	2.7811	2.7509	0.2426	0.4648
$\beta = 1$	C	$\delta = 0.01$	1.0657	1.9898	1.9940	0.0452	0.2103

References

[1] A. Fialová, J. Jurečková, and J. Picek, *Estimating Pareto Tail Index Based on Sample Means.* Revstat (2004) (in print), available at *http://adela.karlin.mff.cuni.cz/~jurecko/janaus.htm.*

[2] J.L. Hodges and E.L. Lehmann, *Estimation of Location Based on Rank Tests.* Ann. Math. Statist. **34** (1963), 598–611.

[3] J. Jurečková, *Tests of Tails Based on Extreme Regression Quantiles.* Statist. Probab. Lett. **49** (2000), 53–61.

[4] J. Jurečková and J. Picek, *A Class of Tests on the Tail Index.* Extremes **4** (2001), 165–183.

[5] J. Komlós, P. Major, and G. Tusnády, *An Approximation of Partial Sums of Independent R.V.'s and the Sample DF. I.* Z. Wahrscheinlichkeitsth. Verw. Geb. **32** (1975), 111–131.

[6] J. Picek and J. Jurečková, *A Class of Tests on the Tail Index Using the Modified Extreme Regression Quantiles.* In J. Antoch and G. Dohnal, editors, ROBUST'2000, pages 217–226. Union of Czech Mathematicians and Physicists, Prague, 2001.

Acknowledgments

The authors thank two referees for their valuable comments and suggestions, which led to a better readability of the text and to further considerations.

J. Jurečková
Department of Probability and Statistics
Charles University
Sokolovská 83
CZ-186 75 Prague 8
Czech Republic
e-mail: `jurecko@karlin.mff.cuni.cz`

J. Picek
Department of Applied Mathematics
Technical University
Hálkova 6
CZ-461 17 Liberec
Czech Republic
e-mail: `jan.picek@vslib.cz`

Statistics for Industry and Technology, 153–164

On Mardia's Tests of Multinormality

A. Kankainen, S. Taskinen and H. Oja

Abstract. Classical multivariate analysis is based on the assumption that the data come from a multivariate normal distribution. The tests of multinormality have therefore received very much attention. Several tests for assessing multinormality, among them Mardia's popular multivariate skewness and kurtosis statistics, are based on standardized third and fourth moments. In Mardia's construction of the affine invariant test statistics, the data vectors are first standardized using the sample mean vector and the sample covariance matrix. In this paper we investigate whether, in the test construction, it is advantageous to replace the regular sample mean vector and sample covariance matrix by their affine equivariant robust competitors. Limiting distributions of the standardized third and fourth moments and the resulting test statistics are derived under the null hypothesis and are shown to be strongly dependent on the choice of the location vector and scatter matrix estimate. Finally, the effects of the modification on the limiting and finite-sample efficiencies are illustrated by simple examples in the case of testing for the bivariate normality. In the cases studied, the modifications seem to increase the power of the tests.

Mathematics Subject Classification (2000). Primary 62H15; Secondary 62F03.

Keywords. Kurtosis, multinormality, robustness, skewness.

1. Introduction

Classical methods of multivariate analysis are for the most part based on the assumption that the data are coming from a multivariate normal population. In the one-sample multinormal case, the regular sample mean vector and sample covariance matrix are complete and sufficient, but on the other hand highly sensitive to outlying observations. It is also well known that the tests and estimates based on the sample mean vector and sample covariance matrix have poor efficiency properties in the case of heavy tailed noise distributions. Testing for departures from multinormality helps to make practical choices between competing methods of analysis.

The research was partially supported by the Academy of Finland.

In the univariate case, standardized third and fourth moments b_1 and b_2 are often used to indicate the skewness and kurtosis. For a random sample $\boldsymbol{x}_1, \ldots, \boldsymbol{x}_n$ from a p-variate distribution with sample mean vector $\bar{\boldsymbol{x}}$ and sample covariance matrix S, Mardia (1970, 1974, 1980) defined the p-variate skewness and kurtosis statistics as

$$b_{1,p} = \text{ave}_{i,j}[(\boldsymbol{x}_i - \bar{\boldsymbol{x}})^T S^{-1}(\boldsymbol{x}_j - \bar{\boldsymbol{x}})]^3$$

and

$$b_{2,p} = \text{ave}_i[(\boldsymbol{x}_i - \bar{\boldsymbol{x}})^T S^{-1}(\boldsymbol{x}_i - \bar{\boldsymbol{x}})]^2,$$

respectively. The statistics $b_{1,p}$ and $b_{2,p}$ are functions of standardized third and fourth moments, respectively. From the definition one easily sees that $b_{1,p}$ and $b_{2,p}$ are invariant under affine transformations. In the univariate case these reduce to the usual univariate skewness and kurtosis statistics b_1 and b_2. Mardia advocated using the skewness and kurtosis statistics to test for multinormality as they are distribution-free under multinormality. See also Bera and John (1983) and Koziol (1993) for the use of the standardized third and fourth moments in the test construction.

Intuitively, as in the univariate case, the skewness and kurtosis statistics $b_{1,p}$ and $b_{2,p}$ compare the variation measured by third and fourth moments to that measured by second moments. The second, third and fourth moments are all highly non-robust statistics, however, and more efficient procedures may be obtained through comparisons of robust and nonrobust measures of variations. It is therefore an appealing idea to study whether it is useful or reasonable to replace the regular sample mean vector $\bar{\boldsymbol{x}}$ and sample covariance matrix S by some affine equivariant robust competitors, location vector estimate $\hat{\boldsymbol{\mu}}$ and scatter matrix estimate $\hat{\Sigma}$.

The paper is organized as follows. In Section 2 the limiting distributions of Mardia's test statistics under the null model of multinormality are discussed. The test statistics are based on the standardized third and fourth moments; the regular mean vector and regular covariance matrix are used in the standardization. Section 3 lists the tools and assumptions for alternative $\sqrt{n}$-consistent affine equivariant location and scatter estimates. The location and scatter estimates are then used in the regular manner to standardize the data vectors. The limiting distributions of the standardized third and fourth moments in this general case are found in Section 4. As an illustration of the theory, the effect of the way of standardization on the limiting and small sample efficiencies in the bivariate case are studied in Sections 5 and 6. The auxiliary lemmas and proofs have been postponed to the Appendix.

2. Limiting Null Distributions of Mardia's Statistics

In this section we recall the well-known results concerning the limiting null distributions of the Mardia's skewness and kurtosis statistics. Due to affine invariance, the distributions of the test statistics do not depend on the unknown $\boldsymbol{\mu}$ and

Σ. Therefore it is not a restriction to assume in the following derivations that $\boldsymbol{x}_1, \ldots, \boldsymbol{x}_n$ is a random sample from the $N_p(\mathbf{0}, I_p)$ distribution. Write

$$\boldsymbol{z}_i = S^{-1/2}(\boldsymbol{x}_i - \bar{\boldsymbol{x}}), \quad i = 1, \ldots, n,$$

for the data vectors standardized in the classical way. Mardia's skewness statistic can be decomposed as

$$b_{1,p} = 6 \sum_{j<k<l} [\text{ave}_i\{z_{ij} z_{ik} z_{il}\}]^2 + 3 \sum_{j \neq k} [\text{ave}_i\{z_{ij}^2 z_{ik}\}]^2 + \sum_j [\text{ave}_i\{z_{ij}^3\}]^2$$

and the Mardia's kurtosis statistic is similarly

$$b_{2,p} = \sum_{j \neq k} \text{ave}_i\{z_{ij}^2 z_{ik}^2\} + \sum_j \text{ave}_i\{z_{ij}^4\}.$$

See, e.g., Koziol (1993). Under multinormality $b_{1,p}$ and $b_{2,p}$ are affine invariant, all the standardized third and fourth moments are asymptotically normal and asymptotically independent and consequently the limiting distributions of

$$n \, \frac{b_{1,p}}{6} \quad \text{and} \quad \sqrt{n} \, \frac{b_{2,p} - p(p+2)}{\sqrt{8p(p+2)}}$$

are a χ^2-distribution with $p(p+1)(p+2)/6$ degrees of freedom and a $N(0,1)$ distribution, respectively.

3. Location and Scatter Estimates

In Mardia's test construction, the data vectors are standardized using the regular sample mean vector $\bar{\boldsymbol{x}}$ and sample covariance matrix S. In this paper we thus consider what happens if these are replaced by other $\sqrt{n}$-consistent affine equivariant location and scatter estimates of $\boldsymbol{\mu}$ and Σ, say, $\hat{\boldsymbol{\mu}}$ and $\hat{\Sigma}$. Next we list assumptions and useful notations for location vector estimate $\hat{\boldsymbol{\mu}}$ and scatter matrix estimate $\hat{\Sigma}$.

First we assume that, for the $N_p(\mathbf{0}, I_p)$ distribution, the influence functions of these location and scatter functionals at $\boldsymbol{z}$ are

$$\gamma(r) \, \boldsymbol{u} \quad \text{and} \quad \alpha(r) \, \boldsymbol{u}\boldsymbol{u}^T - \beta(r) \, I_p,$$

respectively, where $r = ||\boldsymbol{z}||$ and $\boldsymbol{u} = ||\boldsymbol{z}||^{-1}\boldsymbol{z}$. We will later see that the limiting distributions of the modified test statistics depend on the used location and scatter estimates through these functions. For the influence functions of location and scatter functionals, see also Hampel et al. (1986), Croux and Haesbroek (2000) and Ollila et al. (2003a, 2003b).

We further assume that, again under the $N_p(\mathbf{0}, I_p)$ distribution,

$$\boldsymbol{t} = \sqrt{n} \; \text{ave}\{\gamma(r_i)\boldsymbol{u}_i\} = \sqrt{n} \; \hat{\boldsymbol{\mu}} + o_P(1)$$

and

$$\boldsymbol{C} = \sqrt{n} \; \text{ave}\{\alpha(r_i)\boldsymbol{u}_i\boldsymbol{u}_i^T - \beta(r_i)I_p\} = \sqrt{n} \; (\hat{\Sigma} - I_p) + o_P(1).$$

Finally, assume that, for the $N_p(\mathbf{0}, I_p)$ distribution, the expected values $E(\gamma^2(r))$, $E(\alpha^2(r))$ and $E(\beta^2(r))$ are finite. These assumptions imply that the limiting distributions of $\sqrt{n}\hat{\boldsymbol{\mu}}$ and $\sqrt{n}\ vec(\hat{\Sigma} - I_p)$ are p- and p^2-variate normal distributions with zero mean vectors and covariance matrices

$$\frac{E[\gamma^2(r)]}{p} I_p$$

and

$$\frac{E[\alpha^2(r)]}{p(p+2)}[I_{p^2} + I_{p,p} + \mathrm{vec}(I_p)\mathrm{vec}(I_p)^T]$$
$$+\left(E[\beta^2(r)] + \frac{2}{p}E[\alpha(r)\beta(r)]\right)\mathrm{vec}(I_p)\mathrm{vec}(I_p)^T,$$

respectively. (Here $I_{p,p}$ is the so called commutation matrix. See, e.g., Ollila et al., 2003b.)

4. Limiting Distribution of the Standardized Third and Fourth Moments

Consider now any estimators $\hat{\boldsymbol{\mu}}$ and $\hat{\Sigma}$ and write

$$\boldsymbol{z}_i = \hat{\Sigma}^{-1/2}(\boldsymbol{x}_i - \hat{\boldsymbol{\mu}}), \quad i = 1, \ldots, n,$$

for the standardized observations. Denote the standardized third and fourth moments by

$$U_{jkl} = \sqrt{n}\ \mathrm{ave}_i\{z_{ij}z_{ik}z_{il}\},$$
$$V_{jk} = \sqrt{n}\ \mathrm{ave}_i\{z_{ij}^2 z_{ik}^2 - 1\}$$

and

$$V_{jj} = \sqrt{n}\ \mathrm{ave}_i\{z_{ij}^4 - 3\}.$$

The limiting distributions of the standardized third and fourth moments are given by the following theorem. The results easily follow from Lemmas 7.1 and 7.2 stated and proven in the Appendix.

Theorem 4.1. *Under multinormality, the limiting distribution of the vector of all possible third moments U_{jkl} $(j < k < l)$, U_{jjk} $(j \neq k)$ and U_{jjj} is a multivariate normal distribution with zero mean vector and limiting variances and covariances*

$$\begin{aligned} Var(U_{jkl}) &= 1 \\ Var(U_{jjk}) &= 2 + \kappa \\ Var(U_{jjj}) &= 6 + 9\kappa \\ Cov(U_{jjk}, U_{llk}) &= \kappa \\ Cov(U_{jjk}, U_{kkk}) &= 3\kappa \end{aligned}$$

where

$$\kappa = 1 + \frac{E[\gamma^2(r)]}{p} - 2\frac{E[r^3\gamma(r)]}{p(p+2)}.$$

All the other covariances are zero.

Theorem 4.2. *Under multinormality, the limiting distribution of the vector of fourth moments V_{jk} $(j \neq k)$ and V_{jj} is a multivariate normal distribution with zero mean vector and limiting variances and covariances*

$$\begin{aligned}
Var(V_{jk}) &= 4 + \tau_1 + 2\tau_2 \\
Var(V_{jj}) &= 24 + 9\tau_1 + 36\tau_2 \\
Cov(V_{jk}, V_{jl}) &= \tau_1 + \tau_2 \\
Cov(V_{jk}, V_{ml}) &= \tau_1 \\
Cov(V_{jj}, V_{kk}) &= 9\tau_1 \\
Cov(V_{jk}, V_{jj}) &= 3\tau_1 + 6\tau_2 \\
Cov(V_{jk}, V_{ll}) &= 3\tau_1
\end{aligned}$$

where

$$\tau_1 = 4\left[\frac{E[\alpha^2(r)]}{p(p+2)} - 2\frac{E[\alpha(r)\beta(r)]}{p} + E[\beta^2(r)] - \frac{E[\alpha(r)r^4]}{p(p+2)(p+4)} + \frac{E[\beta(r)r^4]}{p(p+2)}\right]$$

and

$$\tau_2 = 2 + 2\frac{E[\alpha^2(r)]}{p(p+2)} - 4\frac{E[\alpha(r)r^4]}{p(p+2)(p+4)}.$$

We close this section with some discussion on the implications of the above results. First note that, if the regular mean vector and sample covariance matrix are used in the standardization, then simply

$$\gamma(r) = r, \quad \alpha(r) = r^2 \text{ and } \beta(r) = 1 \quad \text{which gives} \quad \kappa = \tau_1 = \tau_2 = 0$$

and all the third moments and fourth moments are asymptotically mutually independent, respectively. For other location and scatter statistics this is not necessarily true and also the limiting variances may vary. This means that the limiting distributions and efficiencies of $b_{1,p}$ and $b_{2,p}$ may drastically change. The limiting distribution of $nb_{1,p}/6$ is a weighted sum of chi-square variables with one degree of freedom which makes the calculation of the approximate p-value less practical. The limiting distribution of $\sqrt{n}(b_{2,p} - p(p+2))$ is still normal with mean value zero, but its limiting variance depends on the chosen scatter matrix. In the last two sections we illustrate the impact of the choice of the location vector and scatter matrix estimate on the limiting distribution and efficiency in the bivariate case. The extension to the general p-variate case is straightforward.

5. Tests for Bivariate Normality

Assume now that $\boldsymbol{x}_1, \ldots, \boldsymbol{x}_n$ is a random sample from an unknown bivariate distribution and we wish to test the null hypothesis H_0 of bivariate normality. For limiting power considerations we use the contaminated bivariate normal distribution, the so called Tukey distribution: We say that the distribution of $\boldsymbol{x}$ is a bivariate Tukey distribution $T(\Delta, \boldsymbol{\mu}, \sigma)$ if the pdf of $\boldsymbol{x}$ is

$$f(\boldsymbol{x}) = (1-\Delta)\phi(\boldsymbol{x}) + \Delta\sigma^{-2}\phi((\boldsymbol{x}-\boldsymbol{\mu})/\sigma),$$

where $\phi(\boldsymbol{x})$ is the pdf of $N_2(\mathbf{0}, I_2)$ distribution and $\Delta \in [0,1]$. Alternative sequences $H_1^n : T(\Delta, \boldsymbol{\mu}, 1)$ and $H_2^n : T(\Delta, \mathbf{0}, \sigma)$ with $\Delta = \delta/\sqrt{n}$, $r = ||\boldsymbol{\mu}|| > 0$ and $\sigma > 1$ are then used for the comparisons of the skewness and kurtosis test statistics, respectively.

Write now

$$\boldsymbol{T}_1 = (U_{112}, U_{221}, U_{111}, U_{222})^T \quad \text{and} \quad \boldsymbol{T}_2 = (V_{11}, V_{22}, V_{12})^T$$

for the vectors of the standardized third and fourth moments. Using the results in Section 4, one directly gets the following results.

Corollary 5.1. *Under H_0, the limiting distribution of $\sqrt{n}\boldsymbol{T}_1$ is a 4-variate normal distribution with zero mean vector and covariance matrix*

$$\Omega_1 = \begin{pmatrix} 2 & 0 & 0 & 0 \\ 0 & 2 & 0 & 0 \\ 0 & 0 & 6 & 0 \\ 0 & 0 & 0 & 6 \end{pmatrix} + \kappa \begin{pmatrix} 1 & 0 & 0 & 3 \\ 0 & 1 & 3 & 0 \\ 0 & 3 & 9 & 0 \\ 3 & 0 & 0 & 9 \end{pmatrix}.$$

Corollary 5.2. *Under H_0, the limiting distribution of $\sqrt{n}\boldsymbol{T}_2$ is a 3-variate normal distribution with zero mean vector and covariance matrix*

$$\Omega_2 = \begin{pmatrix} 24 & 0 & 0 \\ 0 & 24 & 0 \\ 0 & 0 & 4 \end{pmatrix} + \tau_1 \begin{pmatrix} 9 & 9 & 3 \\ 9 & 9 & 3 \\ 3 & 3 & 1 \end{pmatrix} + \tau_2 \begin{pmatrix} 36 & 0 & 6 \\ 0 & 36 & 6 \\ 6 & 6 & 2 \end{pmatrix}.$$

It is now possible to construct the limiting distributions of the affine invariant test statistics $b_{1,2}$ and $b_{2,2}$. The comparison of the effect of different standardizations is possible but not easy, however, as the limiting distributions may be of different type. This is avoided if one uses the following test statistic versions:

Definition 5.3. Modified Mardia's skewness and kurtosis statistics in the bivariate case are

$$b_{1,2}^* = \boldsymbol{T}_1{}^T \Omega_1^{-1} \boldsymbol{T}_1 \quad \text{and} \quad b_{2,2}^* = (1,1,2)\boldsymbol{T}_2.$$

Note that, if the sample mean and covariance matrix are used, the definition gives $b_{1,2}^* = b_{1,2}/6$ and $b_{2,2}^* = b_{2,2} - 8$, respectively. In the former case, with general $\hat{\boldsymbol{\mu}}$ and $\hat{\Sigma}$, the affine equivariance property is lost.

The alternative to the regular sample mean vector and sample covariance matrix used in this example is the affine equivariant Oja median (Oja, 1983) and the related scatter matrix estimate based on the Oja sign covariance matrix (SCM). See Visuri et al. (2000) and Ollila et al. (2003b) for the SCM. Note also that the scatter matrix estimate based on the Oja SCM is asymptotically equivalent with the zonoid covariance matrix (ZCM) and can be seen as an affine equivariant extension on mean deviation. For the definition and use of the ZCM, see Koshevoy et al. (2003). We chose these statistics as their influence functions are relatively simple at the $N_2(\mathbf{0}, I_2)$ case, the influence functions are given by

$$\gamma(r) = 2\sqrt{\frac{2}{\pi}}, \quad \alpha(r) = 2\left[2\sqrt{\frac{2}{\pi}}\, r - 1\right] \quad \text{and} \quad \beta(r) = 1.$$

The constants needed for the asymptotic variances and covariances are then

$$\kappa = \frac{4}{\pi} - \frac{1}{2}, \quad \tau_1 = \frac{32}{\pi} - \frac{29}{3} \quad \text{and} \quad \tau_2 = \frac{16}{\pi} - \frac{14}{3}.$$

Consider first the alternative contiguous sequence H_1^n with $\boldsymbol{\mu} = (r, 0, \dots, 0)^T$. Using classical LeCam's third lemma one then easily sees that, under the alternative sequence H_1^n, the limiting distribution of $nb_{1,2}^*$ is a noncentral χ^2-distribution with 4 degrees of freedom and non-centrality parameter $\boldsymbol{\delta}_1^T \Omega_1^{-1} \boldsymbol{\delta}_1$, where $\boldsymbol{\delta}_1$ depends on the location estimate used in the standardization. In the regular case (sample mean vector),

$$\boldsymbol{\delta}_1 = (0,\ 0,\ r^3,\ 0)^T$$

and the Oja median gives

$$\boldsymbol{\delta}_1 = (0,\ r - q(r)2\sqrt{2/\pi},\ r^3 + 3r - q(r)6\sqrt{2/\pi},\ 0)^T.$$

The function $q(r) = ||R(\boldsymbol{\mu})||$ is easily computable and defined through the so-called spatial rank function

$$R(\boldsymbol{\mu}) = E(||\boldsymbol{\mu} - \boldsymbol{x}||^{-1}(\boldsymbol{\mu} - \boldsymbol{x}))$$

with $\boldsymbol{x} \sim N_2(\mathbf{0}, I_2)$. See Möttönen and Oja (1995).

Next we consider the sequence H_2^n. Then

$$Var(b_{2,2}^*) = (1, 1, 2)\ \Omega_2 \begin{pmatrix} 1 \\ 1 \\ 2 \end{pmatrix}$$

and under H_2^n the limiting distribution of $n(b_{2,2}^*)^2/Var(b_{2,2}^*)$ is a noncentral χ^2-distribution with one degree of freedom and non-centrality parameter $[(1,1,2)\boldsymbol{\delta}_2]^2/Var(b_{2,2}^*)$, where $\boldsymbol{\delta}_2$ in this case depends on the scatter estimate used in the standardization. In the regular case (sample covariance matrix) we get

$$\boldsymbol{\delta}_2 = (3(\sigma^2 - 1)^2,\ 3(\sigma^2 - 1)^2,\ (\sigma^2 - 1)^2)^T$$

and the ZCM gives

$$\boldsymbol{\delta}_2 = (3(\sigma^4 - 1) - 12(\sigma - 1),\ 3(\sigma^4 - 1) - 12(\sigma - 1),\ (\sigma^4 - 1) - 4(\sigma - 1))^T$$

In Figure 1 the asymptotical relative efficiencies of the new tests (using the Oja median and the ZCM estimate) relative to the regular tests (using the sample mean and sample covariance matrix) are illustrated for different values of r and for different values of σ. The AREs are then simply the ratios of the non-centrality parameters. The new tests seem to perform very well: The test statistics using robust standardization seem to perform better than the tests with regular location and scatter estimates.

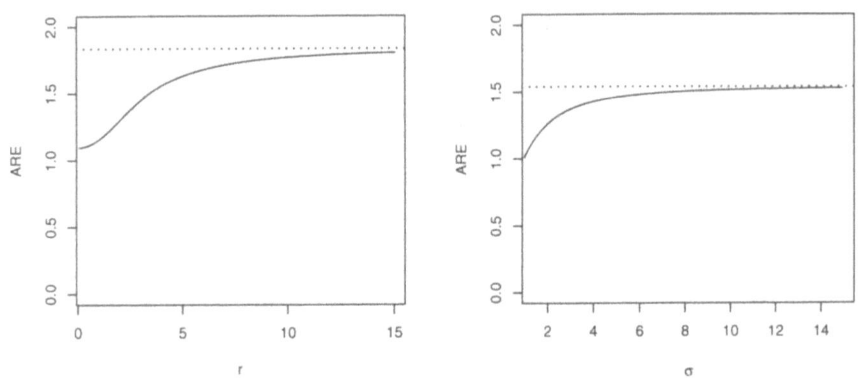

FIGURE 1. Limiting efficiencies of the tests based on the Oja median and the ZCM relative to the regular Mardia's tests for different values of r and σ. The left (right) panel illustrates the efficiency of $b_{1,2}^{*}$ ($b_{2,2}^{*}$).

6. A Real Data Example

In our real data example we consider two bivariate data sets of $n = 50$ observations and the combined data set ($n = 100$). The data sets are shown and explained in Figure 2. We wish to test the null hypothesis of bivariate normality. Again the skewness and kurtosis statistics to be compared are those based on the sample mean vector and sample covariance matrix (regular Mardia's $b_{1,2}$ and $b_{2,2}$) and those based on the Oja median and the ZCM (denoted by $b_{1,2}^{*}$ and $b_{2,2}^{*}$). The values of the test statistics and the approximate p-values were calculated for the three data sets ('Boys', 'Girls', 'Combined'). See Table 1 for the results. As expected, the observed p-values were much smaller in the combined data case.

As discussed in the introduction, the statistics $b_{1,2}$ and $b_{2,2}$ compare the third and fourth central moments to the second ones (given by the sample covariance matrix). The statistics $b_{1,2}^{*}$ and $b_{2,2}^{*}$ use robust location vector and scatter matrix estimates in this comparison. One can then expect that, for data sets with outliers or for data sets coming from heavy tailed distributions, the comparison to robust estimates tends to produce higher values of the test statistics and smaller p-values.

In our examples for the analysis of bivariate data we used the Oja median and the ZCM to standardize the data vectors. In the general p-variate case (with large p and large n) these estimates are, however, computationally highly intensive, and should be replaced by more practical robust location and scatter estimates.

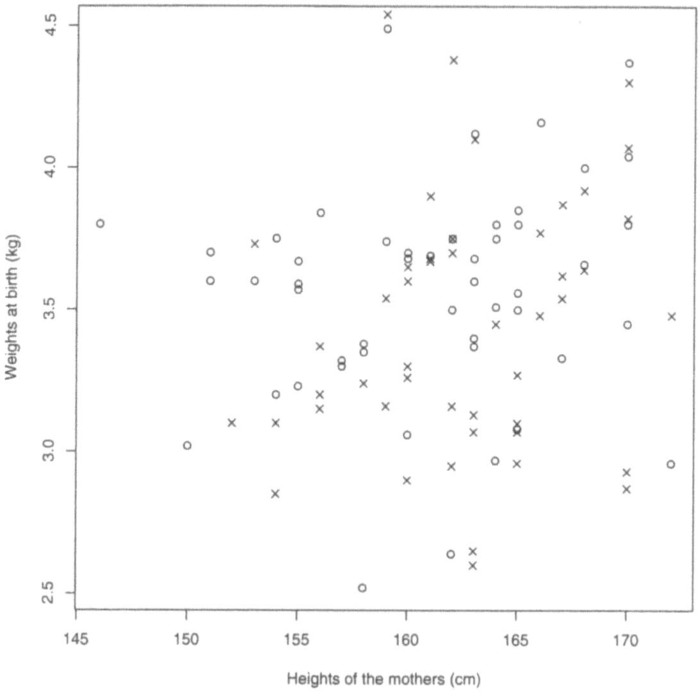

FIGURE 2. The height of the mother and the birth weight of the child measured on 50 boys who have siblings (o) and 50 girls (x) who do not have siblings.

TABLE 1. The observed values of the test statistics with corresponding p-values for the three data sets.

	Boys		Girls		Combined	
	Test statistic	p-value	Test statistic	p-value	Test statistic	p-value
$b_{1,2}$	3.034	0.552	3.376	0.497	9.242	0.055
$b_{1,2}^*$	5.152	0.272	2.336	0.674	11.257	0.024
$b_{2,2}$	0.016	0.900	0.282	0.596	2.275	0.132
$b_{2,2}^*$	0.834	0.361	1.152	0.283	5.166	0.023

7. Appendix: Auxiliary Lemmas

The limiting distributions of the standardized third and fourth moments are given by the following linear approximations.

Lemma 7.1. *Let $\boldsymbol{x}_1, \ldots, \boldsymbol{x}_n$ be a random sample from $N(\mathbf{0}, I_p)$-distribution and let $\boldsymbol{z}_i = \hat{\Sigma}^{-1/2}(\boldsymbol{x}_i - \hat{\boldsymbol{\mu}})$, $i = 1, \ldots, n$, be the standardized observations. Write also $r_i = ||\boldsymbol{x}_i||$ and $\boldsymbol{u}_i = ||\boldsymbol{x}_i||^{-1}\boldsymbol{x}_i$. Then*

$$\begin{aligned}\sqrt{n}\ \text{ave}_i\{z_{ij}z_{ik}z_{il}\} &= \sqrt{n}\ \text{ave}_i\{x_{ij}x_{ik}x_{il}\} + o_P(1)\\ &= \sqrt{n}\ \text{ave}_i\{r_i^3 u_{ij}u_{ik}u_{il}\} + o_P(1) \quad \textit{for} \;\; j < k < l,\end{aligned}$$

$$\begin{aligned}\sqrt{n}\ \text{ave}_i\{z_{ij}^2 z_{ik}\} &= \sqrt{n}\ \text{ave}_i\{x_{ij}^2 x_{ik}\} - t_k + o_P(1)\\ &= \sqrt{n}\ \text{ave}_i\{r_i^3 u_{ij}^2 u_{ik} - \gamma(r_i)u_{ik}\} + o_P(1) \quad \textit{for} \;\; j \neq k\end{aligned}$$

and

$$\begin{aligned}\sqrt{n}\ \text{ave}_i\{z_{ij}^3\} &= \sqrt{n}\ \text{ave}_i\{x_{ij}^3\} - 3t_j + o_P(1)\\ &= \sqrt{n}\ \text{ave}_i\{r_i^3 u_{ij}^3 - 3\gamma(r_i)u_{ij}\} + o_P(1).\end{aligned}$$

Lemma 7.2. *Let $\boldsymbol{x}_1, \ldots, \boldsymbol{x}_n$ be a random sample from $N(\mathbf{0}, I_p)$-distribution and let $\boldsymbol{z}_i = \hat{\Sigma}^{-1/2}(\boldsymbol{x}_i - \hat{\boldsymbol{\mu}})$, $i = 1, \ldots, n$, be the standardized observations. Again, $r_i = ||\boldsymbol{x}_i||$ and $\boldsymbol{u}_i = ||\boldsymbol{x}_i||^{-1}\boldsymbol{x}_i$. Then for $j \neq k$,*

$$\begin{aligned}\sqrt{n}\ \text{ave}_i\{z_{ij}^2 z_{ik}^2 - 1\} &= \sqrt{n}\ \text{ave}_i\{x_{ij}^2 x_{ik}^2 - 1\} - C_{jj} - C_{kk} + o_P(1)\\ &= \sqrt{n}\ \text{ave}_i\{r_i^4 u_{ij}^2 u_{ik}^2 - 1 - \alpha(r_i)[u_{ij}^2 + u_{ik}^2] + 2\beta(r_i)\} + o_P(1)\end{aligned}$$

and

$$\begin{aligned}\sqrt{n}\ \text{ave}_i\{z_{ij}^4 - 3\} &= \sqrt{n}\ \text{ave}_i\{x_{ij}^4 - 3\} - 6C_{jj} + o_P(1)\\ &= \sqrt{n}\ \text{ave}_i\{r_i^4 u_{ij}^4 - 3 - 6\alpha(r_i)u_{ij}^2 + 6\beta(r_i)\} + o_P(1).\end{aligned}$$

Proofs of the lemmas: Note first that

$$\boldsymbol{z}_i = \hat{\Sigma}^{-1/2}(\boldsymbol{x}_i - \hat{\boldsymbol{\mu}}) = \boldsymbol{x}_i - \frac{1}{\sqrt{n}}\boldsymbol{t} - \frac{1}{2\sqrt{n}}C\boldsymbol{x}_i + o(1/\sqrt{n}).$$

Thus

$$z_{ij} = x_{ij} - \frac{1}{\sqrt{n}}t_j - \frac{1}{2\sqrt{n}}\sum_{r=1}^{p} C_{jr}x_{ir} + o(1/\sqrt{n}),$$

$$z_{ij}^2 = x_{ij}^2 - \frac{2}{\sqrt{n}}t_j x_{ij} - \frac{1}{\sqrt{n}}\sum_{r=1}^{p} C_{jr}x_{ir}x_{ij} + o(1/\sqrt{n}),$$

$$z_{ij}^3 = x_{ij}^3 - \frac{3}{\sqrt{n}}t_j x_{ij}^2 - \frac{3}{2\sqrt{n}}\sum_{r=1}^{p} C_{jr}x_{ir}x_{ij}^2 + o(1/\sqrt{n})$$

and

$$z_{ij}^4 = x_{ij}^4 - \frac{4}{\sqrt{n}} t_j x_{ij}^3 - \frac{2}{\sqrt{n}} \sum_{r=1}^{p} C_{jr} x_{ir} x_{ij}^3 + o(1/\sqrt{n}).$$

As the observations come from the $N_p(\mathbf{0}, I_p)$ distribution (all the moments exist), the standard law of large numbers and central limit theorem together with Slutsky's lemma yield the result.

References

[1] A. Bera and S. John, *Tests for Multivariate Normality with Pearson Alternatives.* Commun. Statist. Theor. Meth. **12** (1983), 103–117.

[2] C. Croux, G. Dhaene, and D. Hoorelbeke, *Testing the Information Matrix Equality with Robust Estimators.* Submitted, 2003, available at *http://www.econ.kuleuven.ac.be/public/NDBAE06/public.htm.*

[3] C. Croux and G. Haesbroeck, *Principal Component Analysis Based on Robust Estimators of the Covariance or Correlation Matrix: Influence Functions and Efficiencies.* Biometrika **87** (2000), 603–618.

[4] F.R. Hampel, E.M. Ronchetti, P.J. Rousseeuw, and W.A. Stahel, *Robust Statistics: The Approach Based on Influence Functions.* Wiley, New York, 1986.

[5] G. Koshevoy, J. Möttönen, and H. Oja, *A Scatter Matrix Estimate Based on the Zonotope.* Ann. Statist. **31** (2003), 1439–1459.

[6] J.A. Koziol, *Probability Plots for Assessing Multivariate Normality.* Statistician **42** (1993), 161–173.

[7] K.V. Mardia, *Measures of Multivariate Skewness and Kurtosis with Applications.* Biometrika **57** (1970), 519–530.

[8] K.V. Mardia, *Applications of Some Measures of Multivariate Skewness and Kurtosis in Testing Normality and Robustness Studies.* Sankhyã, Ser B **36** (1974), 115–128.

[9] K.V. Mardia, *Tests of Univariate and Multivariate Normality.* In P.R. Krishnaiah, editor, Handbook of Statistics, vol 1, pages 279–320. North Holland, 1980.

[10] J. Möttönen and H. Oja, *Multivariate Spatial Sign and Rank Methods.* J. Nonparam. Statist. **5** (1995), 201–213.

[11] H. Oja, *Descriptive Statistics for Multivariate Distributions.* Statist. Probab. Lett. **1** (1983), 327–332.

[12] E. Ollila, T.P. Hettmansperger, and H. Oja, *Affine Equivariant Multivariate Sign Methods.* Submitted, 2003a.

[13] E. Ollila, H. Oja, and C. Croux, *The Affine Equivariant Sign Covariance Matrix: Asymptotic Behavior and Efficiencies.* J. Multivariate Anal. **87** (2003b), 328–355.

[14] S. Visuri, V. Koivunen, and H. Oja, *Sign and Rank Covariance Matrices.* J. Statist. Plann. Inference **91** (2000), 557–575.

Acknowledgements

We thank the referees for the critical and constructive comments. Finally, we wish to mention that, after writing the paper, we learned about the paper Croux et al. (2003) with a very similar but more general approach. Croux and his co-writers study the behavior of the information matrix (IM) test when the maximum likelihood estimates are replaced by robust estimates. One of their examples is the modified Jarque-Bera test for univariate skewness and kurtosis. Their conclusion also is that the use of the robust estimates increases the power of the IM test.

A. Kankainen, S. Taskinen and H. Oja
University of Jyväskylä
Department of Mathematics and Statistics
P.O. Box 35 (MaD)
FIN-40014 University of Jyväskylä
Finland
e-mail: {kankaine, slahola, ojahannu}@maths.jyu.fi

Statistics for Industry and Technology, 165–171

Robustness in Sequential Discrimination of Markov Chains under "Contamination"

A. Kharin

Abstract. The problem of robustness is considered for sequential hypotheses testing on the parameters of Markov chains. The exact expressions for the conditional error probabilities, and for the conditional expected sequence lengths are obtained. Robustness analysis under "contamination" is performed. Numerical results are given to illustrate the theory.

Mathematics Subject Classification (2000). Primary 62F35; Secondary 62L10, 62M02.

Keywords. Robustness, sequential test, Markov chain.

1. Introduction

The sequential approach (Wald, 1947) is used in applications, especially in medical trials (Whitehead, 1997), and in statistical quality control, for testing of hypotheses on parameters of models (Siegmund, 1985). One of the applied problems of sequential testing is the problem of discrimination of Markov chains (Durbin et al., 1998; Wu et al., 2001).

Testing of simple hypotheses on parameters of Markov chains is realized by the sequential probability ratio test, which minimizes the expected sequence length. But such a model is usually not adequate to the applied problem. A possibility of some deviations from the model is needed. Sequential testing of composite hypotheses leads to significant increase in the expected sequence length, even for asymptotically optimal tests (Malyutov and Tsitovich, 2000).

An alternative way to work with the deviations from the model of simple hypotheses is to analyze robustness (Huber, 1981; Stockinger and Dutter, 1987) of the test in the case of simple hypotheses under distortions of the model (Kharin, 2002). In the paper we analyze robustness of the sequential hypotheses testing under "contamination" (Huber, 1981).

2. Mathematical Model and Notation

Let a homogeneous Markov chain $x_1, x_2, \ldots,$ be observed, $x_i \in V = \{0, 1, \ldots, M-1\}$, $i \in \mathbb{N}$, with an initial states probabilities vector $\pi = (\pi_i)$, $i \in V$, and with a one-step transition probabilities matrix $P = (p_{ij})$, $i, j \in V$: $\mathrm{P}\{x_1 = i\} = \pi_i$, $\mathrm{P}\{x_n = j \mid x_{n-1} = i\} = p_{ij}$, $i, j \in V, n > 1$. Let us consider two hypotheses on the parameters of the Markov chain: H_0: $\pi = \pi^{(0)}$, $P = P^{(0)}$, with an alternative H_1: $\pi = \pi^{(1)}$, $P = P^{(1)}$, where $\pi^{(0)}$, $\pi^{(1)}$ are given values of the initial states probabilities vector, $P^{(0)}$, $P^{(1)}$, $P^{(0)} \neq P^{(1)}$, are given matrices of transition probabilities for the corresponding hypotheses. Introduce the notation:

$$\lambda_1 = \ln \frac{\pi^{(1)}_{x_1}}{\pi^{(0)}_{x_1}}, \quad \lambda_k = \ln \frac{p^{(1)}_{x_{k-1}x_k}}{p^{(0)}_{x_{k-1}x_k}}, \quad k > 1; \ \Lambda_n = \sum_{k=1}^{n} \lambda_k, \ n \in \mathbb{N}.$$

Let us describe the sequential test for the hypotheses H_0, H_1. Suppose $C_-, C_+ \in \mathbb{Z}$, $C_- < 0$, $C_+ > 0$ be the parameters of the test (thresholds); $N = C_+ - C_- - 1$. The hypothesis H_0 is accepted after n observations, if $\Lambda_n \leq C_-$, the hypothesis H_1 is accepted after n observations, if $\Lambda_n \geq C_+$, else the test should be continued, it means that the $(n+1)$-th observation should be made. The sequence of random vectors $(\Lambda_n, x_n)'$, $n \in \mathbb{N}$, is a Markov chain (Kemeni and Snell, 1959) : by the definition of Λ_n, $n \in \mathbb{N}$, and also because of x_n, $n \in \mathbb{N}$, is a Markov chain,

$$\mathcal{L}\{\Lambda_n, x_n \mid \Lambda_{n-1}, \Lambda_{n-2}, \ldots, \Lambda_1; x_{n-1}, x_{n-2}, \ldots, x_1\} = \mathcal{L}\{\Lambda_n, x_n \mid \Lambda_{n-1}, x_{n-1}\}.$$

Let the parameters $\pi^{(0)}$, $P^{(0)}$, $\pi^{(1)}$, $P^{(1)}$ satisfy the condition:

$$a > 0, \quad m_i, \ m_{ij} \in \mathbb{Z}, \quad i, j \in V, \quad \text{that} \quad \ln \frac{\pi_i^{(1)}}{\pi_i^{(0)}} = m_i a, \quad \ln \frac{p_{ij}^{(1)}}{p_{ij}^{(0)}} = m_{ij} a.$$

In practice, these equations can be approximated with the necessary accuracy by choosing the value of a small enough. Let $t^{(k)}$ be the conditional expected sequence length under the condition that the hypothesis H_k is true, $k \in \{0, 1\}$; let α, β be the conditional error probabilities (error probabilities of the type I and II respectively).

3. Exact Expressions for Conditional Error Probabilities and Expected Sequence Lengths

Let us denote by

$$W^{(k)} = (w_{ij}^{(k)}) = \left(\begin{array}{c|c} \mathbf{I}_2 & \mathbf{0}_{2 \times MN} \\ \hline R^{(k)} & Q^{(k)} \end{array} \right),$$

the matrix of the size $(MN+2)\times(MN+2)$, where the blocks $R^{(k)}$, $Q^{(k)}$ are defined by their components:

$$w^{(k)}_{Mi+s\ Mj+t}=\delta(j-i,m_{st})p^{(k)}_{st},\ i,j\in(C_-,C_+),\ s,t\in V,$$

$$w^{(k)}_{Mi+s\ Mj}=\begin{cases}\sum_{t\in V}\mathbf{1}(i+m_{st}-C_+), & j=C_+, i\in(C_-,C_+), s\in V,\\ \sum_{t\in V}\mathbf{1}(C_- - i-m_{st}), & j=C_-, i\in(C_-,C_+), s\in V,\end{cases}\tag{3.1}$$

where $\delta(\cdot,\cdot)$ means the Kronecker delta function: $\delta(k,l)=1$, if $k=l$, and $\delta(k,l)=0$, otherwise, $k,l\in\mathbb{N}$; $\mathbf{1}(\cdot)$ means the unit step function: $\mathbf{1}(u)=1$, if $u\geq 0$, and $\mathbf{1}(u)=0$, otherwise. Define the vector $\omega^{(k)}=(\omega^{(k)}_i)$, $i\in\{MC_-+1,\dots,MC_+-1\}$ by its components:

$$\omega^{(k)}_{Mi+s}=\delta(i,m_s)\pi^{(k)}_s,\ i\in(C_-,C_+),\ s\in V,\tag{3.2}$$

and define

$$\omega^{(k)}_{MC_-}=\sum_{s\in V}\mathbf{1}(C_- - m_s)\pi^{(k)}_s,\ \omega^{(k)}_{MC_+}=\sum_{s\in V}\mathbf{1}(m_s-C_+)\pi^{(k)}_s.\tag{3.3}$$

Define also the matrices $S^{(k)}=\mathbf{I}_{MN}-Q^{(k)}$, $B^{(k)}=(S^{(k)})^{-1}R^{(k)}$, $k\in\{0,1\}$.

Theorem 3.1. *For the model described above*

$$\alpha=(\omega^{(0)})'B^{(0)}_{(2)}+\omega^{(0)}_{MC_+},\ \beta=(\omega^{(1)})'B^{(1)}_{(1)}+\omega^{(1)}_{MC_-};$$
$$t^{(k)}=(\omega^{(k)})'(S^{(k)})^{-1}\mathbf{1}_{MN}+1.$$

Proof. It is based on the theory of absorbing Markov chains (Kemeni and Snell, 1959). This theory is applied to the sequence

$$\begin{aligned}\xi_n=&\ MC_-\mathbf{1}_{(-\infty,C_-]}\left(\tfrac{\Lambda_n}{a}\right)+MC_+\mathbf{1}_{[C_+,+\infty)}\left(\tfrac{\Lambda_n}{a}\right)\\&+\left(\tfrac{\Lambda_n}{a}M+x_n\right)\mathbf{1}_{(C_-,C_+)}\left(\tfrac{\Lambda_n}{a}\right),\ n\in\mathbb{N},\end{aligned}\tag{3.4}$$

which is a Markov chain with $MN+2$ states. Two of these states, MC_- and MC_+, are the absorbing states. The transition probabilities matrix is given by (3.1). The vector of initial states probabilities is given by (3.2), (3.3). Note that $|S^{(k)}|\neq 0$, $k\in\{0,1\}$, because $S^{(k)}$ is the fundamental matrix (Kemeni and Snell, 1959) of the Markov chain (3.4). □

4. Robustness Analysis

Let us analyze the robustness of the sequential test under a "contamination" (Huber, 1981) in observations. Let the described above hypothetical model be distorted: the initial states probabilities vector and the transition probabilities matrix are given by

$$\bar{\pi}^{(k)}=(1-\varepsilon)\pi^{(k)}+\varepsilon\tilde{\pi}^{(k)},\ \bar{P}^{(k)}=(1-\varepsilon)P^{(k)}+\varepsilon\tilde{P}^{(k)},\tag{4.1}$$

where $\tilde{\pi}^{(k)}$ and $\tilde{P}^{(k)}\neq P^{(k)}$ are the initial states probabilities vector and the transition probabilities matrix for the "contaminating" Markov chain, $\varepsilon\in[0,\frac{1}{2})$ is the probability of "contamination" presence (Huber, 1981).

Define $\tilde{W}^{(k)}$, $\tilde{Q}^{(k)}$, $\tilde{R}^{(k)}$, $\tilde{\omega}^{(k)}$ and $\tilde{\omega}^{(k)}_{MC_{\pm}}$, $k \in \{0,1\}$, similarly to the hypothetical case, replacing $p_{ij}^{(k)}$ and $\pi_i^{(k)}$ with $\tilde{p}_{ij}^{(k)}$ and $\tilde{\pi}_i^{(k)}$ respectively. Define the matrix $\tilde{S}^{(k)} = \mathbf{I}_{MN} - Q^{(k)} - \varepsilon(\tilde{Q}^{(k)} - Q^{(k)})$, $k \in \{0,1\}$.

Theorem 4.1. *If the hypothetical model is distorted according to* (4.1), *then the conditional error probabilities* $\tilde{\alpha}$, $\tilde{\beta}$, *and the conditional expected sequence lengths* $\{\tilde{t}^{(k)}\}$ *differ from the corresponding hypothetical characteristics by the values of the order* $\mathcal{O}(\varepsilon)$*:*

$$\bar{t}^{(k)} - t^{(k)} = \varepsilon\left(\left(\tilde{\omega}^{(k)} - \omega^{(k)}\right)' + \left(\omega^{(k)}\right)'\left(S^{(k)}\right)^{-1}\left(\tilde{Q}^{(k)} - Q^{(k)}\right)\right) \times \left(S^{(k)}\right)^{-1}\mathbf{1}_{MN} + \mathcal{O}(\varepsilon^2), \tag{4.2}$$

$$\bar{\alpha} - \alpha = \varepsilon\Bigg(\left(\omega^{(0)}\right)'\left(\left(S^{(0)}\right)^{-1}\left(\left(\tilde{Q}^{(0)} - Q^{(0)}\right)\left(S^{(0)}\right)^{-1}R^{(0)} + \tilde{R}^{(0)} - R^{(0)}\right)\right)_{(2)} + \left(\tilde{\omega}^{(0)} - \omega^{(0)}\right)' B^{(0)}_{(2)} + \tilde{\omega}^{(0)}_{MC_+} - \omega^{(0)}_{MC_+}\Bigg) + \mathcal{O}(\varepsilon^2), \tag{4.3}$$

$$\bar{\beta} - \beta = \varepsilon\Bigg(\left(\omega^{(1)}\right)'\left(\left(S^{(1)}\right)^{-1}\left(\left(\tilde{Q}^{(1)} - Q^{(1)}\right)\left(S^{(1)}\right)^{-1}R^{(1)} + \tilde{R}^{(1)} - R^{(1)}\right)\right)_{(1)} + \left(\tilde{\omega}^{(1)} - \omega^{(1)}\right)' B^{(1)}_{(1)} + \tilde{\omega}^{(1)}_{MC_-} - \omega^{(1)}_{MC_-}\Bigg) + \mathcal{O}(\varepsilon^2). \tag{4.4}$$

Proof. It is based on the fact that under the distortions (4.1) the transition probabilities matrix and the initial states probabilities vector for the sequence (3.4) have the forms

$$\bar{W}^{(k)} = W^{(k)} + \varepsilon\left(\tilde{W}^{(k)} - W^{(k)}\right), \quad \bar{\omega}^{(k)} = \omega^{(k)} + \varepsilon(\tilde{\omega}^{(k)} - \omega^{(k)}).$$

□

Theorem 4.1 allows to approximate linearly the characteristics of the test under "contamination" using the result of Theorem 3.1. In Theorem 4.1 the coefficients of ε in (4.2)–(4.4) are bounded for all possible values of $\{\tilde{\pi}^{(k)}, \tilde{P}^{(k)}\}$, and this leads to the fact that the linear term of the influence function (Hampel et al., 1986) is bounded.

5. Simulation Results

To illustrate the accuracy of linear approximations (4.2)–(4.4) under the "contamination" (4.1) for the conditional error probabilities and for the conditional expected sequence lengths, we performed statistical simulation.

The experiments are done for the case, where $M = 5$, with the following values of the parameters of the hypothetical and "contaminating" Markov chains:

$$\pi^{(0)} = \left(\frac{1}{4}, \frac{1}{4}, \frac{1}{4}, \frac{1}{8}, \frac{1}{8}\right)', \quad P^{(0)} = \begin{pmatrix} \frac{1}{8} & \frac{1}{8} & \frac{1}{4} & \frac{1}{4} & \frac{1}{4} \\ \frac{1}{4} & \frac{1}{8} & \frac{1}{8} & \frac{1}{4} & \frac{1}{4} \\ \frac{1}{4} & \frac{1}{4} & \frac{1}{8} & \frac{1}{8} & \frac{1}{4} \\ \frac{1}{4} & \frac{1}{4} & \frac{1}{4} & \frac{1}{8} & \frac{1}{8} \\ \frac{1}{4} & \frac{1}{4} & \frac{1}{4} & \frac{1}{8} & \frac{1}{8} \end{pmatrix},$$

$$\pi^{(1)} = \left(\frac{1}{8}, \frac{1}{8}, \frac{1}{4}, \frac{1}{4}, \frac{1}{4}\right)', \quad P^{(1)} = \begin{pmatrix} \frac{1}{4} & \frac{1}{4} & \frac{1}{4} & \frac{1}{8} & \frac{1}{8} \\ \frac{1}{4} & \frac{1}{4} & \frac{1}{8} & \frac{1}{8} & \frac{1}{4} \\ \frac{1}{4} & \frac{1}{8} & \frac{1}{8} & \frac{1}{4} & \frac{1}{4} \\ \frac{1}{8} & \frac{1}{8} & \frac{1}{4} & \frac{1}{4} & \frac{1}{4} \\ \frac{1}{8} & \frac{1}{8} & \frac{1}{4} & \frac{1}{4} & \frac{1}{4} \end{pmatrix},$$

$$\tilde{\pi}^{(0)} = \left(\frac{1}{8}, \frac{1}{8}, \frac{1}{4}, \frac{1}{4}, \frac{1}{4}\right)', \quad \tilde{P}^{(0)} = \begin{pmatrix} \frac{1}{2} & \frac{1}{4} & \frac{1}{8} & \frac{1}{16} & \frac{1}{16} \\ \frac{1}{4} & \frac{1}{2} & \frac{1}{8} & \frac{1}{16} & \frac{1}{16} \\ \frac{1}{16} & \frac{1}{4} & \frac{1}{2} & \frac{1}{8} & \frac{1}{16} \\ \frac{1}{16} & \frac{1}{16} & \frac{1}{8} & \frac{1}{2} & \frac{1}{4} \\ \frac{1}{16} & \frac{1}{16} & \frac{1}{8} & \frac{1}{4} & \frac{1}{2} \end{pmatrix},$$

$$\tilde{\pi}^{(1)} = \left(\frac{1}{4}, \frac{1}{4}, \frac{1}{4}, \frac{1}{8}, \frac{1}{8}\right)', \quad \tilde{P}^{(1)} = \begin{pmatrix} \frac{1}{16} & \frac{1}{16} & \frac{1}{8} & \frac{1}{4} & \frac{1}{2} \\ \frac{1}{16} & \frac{1}{16} & \frac{1}{8} & \frac{1}{2} & \frac{1}{4} \\ \frac{1}{16} & \frac{1}{8} & \frac{1}{2} & \frac{1}{4} & \frac{1}{16} \\ \frac{1}{4} & \frac{1}{2} & \frac{1}{8} & \frac{1}{16} & \frac{1}{16} \\ \frac{1}{2} & \frac{1}{4} & \frac{1}{8} & \frac{1}{16} & \frac{1}{16} \end{pmatrix}.$$

In this example, $a = \ln 2$, and the values of m_i, m_{ij}, $i, j \in \{0, 1, \ldots, 4\}$ can be calculated easily according to Section 2.

Let α_0, β_0 be the requested values of the error probabilities of the type I and II respectively, $\tilde{\alpha}(\varepsilon) = \frac{N_{H_1|H_0}}{N_{H_0}}$, $\tilde{\beta}(\varepsilon) = \frac{N_{H_0|H_1}}{N_{H_1}}$ be the estimates of the error probabilities of the type I and II, which are obtained by the Monte–Carlo simulation under "contamination" with the probability of "contamination" presence ε, where N_{H_l} is the total number of realizations under H_l, $N_{H_k|H_l}$ is the numbers of decisions in favor of H_k from the total number of N_{H_l} realizations ($k, l \in \{0, 1\}$, $k \neq l$). Let $\bar{\alpha}(\varepsilon)$, $\bar{\beta}(\varepsilon)$ be the linear approximations of the conditional error probabilities $\bar{\alpha}$, $\bar{\beta}$, computed according to (4.3), (4.4); $\tilde{t}^{(0)}(\varepsilon)$, $\tilde{t}^{(1)}(\varepsilon)$ be the estimates of the conditional expected sequence lengths, which are obtained by statistical simulation, $\bar{t}^{(0)}(\varepsilon)$, $\bar{t}^{(1)}(\varepsilon)$ be the linear approximations of the characteristics $\bar{t}^{(0)}$, $\bar{t}^{(1)}$, computed according to (4.2).

The thresholds C_+, C_- were computed by the formulae (Wald, 1947):

$$C_+ = \left[\frac{1}{a} \cdot \ln \frac{1 - \beta_0}{\alpha_0}\right], \quad C_- = \left[\frac{1}{a} \cdot \ln \frac{\beta_0}{1 - \alpha_0}\right],$$

where $[\cdot]$ means the integer part. The number of experiments with each parameter set was equal to $N_{H_0} = N_{H_1} = 10000$. The results are presented in the Table 1.

TABLE 1. Results of the experiments.

ε	$\tilde{\alpha}(\varepsilon)$	$\bar{\alpha}(\varepsilon)$	$\tilde{\beta}(\varepsilon)$	$\bar{\beta}(\varepsilon)$	$\tilde{t}^{(0)}(\varepsilon)$	$\bar{t}^{(0)}(\varepsilon)$	$\tilde{t}^{(1)}(\varepsilon)$	$\bar{t}^{(1)}(\varepsilon)$
$\alpha_0 = 0.01,\ \beta_0 = 0.01$								
0.01	0.0146	0.0178	0.0090	0.0090	34.085	34.733	29.282	29.780
0.05	0.0248	0.0268	0.0167	0.0143	36.663	38.512	31.861	33.260
0.2	0.1055	0.0607	0.0744	0.0340	50.010	52.684	43.951	46.310
$\alpha_0 = 0.01,\ \beta_0 = 0.1$								
0.01	0.0139	0.0167	0.0621	0.0675	19.372	19.616	26.642	26.847
0.05	0.0236	0.0250	0.0909	0.0910	21.098	21.736	28.210	28.911
0.2	0.0890	0.0559	0.2110	0.1790	27.428	29.686	33.653	36.547
$\alpha_0 = 0.05,\ \beta_0 = 0.1$								
0.01	0.0631	0.0642	0.0587	0.0642	17.385	17.796	17.186	17.558
0.05	0.081	0.0856	0.0792	0.0856	18.742	19.091	18.160	18.821
0.2	0.1830	0.1660	0.1870	0.1660	21.847	23.947	21.271	23.558
$\alpha_0 = 0.1,\ \beta_0 = 0.1$								
0.01	0.1140	0.1260	0.0589	0.0599	15.636	15.882	12.763	12.971
0.05	0.1520	0.1570	0.0815	0.0793	16.391	16.612	13.557	13.880
0.2	0.2680	0.2750	0.1681	0.1518	18.220	19.349	15.838	17.291

On the basis of the results of experiments one can conclude that the accuracy of the linear approximation w.r.t. ε is satisfactory under "contamination" of the hypothetical model.

Our previous results for the i.i.d. random variables (Kharin, 2002) agree with the results presented in this paper, as a singular case.

For the values of ε large enough, the conditional error probabilities $\tilde{\alpha}(\varepsilon)$, $\bar{\alpha}(\varepsilon)$, $\tilde{\beta}(\varepsilon)$, $\bar{\beta}(\varepsilon)$ differ from the targeted values α_0 and β_0 significantly (see Table 1, the case of $\varepsilon = 0.2$). This fact means that a robust sequential test should be used in this case.

Note in conclusion, that by Theorems 3.1, 4.1, and by the approach proposed in Kharin (2002), the minimax robust sequential test can be constructed.

References

[1] R. Durbin, S. Eddy, A. Krogh, and G. Mitchison, *Biological Sequence Analysis.* Cambridge University Press, 1998.

[2] F. Hampel, E. Ronchetti, P. Rousseeuw, and W. Stahel, *Robust Statistics. The Approach Based on Influence Functions.* Wiley, New York, 1986.

[3] P. Huber, *Robust Statistics.* Wiley, New York, 1981.

[4] A. Kharin, *On Robustifying of the SPRT For a Discrete Model Under "Contaminations".* Austrian J. Statist. **31** (2002), 267–277.

[5] J.G. Kemeni and J.L. Snell, *Finite Markov chains.* Wiley, New York, 1959.

[6] M. Malyutov and I. Tsitovich, *Asymptotically Optimal Sequential Testing of Hypotheses.* Problems Inform. Transmission **36** (2000), 98–112.

[7] D. Siegmund, *Sequential Analysis. Tests and Confidence Intervals.* Springer, New York, 1985.

[8] N. Stockinger and R. Dutter, *Robust Time Series Analysis: a Survey.* Kybernetika **23** (1987), 1–90.

[9] A. Wald, *Sequential Analysis.* Wiley, New York, 1947.

[10] J. Whitehead, *The Design and Analysis of Sequential Clinical Trials.* Wiley, New York, 1997.

[11] T.-J. Wu, Ya-Ch. Hsieh, and L.-A. Li, *Statistical Measures of DNA Sequence Dissimilarity Under Markov Chain Models of Base Composition.* Biometrics **57** (2001), 441–448.

Acknowledgment

The author would like to thank the Organizing Committee of the ICORS'03 for the financial support making participation in the Conference possible.

The author is also thankful to the Referees for useful comments which helped to improve the paper.

A. Kharin
Belarussian State University
Fr. Skoriny av., 4
BY 220050 Minsk
Belarus
e-mail: KharinAY@bsu.by

Statistics for Industry and Technology, 173–182

Robust Box-Cox Transformations for Simple Regression

A. Marazzi and V.J. Yohai

Abstract. The use of the Box-Cox family of transformations is a popular approach to make data behave according to a linear regression model. The regression coefficients, as well as the parameter λ defining the transformation, are generally estimated by maximum likelihood, assuming homoscedastic normal errors. These estimates are nonrobust; in addition, consistency to the true parameters holds only if the assumptions of normality and homoscedasticity are satisfied. We present here a new class of estimates, for the case of simple regression, which are robust and consistent even if the assumptions of normality and homoscedasticity do not hold.

Mathematics Subject Classification (2000). Primary 62J05; Secondary 62G35.

Keywords. Box-Cox transformations, robust estimation, heteroscedasticity.

1. Introduction

The Box-Cox family of transformations has become a widely used tool to make data behave according to a linear regression model. Sakia (1992) has given an excellent review of the work relating to this transformation. The response variable, transformed according to the Box-Cox procedure, is usually assumed to be linearly related to its covariates and the errors normally distributed with constant variance. The regression coefficients, as well as the parameter λ defining the transformation, are generally estimated by maximum likelihood (ML). Unfortunately, near normality and homoscedasticity are hard to attain simultaneously with a single transformation. In addition, the ML-estimate is not consistent under non-normal or heteroscedastic errors and it is not robust.

Carroll and Ruppert (1988) proposed bounded influence estimates based on the normal homoscedastic model to limit the influence of a moderate number of

This work was completed with the support of Grant 2053-066895.01 from the Swiss National Science Foundation, Grant PICT99 0306277 from the Agencia Nacional de Promoción de la Ciencia y la Tecnología, Grant X611 from the Universidad de Buenos Aires, Argentina, and a grant from Fundación Antorchas, Argentina.

outliers. Within their approach, heteroscedasticity was modelled with the help of a weighting system. Moreover, various semiparametric and nonparametric approaches to relax the parametric structure of the response distribution have been studied. References concerning these proposals can be found in Foster et al. (2001), where a semiparametric procedure based on a minimum distance criterion between a semiparametric and a nonparametric estimate of the error distribution is developed. In general, these approaches do not provide effective protection against heavy contamination and heteroscedasticity.

In this paper we present a new class of estimates for the case of simple regression which are robust and consistent even if the assumptions of normality and homoscedasticity do not hold. The new estimates are based on minimization (with respect to λ) of a measure of autocorrelation among the residuals with respect to a robust estimate of the regression coefficients (for given λ). In order to compute the autocorrelation, the residuals have to be ordered according to the values of the regressor. We use the residual autocorrelation as a measure of functional relationship between the residuals and the regressor. This measure is reminiscent of a proposal by Maravall (1983) to detect nonlinearity in a time series.

2. The Model

We consider a bivariate random sample $(x_1, y_1), \ldots, (x_n, y_n)$ and assume that the response and the explanatory variables are linked by the linear relationship

$$y_i^{(\lambda_0)} = \alpha_0 + \beta_0 x_i + h(x_i)u_i, \tag{2.1}$$

where α_0, β_0 and λ_0 are real parameters and $y_i > 0$. The function $h(\cdot)$ is unknown and

$$y^{(\lambda)} = \begin{cases} (y^\lambda - 1)/\lambda & \text{if} \quad \lambda \neq 0, \\ \ln(\lambda) & \text{if} \quad \lambda = 0, \end{cases} \tag{2.2}$$

denotes the usual Box-Cox transformation. We assume that the errors u_i are i.i.d. according to a common cdf F, that u_i is independent of x_i, and $E(u_i) = 0$.

3. The Robust Autocorrelation Estimate

Suppose that (α_n, β_n) are robust estimates of intercept and slope for the simple linear regression model, e.g., MM-estimates (Yohai, 1987). Let $(\alpha_n(\lambda), \beta_n(\lambda))$ be the result of applying these estimates to the responses $y_i^{(\lambda)}$ for a given λ and $s_n(\lambda)$ a robust measure of the residual scale, e.g., the one used to compute the MM-estimate (usually, the initial high breakdown point S-estimate of scale; Rousseeuw and Leroy, 1987). If the residuals are computed using the true parameter λ_0, their conditional mean is close to zero for all values of x. On the other hand, when the residuals are computed using a $\lambda \neq \lambda_0$, there is a functional relationship between the residual conditional mean and x. A suitable value of λ should therefore minimize a measure of dependency between residuals and x. One such measure is the *robust residual autocorrelation* $\rho_n(\lambda)$. Its definition depends on whether there are tied x-values or not.

3.1. The Case Without Tied x-values

Firstly, we suppose that all the values $x_1, \ldots, x_n$ are distinct. In this case $\rho_n(\lambda)$ is given by the following steps:

1. Sort x_i, $i = 1, \ldots, n$ in ascending order and let $i_1, \ldots, i_n$ be the corresponding permuted indices. Thus, $x_{i_1} < x_{i_2} < \cdots < x_{i_n}$.

2. For a given λ, let $r_j(\lambda) = y_{i_j}^{(\lambda)} - \alpha_n(\lambda) - x_{i_j}\beta_n(\lambda)$ $(j = 1, \ldots, n)$ and

$$v_j(\lambda) = \psi\left(\frac{r_j(\lambda)}{s_n(\lambda)}\right), \quad j = 1, \ldots, n.$$

3. Set

$$\rho_n(\lambda) = \frac{1}{n}\sum_{j=1}^{n-1} v_j(\lambda)v_{j+1}(\lambda), \tag{3.1}$$

where $\psi(\cdot)$ is any monotone, odd, and bounded function, e.g., the well-known Huber's function.

3.2. The Case With Tied x-values

We now assume that there are only k distinct x-values and that n_1 observations are equal to the smallest, n_2 to the next smallest, ..., n_k to the largest. Thus, $n_1 + n_2 + \cdots + n_k = n$. In this case, we modify the terms in (3.1) as follows:

1. Replace the first $n_1 - 1$ terms by

$$\frac{(n_1 - 1)}{\binom{n_1}{2}} \sum_{i=1}^{n_1-1} \sum_{j=i+1}^{n_1} v_i(\lambda)v_j(\lambda).$$

2. Replace the term $v_{n_1}(\lambda)v_{n_1+1}(\lambda)$ by

$$\frac{1}{n_1 n_2} \sum_{i=n_1+1}^{n_1+n_2} \sum_{j=1}^{n_1} v_i(\lambda)v_j(\lambda).$$

3. Replace the following $n_2 - 1$ terms by

$$\frac{(n_2 - 1)}{\binom{n_2}{2}} \sum_{i=n_1+1}^{n_1+n_2-1} \sum_{j=i+1}^{n_1+n_2} v_i(\lambda)v_j(\lambda).$$

and so on. Note that this replacement procedure is equivalent to arbitrarily permuting the tied observations and computing their correlations by averaging over the permutations.

The robust autocorrelation is positive when there is dependency between residuals and the regressor x; otherwise it is close to zero. However, the equation $\rho_n(\lambda) = 0$ may have more than one or no solutions. Therefore, we define the *robust autocorrelation estimate* of λ_0 as the global minimum of $\rho_n(\lambda)$. Another proposal for robust autocorrelation can be found in Ma and Genton (2001).

Remark 3.1. We tried to extend the robust autocorrelation estimate to the case of several covariates. For that purpose the residuals were sorted according to the fitted values (in place of the x-values). Unfortunately, simulation results showed that the performance of this proposal is not satisfactory. The reason of this poor performance may be due to the fact that the fitted value corresponding to a value $\lambda \neq \lambda_0$ may be almost uncorrelated with the fitted value associated to $\lambda = \lambda_0$.

4. Consistency

For simplicity, we consider the model given by (2.1) and $y^{(\lambda)} = y^\lambda$. If $\lambda \neq 0$, this model is equivalent to the one given by (2.1) and (2.2). We prove consistency of the robust autocorrelation estimate supposing that $\lambda \in \Lambda = [l_1, l_2]$ with $l_1 > 0$ and $l_2 < \infty$. We make the following assumptions.

(A1) In the model (2.1), $(u_1, x_1), (u_2, x_2), \ldots, (u_n, x_n)$ are i.i.d. and u_i and x_i are independent. The distribution F of u_i is continuous, symmetric, and $E(u_i) = 0$. The distribution G of x_i is continuous.

(A2) h is continuous.

(A3) ψ is continuous, odd, monotone, and bounded.

(A4) (Uniform consistency of the estimates α_n, β_n, and s_n). For all $\lambda \in \Lambda$, there exist real numbers $\varepsilon_1(\lambda) > 0$, $s_0 > 0$ and continuous functions $\alpha(\lambda)$, $\beta(\lambda)$, $s(\lambda)$ such that $\alpha(\lambda_0) = \alpha_0$, $\beta(\lambda_0) = \beta_0$, $s(\lambda) > s_0 > 0$, and

$$\begin{aligned} p\lim_{n\to\infty} \sup_{|\lambda^*-\lambda|<\varepsilon_1(\lambda)} |\alpha_n(\lambda^*) - \alpha(\lambda^*)| &= 0, \\ p\lim_{n\to\infty} \sup_{|\lambda^*-\lambda|<\varepsilon_1(\lambda)} |\beta_n(\lambda^*) - \beta(\lambda^*)| &= 0, \\ p\lim_{n\to\infty} \sup_{|\lambda^*-\lambda|<\varepsilon_1(\lambda)} |s_n(\lambda^*) - s(\lambda^*)| &= 0, \end{aligned}$$

where $p\lim$ denotes convergence in probability.

(A5) (Robust identifiability condition). Let

$$r(\lambda, \alpha, \beta, u, x) = y^\lambda - \alpha - \beta x = (\alpha_0 + \beta_0 x + h(x)u)^{\lambda/\lambda_0} - \alpha - \beta x,$$
$$d(\lambda, \alpha, \beta, s, x) = E(\psi(r(\lambda, \alpha, \beta, u, x)/s)|x),$$

where x and y denote generic values of the covariate and the response. Then, for any $\lambda \neq \lambda_0$, α, β and $s > 0$, there exists an interval $I(\lambda, \alpha, \beta, s)$ such that $P(x \in I(\lambda, \alpha, \beta, s)) > 0$ and for all $x \in I(\lambda, \alpha, \beta, s)$

$$d(\lambda, a(\lambda), \beta(\lambda), s, x) \neq 0.$$

We consider the estimates:

$$\hat\lambda_n = \arg\min_{\lambda\in\Lambda} \rho_n(\lambda), \quad \hat\alpha_n = \alpha_n(\hat\lambda_n), \quad \hat\beta_n = \beta_n(\hat\lambda_n).$$

Theorem 4.1. *Under (A1)-(A5), $\hat\lambda_n \to \lambda_0$, $\hat\alpha_n \to \alpha_0$, and $\hat\beta_n \to \beta_0$ in probability.*

The proof is given in the appendix.

Remark 4.2. Suppose that (A5) does not hold. Then, there exist values $\lambda_1 \neq \lambda_0$, α_1, β_1 and s such that, with probability one, $y^{\lambda_1} = \alpha_1 + \beta_1 x + u$, where u satisfies the condition $E(\psi_s(u)|x) = 0$ with $\psi_s(u) = \psi(u/s)$. But, according to (A1), the model (2.1) satisfies the same condition and, therefore, it is not identifiable. When $h(x) = 1$ and $\beta_0 = 0$, this condition does not hold; in this case λ_0 in model (2.1) is clearly not identifiable.

Remark 4.3. At present, we did not check that the MM-estimates and the scale estimates we are suggesting for α_n, β_n, and s_n satisfy the uniform consistency condition (A4). However, this assumption seems very plausible. The methods of Berrendero and Zamar (2003) could be used to prove this property.

5. Empirical Results

Bivariate observations (x_i, y_i) were generated according to the model

$$y_i^{\lambda_0} = \alpha_0 + \beta_0 x_i + h(x_i) u_i, \quad i = 1, \ldots, n,$$

where the u_i were i.i.d., such that $u_i = be_i$, $E(e_i) = 0$, the distribution of e_i was uniform $U(-0.5, 0.5)$, normal $N(0,1)$, Student t_6, t_3, contaminated normal $0.9N(0,1) + 0.1N(0,25)$, or exponential, and the factor b was chosen so that MAD(u_i)=1/3. In addition, $\alpha_0 = 10$, $\beta_0 = 2$, $\lambda_0 = 0.5$, $n = 100$, $x_i = 0.2 \cdot i$ $(i = 1, \ldots, n)$. Two options for $h(x)$ were used: $h(x) = 3$ (homoscedastic case) and $h(x) = x/2$ (heteroscedastic case). Each experiment was based on 1000 samples. The parameters α_0, β_0 and the function h were chosen in order to have positive responses, clear identifiability, and perceptible homo/heteroscedasticity. The choice of λ_0 was not relevant, because of the equivariance of $\hat{\lambda}_n$ ($\hat{\lambda}_n(y_i^a, \ldots, y_n^a) = \hat{\lambda}_n(y_1, \ldots, y_n)/a$). Various parameter sets provided similar results.

The results are reported in Table 1 and Table 2 and include the average bias of the simulated estimates (bias=$1000 \times \text{mean}(\hat{\lambda}_n - \lambda_0)$) the root of the mean squared error ($\sqrt{\text{mse}}$=$1000 \times (\text{mean}(\hat{\lambda}_n - \lambda_0)^2)^{0.5}$) and the median absolute deviation with respect to λ_0 (mde=$1000 \times \text{median}(\hat{\lambda}_n - \lambda_0)/0.6745$). RAC denotes the robust autocorrelation estimator defined in Section 3, where $\alpha_n(\lambda)$, $\beta_n(\lambda)$, are MM-estimators (with breakdown point 0.5 and efficiency 0.95 for normal errors, as defined in Yohai, 1987) and $s_n(\lambda)$ the robust measure of the residual scale provided by an initial S-estimate (Rousseeuw and Leroy, 1987, p. 135, with tuning constant $c = 1.547$). AC denotes the autocorrelation estimator when $\alpha_n(\lambda)$, $\beta_n(\lambda)$, and $s_n(\lambda)$ are least squares estimates. F is the estimator defined in Foster et al. (2001) and RF is a modification that uses the MM-estimates $\alpha_n(\lambda)$, $\beta_n(\lambda)$ mentioned above in place of least squares estimates. BI denotes the bounded influence estimator as defined in Carroll and Ruppert (1988, p.186, with tuning constant $a = 1.5$ and based on 10 iterations). ML denotes the maximum likelihood estimator.

TABLE 1. Simulation results: homoscedastic case.

		RAC	AC	F	RF	BI	ML
	bias	0	0	0	0	-2	0
Unif	$\sqrt{\text{mse}}$	22	20	22	21	23	18
	mde	22	21	22	21	23	18
	bias	-1	-1	0	0	-3	0
Gauss	$\sqrt{\text{mse}}$	28	26	27	26	26	23
	mde	28	26	28	27	24	23
	bias	0	0	1	1	-4	0
t6	$\sqrt{\text{mse}}$	29	30	29	28	26	29
	mde	30	31	30	29	27	29
	bias	-2	-1	-1	-1	-6	1
t3	$\sqrt{\text{mse}}$	31	46	30	29	27	54
	mde	31	36	28	28	27	45
	bias	0	-2	-1	-1	-4	0
CntG	$\sqrt{\text{mse}}$	30	48	31	30	27	54
	mde	29	46	31	30	26	56
	lme	-2	-1	-1	-1	2	-26
Exp	$\sqrt{\text{mse}}$	28	38	24	22	26	44
	mde	27	36	23	22	27	44

In the cases of homoscedastic short tailed error distributions, the performance of the various estimators was similar. For long tailed distributions, BI, RAC and RF provided effective and similar protections (RF being slightly superior for exponential errors). In the heteroscedastic symmetric cases, only the autocorrelation estimates RAC and AC were not biased and the robustness of RAC was very satisfying. Unfortunately, all estimates (with the exception of AC) were strongly biased in the asymmetric heteroscedastic case.

Remark 5.1. We used FORTRAN programs loaded into S-plus to evaluate the initial estimates $\alpha_n(\lambda)$, $\beta_n(\lambda)$, and $s_n(\lambda)$ on a grid of 101 equally spaced points between 0.05 and 1.25; linear interpolation was used for other values of λ. On a 733 MHz Pentium III, the computing time (seconds) to obtain the initial estimates on the grid was 0.22 ($n = 20$), 1.60 ($n = 50$), 10.77 ($n = 100$), 78.14 ($n = 200$). The computing time for minimizing $\rho(\lambda)$ was 0.05 ($n = 20$), 0.08 ($n = 50$), 0.15 ($n = 100$), 0.27 ($n = 200$).

6. Appendix

For any $\lambda \in \Lambda$, u, u^*, x, and $\varepsilon > 0$ we define

$$t(\lambda, u, u^*, x, \varepsilon) = \inf_{D(\lambda,\varepsilon,x)} \psi(r(\lambda', \alpha', \beta', u, x)/s')\psi(r(\lambda', \alpha', \beta', u^*, x')/s'),$$

TABLE 2. Simulation results: heteroscedastic case.

		RAC	AC	F	RF	BI	ML
	bias	1	2	-34	-34	-120	-136
Unif	$\sqrt{\text{mse}}$	45	40	62	54	124	139
	mde	44	40	64	53	179	205
	bias	2	3	-33	-32	-142	-192
Gauss	$\sqrt{\text{mse}}$	46	50	62	54	146	195
	mde	46	49	61	52	207	285
	bias	5	6	-30	-30	-149	-222
t6	$\sqrt{\text{mse}}$	49	59	63	55	153	226
	mde	50	58	61	54	221	332
	bias	1	6	-36	-34	-158	-234
t3	$\sqrt{\text{mse}}$	47	97	66	56	162	252
	mde	44	71	68	56	233	365
	bias	3	8	-36	-33	-146	-216
CntG	$\sqrt{\text{mse}}$	48	122	68	58	150	242
	mde	45	93	67	58	216	339
	lme	22	6	-56	-54	-171	-313
Exp	$\sqrt{\text{mse}}$	70	80	87	74	179	319
	mde	71	72	94	82	255	467

where $D(\lambda, \varepsilon, x)$ is the set of all λ', α', β', s', x' that satisfy the following conditions:

$$|\lambda' - \lambda| \le \varepsilon, |\alpha' - \alpha(\lambda')| \le \varepsilon, |\beta' - \beta(\lambda')| \le \varepsilon, |s' - s(\lambda')| \le \varepsilon, |x' - x| \le \varepsilon.$$

Lemma 6.1. *The following properties hold:*

(i) $d(\lambda, \alpha, \beta, s, x)$ *is continuous;*
(ii) $d(\lambda_0, \alpha_0, \beta_0, s, x) = E(\psi(h(x)u/s)|x) = 0$.

Proof. (i) follows from the Dominated Convergence Theorem and (ii) it is immediate.

Lemma 6.2. *Let, u, u^*, x be independent variables such that u and u^* have distribution F and x distribution G. Then, for a given $\lambda \neq \lambda_0$, there exists $\varepsilon_2(\lambda) \le \varepsilon_1(\lambda)$, such that $E(t(\lambda, u, u^*, x, \varepsilon_2(\lambda))) > 0$.*

Proof. By (A5),

$$P(E(\psi(r(\lambda, \alpha(\lambda), \beta(\lambda), u, x)/s(\lambda))|x) \neq 0) > 0$$

and then

$$E(t(\lambda, u, u^*, x, 0)) = E(E^2(\psi(r(\lambda, \alpha(\lambda), \beta(\lambda), u, x)/s(\lambda))|x)) > 0.$$

The Lemma follows from the continuity of $t(\lambda, u, u^*, x, \varepsilon)$ and the Dominated Convergence Theorem.

Lemma 6.3. *Let $x_1, x_2, \ldots, x_n$ be a random sample of a distribution G, and let $x_{(1)} \leq x_{(2)} \leq \cdots \leq x_{(n)}$ be the ordered sample. Define $d_i = x_{(i+1)} - x_{(i)}$, $i = 1, \ldots, n-1$ and $m_n(\varepsilon) = \#\{i : d_i > \varepsilon\}/n$. Then, for any $\varepsilon > 0$, $m_n(\varepsilon) \to 0$ a.s. as $n \to \infty$.*

Proof. Take $\delta > 0$ and let M such that $P(|x| > M) < \delta$. Then

$$\begin{aligned} m_n(\varepsilon) &\leq \frac{\#\{i : d_i > \varepsilon, |x_i| > M\}}{n} + \frac{\#\{i : d_i > \varepsilon, |x_i| \leq M\}}{n} \\ &\leq \frac{\#\{i : |x_i| > M\}}{n} + \frac{1}{n}\frac{2M}{\varepsilon}. \end{aligned}$$

Therefore,

$$\limsup_{n\to\infty} m_n(\varepsilon) \leq \lim_{n\to\infty} \frac{\#\{i : |x_i| > M\}}{n} < \delta.$$

Since this inequality holds for any $\delta > 0$, the Lemma follows.

Proof of Theorem 4.2

To prove that $p\lim_{n\to\infty} \hat{\lambda}_n = \lambda_0$, it will be enough to show that:

(a) If $\delta > 0$ and $L = \{|\lambda - \lambda_0| > \delta\}$, then there exists $z > 0$, such that

$$p\lim \inf_{n\to\infty} \inf_{\lambda \in L} \rho_n(\lambda) > z;$$

(b) $p\lim \rho_n(\lambda_0) = 0$.

We start proving (a). Since we are assuming that the distribution G of x is continuous we can assume that all the x_i's are different. We can write

$$\rho_n(\lambda) = \frac{1}{n}\sum_{i=1}^{n-1} \psi\left(\frac{r(\lambda, \alpha_n(\lambda), \beta_n(\lambda), u_{(i)}, x_{(i)})}{s_n(\lambda)}\right)\psi\left(\frac{r(\lambda, \alpha_n(\lambda), \beta_n(\lambda), u_{(i+1)}, x_{(i+1)})}{s_n(\lambda)}\right)$$

where $x_{(i)} = x_{j_i}$, $u_{(i)} = u_{j_i}$ and $j_1, j_2, \ldots, j_n$ are the indices such that $x_{j_1} < x_{j_2} < \cdots < x_{j_n}$. For any $\lambda \neq \lambda_0$, let $\varepsilon_2(\lambda)$ be defined as in Lemma 6.2. According to the Heine-Borel theorem, we can find $\lambda_1, \ldots, \lambda_k$ in L, such that $L \subset \cup_{j=1}^k L_j$, where $L_j = \{\lambda : |\lambda - \lambda_j| < \varepsilon_2(\lambda)\}$. It is therefore enough to show that there exist numbers $z_j > 0$, $j = 1, \ldots, k$, such that

$$p\lim \inf_{n\to\infty} \inf_{\lambda \in Lj} \rho_n(\lambda) > z_j, \quad j = 1, \ldots, k.$$

Let $I = \{i : x_{(i+1)} - x_{(i)} > \varepsilon_2(\lambda)\}$ and $m_n = \#I/n$. In addition, for a given j let

$$H_{n1} = \left\{ \sup_{|\lambda - \lambda_j| \leq \varepsilon_1(\lambda_j)} |\alpha_n(\lambda) - \alpha(\lambda)| < \varepsilon_2(\lambda_j) \right\},$$

$$H_{n2} = \left\{ \sup_{|\lambda - \lambda_j| \leq \varepsilon_1(\lambda_j)} |\beta_n(\lambda) - \beta(\lambda)| < \varepsilon_2(\lambda_j) \right\},$$

$$H_{n3} = \left\{ \sup_{|\lambda - \lambda_j| \leq \varepsilon_1(\lambda_j)} |s_n(\lambda) - s(\lambda)| < \varepsilon_2(\lambda_j) \right\},$$

and

$$H_n = \cap_{l=1}^{3} H_{nl}^C.$$

Then, by (A4)

$$\lim_{n\to\infty} P(H_n) = 1. \tag{6.1}$$

In H_n, we have:

$$\begin{aligned}
&\inf_{\lambda\in L_j} \rho_n(\lambda)\\
> \;& \frac{1}{n}\sum_{i\notin I} t(\lambda_i, u_{(i)}, u_{(i+1)}, x_{(i)}, \varepsilon_2(\lambda_j)) +\\
& \frac{1}{n}\sum_{i\in I} \psi\left(\frac{r(\lambda,\alpha_n(\lambda),\beta_n(\lambda),u_{(i)},x_{(i)})}{s_n(\lambda)}\right)\psi\left(\frac{r(\lambda,\alpha_n(\lambda),\beta_n(\lambda),u_{(i+1)},x_{(i+1)})}{s_n(\lambda)}\right)\\
\geq \;& \frac{1}{n}\sum_{i\notin I} t(\lambda_j, u_{(i)}, u_{(i+1)}, x_{(i)}, \varepsilon_2(\lambda_j)) - \frac{m_n K^2}{n}\\
\geq \;& \frac{1}{n}\sum_{i=1}^{n-1} t(\lambda_j, u_{(i)}, u_{(i+1)}, x_{(i)}, \varepsilon_2(\lambda_j)) - \frac{2m_n K^2}{n}.
\end{aligned}$$

where $K = \sup\psi$. Therefore, by Lemma 6.3 and (6.1),

$$\begin{aligned}
\lim\inf_{n\to\infty}\inf_{\lambda\in L_j} \rho_n(\lambda) \;&>\; p\lim_{n\to\infty}\frac{1}{n}\sum_{i=1}^{n-1} t(\lambda_j, u_{(i)}, u_{(i+1)}, x_{(i)}, \varepsilon_2(\lambda_j))\\
&=\; p\lim_{n\to\infty}\frac{1}{n}\sum_{i=1}^{n-1} t(\lambda_j, u_i, u_{r_{i+1}}, x_i, \varepsilon_2(\lambda_j)) \qquad (6.2)
\end{aligned}$$

where $(r_1,\ldots,r_n)$ is the inverse permutation of $(j_1,\ldots,j_n)$. Since this permutation depends only on the x_i's, but not on the u_i's, we have

$$E\left(\frac{1}{n}\sum_{i=1}^{n-1} t\left(\lambda_j, u_i, u_{r_{i+1}}, x_i, \varepsilon_2(\lambda_i)\right)\right) = E(t(\lambda_j, u, u^*, x, \varepsilon_2(\lambda_j))), \tag{6.3}$$

where u, u^* and x are independent random variables, the first two with distribution F and the third with distribution G. In addition,

$$\begin{aligned}
&\mathrm{Var}\left(\frac{1}{n}\sum_{i=1}^{n-1} t(\lambda_j, u_i, u_{r_{i+1}}, x_i, \varepsilon_2(\lambda_j))\right)\\
=\;& \frac{n-1}{n^2}[\mathrm{Var}(t(\lambda_j, u, u^*, x, \varepsilon_2(\lambda_j))\\
&+ \mathrm{Cov}(t(\lambda_j, u, u^*, x, \varepsilon_2(\lambda_j))t(\lambda_j, u^*, u^{**}, x^*, \varepsilon_2(\lambda_j)))]
\end{aligned}$$

where u, u^*, u^{**}, x, x^* are independent. Then

$$\lim_{n\to\infty}\mathrm{Var}\left(\frac{1}{n}\sum_{i=1}^{n-1} t(\lambda_j, u_i, u_{r_{i+1}}, x_i, \varepsilon_2(\lambda_j))\right) = 0. \tag{6.4}$$

Therefore, by (6.2), (6.3), (6.4), Chebychev's inequality, and Lemma 6.2, we get

$$p\lim_{n\to\infty}\frac{1}{n}\sum_{i=1}^{n-1}t(\lambda_j,u_i,u_{r_{i+1}},x_i,\varepsilon_2(\lambda_j)))=E(t(\lambda_j,u,u^*,x,\varepsilon_2(\lambda_j))>0,$$

which proves (a).

The proof of part (b) is similar to the proof of part (a). The main difference is that we now use Lemma 6.1 (ii) instead of Lemma 6.2. This proves $p\lim_{n\to\infty}\hat{\lambda}_n=\lambda_0$. Using (A4) we get $p\lim_{n\to\infty}\hat{\alpha}_n=\alpha_0$ and $p\lim_{n\to\infty}\hat{\beta}_n=\beta_0$.

References

[1] J.R. Berrendero and R.H. Zamar, *A Note on the Uniform Asymptotic Normality of Location M-Estimates.* Submitted, 2003, available at *http://hajek.stat.ubc.ca/~ruben/website/publications.htm.*

[2] R.J. Carrol and D. Ruppert, *Transformation and Weighting in Regression.* Chapman and Hall, New York, 1988.

[3] A.M. Foster, L. Tial, and L.W. Wei, *Estimation for the Box-Cox Transformation Model Without Assuming Parametric Error Distribution.* J. Amer. Statist. Assoc. **96** (2001), 1097–1101.

[4] Y. Ma and M.G. Genton, *Highly Robust Estimation of the Autocovariance Function.* J. Time Ser. Anal. **78** (2001), 663–684.

[5] A. Maravall, *An Application of Nonlinear Time Series Forecasting.* J. Bus. Econom. Statist. **1** (1983), 66–74.

[6] P.J. Rousseeuw and A.M. Leroy, *Robust Regression and Outlier Detection.* Wiley, New York, 1987.

[7] R.M. Sakia, *The Box-Cox Transformation Technique: A Review.* Statistician **41** (1992), 169–178.

[8] V. Yohai, *High Breakdown-Point and High Efficiency Robust Estimates For Regression.* Ann. Statist. **15** (1987), 642–656.

Acknowledgment

Many thanks to Danielle Meylan and Alex Randriamiharisoa for their programming support.

A. Marazzi
Université de Lausanne, Institut de médecine sociale et préventive
Bugnon 17
CH-1005 Lausanne, Switzerland
e-mail: `alfio.marazzi@unil.ch`

V.J. Yohai
Universidad de Buenos Aires, Departamento de Matematica
Ciudad Universitaria, Pabellon 1
1428 Buenos Aires, Argentina
e-mail: `vyohai@dm.com.ar`

Statistics for Industry and Technology, 183–194

Consistency of the Least Weighted Squares Regression Estimator

L. Mašíček

Abstract. The paper deals with the least weighted squares estimator which is robust and generalizes classical least trimmed squares. We provide conditions under which this estimator is consistent and we prove consistency for general regression.

Mathematics Subject Classification (2000). Primary 62F35; Secondary 62J05.

Keywords. Robust regression, the least trimmed squares, the least weighted squares, weak consistency.

1. Introduction

Let us consider the following regression model

$$Y_i = X_i^T \beta_0 + Z_i \qquad \text{for } i = 1, \ldots, n \tag{1.1}$$

where $X_i = (X_{i1}, \ldots, X_{ip})^T$ is the $p \times 1$ column vector of random explanatory variables, β_0 is the $p \times 1$ column vector of unknown regression coefficients and Z_i are the random fluctuations with $\mathrm{E}Z_i = 0$. Moreover, the sequence of random vectors $X_1, \ldots, X_n$ is independent and identically distributed (IID), the sequence of random variables $Z_1, \ldots, Z_n$ is IID and both sequences are mutually independent.

For any $\beta \in R^p$ denote the ith residual as

$$r_i(\beta) := (Y_i - X_i^T \beta) = Z_i - X_i^T(\beta - \beta_0) \tag{1.2}$$

and the hth order statistics of the squared residuals by $r^2_{(h)}(\beta)$, i.e.,

$$0 \le r^2_{(1)}(\beta) \le r^2_{(2)}(\beta) \le \cdots \le r^2_{(n)}(\beta). \tag{1.3}$$

Denote the order statistics of the absolute value of the residuals by $r_{|h|}(\beta) := \sqrt{r^2_{(h)}(\beta)}$ (i.e., also the square root of the hth order statistics of squared residuals).

Research was supported by grant of GA ČR 402/03/0084 and MSM113200008.

Now we can define *the least weighted squares estimator* (LWS) as

$$\hat{\beta}_n = \hat{\beta}_n^{LWS} := \arg\min_{\beta \in R^p} \sum_{h=1}^{n} w\left(\frac{h-1}{n}\right) r_{(h)}^2(\beta), \tag{1.4}$$

where $w : [0,1] \to R$ is a given *weight function.* Typically we suppose, that w is nonincreasing (i.e., observations with larger residuals have smaller weight). Without loss of generality we suppose $w(1) = 0$.

The least weighted squares estimator was developed by Víšek (2000) and it generalizes classical *least trimmed squares* (e.g., Rousseeuw, 1984; Rousseeuw and Leroy, 1987) which we get for the choice $w(\alpha) = \mathrm{I}\{\alpha < \overline{\alpha}\}$ where $\mathrm{I}\{\dots\}$ is an indicator function and $\overline{\alpha} \in (0,1)$. The main reason for developing this estimator was to improve applicability. In the least trimmed squares estimator we can adjust only one constant but in the LWS estimator we can choose the whole weight function. This gives us a chance to increase efficiency or decrease gross error sensitivity. The case of a location parameter was discussed by Mašíček (2002) and $n^{1/4}$-consistency was proved.

The LWS estimator has several nice properties. First of all the breakdown point can be computed immediately from the weight function. If $w(\alpha) > 0$ for $\alpha < \overline{\alpha}$ and $w(\alpha) = 0$ for $\alpha > \overline{\alpha}$ then the breakdown point of the LWS estimator equals $\min\{1 - \overline{\alpha}, \overline{\alpha}\}$. This means that the breakdown point is under control and we can choose it arbitrarily up to 0.5. Finally, the LWS estimator is regression and scale equivariant.

In the next section we rewrite the function inside the minimization (1.4) in the form of a statistical functional (i.e., as a function of the empirical distribution function). In Section 3 we provide a criterion for consistency of the LWS estimator and the main theorem is formulated. Section 4 provides a detailed proof.

2. LWS Estimator as a Statistical Functional

Let us denote the function to be minimized in (1.4) as

$$\mathrm{MF}_n(\beta) := \sum_{h=1}^{n} w\left(\frac{h-1}{n}\right) r_{(h)}^2(\beta). \tag{2.1}$$

For $x \in R^p$ and $z \in R$ denote by $F(x,z)$ the distribution function (d.f.) of the $(p+1)$-dimensional vector $(X_i, Z_i) = (X_{i1} \dots X_{ip}, Z_i)$ and denote by $F_n(x,z)$ the corresponding empirical distribution function (e.d.f.) based on random vectors $(X_1, Z_1), \dots, (X_n, Z_n)$. We suppose distribution functions to be left continuous.

Now we rewrite $\mathrm{MF}_n(\beta)$ and the LWS estimator into the form of a statistical functional, i.e., as a function of the e.d.f. F_n as follows. The residuals in (2.1) can be expressed as $r_{(h)}^2(\beta) = \sum_{i=1}^{n} r_i^2(\beta)\, \mathrm{I}\left\{|r_i(\beta)| = r_{|h|}(\beta)\right\}$. Hence we can reorder the summation in (2.1):

$$\frac{1}{n}\mathrm{MF}_n(\beta) = \sum_{h=1}^{n} w_{h,n} \left[\frac{1}{n} \sum_{i=1}^{n} \left(Z_i - X_i^T t\right)^2 \mathrm{I}\left\{ \left|Z_i - X_i^T t\right| \le r_{|h|}(\beta_0 + t) \right\} \right], \quad (2.2)$$

where $t := (\beta - \beta_0)$ and $w_{h,n} := \left[w\left(\frac{h-1}{n}\right) - w\left(\frac{h}{n}\right) \right]$ (let us recall that $w(1) = 0$). Since we are working with continuous random errors we need not take into account the case of $r_{|h|}(\beta) = r_{|l|}(\beta)$ for $h \neq l$.

Now we define for an arbitrary $(p+1)$-dimensional d.f. $G(x, z)$ (where $x \in R^p$ and $z \in R$) the following statistical functional

$$\mathrm{MF}_\alpha(\beta_0 + t, G) := \int (z - x^T t)^2 \mathrm{I}\left\{ |z - x^T t| \le G_t^{-1}(\alpha) \right\} dG(x, z) \quad (2.3)$$

where G_t is the d.f. of random variable $|Z_G - X_G^T t|$ and the random vector (X_G, Z_G) has d.f. G. If we choose G equal to the e.d.f. F_n we will get $G_t^{-1}(\alpha) = F_{n,t}^{-1}(\alpha) = r_{|h|}(\beta_0 + t)$ for $\alpha \in \left(\frac{h-1}{n}, \frac{h}{n}\right]$ where $F_{n,t}$ denotes the e.d.f. based on the absolute values of the residuals (i.e., based on $|Z_1 - X_1^T t|, \ldots, |Z_n - X_n^T t|$). Hence we obtain

$$\mathrm{MF}_\alpha(\beta_0 + t, F_n) = \frac{1}{n} \sum_{i=1}^{n} \left(Z_i - X_i^T t\right)^2 \mathrm{I}\left\{ \left|Z_i - X_i^T t\right| \le r_{|h|}(\beta_0 + t) \right\} \quad (2.4)$$

for $\alpha \in \left(\frac{h-1}{n}, \frac{h}{n}\right]$, which is exactly the term in the brackets in (2.2). Now we can define

$$\mathrm{MF}(\beta, G) := \int_0^1 \mathrm{MF}_\alpha(\beta, G)\, dw^*(\alpha) \quad (2.5)$$

where $w^*(\alpha) := w(0) - w(\alpha)$. Substituting $G := F_n$ in (2.5) we obtain equation $\mathrm{MF}(\beta, F_n) = \frac{1}{n}\mathrm{MF}_n(\beta)$. This is because $\mathrm{MF}_\alpha(\beta, F_n) = \mathrm{MF}_{h/n}(\beta, F_n)$ for $\alpha \in \left(\frac{h-1}{n}, \frac{h}{n}\right]$ and hence

$$\mathrm{MF}(\beta, F_n) = \sum_{h=1}^{n} w_{h,n} \mathrm{MF}_{h/n}(\beta, F_n), \quad (2.6)$$

which is equal to (2.2) (see also (2.4)).

Now we can define the statistical functional $T(G)$

$$T(G) := \arg\min_{\beta \in R^p} \mathrm{MF}(\beta, G). \quad (2.7)$$

Obviously $T(F_n) = \hat{\beta}_n^{LWS}$, i.e., T represents the LWS estimator.

3. The Weak Consistency of the LWS Regression Estimator

The following assumptions will be needed throughout the paper.

A1: The weight function w is nonincreasing, bounded and its first derivative exists almost everywhere. Moreover, there exists $0 < \overline{\alpha} < 1$ such that $w(\alpha) > 0$ for $\alpha \in (0, \overline{\alpha})$ and $w(\alpha) = 0$ for $\alpha \in (\overline{\alpha}, 1)$.

A2: The random errors Z_i have continuous distribution with $\mathrm{E}Z_i^2 < \infty$, distribution function F_Z and density f_Z. The density is bounded, symmetric, strictly decreasing on $(0, \infty)$, $f_Z(x) > 0$ for $x \in R$ and f_Z' exists everywhere.

A3: The random vectors of explanatory variables X_i have finite second moments and there exist $\delta_1 > 0$ and $\delta_2 > 0$ such that

$$P\left(|X_i^T t| \leq \delta_1\right) \leq \overline{\alpha} - \delta_2 \qquad \text{for } t \in H_1, \tag{3.1}$$

where $H_1 := \{t \in R^p : \|t\| = 1\}$.

Theorem 3.1 (Weak consistency of the LWS regression estimator). *Let conditions* **A1**, **A2** *and* **A3** *be satisfied. Then the LWS estimator is a weak consistent estimator of* β_0, *i.e.,* $\hat{\beta}_n^{LWS} \to_P \beta_0$.

The proof of Theorem 3.1 is in Section 4. Let us now look deeply at the conditions from Theorem 3.1.

Condition **A1** implies that the breakdown point of our estimator is positive and equal to $\min\{1 - \overline{\alpha}, \overline{\alpha}\} > 0$.

In condition **A2** we suppose symmetric and one peak density of the random errors. This condition is very important for consistency. Suppose just one dimensional data with symmetric density of observations which has two sharp peaks – one around -1 and one around 1. I.e. approximately one half of the data is around -1 and one half around 1. If we apply the LWS estimator with $w(\alpha) = \mathrm{I}\{\alpha < 0.5\}$ we get a value close to -1 or 1. That is because LWS wants to fit 50% of data. But we want to obtain a value close to zero, which is the expected value of the observations.

It can be shown that **A3** is equivalent to the following condition: the random variables X_i have finite second moments and there exists $\varepsilon > 0$ such that

$$P\left(|X_i^T t| = 0\right) \leq \overline{\alpha} - \varepsilon \qquad \text{for } t \in H_1. \tag{3.2}$$

The last inequality, together with the existence of second moments, implies that the $p \times p$ matrix $\mathrm{E}X_i X_i^T$ is positive definite ($\mathrm{E}X_i X_i^T$ being a positive definite matrix is necessary for the consistency of the classical least squares). But condition **A3** is stronger and is important for the consistency of the LWS estimator. Because $w(\alpha) = 0$ for $\alpha > \overline{\alpha}$ we try to fit only $\lfloor n\overline{\alpha} \rfloor$ observations ($\lfloor x \rfloor$ denotes the integral part of $x \in R$). Suppose there exists $t \in R^p$ such that $P\left(|X_i^T t| = 0\right) > \overline{\alpha}$. Hence for large dataset (i.e., large n) there exists (with high probability) a subgroup of observations which contains at least $\lfloor n\overline{\alpha} \rfloor$ observations and their matrix of the explanatory variables is singular. Hence we have a problem to prove consistency.

See also Davies (1990) for the definition of $\lambda_n(\alpha)$ (page 1657). The function $\lambda_n(\alpha)$ measures the worst possible conditioning of any $\lfloor n\alpha \rfloor$ subset of the explanatory variables for the linear regression model with fixed explanatory variables. We can define an analogical measure in the case of random explanatory variables as follows. Let $\delta(\alpha)$ is the largest possible δ which satisfies $P\left(|X_i^T t| \leq \delta\right) \leq \alpha$ for any $t \in H_1$ ($\delta(\alpha)$ is equal to the largest possible δ_1 in (3.1) for $\alpha := \overline{\alpha} - \delta_2$). Condition **A3** implies $\delta(\alpha) > 0$ for some $\alpha < \overline{\alpha}$.

4. Proof of the Weak Consistency of the LWS Regression Estimator

To prove consistency we approximate the function $\mathrm{MF}(\beta, F_n)$ by $\mathrm{MF}(\beta, F)$. If we knew that $\mathrm{MF}(\beta, F)$ attains a global minimum for $\beta = \beta_0$ and is increasing in the neighborhood of β_0, then we would know that $\mathrm{MF}(\beta, F_n)$ has (for large n, with high probability) global minimum close to β_0. The following lemma proves the fact that $\mathrm{MF}(\beta, F)$ has minimum for $\beta = \beta_0$.

Lemma 4.1. *Under conditions* **A1**, **A2** *and* **A3** *for any* $\eta > 0$ *there exists* $\delta > 0$ *such that*

$$MF(\beta, F) \geq MF(\beta_0, F) + \delta \tag{4.1}$$

for any $\beta \in R^p$, $\|\beta - \beta_0\| \geq \eta$.

Proof of Lemma 4.1 Because the LWS estimator is regression equivariant, we may put $\beta_0 = 0$. It is sufficient to prove that $\mathrm{MF}(\beta, F)$ is continuous function of β and that $\mathrm{MF}_\alpha(\beta, F)$ strictly increasing in 0 in any direction, i.e.,

$$\mathrm{MF}_\alpha(0, F) < \mathrm{MF}_\alpha(\beta, F) \qquad \text{for } 0 < \alpha < 1,\ \beta \neq 0, \tag{4.2}$$

$$\mathrm{MF}_\alpha(\beta, F) < \mathrm{MF}_\alpha(K\beta, F) \qquad \text{for } 0 < \alpha < 1,\ \beta \neq 0,\ K > 1. \tag{4.3}$$

Both inequalities together with (2.5) imply the same property for $\mathrm{MF}(\beta, F)$, because $\mathrm{MF}(\beta, F)$ is an average of $\mathrm{MF}_\alpha(\beta, F)$ (see (2.5)). Hence $\mathrm{MF}(\beta, F)$ is strictly increasing in 0 in any direction. Because $\mathrm{MF}(\beta, F)$ is also continuous the inequality (4.1) holds. Hence the proof is completed by showing continuity of $\mathrm{MF}(\beta, F)$ and inequalities (4.2) and (4.3).

To prove the continuity of $\mathrm{MF}(\beta, F)$ we substitute (2.3) into (2.5) and rewrite $\mathrm{MF}(\beta, F)$ into the form

$$\mathrm{MF}(\beta_0 + t, F) = \int w\left(F_t\left(|z - x^T t|\right)\right)(z - x^T t)^2\, dF(x, z). \tag{4.4}$$

Hence we can easily verify $\lim_{\beta' \to \beta} \mathrm{MF}(\beta', F) = \mathrm{MF}(\beta, F)$ and we can also express $\frac{\partial}{\partial \beta}\mathrm{MF}(\beta, F)$ (the formula for $\frac{\partial}{\partial \beta}\mathrm{MF}(\beta, F)$ in case of a location parameter is in Mašíček, 2002).

We prove inequalities (4.2) and (4.3) using stochastical ordering (for the definition see appendix, Lemma 5.1 and 5.2). Let X and Z be two independent random variables with conjugated d.f. $F(x, z) = F_X(x)F_Z(z)$ ($x \in R^p$ and $z \in R$) where $F_X(x)$ and $F_Z(z)$ are marginal d.f.. Let us express $\mathrm{MF}_\alpha(\beta, F)$ for fixed $\alpha \in (0, 1)$ and $\beta \in R^p$. By definition (2.3) the value $\mathrm{MF}_\alpha(\beta, F)$ depends only on the distribution of the random variable $|Z - X^T\beta|$. Because X and Z are independent and Z is symmetric it is easily seen that $(X, Z) =_D (X, Z\,\mathrm{sign}(X^T\beta))$ and hence (denote $=_D$ equation of distributions)

$$|Z - X^T\beta| = |Z\,\mathrm{sign}(X^T\beta) - |X^T\beta|| =_D |Z - |X^T\beta||. \tag{4.5}$$

Let us introduce random variables

$$V_\beta := |X^T\beta|, \qquad U_\beta := |Z - V_\beta|. \tag{4.6}$$

Obviously $\mathrm{MF}_\alpha(\beta, F)$ depends only on the distribution of U_β. It is easy to check (see (2.3)) that

$$\mathrm{MF}_\alpha(\beta, F) = \mathrm{E}(U'_\beta)^2 \tag{4.7}$$

where $U'_\beta := U_\beta \cdot \mathrm{I}\{U_\beta \le u_\beta(\alpha)\}$ and $u_\beta(\alpha)$ is defined by equation

$$P\left(U_\beta \le u_\beta(\alpha)\right) = \alpha. \tag{4.8}$$

Notice F_β is the d.f. of the random variable U_β and $u_\beta(\alpha) = F_\beta^{-1}(\alpha)$. Let us recall that F_β is the d.f. of $|Z - X^T\beta|$ (see (2.3) – the definition of G_t).

Let us denote by $F_{\beta*}$ the d.f. of U'_β. Distribution function $F_{\beta*}$ can be easily expressed using F_β

$$F_{\beta*}(u) = \min\{F_\beta(u) + (1-\alpha), 1\} \qquad \text{for } u > 0. \tag{4.9}$$

This holds because U'_β is equal to U_β cut off at its α-quantile. For $u \le 0$ obviously $F_\beta(u) = F_{\beta*}(u) = 0$.

Now we prove inequalities (4.2) and (4.3). Let $\beta_1, \beta_2 \in R^p$ be fixed such that

$$V_{\beta_1} \prec V_{\beta_2}, \qquad V_{\beta_1} \ne_D V_{\beta_2}, \tag{4.10}$$

i.e., V_{β_2} is stochastically larger than V_{β_1} and both random variables do not have the same distribution. If we prove

$$\mathrm{MF}_\alpha(\beta_1, F) < \mathrm{MF}_\alpha(\beta_2, F) \tag{4.11}$$

the inequalities (4.2) and (4.3) will follow. The first inequality (4.2) is a special case of (4.11) for $\beta_1 := 0$ and $\beta_2 := \beta \ne 0$. It follows immediately that $V_{\beta_1} \prec V_{\beta_2}$ (because $V_{\beta_1} = V_0 = 0$ and $V_{\beta_2} = V_\beta \ge 0$) and $V_{\beta_1} \ne_D V_{\beta_2}$ (suppose $V_\beta =_D V_0 \equiv 0$ hence $X^T\beta = 0$ a.s. which is in contradiction to assumption **A3**). The second inequality (4.3) is obtained for $\beta_1 := \beta$ and $\beta_2 := K\beta$, where $\beta \ne 0$ and $K > 1$.

It remains to prove (4.11). The d.f. of U_β for $u > 0$ and $\beta \in R^p$ is

$$\begin{aligned} F_\beta(u) &= P\left(U_\beta < u\right) = P\left(-u < Z - V_\beta < u\right) \qquad (4.12)\\ &= \mathrm{E}P\left(-u < Z - v < u | V_\beta = v\right) = \mathrm{E}\left[F_Z(V_\beta + u) - F_Z(V_\beta - u)\right]. \end{aligned}$$

The last equality holds by independency of Z and V_β.

Fix $u > 0$. Observe that the function $g(x) := F_Z(x+u) - F_Z(x-u)$ is strictly decreasing in x for $x > 0$ because

$$\frac{\partial g(x)}{\partial x} = \frac{\partial}{\partial x}[F_Z(x+u) - F_Z(x-u)] = f_Z(x+u) - f_Z(x-u) < 0 \tag{4.13}$$

for $x > 0$. The last inequality follows by assumption **A2** ($|x+u| > |x-u|$ and f_Z being symmetric and strictly decreasing on $(0, \infty)$). By Lemma 5.2 (implication 1. ⇒ 4.), (4.10) and (4.12) is

$$F_{\beta_1}(u) = \mathrm{E}g(V_{\beta_1}) > \mathrm{E}g(V_{\beta_2}) = F_{\beta_2}(u) \qquad \text{for } u > 0. \tag{4.14}$$

Combining (4.9) and (4.14) implies that $F_{\beta_1 *}(u) \ge F_{\beta_2 *}(u)$ for $u > 0$, i.e., $U'_{\beta_1} \prec U'_{\beta_2}$ (see Lemma 5.1, implication 3. ⇒ 1.). Using $F_\beta(0^+) < \overline{\alpha}$ (see condition **A3** and definition of F_β) we can assert that for u from a right neighborhood of zero is $F_{\beta_1 *}(u) > F_{\beta_2 *}(u)$. Applying Lemma 5.2, implication 2. ⇒ 3. (substitute

$X := U'_{\beta_1}$, $Y := U'_{\beta_2}$ and $h(x) := \text{sign}(x)\, x^2$) yields

$$\text{MF}_\alpha(\beta_1, F) = \text{E}\left(U'_{\beta_1}\right)^2 = \text{E}h\left(U'_{\beta_1}\right) < \text{E}h\left(U'_{\beta_2}\right) = \text{E}\left(U'_{\beta_2}\right)^2 = \text{MF}_\alpha(\beta_2, F), \tag{4.15}$$

because $U'_{\beta_1} \geq 0$ and $U'_{\beta_2} \geq 0$, i.e., (4.11) holds. □

Proof of Theorem 3.1 To prove the weak consistency we will use Prochorov's metric defined for any two d-dimensional distribution functions G and H as follows

$$\pi(G, H) = \inf\left\{\delta : P_G(V \in A) \leq P_H(V \in A^\delta) + \delta, A - \text{Borel subset } R^d\right\} \tag{4.16}$$

where A^δ is the set of points in R^d which have distance from A smaller than δ and $P_G(V \in A)$ denotes the probability of $[V \in A]$ if V has distribution function G (for details see Jurečková and Sen, 1996, Section 2.5.3).

Let us fix a probability space $(\Omega, \mathcal{A}, P)$ and denote F_n the e.d.f. based on random vectors $(X_1, Z_1), \ldots, (X_n, Z_n)$. Observe that $F_n \to F$ pointwise and almost surely (by the SLLN), and hence weakly. Because Prochorov's metric metricize weak convergence we obtain $\pi(F_n, F) \to 0$ a.s. and hence

$$\pi(F_n, F) \to_P 0. \tag{4.17}$$

The LWS estimator is regression equivariant, hence we may put $\beta_0 = 0$. Suppose the constant $\delta \in (0, 1)$ is small but fixed. In the next part of the proof we will need several upper bounds on δ. These bounds will depend only on the weight function w and the d.f. F (it would be complicate to express δ now therefore we express it at the end of the proof).

Denote by $K_z := \text{E}Z_i^2 + 1$. Convergence (4.17) together with the weak law of large numbers gives us the following: there exists a sequence $B_n \subset \Omega$ such that $P(B_n) \to 1$ and for any $\omega \in B_n$ and any Borel $A \subset R^{p+1}$ the following inequalities hold

$$P((X, Z) \in A) \leq P_n((X, Z) \in A^\delta) + \delta \tag{4.18}$$

$$P_n((X, Z) \in A) \leq P((X, Z) \in A^\delta) + \delta \tag{4.19}$$

$$\frac{1}{n}\sum_{i=1}^{n} Z_i^2 \leq K_z \tag{4.20}$$

where P denotes the probability measure based on d.f. F (i.e., (X, Z) has d.f. F) and P_n denotes the empirical probability measure based on F_n (i.e., (X, Z) has d.f. F_n).

We split the rest of the proof into two steps.

1. We find a constant K_1 (which does not depend on δ) such that for any $\omega \in B_n$

$$\text{MF}(\beta, F_n) > \text{MF}(0, F_n) \qquad \text{for } \|\beta\| \geq K_1.$$

2. We find a constant K_2 (which does not depend on δ) such that for any $\omega \in B_n$

$$|\text{MF}(\beta, F_n) - \text{MF}(\beta, F)| \leq \delta K_2 + \frac{1}{n} K_2 \qquad \text{for } \|\beta\| \leq K_1.$$

These two results together with Lemma 4.1 conclude the proof. Fix $\eta > 0$. By Lemma 4.1 there exists $\delta_1 > 0$ such that for and β with $\|\beta\| \geq \eta$ it holds

$$\mathrm{MF}(\beta, F) \geq \mathrm{MF}(0, F) + \delta_1. \tag{4.21}$$

The second step together with (4.21) implies

$$\begin{aligned} \mathrm{MF}(\beta, F_n) &\geq \mathrm{MF}(\beta, F) - \delta K_2 - \frac{1}{n} K_2 \geq \mathrm{MF}(0, F) + \delta_1 - \delta K_2 - \frac{1}{n} K_2 \\ &\geq \mathrm{MF}(0, F_n) + \delta_1 - 2\delta K_2 - 2\frac{1}{n} K_2 \end{aligned} \tag{4.22}$$

for any β with $\eta \leq \|\beta\| \leq K_1$. The first and the third inequalities come from step 2 and the second one from (4.21). Choosing $\delta \leq \delta_1/(6K_2)$ and $n_1 := 6K_2/\delta_1$ (notice δ_1 and K_2 do not depend on δ) we obtain

$$\mathrm{MF}(\beta, F_n) > \mathrm{MF}(0, F_n) \qquad \text{for } \eta \leq \|\beta\| \leq K_1,\ n > n_1 \text{ and } \omega \in B_n. \tag{4.23}$$

The last inequality, together with step 1, implies

$$\mathrm{MF}(\beta, F_n) > \mathrm{MF}(0, F_n) \qquad \text{for } \|\beta\| \geq \eta,\ n > n_1 \text{ and } \omega \in B_n. \tag{4.24}$$

Hence for $n > n_1$ and for any $\omega \in B_n$, $\|\hat{\beta}_n\| \leq \eta$ because $\hat{\beta}_n$ is defined by minimization of $\mathrm{MF}(\beta, F_n)$ (see (1.4)). Together with convergence of $P(B_n) \to 1$ we obtain weak consistency of $\hat{\beta}_n$ because η was an arbitrary positive number. Now we should prove steps 1 and 2.

Fix $\omega \in B_n$ for the rest of the proof. Then there are two distributions – the empirical (i.e., P_n where (X, Z) has d.f. F_n) and the theoretical (i.e., P where (X, Z) has d.f. F).

Let us prove step 1, i.e., we find a constant K_1 (which does not depend on δ or ω) such that for any β with $\|\beta\| \geq K_1$ and $\omega \in B_n$

$$\mathrm{MF}(\beta, F_n) > \mathrm{MF}(0, F_n). \tag{4.25}$$

To prove this we show that $\mathrm{MF}(0, F_n)$ is bounded on B_n and that $\mathrm{MF}(\beta, F_n)$ is increasing with $\|\beta\|$.

The upper bound for $\mathrm{MF}(0, F_n)$ comes from (4.20) and from condition **A1** (w is nonincreasing)

$$\mathrm{MF}(0, F_n) = \frac{1}{n} \sum_{h=1}^{n} w\left(\frac{h-1}{n}\right) r_{(h)}^2(0) \leq w(0) \cdot \frac{1}{n} \sum_{i=1}^{n} Z_i^2 \leq w(0) K_z \tag{4.26}$$

for any $\omega \in B_n$.

Now we find a lower bound for $\mathrm{MF}(\beta, F_n)$. By condition **A3** there exist $\delta_2 > 0$ and $\delta_3 > 0$ such that for $\beta \neq 0$

$$P\left(|X^T \beta| \leq 2\delta_3 \|\beta\|\right) \leq \overline{\alpha} - 5\delta_2. \tag{4.27}$$

We can choose δ_3 in such a way that $\delta_3 \leq \delta_2$. Supposing $\delta \leq \delta_3$ (notice δ_3 does not depend on δ) and applying (4.19) for $A := \{(x, z) \in R^{p+1} : |x^T \beta| < \delta_3 \|\beta\|\}$

we obtain

$$P_n\left(|X^T\beta| < \delta_3\|\beta\|\right) = P_n\left(X \in A\right) \le P\left(X \in A^\delta\right) + \delta \qquad (4.28)$$
$$= P\left(|X^T\beta| \le (\delta_3+\delta)\|\beta\|\right) + \delta \le P\left(|X^T\beta| \le 2\delta_3\|\beta\|\right) + \delta_3$$

because $A^\delta = \{(x,z) \in R^{p+1} : |x^T\beta| \le (\delta_3+\delta)\|\beta\|\}$. Combining (4.27), (4.28) and $\delta_3 \le \delta_2$ yields

$$P_n\left(|X^T\beta| < \delta_3\|\beta\|\right) \le \overline{\alpha} - 5\delta_2 + \delta_3 \le \overline{\alpha} - 4\delta_2, \qquad (4.29)$$

i.e., for any $\beta \in R^p$ and $\omega \in B_n$ there is no more than $\lfloor n(\overline{\alpha} - 4\delta_2)\rfloor$ observations which have $|X_i^T\beta| < \delta_3\|\beta\|$ (denote $\lfloor x \rfloor$ the integral part of $x \in R$).

Let us take K_3 sufficiently large such that

$$P\left(|Z| > K_3 - 1\right) \le \delta_2. \qquad (4.30)$$

Notice that K_3 depends only on d.f. F. Inequality (4.19) for $A := \{(x,z) : |z| > K_3 - \delta\}$, together with (4.30), implies that for any $\omega \in B_n$ (recall $\delta \le \delta_2$ and $\delta < 1$)

$$P_n\left(|Z| > K_3\right) \le P\left(|Z| > K_3 - \delta\right) + \delta \le \delta_2 + \delta \le 2\delta_2, \qquad (4.31)$$

i.e., no more than $\lfloor 2n\delta_2 \rfloor$ random errors have $|Z_i| > K_3$.

Notice that for any β there are no more than $\lfloor n(1 - (\overline{\alpha} - \delta_2))\rfloor$ observations which do not have weights greater than or equal to $w(\overline{\alpha} - \delta_2) > 0$, because w is nonincreasing and hence at least $\lfloor n(\overline{\alpha} - \delta_2)\rfloor$ weights are greater than or equal to $w(\overline{\alpha} - \delta_2)$. Together we get that for any β and $\omega \in B_n$ at least n^* observations satisfy $|X_i^T\beta| \ge \delta_3\|\beta\|$ and $|Z_i| \le K_3$, and have weight greater than or equal to $w(\overline{\alpha} - \delta_2)$. The number of these observations can be expressed as

$$n^* := n - \lfloor n(\overline{\alpha} - 4\delta_2)\rfloor - \lfloor 2n\delta_2 \rfloor - \lfloor n(1 - (\overline{\alpha} - \delta_2))\rfloor \ge n\delta_2. \qquad (4.32)$$

Suppose $\delta_3\|\beta\| \ge K_3$. Hence at least n^* terms in (2.1) are greater than or equal to $w(\overline{\alpha} - \delta_2)(\delta_3\|\beta\| - K_3)^2$. This implies

$$\mathrm{MF}(\beta, F_n) = \frac{1}{n}\mathrm{MF}_n(\beta) \ge \frac{1}{n}n^* w(\overline{\alpha} - \delta_2)(\delta_3\|\beta\| - K_3)^2 \qquad (4.33)$$
$$\ge \delta_2 w(\overline{\alpha} - \delta_2)(\delta_3\|\beta\| - K_3)^2.$$

Combining (4.26) and (4.33) implies that there exists a sufficiently large constant K_1 such that for any β with $\|\beta\| \ge K_1$ and $\omega \in B_n$

$$\mathrm{MF}(\beta, F_n) \ge \delta_2 w(\overline{\alpha} - \delta_2)(\delta_3\|\beta\| - K_3)^2 > w(0)K_z \ge \mathrm{MF}(0, F_n) \qquad (4.34)$$

which completes step 1. Notice K_1 does not depend on δ or ω.

It remains to prove step 2, i.e., to prove that on B_n the value $\mathrm{MF}(\beta, F_n)$ is close to $\mathrm{MF}(\beta, F)$ for any β with $\|\beta\| \le K_1$. Because MF is an average of MF_α (see (2.3) and (2.5)) it is sufficient to prove that $\mathrm{MF}_\alpha(\beta, F_n)$ is close to $\mathrm{MF}_\alpha(\beta, F)$ for $\alpha \le \overline{\alpha}$.

Fix β with $\|\beta\| \le K_1$. We start with looking for an upper bound for the squared residuals on B_n. Let us take sufficiently large K_4 and K_5 such that

$$P\left(|Z| > K_4 - 1\right) \le (1 - \overline{\alpha})/4, \qquad P\left(\|X\| > K_5 - 1\right) \le (1 - \overline{\alpha})/4. \qquad (4.35)$$

Notice that K_4 and K_5 depend only on d.f. F. Suppose $\delta \leq (1-\overline{\alpha})/4$. Inequalities (4.35) imply for any $\omega \in B_n$ (see (4.19) and recall $\delta < 1$)

$$P_n(|Z| \leq K_4) = [1 - P_n(|Z| > K_4)] \geq [1 - P(|Z| > K_4 - \delta) - \delta] \tag{4.36}$$
$$\geq 1 - (1-\overline{\alpha})/4 - \delta \geq (1+\overline{\alpha})/2,$$
$$P_n(\|X\| \leq K_5) = [1 - P_n(\|X\| > K_5)] \geq [1 - P(\|X\| > K_5 - \delta) - \delta] \tag{4.37}$$
$$\geq 1 - (1-\overline{\alpha})/4 - \delta \geq (1+\overline{\alpha})/2.$$

Inequalities (4.35) also imply

$$P(|Z| \leq K_4) \geq (1+\overline{\alpha})/2, \qquad P(\|X\| \leq K_5) \geq (1+\overline{\alpha})/2. \tag{4.38}$$

Denote $K_6 := K_4 + K_5 K_1$. The inequality $\|\beta\| \leq K_1$, together with (4.36), (4.37) and (4.38), implies (for any $\omega \in B_n$)

$$P_n\left(|Z - X^T\beta| \leq K_6\right) \geq \overline{\alpha}, \qquad P\left(|Z - X^T\beta| \leq K_6\right) \geq \overline{\alpha}. \tag{4.39}$$

The first inequality in (4.39) gives us that for any β with $\|\beta\| \leq K_1$ and for any $\omega \in B_n$ at least $\lfloor n\overline{\alpha} \rfloor$ observations have residuals $(Z_i - X_i^T\beta)$ bounded by a constant K_6. The second one provides an upper bound on the theoretical quantile.

Let us recall that F_β is the d.f. of the random variable $|Z_i - X_i^T\beta|$ and $F_{n,\beta}$ the corresponding e.d.f. Inequalities (4.39) imply $F_{n,\beta}(K_6) \geq \overline{\alpha}$ and $F_\beta(K_6) \geq \overline{\alpha}$. Hence we can write

$$r_{|h|}(\beta) = F_{n,\beta}^{-1}(h/n) \leq K_6, \qquad F_\beta^{-1}(\alpha) \leq K_6 \tag{4.40}$$

for $h \leq \lfloor n\overline{\alpha} \rfloor$, $\alpha \leq \overline{\alpha}$ and β with $\|\beta\| \leq K_1$. Notice that we are still working on B_n, i.e., the first inequality in (4.40) holds for any $\omega \in B_n$.

Now we prove that $F_{n,\beta}$ is close to F_β. Applying (4.19) yields

$$F_{n,\beta}(u) = P_n\left(|Z - X^T\beta| < u\right) \leq P\left(|Z - X^T\beta| < u + \delta(1+\|\beta\|)\right) + \delta \tag{4.41}$$
$$= F_\beta(u) + P\left(u \leq |Z - X^T\beta| < u + \delta(1+\|\beta\|)\right) + \delta$$

for $u > 0$. Independency of Z and X and inequality $\|\beta\| \leq K_1$ imply

$$P\left(u \leq |Z - X^T\beta| < u + \delta(1+\|\beta\|)\right) = \mathrm{E}P[u \leq |Z - X^T\beta| < u + \delta(1+\|\beta\|)|X]$$
$$= \mathrm{E}\int \mathrm{I}\left\{u \leq |z - X^T\beta| < u + \delta(1+\|\beta\|)\right\} f_Z(z)\, dz$$
$$\leq 2\delta(1+\|\beta\|)M_f \leq 2\delta(1+K_1)M_f \tag{4.42}$$

where $M_f := \sup\{f_Z(z) : z \in R\} < \infty$. Combining (4.41) and (4.42) yields

$$F_{n,\beta}(u) \leq F_\beta(u) + \delta(1 + 2(1+K_1)M_f) = F_\beta(u) + \delta K_7 \tag{4.43}$$

for any β with $\|\beta\| \leq K_1$ where $K_7 := (1 + 2(1+K_1)M_f)$. Notice K_7 does not depend on δ or ω. In the same way we express the lower bound for $F_{n,\beta}(u)$ (using (4.18)) and obtain ($\|.\|_\infty$ denotes the supremum norm)

$$\|F_{n,\beta} - F_\beta\|_\infty \leq \delta K_7 \tag{4.44}$$

for any β with $\|\beta\| \leq K_1$ and any $\omega \in B_n$.

Fix $\alpha \leq \overline{\alpha}$. In the rest of the proof we use the same method as in Lemma 4.1. We define a random variable U_β which has d.f. F_β and random variable U'_β which has d.f. $F_{\beta*}$ where

$$F_{\beta*}(u) = \min\left\{F_\beta(u) + (1-\alpha), 1\right\} \qquad \text{for } u > 0 \tag{4.45}$$

and $F_{\beta*}(u) = 0$ for $u \leq 0$. Notice that $F_{\beta*}(u) = 1$ for $u \geq K_6$ because $F_\beta(K_6) \geq \alpha$ (see (4.39) or (4.40)). Hence we can write

$$\mathrm{MF}_\alpha(\beta, F) = \mathrm{E}\left(U'_\beta\right)^2 = \int_0^\infty (1 - F_{\beta*}(\sqrt{z}))\, dz = \int_0^{K_6^2} (1 - F_{\beta*}(\sqrt{z}))\, dz. \tag{4.46}$$

In the same way we express $\mathrm{MF}_\alpha(\beta, F_n)$. Denote by $F_{n,\beta*}$ the e.d.f. based on values $|Z_i - X_i^T\beta| \cdot \mathrm{I}\left\{|Z_i - X_i^T\beta| \leq r_{|h|}(\beta)\right\}$ where $h := \lfloor n\alpha \rfloor$. Because $F_{n,\beta}$ is not continuous we should express $F_{n,\beta*}$ more carefully. Noticing $r_{|h|}(\beta)$ is the α-quantile of $F_{n,\beta}$ gives

$$F_{n,\beta*}(u) = \min\left\{F_{n,\beta}(u) + \lfloor n(1-\alpha)\rfloor/n, 1\right\} \qquad \text{for } u > 0 \tag{4.47}$$

and $F_{n,\beta*}(u) = 0$ for $u \leq 0$. Again $F_{n,\beta*}(u) = 1$ for $u \geq K_6$ and $\omega \in B_n$ because $F_{n,\beta}(K_6) \geq \alpha$ (see (4.39) or (4.40)). Hence

$$\mathrm{MF}_\alpha(\beta, F_n) = \int_0^\infty (1 - F_{n,\beta*}(\sqrt{z}))\, dz = \int_0^{K_6^2} (1 - F_{n,\beta*}(\sqrt{z}))\, dz. \tag{4.48}$$

Combining (4.44), (4.45) and (4.47) gives

$$\|F_{n,\beta*} - F_{\beta*}\|_\infty \leq \delta K_7 + 1/n. \tag{4.49}$$

Last inequality, together with (4.46) and (4.48), implies

$$|\mathrm{MF}_\alpha(\beta, F_n) - \mathrm{MF}_\alpha(\beta, F)| = \left|\int_0^{K_6^2} (F_{n,\beta*}(z) - F_{\beta*}(z))\, dz\right| \tag{4.50}$$

$$\leq \int_0^{K_6^2} |F_{n,\beta*}(z) - F_{\beta*}(z)|\, dz \leq \delta K_8 + K_9/n$$

for $\|\beta\| \leq K_1$ and $\omega \in B_n$ where $K_8 := K_6^2 K_7$ and $K_9 := K_6^2$ which concludes the proof of step 2 because we can take $K_2 := \max\{w(0)K_8, w(0)K_9\}$ (see (2.5)). Notice that K_2 depends only on the weight function w and the d.f. F.

Finally notice that it is sufficient to take

$$\delta := \min\{1, \delta_1/(6K_2), \delta_3, (1-\overline{\alpha})/4\} \in (0,1) \tag{4.51}$$

where constants δ_1, K_2, δ_3 and $\overline{\alpha}$ depend only on the weight function w and on the distribution of random errors and explanatory variables (i.e., on the d.f. F).

□

5. Appendix

Let X and Y be two random variables. We say that Y is *stochastic larger* than X (denote $X \prec Y$) if and only if $\mathrm{E}h(X) \leq \mathrm{E}h(Y)$ for any nondecreasing function h for which both expected values exist and are finite.

Lemma 5.1. (*stochastic ordering* 1) *Let X and Y be two random variables with distribution functions F_X and F_Y. Then the following conditions are equivalent:*

1. $\mathrm{E}h(X) \leq \mathrm{E}h(Y)$ *for any nondecreasing function h for which both expected values exist and are finite (i.e., $X \prec Y$),*
2. $\mathrm{E}g(X) \geq \mathrm{E}g(Y)$ *for any nonincreasing function g for which both expected values exist and are finite,*
3. $F_X(z) \geq F_Y(z)$ *for $z \in R$.*

Lemma 5.2. (*stochastic ordering* 2) *Let X and Y be two random variables with distribution functions F_X and F_Y. Suppose that $X \prec Y$, then the following conditions are equivalent:*

1. $X \neq_D Y$,
2. *there exists $z_0 \in R$ such that $F_X(z_0) \neq F_Y(z_0)$,*
3. $\mathrm{E}h(X) < \mathrm{E}h(Y)$ *for any increasing function h, for which both expected values exist and are finite,*
4. $\mathrm{E}g(X) > \mathrm{E}g(Y)$ *for any decreasing function g, for which both expected values exist and are finite.*

References

[1] L. Davies, *The Asymptotics of S-estimators in the Linear Regression Model.* Ann. Statist. **18** (1990), 1651–1675.

[2] J. Jurečková and P.K. Sen, *Robust Statistical Procedures.* Wiley, New York, 1996.

[3] L. Mašíček, *Konzistence odhadu LWS pro parametr polohy (Consistency of LWS Estimator for Location Parameter).* In J. Antoch, G. Dohnal, and J. Klaschka, editors, ROBUST'2002, pages 240–246, 2002.

[4] P.J. Rousseeuw, *Least Median of Squares Regression.* J. Amer. Statist. Assoc. **79** (1984), 871–880.

[5] P.J. Rousseeuw and A.M. Leroy, *Robust Regression and Outlier Detection.* Wiley, New York, 1987.

[6] J.Á. Víšek, *Regression with High Breakdown Point.* In J. Antoch and G. Dohnal, editors, ROBUST'2000, pages 324–356, 2000.

[7] J.Á. Víšek, *A New Paradigm of Point Estimation.* In K. Kupka, editor, Proceedings of "Data Analysis 2000 – Modern Statistical Methods – Modelling, Regression, Classification and Data Mining", pages 195–230. Trilobyte, Pardubice, 2000.

L. Mašíček, Department of Probability and Statistics, Faculty of Mathematics and Physics
Charles University, Sokolovská 83, 186 75 Prague, Czech Republic
e-mail: `masicek@karlin.mff.cuni.cz`, `libor@matfyz.cz`

Statistics for Industry and Technology, 195–206

Algorithms for Robust Model Selection in Linear Regression

S. Morgenthaler, R.E. Welsch and A. Zenide

Abstract. In modelling situations, we often have a choice. We can add to model complexity to describe some unusual observations or groups of observations or we can remove those observations and fit the rest with a different, perhaps simpler, model. Or, we might put weights on observations and/or variables and use all of the data or all of the model complexity or some combination. To decide what is unusual generally requires some sort of model, but that very model may be what is causing some observations to be seen as unusual or, in fact, masking unusual observations.

When confronted both by unnecessary model complexity and by unusual observations, a model selection technique must be able to identify the correct model structure as well as unusual observations. This paper proposes some new algorithms that are designed to address these problems and compares them with currently available approaches such as robust C_p and robust cross-validated selection.

Mathematics Subject Classification (2000). Primary 62F35; Secondary 62G09.

Keywords. Variable selection, outliers, cross-validation, robust regression.

1. Introduction

Given observations $\boldsymbol{y} = (y_1, \ldots, y_n) \in \mathbb{R}^n$ and corresponding instances of explanatory variables $\boldsymbol{x}_1 \in \mathbb{R}^p, \ldots, \boldsymbol{x}_n \in \mathbb{R}^p$, we consider linear predictors

$$\widehat{y}_i = \hat{\beta}_0 + \hat{\boldsymbol{\beta}}' \boldsymbol{x}_i \qquad i = 1, \ldots, n, \tag{1.1}$$

with regression coefficients $\beta_0, \beta_1, \ldots, \beta_p$. We assume that the observations are realizations of

$$Y_i = E(Y_i|\boldsymbol{x}_i) + \varepsilon_i, \qquad i = 1, \ldots, n, \tag{1.2}$$

The work described in this paper was supported by a fellowship for young researchers awarded by the the Swiss National Science Foundation and by the Center for Computational Research in Economics and Management Science at the Massachusetts Institute of Technology Sloan School of Management.

where $\varepsilon_1, \ldots, \varepsilon_n$ are independent and have common mean 0 and variance σ^2. A model selection technique selects a best subset among the p explanatory variables such that an accurate linear predictor of type (1.1) is obtained if non-zero regression coefficients are only allowed for the selected variables. This is a fundamental problem of statistics, which is of great importance beyond the linear regression setting. Assuming a Gaussian distribution for the errors ε_i many asymptotically equivalent criteria have been proposed, among them (in chronological order):

$$\begin{aligned}
\text{MS/df(subset)} &= \text{RSS}_{\text{subset}}/(n - q_{\text{subset}})^2 \quad \text{(Tukey, 1967)} \\
\text{FPE(subset)} &= \text{RSS}_{\text{subset}}\,(n + q_{\text{subset}})/(n - q_{\text{subset}}) \quad \text{(Akaike, 1969)} \\
\text{PRESS(subset)} &= \text{leave-out-1 cross-validated RSS}_{\text{subset}} \quad \text{(Allen, 1971)} \\
&= \sum_{i=1}^{n} \left((y_i - \widehat{y}_i)^2/(1 - h_{ii})^2\right) \\
\text{AIC(subset)} &= n \log(\text{RSS}_{\text{subset}}) + 2\, q_{\text{subset}} \quad \text{(Akaike, 1973)}, \\
\text{AIC}_c\text{(subset)} &= n \log(\text{RSS}_{\text{subset}}) + 2\,(q_{\text{subset}} + 1) \times \\
&\qquad n/(n - q_{\text{subset}} - 2) \quad \text{(Sugiura, 1978)},
\end{aligned}$$

where $q_{\text{subset}} =$ (size of the subset + 1) counts the number of non-zero regression coefficients in (1.1). These criteria only depend on the subset and in particular do not require knowledge of σ^2. The basic ingredient for their computation is the residual sum of squares $\text{RSS}_{\text{subset}} = \sum_{i=1}^{n} (y_i - \widehat{y}_i)^2$. In the case of PRESS, the leave-out-1 cross-validation estimate refers to $\sum_{i=1}^{n} (y_i - \widehat{y}_{i,\text{cv}})^2$, with $\widehat{y}_{i,\text{cv}}$ denoting the prediction computed without using the ith observation. When using the least-squares estimator, the corresponding residuals can be directly computed with the help of $h_{ii} = \boldsymbol{x}_i'(\sum_{l=1}^{n} \boldsymbol{x}_l\, \boldsymbol{x}_l')^{-1}\, \boldsymbol{x}_i$, which on average are equal to q_{subset}/n. The MS/df criterion was proposed by J.W. Tukey in reaction to a remark in Anscombe (1967) and the finite sample correction to AIC was popularized and generalized by Hurvich and Tsai (1989). As shown in Shibata (1981), these criteria are asymptotically efficient in the sense that the variance of the prediction error achieved by the chosen predictor converges to the best variance possible. The first three criteria are all approximately proportional to $\text{RSS}_{\text{subset}}\,(1 + 2\, q_{\text{subset}}/n)$.

If we assume that the mean function $E(Y_i|\boldsymbol{x}_i)$ satisfies

$$Y_i = \mu + \boldsymbol{\theta}'\boldsymbol{x}_i + \varepsilon_i, \qquad i = 1, \ldots, n, \tag{1.3}$$

for some unknown $\mu \in \mathbb{R}$ and $\boldsymbol{\theta} \in \mathbb{R}^p$, then a different formulation of the model selection problem asks for identification of the "true subset", that is the index set of the non-zero components of $\boldsymbol{\theta}$. Related to this are the notions of overfitting, underfitting (parsimony) and asymptotic consistency. All the criteria mentioned above are asymptotically inconsistent and tend to select too large a subset (they overfit). PRESS can be rendered consistent by fitting on fewer than $n-1$ observations and thus keeping in reserve for validation more than 1 observation (leave-out-more-more-than-1, cross-validation, see Shao, 1993). In fact, the validation set should be $O(n)$.

In practice, the errors are rarely generated by a Gaussian distribution and this invalidates the theoretical arguments behind both the least squares estimates used in (1.1) and the selection criteria based on the residual sum of squares. Robust model selection refers to the problem of finding a good subset – either for prediction or in order to identify correctly the true model – in a situation where the observations (both $\mathbf{x}$ and y) may contain outliers (gross errors) and with errors exhibiting non-Gaussian behavior. A possible solution to this problem is the construction of robust versions of the model selection criteria, applicable for heavy-tailed error distributions and based on the replacement of $\widetilde{\beta}_0$ and $\widetilde{\boldsymbol{\beta}}$ in (1.1) with robust estimates. A nonexhaustive list of related papers contains Ronchetti (1985), who used the AIC statistic, Hurvich and Tsai (1990), who considered least-absolute-deviations regression and AIC$_c$, Ronchetti and Staudte (1994), who used the C_p statistic (see Mallows, 1973) and Ronchetti et al. (1997), who used cross-validation. A Bayesian approach is contained in Hoeting et al. (1996).

It is evident that outliers can create insoluble problems due to the heterogeneous apportioning of the information in regression problems. If we are asked to uncover unusual observations using only the available data, then we have to rely on the rule of the majority; that is, an observation is unusual because most of the others are quite different. Outliers in data generated by (1.2) or (1.3), however, are outliers because of the outlyingness of the errors and these we do not observe directly. Instead we have to rely on residuals $y_i - \widehat{y}_i$ and these in turn depend on our estimation of β_0 and $\boldsymbol{\beta}$. Each residual thus does not only depend on its associated error, but rather on the errors of all the data points used in the estimation. If an observation is exceptionally informative for one of the regression parameters, it will necessarily have a small residual and its outlyingness is directly linked to the presence or absence of the corresponding parameter in the selected model. In a simple linear regression, for example, with $x_1 = \cdots = x_{n-1} = 0$ and $x_n = 1$, the last observation can never be judged internally. If it is an outlier, we simply have had bad luck. Robust methods cannot resolve this predicament and can thus only be applied in situations where sufficient redundancy is assured.

In some sense, uncovering outliers and selecting a subset of the variables are interchangeable problems, a fact that can be made explicit by adding a dummy variable for each observation. If we had a model selection method that could deal with the case $p > n$, then we could select not only among the original p explanatory variables, but also from the n dummy variables. Each dummy variable ending up in the final selection corresponds to an observation that is being treated specially. It is also important to realize that outlyingness can only be defined with respect to a model.

In this paper, we will explore methods that play on this duality between variable selection and data selection. Sampling or selecting among the n observations is a widely used statistical technique that can be adapted to robust regression and regression diagnostics. It has been applied to inference problems (bias estimation, variance estimation, confidence intervals) (cross validation, bootstrap, jackknife), the detection or nomination of outliers (Hawkins et al., 1984, Billor et al., 1989 and Fernholz et al., 2004), to the computation of high-breakdown estimators (Rousseeuw, 1984 and Rousseeuw and Leroy, 1987) and for general diagnostic purposes (Atkinson and Riani, 2000). We will use sampling for variable selection as well.

2. Building Blocks

Variable selection in a linear model is a challenge even in the best of circumstances when

$$\boldsymbol{y} = \beta_0 + \beta_1 \boldsymbol{x}_1 + \cdots + \beta_p \boldsymbol{x}_p + \boldsymbol{\varepsilon},$$

with the random errors $\varepsilon_1, \ldots, \varepsilon_n$ uncorrelated and following a normal distribution and the design $\boldsymbol{x}_1, \ldots, \boldsymbol{x}_p$ carefully chosen. Many different methods for selection among the variables $\{\boldsymbol{x}_1, \ldots, \boldsymbol{x}_p\}$ have been proposed and analyzed. Most are based on the minimization of the estimated prediction error of the selected model. No best method, universally agreed upon, exists even in this simplistic setting.

The presence of outliers among the observed errors impacts variable selection profoundly. If we simply apply methods designed for the Gaussian case, we may make a bad choice in the sense that a single outlier can lead us to exclude one or several important variables from the final model and substitute variables with little predictive power. This traditional robustness problem is intimately linked with the difficulties caused by the presence of outlying points in the design space. Such singular points can also have an overwhelming influence. They greatly enlarge the space spanned by the design and if we search for a model with good predictive power over the whole span then these points cannot simply be ignored. On the other hand, if a wild design point is afflicted with a very large error, then it may be impossible to estimate a predictor that works well over the whole span. This dilemma is very different from cases where replicated observations that allow a clear identification of the potential outliers are available. Sometimes the best one can do is to delete wild design points and to concentrate on the prediction within the reduced span. The developer of a procedure for variable selection in the presence of outliers must respond to these challenges. The choices made by the developer should be transparent and understandable to the user.

Two methodologies are discussed in the literature. On one hand, pretreat the data with the aim of flagging potential outliers. The technique most often cited for this purpose is LMS (least-median-squares) (Rousseeuw, 1984) applied to the model using all available variables. This method is easy to explain and finds outliers even when they are masked by other outlying observations. Note that all available explanatory variables are used in the initial outlier screening. On the other hand, one may simultaneously nominate outliers and select variables. This is more appealing because it respects the dualism of these tasks. If we combine the data into the matrix $[\boldsymbol{x}_1|\cdots|\boldsymbol{x}_p|\boldsymbol{y}]$, variable selection refers to the selection of some among the first p columns, whereas outlier nomination refers to the flagging of some of the rows.

LMS is a robust regression fit most often implemented via a random search based on repeatedly drawing $(p+1)$ rows at random, computing the corresponding fit and the value of the median squared residual over all observations. The fit with the minimal value of this criterion is finally chosen. Potential outliers are observations with large absolute residual relative to the final fit. One can modify this procedure by updating an outlier score per observation based on the size of its residual in the randomly drawn fits. A host of outlier nomination procedures are discussed in the literature and could be used for this purpose. The impact of the pretreatment step on the second stage, the variable selection, is unclear. Sometimes the variable selection step is based uniquely on the "cleaned data", that is the data where the putative outliers have been removed. In this case, one has to be careful not to remove highly informative observations. It is preferable, if such outliers are kept in play at the second stage, since the flagged observations in any case probably also contain valuable information.

The parallel search of a model and of potential outliers is naturally implemented by sampling. In the base procedure, we select a random sample of the observations and of the variables and fit the corresponding model to the sampled observations by least squares. We then compute a robust estimate of the prediction error based on the residuals of the observations that have not been sampled. Computing the best fit, the one minimizing the estimated prediction error, over repetitions of the sampling leads to a robust model choice as long as the subsamples are not too large. If large subsamples are chosen, robust fitting should be considered. This basic procedure can be modified in many directions. We can search the model space in a more systematic fashion, for example in the manner of a forward/backward method in which models are compared to models with one additional or one less variable. Furthermore, this comparison can be sharpened by computing each of the models under comparison for all subsamples. We can also choose the subsamples of the observations in a balanced manner, ensuring, for example, that all observations are equally often represented or that all couples of observations are equally often represented, and so on. Another idea consists in replacing simple random sampling by weighted sampling with the weights readjusted after each step. The weights on the observations can take into account two aspects, they increase with decreasing outlyingness score and they increase with increasing information content. The weights on the variables should ideally increase with increasing predictive power of the variable and point us towards the good models. This can in general not be achieved with simple multinomial sampling and requires more sophisticated approaches.

Robust variable selection can thus be formulated as an optimization problem where a robust estimate of the prediction error is to be minimized over all possible models. The impact of the choice of this estimate on the quality of the procedure, the behavior of the procedure in the case of unbalanced designs, etc. is unclear. Similarly, there are a variety of algorithms for solving such problems and in this area too, not enough is known about the relative merits of various procedures. The results of extended simulation studies would, in our opinion, be a welcome addition to the existing literature.

3. Penalty Methods

In this section we explore the idea that uncovering outliers and selecting a subset of variables are, in some sense, interchangeable. Our matrix of "explanatory" variables becomes

$$\begin{array}{ccccccc} x_{11} & \cdots & x_{1,p+1} & 1 & 0 & 0 & \cdots & 0 \\ x_{21} & \cdots & x_{2,p+1} & 0 & 1 & 0 & \cdots & 0 \\ x_{31} & \cdots & x_{3,p+1} & 0 & 0 & 1 & \cdots & 0 \\ \vdots & \cdots & \vdots & \vdots & \vdots & \vdots & \cdots & \vdots \\ x_{n1} & \cdots & x_{n,p+1} & 0 & 0 & 0 & \cdots & 0 \end{array}$$

where the last column has been omitted to remove the intercept singularity. Since the number of columns now exceeds the number of rows, many traditional regression and selection procedures cannot be applied. However, penalty methods are useful in such situations and can provide some insight into our model and possible outliers. We now assume that the x-variables have been centered and scaled and that the "dummy" columns have been centered. The new matrix, of explanatory variables, which we call $\mathbf{Z}$, will be n by $n+p-1$ and our problem becomes

$$\min_{\beta} \left\{ \sum_{i=1}^{n} \left(y_i - \beta_0 - \sum_{j=1}^{n+p-1} z_{ij}\beta_j \right)^2 + k \sum_{j=1}^{n+p-1} \beta_j^2 \right\} \tag{3.1}$$

with β_0 estimated by $\bar{y}$. The penalty parameter, k, will control the complexity of the problem, i.e., the degrees of freedom used up in the estimation process. We can never use more than n degrees of freedom, even though we have $n+p-1$ possibilities. For k large enough, (3.1) is a well posed problem. The complexity or effective number of parameters (Hastie et al., 2001) is given by

$$\text{trace } \mathbf{Z}(\mathbf{Z}'\mathbf{Z} + k\mathbf{I})^{-1}\mathbf{Z}'. \tag{3.2}$$

There are many variations on (3.1). If we use the L1 norm for the penalty term, we obtain the "Lasso" of Tibshirani (1996). This is potentially useful in our case since we expect relatively few coefficients to be non-zero, but will not be explored in this paper. We use the L2 norm for the estimation term in (3.1) because the row dummies in the penalty term are designed to provide some robustness to outlying observations. A more robust loss function (e.g., L1) could be used but at increased computational cost.

Fan and Li (2001) consider a wide variety of penalty functions in the selection context. When extended to our selection problem with both real and dummy explanatory variables, the SCAD penalty functions of Fan and Li would allow large outliers not to be penalized (row fully removed without bias) and also allow rows with small outliers to have their dummy coefficients set to zero. We are planning to see if their methods will improve upon the preliminary results presented here.

Penalized methods require that we choose k. When the number of parameters is less than the number of observations there are many ways to do this, but a common choice is cross-validation. When the number of parameters exceeds the number of observations, there is almost no other choice.

Cross-validated penalized methods seem to provide excellent results for prediction (Frank and Friedman, 1996) where there is no need to actually "select" variables. In order to compare our results to a known model with known outlying observations, we need to select the real explanatory variables as well as those dummy variables that try to fit the outlying observations.

We choose to simply look at the t-statistics for the estimated coefficients

$$t_j(k) = \frac{\hat{\beta}_j(k)}{\hat{\sigma}\sqrt{(\mathbf{Z}'\mathbf{Z} + k\mathbf{I})^{-1}\mathbf{Z}'\mathbf{Z}(\mathbf{Z}'\mathbf{Z} + k\mathbf{I})^{-1}_{jj}}}$$

with $\hat{\sigma}$ estimated robustly by

$$\hat{\sigma} = 1.483 \text{ med} \mid y_i - \mathbf{z}_i^T \hat{\boldsymbol{\beta}}(k) - \text{med} \mid y_i - \mathbf{z}_i^T \hat{\boldsymbol{\beta}}(k) \mid \mid .$$

When $|t_j(k)|$ is greater than 2, we take the variable to be in the model. This can be further refined by the use of degrees of freedom, such as n minus the complexity (3.2), because the complexity is a measure of the degrees of freedom "used up" by the model.

In practice, the penalty method works well for cases where gross errors and leverage are not serious problems. However, the penalty $||\boldsymbol{\beta}||^2$ grows with the number of outliers and interferes with the selection of the "real" variables. One possible solution is to have two penalty functions – one related to the row dummy variables and the other to the real explanatory variables. This requires double cross-validation and increased computational complexity.

Our approach is to use the penalty method in five stages:

1. The solution to the optimization problem (3.1) is found using robust (MAD) cross-validation to determine k. We used ten-fold cross validation which leads to large training sets. Smaller training sets as suggested by Shao (1993) did not dramatically change our results.
2. All rows with dummy variables having $|t| > 2$ are removed.
3. The model (3.1) is optimized again without these rows and a new value of k is found by cross-validation which we will call k^*. This is considered to be a "robust" value of k.
4. All rows are again used to optimize (3.1) with k fixed at k^*. Rows with dummy variables having $|t| > 4$ are now removed.
5. The optimization problem (3.1) is now solved for the remaining rows with k found by cross-validation. Explanatory variable selection is done using $|t| > 2$ as the selection criterion.

The above procedure provides a fairly severe initial "pruning" of the data where the pruning is done based on the simultaneous weighting (in the principal component coordinate system) of both the real explanatory variables and row dummies. Then k^* is found as a working guess at the ratio of the likelihood variance to the prior variance (in the ridge or Bayes regression sense). In effect, this keeps k from getting too small due to a large $||\boldsymbol{\beta}||^2$ caused by a large number of outliers. Final row removal using $|t| > 4$ is done to maintain efficiency for the Gaussian case.

4. Sampling Methods

Ronchetti et al. (1997) have addressed variable selection and robustness using bounded influence methods whose breakdown decays as p increases. To maintain high breakdown for larger values of p, sampling methods (rather than true optimization methods) appear to be essential.

The traditional sampling approach is to argue that we can, with high probability, find a "good" subset of the data where our robust objective function would be minimized. Of course, we may never get such a data set in our sample, and the true minimum would be missed and our estimates would be wrong. These sampling methods (e.g., Rousseeuw and Leroy, 1987) have, so far, not been used directly in connection with variable selection. They have been used as a way to "clean" or "pre-screen" the data on the full model (all variables present) before variable selection is applied (Ronchetti et al., 1997) and to determine which outliers to consider in combination with a Bayesian selection method (Hoeting et al., 1996).

Since robust sampling methods like least median of squares are already computationally intensive, the idea of using all possible subsets selection with such methods on large problems is not computationally feasible.

4.1. Model Selection By Random Draws

In this section we study the basic algorithm using simple random sampling. The underlying idea is to draw models and observations at random, to fit this model by least-squares to these observations and to evaluate the predictive quality of the fitted model on the remaining observations. This procedure is summarized in Table 1. With p variables, there are 2^p potential models and with N drawings, each variable will be included in about $N/2$ of the drawings, each pair of variables will be included in about $N/4$ of the drawings, and so on. Each of the models is expected to be sampled $N/2^p$ times. The effectiveness of such a scheme deteriorates when p is large, or else N will have to be increased correspondingly. Robustness can be achieved easily and simply by choosing a robust measure of the predictive quality. In our implementation, we have used the median absolute deviation (MAD) of the prediction residuals.

The crucial question is how to use the result from this algorithm for the final variable selection. For the formal description, we need to introduce some notation.

TABLE 1. Algorithm for variable selection based on randomly drawn subsamples and models.

Step	Action
Initialization	Choose values for the number of draws $N > 0$, for the number of best models stored $n_b > 0$ and for the number of observations randomly chosen during the model fitting $n > n_s > p$.
Drawing	Draw a random model by selecting each variable at random with probability 1/2 and choose a random subset of size n_s from the n observations.
Evaluate	Compute the least squares fit of the model drawn with the selected data and compute the predictive residuals on the unused data; as a measure of prediction quality compute the MAD of these residuals (predictive MAD).
Summarize	Keep score of the n_b models with the lowest predictive MAD.

Let i be an index for the best models found ($i = 1, \ldots, n_b$) and let $m(i,j)$ code for the variables included in the model, in the sense that

$$m(i,j) = \begin{cases} 0, & \text{if variable } j \text{ is not included in model } i\,; \\ 1, & \text{if variable } j \text{ is included in model } i\,; \end{cases}$$

($i = 1, \ldots, n_b; j = 1, \ldots, p$). Furthermore, let pMAD(i) be the predictive MAD for model i. We wish to interpret the predictive MAD as an indicator for the predictive power of a model. However, in doing so, we must keep in mind that this indicator is highly variable and it would, for example, be naive to expect that the overall best model, that is the one with the lowest predictive MAD, is a good final selection. Let $X_1, \ldots, X_m$ be independent and identically distributed normal random variables with expectation 0 and variance σ^2 and let MAD $= med(|X_1|, \ldots, |X_m|)$. It follows that the asymptotic expectation is $\tau = 0.674\,\sigma$, whereas the asymptotic standard deviation is $1.2\,\tau m^{-1/2}$. For small values of m, the relative uncertainty in the estimation of τ as measured by plus/minus one standard deviation is huge. The relative error when $m = 20$, for example, is 50%. For $m = 100$ it still is more than 20% and when $m = 500$ we should expect a 10% error. Depending on the number of observations used in the evaluation of the predictive residuals $m = n - n_s$, a change of the predictive MAD by a factor of two is thus no indication of a real difference in the underlying predictive power.

In order to take account of this variability, we keep a number of the best models rather than a single one. The simplest method for determining the final selection is based on a summary of the form $m(+,j) = \sum_{i=1}^{n_b} m(i,j)$. The ordering of $m(+,j)$ can be taken as an indication for the relative importance of the variables. Under the assumption that none of the variables is needed, $m(+,j)$ has a binomial distribution with parameters n_b and 1/2 and – based on the normal approximation and the Bonferroni inequality – an observed value larger than $n_b * (0.5 + \Phi^{-1}(1 - 0.05/p)/(2\sqrt{n_b})$ is indicative of a significant departure.

Modifications of the basic scheme, for example, by periodically nominating as outliers the observations contributing above average to the worst models, may improve the performance, but we have not pursued such ideas.

5. Some Comparisons

We have discussed a variety of methods for the selection of cases and explanatory variables in the regression setting. To compare all possible combinations of case selection (and/or weighting) with variable selection is beyond the scope of this paper. Our benchmark is the experimental set-up in Ronchetti et al. (1997). The situations considered in this paper are regressions with six variables and four error distributions, namely Gaussian (e1), 7% wild (e2), slash (e3) and 10% asymmetric wild (e4). The reader is invited to consult the original paper for details. We consider model p2, where the last variable x_6 is superfluous, and $n = 60$ observations. In Table 2, we compare our methods (the rows called "draws" and "pen") with their

TABLE 2. This table shows the number of times that the correct model variables have been found in 200 simulations for $n = 60$ and model p2 with the methods discussed above. The row labelled draws shows the number of correct models obtained when using a cutoff value of 0.51, which is slightly smaller than the value proposed in the text, which for $n_b = 200$ is equal to 0.53.

	$n = 60$, model p2							
	without leverage				with leverage			
	e1	e2	e3	e4	e1	e2	e3	e4
LS	188	47	0	3	186	44	0	1
rob	157	160	10	167	169	166	7	172
draws	160	141	9	141	170	150	15	173
pen	164	166	10	141	80	86	45	66
LMSA	139	141	129	141	52	42	60	44

bounded influence method ("rob") and the all possible subsets (LS) method of Shao (1993). Since bounded influence methods have lower breakdown in higher dimensions, we have also included "LMSA" which uses least median of squares to find a "clean" subset of the data with all explanatory variables present. The clean set is defined by eliminating those observations with residuals greater than the 80th percentile in absolute value. Variable selection is accomplished by first ordering all possible subsets (models) using the single row deletion PRESS statistic. Since the PRESS statistic is subject to variation, we look at the lowest values of PRESS until the first "big" jump in PRESS values occurs. A big jump is defined as a jump bigger than 90% of all the jumps. Those variables included in models with PRESS values below the jump is taken as the final selected model.

The penalty method that just solves equation (3.1) and then selects variables with $|t| > 2$ (no steps 2 through 4) does very well in the "nice" case (e1, p2, no leverage) giving Correct 182 and Extra 18. Hence, for clean data, it would appear to be a viable (and cheaper) selection tool when compared to all possible subsets (LS) which gives the correct model in 188 out of 200. However, this simple penalty method does not do well with contaminated data (nor does LS for that matter) and the iteration steps 2 to 4 are required.

6. Conclusions

For the number of variables and observations considered here, sampling variables and observations simultaneously provides good performance in all cases, except when the errors follow a slash distribution. It is, of course, a computationally intensive method and how it will scale to larger problems remains to be seen. We plan to investigate various combinations of sampling and penalty methods to see if reasonable performance can be achieved on larger scale problems. In bioinformatics and data-mining there are many applications where p exceeds n and n is not too large. We are hopeful that the methods discussed in this paper will be particularly useful in these areas.

References

[1] H. Akaike, *Fitting Autoregressive Models for Prediction.* Ann. Inst. Statist. Math. **21** (1969), 243–247.

[2] H. Akaike, *Information Theory and an Extension of the Maximum Likelihood Principle.* In B.N. Petrov and F. Caski, editors, 2nd International Symposium on Information Theory, pages 267–281. Akademia Kiado, Budapest, 1973.

[3] F.J. Anscombe, *Topics in the Investigation of Linear Relations Fitted by the Method of Least Squares* (with discussion). J. Roy. Statist. Soc. Ser. B Stat. Methodol. **29** (1967), 1–52.

[4] A.C. Atkinson and M. Riani, *Robust Diagnostic Regression Analysis.* Springer Verlag, New York, 2000.

[5] N. Billor, A. Hadi, and P. Velleman, *Blocked Adaptive Computationally-efficient Outlier Nomination.* Comput. Statist. Data Anal. **34** (1989), 279–298.

[6] J. Fan and R. Li, *Variable Selection via Nonconcave Penalized Likelihood and its Oracle Properties.* J. Amer. Statist. Assoc. **96** (2001), 1348–1360.

[7] L.T. Fernholz, S. Morgenthaler, and J.W. Tukey, *An Outlier Nomination Method based on the Multihalver.* Statist. Probab. Lett. **21** (2004), in press.

[8] I. Frank and J. Friedman, *A Statistical View of Some Chemometrics Regression Tools* (with discussion). Technometrics **35** (1993), 109–148.

[9] T. Hastie, R. Tibshirani, and J. Friedman, *The Elements of Statistical Learning: Data Mining, Inference, and Prediction.* Springer, New York, 2001.

[10] D.M. Hawkins, D. Bradu, and G.V. Kass, *Location of Several Outliers in Multiple Regression Data Using Elemental Sets.* Technometrics **26** (1984), 197–208.

[11] J. Hoeting, A.E. Raftery, and D. Madigan, *A Method for Simultaneous Variable Solution and Outlier Identification in Linear Regression.* Comput. Statist. Data Anal. **22** (1996), 251–270.

[12] C.M. Hurvich and C.L. Tsai, *Regression and Time Series Model Selection in Small Samples.* Biometrika **76** (1989), 297–307.

[13] C.M. Hurvich and C.L. Tsai, *Model Selection for Least Absolute Deviations Regression in Small Samples.* Statist. Probab. Lett. **9** (1990), 259–265.

[14] C.L. Mallows, *Some Comments on Cp.* Technometrics **15** (1973), 661–675.

[15] E. Ronchetti, *Robust Model Selection in Regression.* Statist. Probab. Lett. **3** (1985), 21–23.

[16] E. Ronchetti and R.G. Staudte, *A Robust Version of Mallow's Cp.* J. Amer. Statist. Assoc. **89** (1994), 550–559.

[17] E. Ronchetti, C. Field, and W. Blanchard, *Robust Linear Model Selection by Cross-Validation.* J. Amer. Statist. Assoc. **92** (1997), 1017–1032.

[18] P.J. Rousseeuw, *Least Median of Squares Regression.* J. Amer. Statist. Assoc. **79** (1984), 871–880.

[19] P.J. Rousseeuw and A.M. Leroy, *Robust Regression and Outlier Detection.* Wiley, New York, 1987.

[20] J. Shao, *Linear Model Selection by Cross-Validation.* J. Amer. Statist. Assoc. **88** (1993), 486–494.

[21] R. Shibata, *An Optimal Selection of Regression Variables.* Biometrika **68** (1981), 45–54.

[22] N. Sugiura, *Further Analysis of the Data by Akaike's Information Criterion and The Finite Corrections.* Comm. Statist. Theory Methods **7** (1978), 13–26.

[23] R. Tibshirani, *Regression Shrinkage and Selection via the Lasso.* J. Roy. Statist. Soc. Ser. B Stat. Methodol. **5** (1996), 267–288.

[24] J.W. Tukey, Discussion of *Topics in the Investigation of Linear Relations Fitted by the Method of Least Squares* by F. J. Anscombe. J. Roy. Statist. Soc. Ser. B Stat. Methodol. **29** (1967), 47–48.

Acknowledgements

We would like to acknowledge the computing assistance of Lauren McCann, Ph.D. student at MIT, and the technical typing of Christine Liberty at MIT.

S. Morgenthaler and A. Zenide
Ecole Polytechnique Fédérale
Institute of Mathematics
CH-1015 Lausanne
Switzerland
e-mail: {stephan.morgenthaler, andrei.zenide}@epfl.ch

R.E. Welsch
Massachusetts Institute of Technology
Sloan School of Management
77 Massachusetts Avenue, E53-383
Cambridge, MA 02142
USA
e-mail: rwelsch@mit.edu

Statistics for Industry and Technology, 207–219

Analyzing the Number of Samples Required for an Approximate Monte-Carlo LMS Line Estimator

D.M. Mount, N.S. Netanyahu and E. Zuck

Abstract. This paper analyzes the number of samples required for an approximate Monte-Carlo least median of squares (LMS) line estimator. We provide a general computational framework, followed by detailed derivations for several point distributions and subsequent numerical results for the required number of samples.

Mathematics Subject Classification (2000). Primary 62J05; Secondary 68W40.
Keywords. Least median of squares (LMS) line estimator, Monte-Carlo LMS, approximation algorithms.

1. Introduction and Problem Definition

We are given a set of n observations in the plane (x_i, y_i), for $i = 1, \ldots, n$, and consider the linear regression model (with intercept)

$$y_i = (ax_i + b) + r_i,$$

where a and b are the slope and y-intercept of the regression line and r_i is the residual associated with the observation (x_i, y_i). We consider robust regression, in which the goal is to estimate the line parameters (a, b) that fit the bulk of the data, even when outliers are present. We assume that an upper bound on the fraction of *outliers* q, $0 \leq q < 0.5$, is known, so that the number of outliers is at most qn. We make no assumptions about the distribution of these outliers. The remaining points, of which there are at least $(1-q)n$ are called *inliers*, and are assumed to lie near the line (see further specifications below).

A popular approach for robust estimation of this type is Rousseeuw's *least median of squares* (LMS) line estimator (Rousseeuw, 1984). The estimator com-

This material is based upon work supported by the National Science Foundation under Grant No. 0098151.

putes the line of fit that minimizes the median squared residual (or any order statistic). Define a *strip* to be the region bounded between two nonvertical, parallel lines, and define the *vertical width* of a strip to be the vertical distance between these lines. LMS is computationally equivalent to determining a strip of minimum vertical width that contains at least 50% of the points. Thus, LMS estimation problem can be reduced into a geometric optimization problem.

A number of algorithms have been derived for the computation of the LMS estimator. Stromberg (1993) gave an exact algorithm for computing the LMS hyperstrip in d-dimensional space. His algorithm runs in $O(n^4 \log n)$ time for simple regression. A slightly improved algorithm which runs in $O(n^4)$ time is due to Agulló (1997). The best algorithm known for finding the LMS strip (in the plane) is the topological plane-sweep algorithm due to Edelsbrunner and Souvaine (1990). It runs in $O(n^2)$ time and requires $O(n)$ space. However, even quadratic running time is unacceptably high for many applications involving large data sets.

In practice, simple Monte-Carlo approximation algorithms are often used. The *feasible solution algorithm* due to Hawkins (1993) is one such algorithm. Additional Monte-Carlo-based algorithms are described in Rousseeuw and Leroy (1987) and Rousseeuw and Hubert (1997). Some number of pairs of distinct points are sampled at random. Each pair of points determines the slope of a line. The optimal intercept is then computed by a process called *intercept adjustment*, where the optimal intercept is determined by reducing the simple regression problem to an LMS location estimation problem. This can be done in $O(n \log n)$ time (see Rousseeuw and Leroy, 1987). Thus for a fixed number of subsamples, it is possible to determine the best LMS line approximation in $O(n \log n)$ time. This is equivalent to finding the narrowest strip containing (at least) half of the points over all the strips of slopes defined by the subsamples. The midline of the narrowest strip is the LMS line approximation.

Although inliers are usually assumed to lie exactly on the line, this is rarely the case in practice. We assume that the distribution of residuals is known and is independent of the x-coordinate. The main issue considered is how many pairs of points should be sampled to guarantee that, with sufficiently high probability, this Monte-Carlo estimate returns a sufficiently close approximation to the LMS line of fit? The meaning of "sufficiently close" is based on the formulation given by Mount et al. (1997). Let w^* denote the vertical width of the narrowest strip containing (at least) 50% of the points. Given an approximation error bound, $\epsilon \geq 0$, we say that any strip that contains (at least) 50% of the points and is of vertical width at most $(1+\epsilon)w^*$ is an *ϵ-approximate solution* to the LMS problem. Given a *confidence probability* p_c, our goal is to compute the minimum number of random samples needed, so that the Monte-Carlo algorithm returns an ϵ-approximate solution to the LMS problem with probability at least p_c. This analysis is based on some assumptions about the distribution of inliers, which are given in the next section.

2. Assumptions on the Point Distribution

Suppose that the line in question is given by $y = ax + b$. As previously noted, we make some concrete assumptions about the distribution of the inliers. Specifically, we assume that the x-coordinates of the points are uniformly distributed over an interval $[x_0, x_1]$, and that the y-coordinate for a point with x-coordinate x is $ax+b+z$, where the deviate z is a random variable having a certain probability density function (pdf). In Section 4 we consider specifically Gaussian and exponential pdf's. (Assuming normally distributed residuals is fairly common. The exponential pdf was chosen as a case study for a one-sided distribution.) In the full paper we also consider the uniform pdf (Mount et al., 2003).

Because the approximation bound is based on the ratio of vertical widths of two parallel strips, we may apply any affine transformation to the plane that preserves such ratios. In particular, in order to simplify the analysis, we transform the x-coordinates to lie within the interval $[0, 1]$, and we transform the y-coordinates by subtracting $ax - b$, so that the line of fit is mapped to the x-axis, and finally we scale the y-coordinates so that the standard deviation of residuals is some desired value. After this transformation, the optimal slope is zero. It is easy to verify that all these transformations preserve the ratio of vertical widths of parallel strips. Our analysis holds in the limit as n tends to infinity.

3. Sketch of Computational Procedure

Our analyzes for the various residual distributions are all based on a common approach. Any sampled pair of points defines a line with some slope s. Based on our knowledge of the distributions of the x- and y-coordinates of the inliers, it will be possible to derive the pdf of s, for pairs of inliers. In later sections we provide this derivation, but for now, let $f_S(s)$ denote this slope pdf.

Consider a parallel strip whose sides are of slope s, whose central line has y-intercept b, and whose vertical width is w. Let $A_b(s, w)$ denote the probability mass of this strip relative to the given point distribution. We will omit b when it is clear from context. The algorithm computes a strip of slope s that contains at least half of the points. Since we know nothing about outliers, if we want to guarantee that at least half of all the n points lie within a given strip, the strip should contain a fraction of at least $1/(2(1-q))$ of the $(1-q)n$ inliers. Throughout, we let $\mathcal{F}_I = 1/(2(1-q))$ denote this fraction of inliers. (Recall that $q < 0.5$.) Define the *slope-width function*, $w(s)$, to be the vertical width of the narrowest strip that has this fraction of the inlier probability mass. That is,

$$w(s) = \operatorname{argmin}_w \exists b \ (A_b(s, w) \geq \mathcal{F}_I). \tag{3.1}$$

For most reasonable residual distributions, we would expect that $w(s)$ increases monotonically relative to its optimum value of $w(0)$, denoted w^*. (See Figure 1.) This can be shown for the distributions presented here, by a straightforward perturbation argument, which will be presented in the full version of the

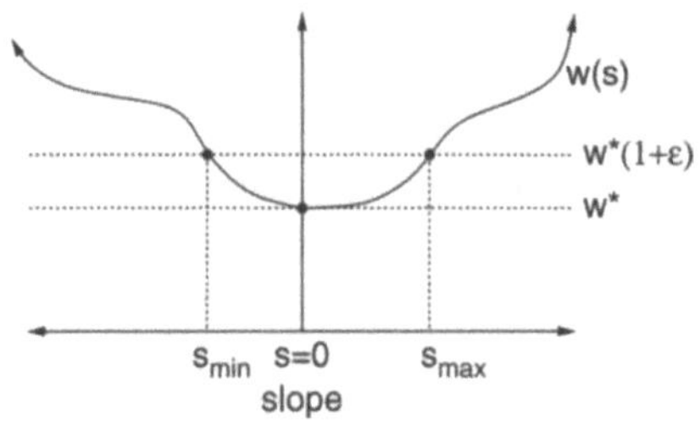

FIGURE 1. An illustration of the slope-width function $w(s)$ and minimum and maximum allowable slopes.

paper (Mount et al., 2003). Let $[s_{\min}, s_{\max}]$ denote the largest interval such that, for all $s \in [s_{\min}, s_{\max}]$, $w(s) \leq (1+\epsilon)w^*$. A slope s is *ϵ-good* if it lies within this interval. A sampled point pair is *ϵ-good* if the line passing through this pair has an ϵ-good slope. Recalling the $f_S(s)$ is the slope pdf, the probability of randomly sampling an ϵ-good point pair, denoted p_g, is

$$p_g = \int_{s_{\min}}^{s_{\max}} f_S(s)ds. \tag{3.2}$$

Finally, given that with probability p_c at least one good sample should be chosen, it follows that if N is the number of point pairs to be sampled, the probability of generating no good pair in N trials should be at most $1 - p_c$. Since we assume that sampling is performed independently, and since the probability of selecting a pair of inliers is $(1-q)^2$, N should be selected such that

$$(1-(1-q)^2 p_g)^N \leq 1 - p_c.$$

Solving for N, we obtain the desired bound on the number of samples.

$$N \geq \frac{\log(1-p_c)}{\log(1-(1-q)^2 p_g)}.$$

The remainder of the analysis reduces to computing the slope density function $f_S(s)$, computing the slope-width function $w(s)$, inverting this function to compute $s_{\min}$ and $s_{\max}$, and finally solving the integral of Eq. (3.2). In the rest of the paper we will provide detailed derivations of these results for two common residual pdf's.

4. Derivation of Slope Distribution

Let $p_i = (x_i, y_i)$, $p_j = (x_j, y_j)$ denote two distinct inlying points sampled at random. According to the point distribution model described in Section 2, x_i, x_j are independent, uniformly distributed in the interval $[0,1]$, that is, $x_i, x_j \sim \mathcal{U}(0,1)$ and y_i, y_j are independent, identically distributed (i.i.d.) according to some specified pdf. We make the general position assumption that the x-coordinates of the points are distinct. The slope of the line joining p_i and p_j is

$$s = \frac{y_j - y_i}{x_j - x_i}. \tag{4.1}$$

Thus to find $f_S(s)$, it suffices to find the pdf's of $(y_j - y_i)$ and $(x_j - x_i)$, and then find the pdf of their quotient. The latter can be found, using the following result from basic probability theory (Fisz, 1963; Papoulis, 1965). Let $S = V/U$ denote the quotient of two random variables V and U. The pdf of S is given by $f_S(s) = \int_{-\infty}^{\infty} f_{VU}(us, u)|u|du$, where $f_{VU}(.)$ is the joint pdf of V and U. Thus if V and U are independent,

$$f_S(s) = \int_{-\infty}^{\infty} f_V(us) f_U(u)|u|du. \tag{4.2}$$

Let U and V denote, respectively, the random variables $x_j - x_i$ and $y_j - y_i$. The pdf of U is given by the convolution of the pdf's corresponding to $\mathcal{U}(0,1)$ and $\mathcal{U}(-1,0)$. It is easy to show that

$$f_U(u) = \begin{cases} 1+u & -1 \leq u \leq 0 \\ 1-u & 0 \leq u \leq 1 \\ 0 & \text{otherwise.} \end{cases} \tag{4.3}$$

In the following subsections we derive the pdf of V for two common distributions. Also, we derive for each case the resulting pdf of the slope, using (4.2).

4.1. The Gaussian Case

We assume, without loss of generality, that $y_i, y_j \sim \mathcal{N}(0,1)$. Thus $-y_i \sim \mathcal{N}(0,1)$, as well. Based on the fact that the sum of (two) independent, normally distributed random variables has a normal pdf whose expected value is the sum of the expected values and whose variance is the sum of the variances, we have $y_j - y_i \sim \mathcal{N}(0, \sqrt{2})$. In other words, $f_V(v) = (1/(2\sqrt{\pi}))\ \exp(-v^2/4)$. Employing (4.2) we thus obtain

$$f_S(s) = \frac{1}{2\sqrt{\pi}} \int_{-1}^{1} e^{-k(s)u^2} f_U(u)|u|du, \tag{4.4}$$

where $k(s) \equiv s^2/4$. Since in view of (4.3) the integrand is an even function of u, we may rewrite the above equation as:

$$f_S(s) = \frac{1}{\sqrt{\pi}} \int_{0}^{1} e^{-k(s)u^2}(1-u)udu. \tag{4.5}$$

Solving the definite integral yields the following bottom-line expression:

$$f_S(s) = \frac{1}{2\sqrt{\pi}k(s)} \left(1 - 0.5\sqrt{\pi/k(s)}\ \mathrm{erf}(\sqrt{k(s)})\right), \tag{4.6}$$

where $\mathrm{erf}(x) = (2/\sqrt{\pi}) \int_0^x e^{-t^2} dt$. (See Mount et al., 2003, for a detailed derivation.) Figure 2(a) gives a plot of $f_S(s)$ for the Gaussian case. Note that although the denominator of $f_S(s)$ approaches zero as $s \to 0$, the function is well defined. Specifically, it can be shown that the limit of $f_S(s)$ as $s \to 0$ is equal to $1/(6\pi)$.

4.2. The Exponential Case

We now consider the exponential case. Without loss of generality, we assume that $y_i, y_j \sim \mathcal{E}(1)$, that is, y_i, y_j are independent, exponentially distributed with $\lambda = 1$. (This one-sided distribution is a special case of the gamma and Weibull distributions.) In other words, $y_j \sim e^{-y}H(y)$ and $-y_i \sim e^{y}H(-y)$, where $H(y)$ denotes a step function, such that $H(y) = 1$ for $y \geq 0$ and $H(y) = 0$ otherwise. It can be shown that the convolution of these pdf's results in a Laplace pdf (see Mount et al., 2003):

$$y_j - y_i \sim \frac{1}{2}e^{-|v|}. \tag{4.7}$$

Employing (4.2) we now obtain

$$f_S(s) = \frac{1}{2}\int_{-1}^{1} e^{-|us|} f_U(u)|u|du. \tag{4.8}$$

As before, since the integrand is an even function of u, we may rewrite this as:

$$f_S(s) = \int_{0}^{1} e^{-|s|u}(1-u)udu. \tag{4.9}$$

Solving the definite integral yields the desired slope pdf:

$$f_S(s) = \frac{1}{|s|^3}\left((|s|+2)e^{-|s|} + |s| - 2\right). \tag{4.10}$$

(See Mount et al., 2003, for a detailed derivation.) Although the denominator is equal to zero at $s = 0$, $f_S(s)$ is well-defined at the origin. In particular, it can be shown (by a Taylor expansion of the numerator up to terms of degree 4) that the limit of $f_S(s)$ as $s \to 0$ is equal to 1/6. Figure 2(b) gives a plot of $f_S(s)$ for the exponential case.

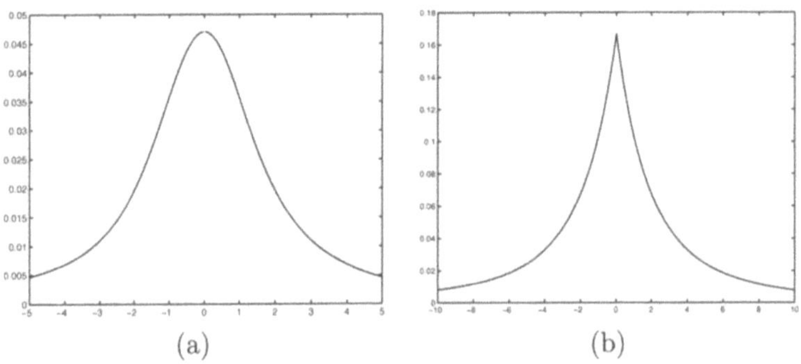

FIGURE 2. Slope pdf: (a) Gaussian case, (b) exponential case.

5. Derivation of the Slope-Width Function

In this section we compute the slope-width function $w(s)$, defined in Section 3.

5.1. The Gaussian Case

We consider first the special case $s = 0$ for which $w(s) = w(0) = w^*$. We make the observation that the desired parallelogram (a horizontal rectangle of width w^* in this case) must be symmetric with respect to the x-axis. By definition of the pdf of the y-coordinates, it is easy to show that the "area" (that is, probability mass) of any other horizontal rectangle of width w^* will be smaller than $A(0, w^*) = A_0(0, w^*)$. Thus

$$A(0, w^*) = \int_{-w^*/2}^{w^*/2} \int_0^1 f_Y(y) dx dy = \frac{1}{\sqrt{2\pi}} \int_{-w^*/2}^{w^*/2} e^{-y^2/2} dy, \qquad (5.1)$$

that is,

$$A(0, w^*) = \sqrt{\frac{2}{\pi}} \int_0^{w^*/2} e^{-y^2/2} dy. \qquad (5.2)$$

Given $\mathcal{F}_I$, we can find w^* (the solution to the implicit equation $A(0, w^*) = \mathcal{F}_I$), by using standard numerical analysis techniques (Atkinson, 1978), such as the Gaussian quadrature (to approximate the various definite integrals) and the iterative bisection method for root-finding of an equation. For $\mathcal{F}_I = 0.5$, for example, we obtain $w^* \approx 1.349$.

We now assume, without loss of generality, that $s > 0$, and consider a parallelogram of slope s and width w. It can be shown that among all such parallelograms, the one that is symmetric with respect to the line $y = sx - s/2$ has maximum area. Specifically this implies that $y = sx - s/2$ is also the midline of the narrowest strip containing half of the data points. (Moving any other strip in the direction of the above midline increases the probability mass; thus the width of any other strip is not a local minimum, in the sense that if the probability mass exceeds, say, 0.5, it is possible to locally shrink that strip so that it contains half of the data.) Henceforth, let $A(s, w) = A_{-s/2}(s, w)$. The non-vertical sides of the desired parallelogram are thus given by the line equations: $y_I = sx - (s - w)/2$ and $y_{II} = sx - (s + w)/2$. We distinguish between two cases: (1) $s \leq w$ and (2) $s > w$ (see Figures 3(a) and 3(b), respectively). Due to space limitations, we only discuss here the first case. (See Mount et al., 2003, for further details.)

We decompose the desired parallelogram into the horizontal rectangle whose sides are $y = \pm(w - s)/2$, and the triangles above and below this rectangle. The area (i.e., probability mass) of the two triangles is clearly the same. Thus the total probability mass of the parallelogram is given by

$$A(s, w) = \int_{-(w-s)/2}^{(w-s)/2} f_Y(y) dy + 2 \int_{-(w+s)/2}^{-(w-s)/2} x_{II}(y) f_Y(y) dy, \qquad (5.3)$$

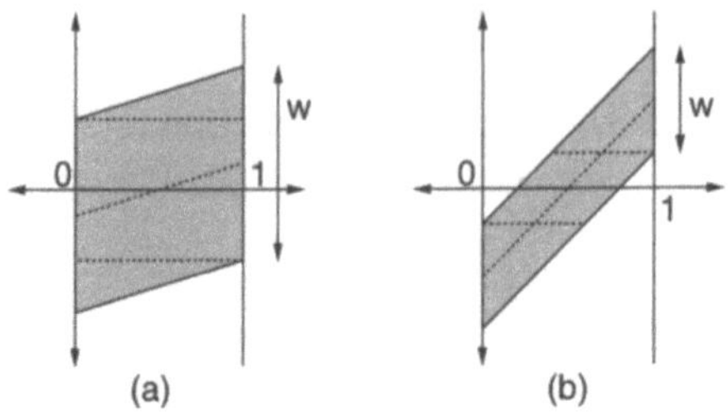

FIGURE 3. The low and high slope cases for the Gaussian distribution.

where $x_{II}(y) = (1/s)y + (1 + w/s)/2$, which yields

$$A(s,w)=\int_{-(w-s)/2}^{(w-s)/2} f_Y(y)dy+\frac{2}{s}\int_{-(w+s)/2}^{-(w-s)/2} yf_Y(y)dy+\left(1+\frac{w}{s}\right)\int_{-(w+s)/2}^{-(w-s)/2} f_Y(y)dy.$$

Solving the second definite integral leads to the following expression for $A(s, w)$:

$$A(s,w) = \sqrt{\frac{2}{\pi}}\int_0^{(w-s)/2} e^{-y^2/2}dy - \sqrt{\frac{8}{\pi}}\left(\frac{1}{s}\right) e^{-(w^2+s^2)/8} \sinh\left(\frac{ws}{4}\right) + \left(1+\frac{w}{s}\right)\frac{1}{\sqrt{2\pi}}\int_{-(w+s)/2}^{-(w-s)/2} e^{-y^2/2}dy.$$

Using standard numerical analysis techniques, as before, we can obtain numerical solutions for the implicit equation $A(s, w(s)) = \mathcal{F}_I$. That is, for a given $\mathcal{F}_I$ and s, we can look up $w(s)$ and vice versa.

Figures 4(a), 4(b) show plots of the strip width as a function of slope, for the Gaussian case and the exponential case, respectively, for various values of $\mathcal{F}_I$.

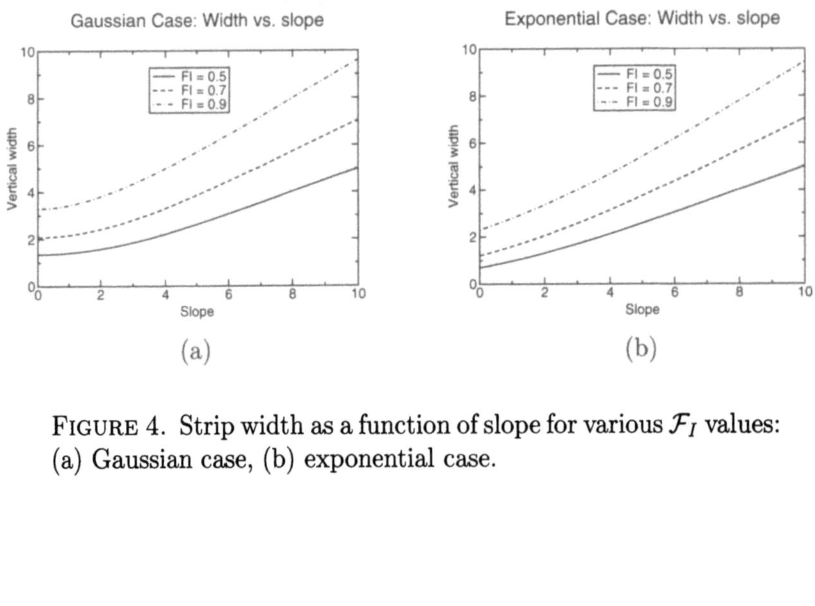

FIGURE 4. Strip width as a function of slope for various $\mathcal{F}_I$ values: (a) Gaussian case, (b) exponential case.

5.2. The Exponential Case

We consider first the special case $s = 0$. Given the one-sided pdf of the y-coordinates, it is obvious that the lower edge of the desired strip (of width w^*) coincides in this case with the x-axis. Thus letting $A(0, w^*) = A_0(0, w^*)$ we have

$$A(0, w^*) = \int_0^{w^*} e^{-y} dy = 1 - e^{-w^*}, \tag{5.4}$$

that is, for a given $\mathcal{F}_I$, we have

$$w^* = \ln\left(\frac{1}{1 - \mathcal{F}_I}\right). \tag{5.5}$$

Because of the lack of symmetry, determining the optimal strip for $s \neq 0$ is much less obvious than the Gaussian case. To do so, we will derive the probability mass associated with a differential strip of slope s, that is, the probability density of a line of slope s as a function of its y-intercept. Consider the family of lines $y = sx + y_{\text{int}}$, where the intercept y_{int} is a random variable. (Assume that $s > 0$. An analogous analysis to the one below applies to $s < 0$.) To obtain the distribution of y_{int}, that is, $F(z) = \mathcal{P}(y_{\text{int}} < z)$, we distinguish between the two cases: (1) $z \geq 0$ and (2) $-s < z < 0$. The distribution is 0 for $z < -s$, due to the underlying exponential pdf of the y-coordinates, and is essentially equal to the probability mass of the strip defined by $y = sx + z$ and $y = sx - s$ (see Figure 5). We have for $z > 0$

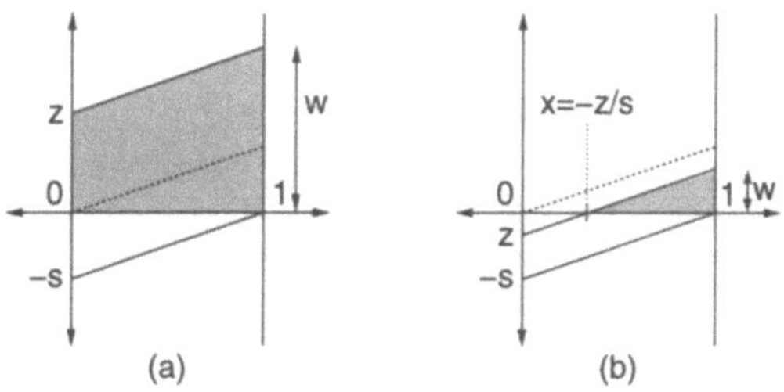

FIGURE 5. Regions for probability mass computation due to (a) (5.6) and (b) (5.7).

$$F(z) = \mathcal{P}(y_{\text{int}} < z) = \int_0^1 \left(\int_0^{sx+z} e^{-y} dy\right) dx = \int_0^1 \left(1 - e^{-sx-z}\right) dx,$$

which yields

$$F(z) = \mathcal{P}(y_{\text{int}} < z) = 1 + \frac{1}{s}\left(e^{-z-s} - e^{-z}\right). \tag{5.6}$$

For $-s < z \leq 0$ we have

$$F(z) = \int_{-z/s}^1 \left(\int_0^{sx+z} e^{-y} dy\right) dx = \int_{-z/s}^1 \left(1 - e^{-sx-z}\right) dx,$$

which yields

$$F(z) = 1 + \frac{z}{s} + \frac{1}{s}\left(e^{-z-s} - 1\right). \tag{5.7}$$

Differentiating the above distribution with respect to z provides the probability density of a line having a slope s and y-intercept z. We obtain

$$f_Z(z) = \begin{cases} \frac{1}{s}\left(1 - e^{-z-s}\right) & -s \leq z \leq 0 \\ \\ \frac{1}{s}\left(e^{-z} - e^{-z-s}\right) & z \geq 0. \end{cases} \tag{5.8}$$

It can easily be seen that the above pdf is unimodal. The mode occurs at $z = 0$ and its value is $f_Z(0) = m$, where $m = (1/s)\left(1 - e^{-s}\right)$.

Having established (for a given slope $s > 0$) the pdf of a line's intercept and its unimodality, we make the observation that the width of the desired strip is given by $w(s) = y_1 - y_0$, for some $y_0 < 0$, $y_1 > 0$, such that:

(1) $f_Z(y_0) = f_Z(y_1)$, and
(2) $F(y_1) - F(y_0) = A(s, w(s)) = \mathcal{F}_I$.

In this context, it is understood that $A(s, w(s))$ applies to the strip bounded by the y-intercepts y_0 and y_1. The first observation arises as a local-minimality constraint on the strip lying between the intercepts $[y_0, y_1]$. (If this did not hold, a differential argument shows that an infinitesimal shift of the strip, either up or down, would increase the probability mass without increasing the strip's width, contradicting the strip's minimality.) The second fact simply states that the strip must contain a fraction $\mathcal{F}_I$ of the probability mass. From (5.8) and the first constraint we obtain

$$1 - e^{-y_0-s} = e^{-y_1} - e^{-y_1-s}, \tag{5.9}$$

and substituting $(y_0 + w)$ for y_1 yields

$$e^{-y_0}\left(e^{-s} + e^{-w} - e^{-w-s}\right) = 1. \tag{5.10}$$

From (5.6), (5.7), we have

$$F(y_0) = 1 - \frac{1}{s} + \frac{y_0}{s} + \frac{1}{s}\; e^{-y_0-s}$$

and

$$F(y_1) = 1 + \frac{1}{s}\left(e^{-s} - 1\right)\; e^{-y_1}.$$

From the last two equations and the second constraint we obtain

$$A(s, w(s)) = \frac{1}{s}\left((e^{-s} - 1)e^{-y_1} + 1 - y_0 - e^{-s}e^{-y_0}\right) = \mathcal{F}_I.$$

Combining the latter equation with (5.9) and (5.10) yields

$$A(s, w(s))\; s = -y_0,$$

or

$$A(s, w(s))\; s = -\ln\left(e^{-s} + e^{-w} - e^{-w-s}\right) = -\mathcal{F}_I s.$$

Hence we get the following closed-form expression

$$w(s) = \ln\left(\frac{1+e^{-s}}{e^{-\mathcal{F}_I s} - e^{-s}}\right). \tag{5.11}$$

Finding s from a given w does not yield a closed-form solution. As in the previous subsection, a standard iterative root-finding technique was used.

6. Numerical Results

We show the concrete results of the analysis for the Gaussian and exponential distributions. Figure 6 shows the probability that single random sampled pair is *good* as a function of ϵ. Recall that a sampled pair is good if there is a strip whose slope is determined by the pair, that contains a fraction of $\mathcal{F}_I$ of the inliers, and whose width is at most $(1+\epsilon)$ times the optimum width. Recall that $\mathcal{F}_I = 1/(2(1-q))$, where q is the expected fraction of outliers, and hence the values $\mathcal{F}_I = \{0.5, 0.7, 0.9\}$ shown in the plots correspond to outlier fractions of $q = \{0, 0.29, 0.44\}$, respectively. Figure 7 shows the required number of samples as a function of ϵ for various confidence levels p_c, and for the fixed value $\mathcal{F}_I = 0.5$.

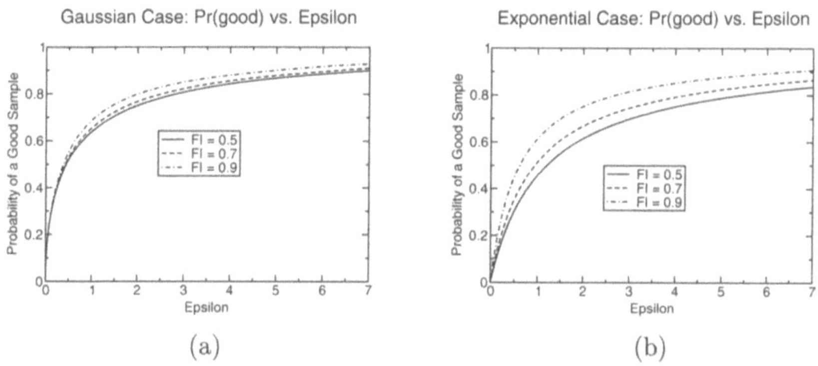

FIGURE 6. Probability that a single random sample is good as a function of ϵ for various $\mathcal{F}_I$ values: (a) Gaussian case, (b) exponential case.

References

[1] J. Agulló, *Exact Algorithms for Computing the Least Median of Squares Estimate in Multiple Linear Regression.* In Y. Dodge, editor, L_1-Statistical Procedures and Related Topics, IMS Lecture Notes Monograph Series **31**, pages 133–146. Institute of Mathematical Statistics, 1997.

[2] K.E. Atkinson, *An Introduction to Numerical Analysis.* John Wiley, New York, 1978.

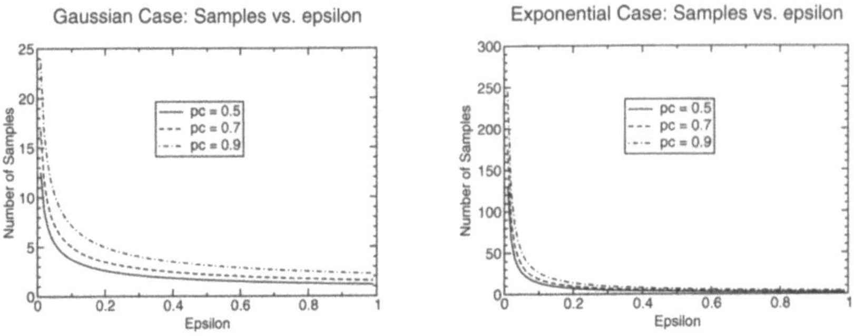

(a) (b)

FIGURE 7. Required number of samples as a function of ϵ for $\mathcal{F}_I = 0.5$ and for various confidence levels p_c: (a) Gaussian case, (b) exponential case.

[3] G. Dahlquist, A. Björk, and N. Anderson, *Numerical Methods.* Prentice-Hall, Englewood Cliffs, NJ, 1974.

[4] H. Edelsbrunner and D.L. Souvaine, *Computing Median-of-Squares Regression Lines and Guided Topological Sweep.* J. Amer. Statist. Assoc. **85** (1990), 115–119.

[5] M. Fisz, *Probability Theory and Mathematical Statistics.* John Wiley, New York, 1963.

[6] D.M. Hawkins, *The Feasible Set Algorithm for Least Median of Squares Regression.* Comput. Statist. Data Anal. **16** (1993), 81–101.

[7] D.M. Mount, N.S. Netanyahu, K. Romanik, R. Silverman, and A.Y. Wu, *Practical Approximation Algorithm for the LMS Line Estimator.* In Proceedings of the Eighth ACM-SIAM Symposium on Discrete Algorithms, pages 464–472. New Orleans, Louisiana, 1997.

[8] D.M. Mount, N.S. Netanyahu, and E. Zuck, *Analyzing The Number of Samples Required for an Approximate Monte-Carlo LMS Line Estimator.* Technical Report, 2003.

[9] A. Papoulis, *Probability, Random Variables, and Stochastic Processes.* McGraw-Hill, Tokyo, 1965.

[10] P.J. Rousseeuw and M. Hubert, *Recent Developments in PROGRESS.* In Y. Dodge, editor, L_1-Statistical Procedures and Related Topics, IMS Lecture Notes Monograph Series **31**, pages 201–214. Institute of Mathematical Statistics, 1997.

[11] P.J. Rousseeuw and A.M. Leroy, *Robust Regression and Outlier Detection.* Wiley & Sons, New York, 1987.

[12] P.J. Rousseeuw, *Least Median of Squares Regression.* J. Amer. Statist. Assoc. **79** (1984), 871–880.

[13] A.J. Stromberg, *Computing the Exact Least Median of Squares Estimate and Stability Diagnostics in Multiple Linear Regression.* SIAM J. Sci. Comput. **14** (1993), 1289–1299.

Acknowledgments

We thank Jeremy Schiff for several insightful discussions and Adi Sweete for his help in producing the numerical results contained in this paper.

D.M. Mount
Department of Computer Science
University of Maryland
College Park, MD 20742
U.S.A.
e-mail: mount@cs.umd.edu

N.S. Netanyahu
Department of Computer Science
Bar-Ilan University
Ramat-Gan 52900
Israel
and
Center for Automation Research
University of Maryland
College Park, MD 20742
U.S.A.
e-mail: nathan@macs.biu.ac.il, nathan@cfar.umd.edu

E. Zuck
Department of Computer Science
Bar-Ilan University
Ramat-Gan 52900
Israel
e-mail: zuckel@macs.biu.ac.il

Statistics for Industry and Technology, 221–233

Visualizing 1D Regression

D.J. Olive

Abstract. Regression is the study of the conditional distribution of the response y given the predictors $\boldsymbol{x}$. In a 1D regression, y is independent of $\boldsymbol{x}$ given a single linear combination $\boldsymbol{\beta}^T\boldsymbol{x}$ of the predictors. Special cases of 1D regression include multiple linear regression, binary regression and generalized linear models. If a good estimate $\hat{\boldsymbol{b}}$ of some non-zero multiple $c\boldsymbol{\beta}$ of $\boldsymbol{\beta}$ can be constructed, then the 1D regression can be visualized with a scatterplot of $\hat{\boldsymbol{b}}^T\boldsymbol{x}$ versus y. A resistant method for estimating $c\boldsymbol{\beta}$ is presented along with applications.

Mathematics Subject Classification (2000). Primary 62G35; Secondary 62F35.

Keywords. Diagnostics, generalized linear models, minimum covariance determinant, response transformations, single index models.

1. Introduction

Regression is the study of the conditional distribution $y|\boldsymbol{x}$ of the response y given the $(p-1)\times 1$ vector of nontrivial predictors $\boldsymbol{x}$. In a *1D regression* (or regression with 1-dimensional structure), y is conditionally independent of $\boldsymbol{x}$ given a single linear combination $\boldsymbol{\beta}^T\boldsymbol{x}$ of the predictors, written

$$y \perp\!\!\!\perp \boldsymbol{x}|\boldsymbol{\beta}^T\boldsymbol{x}. \tag{1.1}$$

A *1D regression model* has the form

$$y = g(\alpha + \boldsymbol{\beta}^T\boldsymbol{x}, e) \tag{1.2}$$

where g is a bivariate (inverse link) function and e is a zero mean error that is independent of $\boldsymbol{x}$. See Li and Duan (1989) and Cook and Weisberg (1999a, p. 414).

The above class of models is very rich. A *single index model* uses

$$y = g(\alpha + \boldsymbol{\beta}^T\boldsymbol{x}, e) \equiv m(\alpha + \boldsymbol{\beta}^T\boldsymbol{x}) + e, \tag{1.3}$$

This research was supported by NSF grant DMS 0202922.

and the *multiple linear regression model* is an important special case where m is the identity function: $m(\alpha + \boldsymbol{\beta}^T \boldsymbol{x}) = \alpha + \boldsymbol{\beta}^T \boldsymbol{x}$. Another important special case of 1D regression is the *response transformation model* where

$$g(\alpha + \boldsymbol{\beta}^T \boldsymbol{x}, e) = t^{-1}(\alpha + \boldsymbol{\beta}^T \boldsymbol{x} + e) \tag{1.4}$$

and t^{-1} is a one to one (typically monotone) function so that $t(y) = \alpha + \boldsymbol{\beta}^T \boldsymbol{x} + e$. Generalized linear models (GLM's) are also a special case of 1D regression.

Some notation from the regression graphics literature will be useful. *Dimension reduction* can greatly simplify our understanding of the conditional distribution $y|\boldsymbol{x}$. If a 1D regression model is appropriate, then the $(p-1)$-dimensional vector $\boldsymbol{x}$ can be replaced by the 1-dimensional scalar $\boldsymbol{\beta}^T \boldsymbol{x}$ with no loss of information. A *sufficient summary plot* (SSP) is a plot that contains all the sample regression information about the conditional distribution of $y|\boldsymbol{x}$. For 1D regression, if $y \perp\!\!\!\perp \boldsymbol{x}|\boldsymbol{\beta}^T \boldsymbol{x}$ then $y \perp\!\!\!\perp \boldsymbol{x}|c\boldsymbol{\beta}^T \boldsymbol{x}$ for any constant $c \neq 0$. The quantity $c\boldsymbol{\beta}^T \boldsymbol{x}$ is called a *sufficient predictor* (SP), and a plot of the SP versus y is a SSP. If a consistent estimator $\hat{\boldsymbol{b}}$ of $c\boldsymbol{\beta}$ can be found for some nonzero c, then an *estimated sufficient summary plot* (ESSP) is a plot of the *estimated sufficient predictor* (ESP) $\hat{\boldsymbol{b}}^T \boldsymbol{x}$ versus y.

Additional notation is needed before giving theoretical results. Let $\boldsymbol{x}$, $\boldsymbol{a}$, $\boldsymbol{t}$, and $\boldsymbol{\beta}$ be $(p-1) \times 1$ vectors where only $\boldsymbol{x}$ is random. The predictors $\boldsymbol{x}$ satisfy the condition of *linearly related predictors* with 1D structure (Cook and Weisberg, 1999a, p. 431) if

$$E[\boldsymbol{x}|\boldsymbol{\beta}^T \boldsymbol{x}] = \boldsymbol{a} + \boldsymbol{t}\boldsymbol{\beta}^T \boldsymbol{x}. \tag{1.5}$$

Notice that $\boldsymbol{\beta}$ is a fixed $(p-1) \times 1$ vector. If $\boldsymbol{x}$ is elliptically contoured (EC) with 1st moments, then the assumption of linearly related predictors holds. See Cook (1998a, p. 130).

Following Cook (1998a, pp. 143–144), assume that there is an objective function

$$L_n(a, \boldsymbol{b}) = \frac{1}{n} \sum_{i=1}^{n} L(a + \boldsymbol{b}^T \boldsymbol{x}_i, y_i) \tag{1.6}$$

where $L(u, v)$ is a bivariate function that is a convex function of the first argument u. Assume that the estimate $(\hat{a}, \hat{\boldsymbol{b}})$ of $(a, \boldsymbol{b})$ satisfies

$$(\hat{a}, \hat{\boldsymbol{b}}) = \arg\min_{a, \boldsymbol{b}} L_n(a, \boldsymbol{b}). \tag{1.7}$$

For example, the ordinary least squares (OLS) estimator uses

$$L(a + \boldsymbol{b}^T \boldsymbol{x}, y) = (y - a - \boldsymbol{b}^T \boldsymbol{x})^2.$$

Maximum likelihood type estimators such as those used to compute GLM's and Huber's M-estimator also work, as does the Wilcoxon rank estimator. Assume that the population analog $(\alpha, \boldsymbol{\eta})$ is the unique minimizer of $E[L(a + \boldsymbol{b}^T \boldsymbol{x}, y)]$ where the expectation exists and is with respect to the joint distribution of $(y, \boldsymbol{x}^T)^T$. For

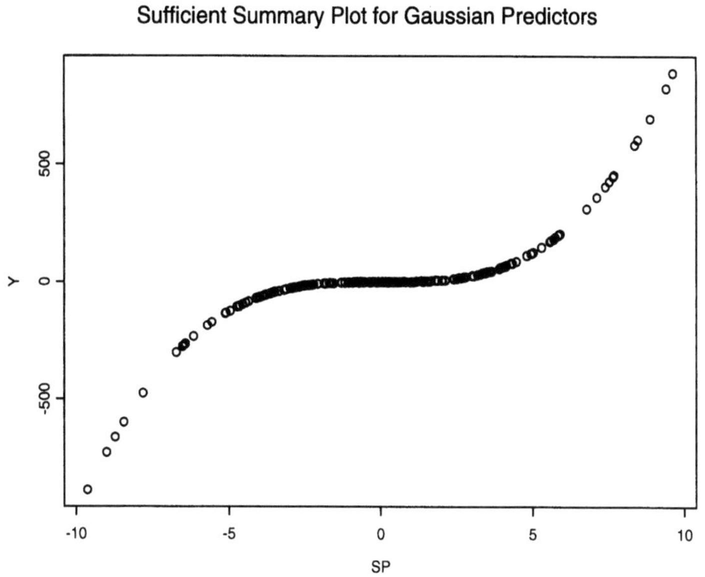

FIGURE 1. SSP for $m(u) = u^3$.

example, $(\alpha, \boldsymbol{\eta})$ is unique if $L(u, v)$ is strictly convex in its first argument. The following result is a useful extension of Brillinger (1977, 1983).

Theorem 1. (Li and Duan, 1989, p. 1016): *Assume that the $\boldsymbol{x}$ are linearly related predictors, that $(y_i, \boldsymbol{x}_i^T)^T$ are iid observations from some joint distribution and that $\text{Cov}(\boldsymbol{x}_i)$ exists and is positive definite. Assume $L(u, v)$ is convex in its first argument and that $\boldsymbol{\eta}$ is unique. Assume that $y \perp\!\!\!\perp \boldsymbol{x} | \boldsymbol{\beta}^T \boldsymbol{x}$. Then $\boldsymbol{\eta} = c\boldsymbol{\beta}$ for some scalar c.*

Remark 1. If $\hat{\boldsymbol{b}}$ is a consistent estimator of $\boldsymbol{\eta} \equiv \boldsymbol{\beta}_b$, then certainly

$$\boldsymbol{\beta}_b = c\boldsymbol{\beta} + \boldsymbol{u}_g$$

where $\boldsymbol{u}_g = \boldsymbol{\beta}_b - c\boldsymbol{\beta}$ is the bias vector. If the conditions of Theorem 1 hold, then $\boldsymbol{u}_g = \boldsymbol{0}$. Under additional conditions, $(\hat{a}, \hat{\boldsymbol{b}}^T)^T$ is asymptotically normal (see Li and Duan, 1989, p. 1031). In particular, the OLS estimator frequently has a $\sqrt{n}$ convergence rate. Often if no strong nonlinearities are present among the predictors, the bias vector is small enough so that $\hat{\boldsymbol{b}} \approx c\boldsymbol{\beta}$. See Cook and Weisberg (1999a, p. 431–441) for checking whether the predictors are linearly related and whether a 1D regression model is appropriate.

A very useful result is that if $y = m(x)$ for some function m, then m can be visualized with both a plot of x versus y and a plot of $c\,x$ versus y if $c \neq 0$. In fact, there are only three possibilities: if $c > 0$, then the two plots are nearly identical. If $c < 0$, then the plot appears to be flipped about the vertical axis. If $c = 0$, then the plot is a dot plot. Similarly, if $y_i = g(\alpha + \boldsymbol{\beta}^T \boldsymbol{x}_i, e_i)$, then the plot

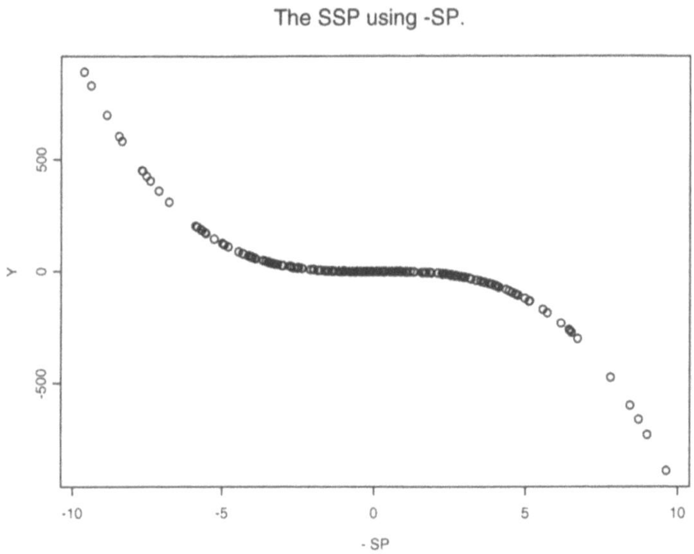

FIGURE 2. Another SSP for $m(u) = u^3$.

of $\boldsymbol{\beta}^T\boldsymbol{x}$ versus y and the plot of $c\boldsymbol{\beta}^T\boldsymbol{x}$ versus y will be nearly identical in overall shape if $c > 0$.

Example 1. Suppose that $\boldsymbol{x}_i \sim N_3(\boldsymbol{0}, \boldsymbol{I}_3)$ where $\boldsymbol{I}_3$ is the 3×3 identity matrix, and

$$y = m(\boldsymbol{\beta}^T\boldsymbol{x}) + e = (x_1 + 2x_2 + 3x_3)^3 + e \quad \text{with} \quad \mathrm{n} = 100.$$

Then a 1D regression model holds with $\boldsymbol{\beta} = (1, 2, 3)^T$. Figure 1 shows the sufficient summary plot of $\boldsymbol{\beta}^T\boldsymbol{x}_i$ versus y_i, and Figure 2 shows the sufficient summary plot of $-\boldsymbol{\beta}^T\boldsymbol{x}_i$ versus y_i. Notice that the functional form m appears to be cubic in both plots and that both plots can be smoothed by eye or with a scatterplot smoother.

Remark 2. The OLS estimator $(\hat{a}_o, \hat{\boldsymbol{b}}_o)$ is obtained from the usual multiple linear regression of y_i on $\boldsymbol{x}_i$, but *we are not assuming that the multiple linear regression model holds*; however, we are hoping that the 1D regression model $y \perp\!\!\!\perp \boldsymbol{x}|\boldsymbol{\beta}^T\boldsymbol{x}$ is a useful approximation to the data and that $\hat{\boldsymbol{b}}_o \approx c\boldsymbol{\beta}$ for some nonzero constant c. Theorem 1 provides some conditions for the above approximation to hold. Notice that if the multiple linear regression model does hold and if the errors e_i are such that OLS is a consistent estimator, then $c = 1$, and $\boldsymbol{u}_g = \boldsymbol{0}$.

The following result, perhaps first noted by Brillinger (1977, 1983), is called the *1D Estimation Result* by Cook and Weisberg (1999a, p. 432): let $(\hat{a}_o, \hat{\boldsymbol{b}}_o)$ denote the OLS estimate obtained from the OLS multiple linear regression of y on $\boldsymbol{x}$. The *OLS view* is a plot of $\hat{\boldsymbol{b}}_o^T\boldsymbol{x}$ versus y. If the 1D regression model is appropriate and if no strong nonlinearities are present among the predictors, then *the OLS view will frequently be a useful estimated sufficient summary plot.* Hence the OLS predictor $\hat{\boldsymbol{b}}_o^T \boldsymbol{x}$ is a useful ESP.

Three additional methods considered in this paper that have proven useful for estimating the ESP are sliced inverse regression (SIR), principal Hessian directions (PHD), and sliced average variance estimation (SAVE). See Cook and Li (2002) for a discussion of when these methods can fail. These methods frequently perform well if there are no strong nonlinearities present in the predictors. All of these methods fail if $c = 0$ or if the bias vector $\boldsymbol{u}_g$ is "large" compared to $c\boldsymbol{\beta}$. For example, the OLS view can fail if the sufficient summary plot of $\boldsymbol{\beta}^T \boldsymbol{x}$ versus y is approximately symmetric, and all of these methods can perform poorly if outliers are present. Gather et al. (2002) show that a single outlier can cause SIR to fail. Some useful references for SIR and related methods include Cook (1998ab, 2000), Cook and Weisberg (1999ab), Fung et al. (2002), Li (1991) and Stoker (1986).

Ellipsoidal trimming is a method for estimating the ESP that can reduce the bias $\boldsymbol{u}_g$. See Cook and Nachtsheim (1994) and Cook (1998a, p. 152). To perform ellipsoidal trimming, an estimator $(T, \boldsymbol{C})$ is computed where T is a $(p-1) \times 1$ multivariate location estimator and $\boldsymbol{C}$ is a $(p-1) \times (p-1)$ symmetric positive definite dispersion estimator. Then the ith squared Mahalanobis distance is the scalar

$$D_i^2 = (\boldsymbol{x}_i - T)^T \boldsymbol{C}^{-1} (\boldsymbol{x}_i - T) \tag{1.8}$$

for each vector of observed predictors $\boldsymbol{x}_i$. If the ordered distance $D_{(j)}$ is unique, then j of the $\boldsymbol{x}_i$'s are in the ellipsoid

$$\{\boldsymbol{x} : (\boldsymbol{x} - T)^T \boldsymbol{C}^{-1} (\boldsymbol{x} - T) \leq D_{(j)}^2\}. \tag{1.9}$$

The ith case $(y_i, \boldsymbol{x}_i^T)^T$ is trimmed if $D_i > D_{(j)}$. Then an estimator of $c\boldsymbol{\beta}$ is computed from the untrimmed cases. For example, if $j \approx 0.9n$, then about 10% of the cases are trimmed, and OLS could be used on the remaining cases.

The following procedure was suggested by Olive (2002). First compute $(T, \boldsymbol{C})$ using the *Splus* function `cov.mcd` (see Rousseeuw and Van Driessen, 1999). Trim the K% of the cases with the largest Mahalanobis distances, and then compute the OLS estimator $(\hat{\alpha}_K, \hat{\boldsymbol{\beta}}_K)$ from the untrimmed cases. Use $K = 0$, 10, 20, 30, 40, 50, 60, 70, 80, and 90 to generate ten plots of $\hat{\boldsymbol{\beta}}_K^T \boldsymbol{x}$ versus y using all n cases. These plots will be called "OLS trimmed views." Notice that $K = 0$ corresponds to the OLS view. The *best OLS trimmed view* is the trimmed view with a smooth mean function and the smallest variance function and is the estimated sufficient summary plot. If $K^* = E$ is the percentage of cases trimmed that corresponds to the best trimmed view, then $\hat{\boldsymbol{\beta}}_E^T \boldsymbol{x}$ is the estimated sufficient predictor.

Example 2. For the data in Example 1, the OLS view is similar to Figure 1 except the plot is not quite as smooth and the horizontal scale is multiplied by $c \approx 42$. The best trimmed view appears to be identical to Figure 1 except that the horizontal scale is multiplied by $c \approx 12.5$. The OLS view used $\hat{\boldsymbol{\beta}}_0 = (41.68, 87.40, 120.83)^T \approx 42\boldsymbol{\beta}$ while the best trimmed view used $\hat{\boldsymbol{\beta}}_{50} = (12.61, 25.07, 37.26)^T \approx 12.5\boldsymbol{\beta}$.

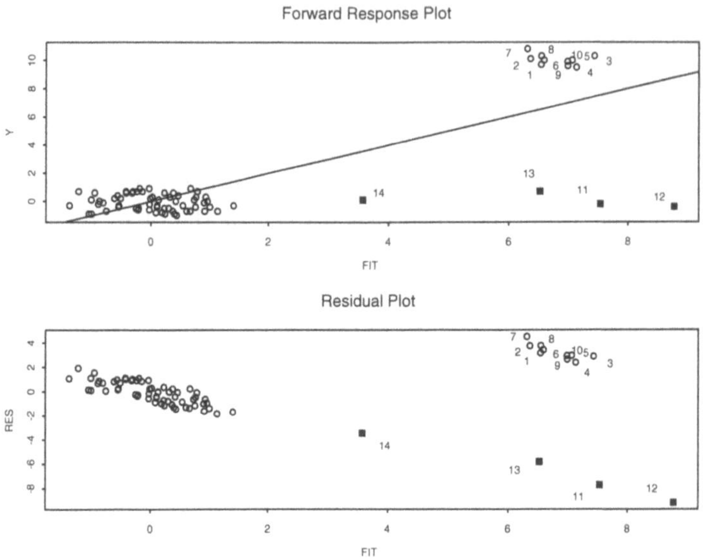

FIGURE 3. Plots for HBK Data.

This section has reviewed the existing literature on 1D regression. Section 2 shows that 1D regression can provide useful diagnostics when g is known. Section 3 considers estimating g when $g = g_\lambda$ and $\lambda \in \{\lambda_1, ..., \lambda_k\}$ where k is a small integer. Section 4 suggests using ellipsoidal trimming with methods other than OLS. This technique gives a resistant version of SIR and shows that the *Splus* function `lmsreg` can be very useful for finding certain types of curvature.

2. 1D Regression Diagnostics

In this section, we suggest that when g is known, an estimated sufficient summary plot should be used in addition to the usual diagnostics for checking the model. Assume that the 1D regression model is $y_i = g(\alpha + \boldsymbol{\beta}^T \boldsymbol{x}_i, e_i)$ for $i = 1, \ldots, n$ where the e_i are iid with zero mean and variance $V(e_i) = \sigma^2$.

Remark 3. If $y \perp\!\!\!\perp \boldsymbol{x} | \boldsymbol{\beta}^T \boldsymbol{x}$ then $y \perp\!\!\!\perp \boldsymbol{x} | a + c\boldsymbol{\beta}^T \boldsymbol{x}$ for any constants a and $c \neq 0$. Hence if $\hat{\boldsymbol{b}}^T \boldsymbol{x}$ is an ESP, so is $a + \hat{\boldsymbol{b}}^T \boldsymbol{x}$.

A good example is the multiple linear regression (MLR) model $y_i = \alpha + \boldsymbol{\beta}^T \boldsymbol{x}_i + e_i$. Let $(\hat{a}, \hat{\boldsymbol{b}})$ be a MLR estimator of $(\alpha, \boldsymbol{\beta})$. Then the fitted values are $\hat{y}_i = \hat{a} + \hat{\boldsymbol{b}}^T \boldsymbol{x}_i$, and the residuals are $r_i = y_i - \hat{y}_i$. The most used residual plot is a plot of $\hat{y}_i$ versus r_i, and the *forward response plot* is a plot of the fitted values $\hat{y}_i$ versus the response y_i.

Remark 3 shows that the forward response plot is an ESSP. Let the scalars $w_i = \alpha + \boldsymbol{\beta}^T \boldsymbol{x}_i$. Ignoring the errors gives $y_i = w_i$ which is the equation of the *identity line* that has unit slope and zero intercept. Hence if the MLR model is appropriate and if $(\hat{a}, \hat{\boldsymbol{b}})$ is a good estimator of $(\alpha, \boldsymbol{\beta})$, then the plotted points in the forward response plot should scatter about the identity line. The vertical deviations from the identity line are the residuals r_i since these deviations are $y_i - \hat{y}_i$. When the OLS estimator is used, the coefficient of determination R^2 is equal to the squared correlation of y_i and $\hat{y}_i$. See Chambers et al. (1983, p. 280).

High leverage outliers challenge conventional numerical MLR diagnostics such as Cook's distance (Cook, 1977), but, as shown in Example 3 below, can often be detected using the forward response and residual plots. Using trimmed views (see section 4) is also effective for detecting outliers and other departures from the MLR model.

Example 3. In the well known artificial HBK data set (Hawkins et al., 1984), the first 10 cases are outliers while cases 11-14 are good leverage points. This data set has $n = 75$ cases and $p - 1 = 3$ nontrivial predictors. Figure 3 shows the residual and forward response plots based on the OLS estimator. The highlighted cases have Cook's distance $> \min(0.5, 2p/n)$, and the identity line is shown in the ESSP.

Now suppose that model (1.2) holds where g is known. If the estimator $(\hat{a}, \hat{\boldsymbol{b}})$ satisfies $\hat{\boldsymbol{b}} \approx c\boldsymbol{\beta}$, then the plot of $\hat{a} + \hat{\boldsymbol{b}}^T \boldsymbol{x}$ or $\hat{\boldsymbol{b}}^T \boldsymbol{x}$ versus y can be used to visualize g provided that $c \neq 0$. Since g is known, the classical (e.g., maximum likelihood) estimator for $\boldsymbol{\beta}$ should be used since then $c = 1$ and the bias vector should be small for large sample size n. Often adding a parametric fit and a lowess smooth to the plot will be useful.

Plots are also useful for additive error models

$$y_i = m(x_{i,1}, \ldots, x_{i,p-1}) + e_i = m(\boldsymbol{x}_i^T) + e_i = m_i + e_i.$$

Many anova, categorical, nonlinear regression, nonparametric regression, and multi-index models have this form. For these models $\hat{y}_i = \hat{m}_i$ and the residuals $r_i = y_i - \hat{y}_i$. In the fit-response plot (FY plot) of $\hat{y}_i$ versus y_i, the plotted points should scatter about the identity line, and the vertical deviations from the identity line are equal to the residuals.

3. 1D Regression Model Selection

In this section, we assume that a 1D regression model holds with $g = g_{\lambda_o}$ where $\lambda_o \in \Lambda = \{\lambda_1, \ldots, \lambda_k\}$ and $k > 1$ is a small integer. To estimate λ_o, make an ESSP for each of the k possible models. Examples include choosing a frequentist or a Bayesian model; a proportional hazards model or one of several competing 1D survival models; a logistic, probit or complementary log-log model in binary regression; a full or sub model in variable selection.

A good example for illustration is the response transformation model

$$t_{\lambda_o}(y_i) \equiv y_i^{(\lambda_o)} = \alpha_0 + \boldsymbol{\beta}_0^T \boldsymbol{x}_i + e_i \tag{3.1}$$

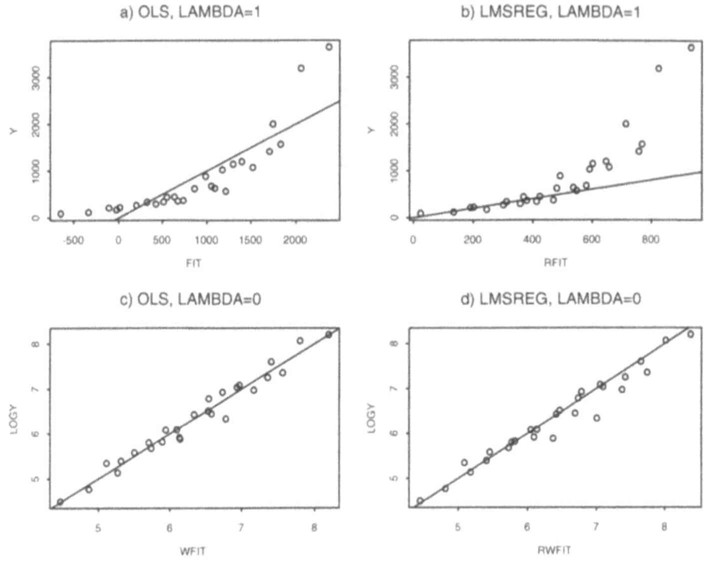

FIGURE 4. OLS and LMSREG Suggest Using log(y) for the Textile Data.

where the response variable $y_i > 0$ and the *power transformation family*

$$t_\lambda(y) \equiv y^{(\lambda)} = \frac{y^\lambda - 1}{\lambda} \tag{3.2}$$

for $\lambda \neq 0$ and $y^{(0)} = \log(y)$. Assume $\lambda_o \in \Lambda = \{0, \pm 1/4, \pm 1/3, \pm 1/2, \pm 2/3, \pm 1\}$.

The literature for estimating λ_o is enormous, and at least two papers using results from 1D regression have appeared. Let the OLS estimator $(\hat{a}_o, \hat{\boldsymbol{b}}_o)$ be computed from the multiple linear regression of y_i on $\boldsymbol{x}_i$. Then Cook and Weisberg (1994) suggest that the inverse response plot of y versus $\hat{y}_{OLS}$ will often show t_{λ_o}. Hence the forward response plot of $\hat{y}_{OLS}$ versus y will show $t^{-1}_{\lambda_o}$. If t_λ is the appropriate transformation, Cook and Olive (2001) suggest that a plot of $\hat{y}_{OLS}$ versus $t_\lambda(y)$ will follow the identity line.

These ideas suggest a graphical method for selecting response transformations that can be used with any good MLR estimator. Let $w_i = t_\lambda(y_i)$ for $\lambda \neq 1$, and let $w_i = y_i$ if $\lambda = 1$. Next, perform the multiple linear regression of w_i on $\boldsymbol{x}_i$ and make the forward response plot of $\hat{w}_i$ versus w_i. If the plotted points follow the identity line, then take $\lambda_o = \lambda$. One plot is made for each of the eleven values of $\lambda \in \Lambda$, and if more than one value of λ works, contact subject matter experts and use the simplest or most reasonable transformation. (Note that this procedure can be modified to create a graphical diagnostic for a numerical estimator by adding the estimate $\hat{\lambda}$ of λ_o to Λ.)

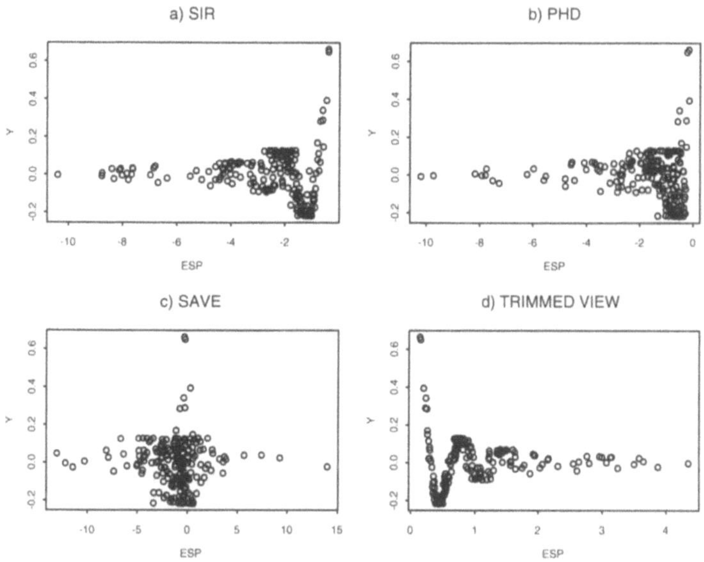

FIGURE 5. Estimated Sufficient Summary Plots.

Example 4. A textile data set is given in Box and Cox (1964) where samples of worsted yarn with different levels of the three factors were given a cyclic load until the sample failed. The goal was to understand how $y =$ *the number of cycles to failure* was related to the predictor variables *length, amplitude* and *load.* Figure 4 shows the forward response plots for two MLR estimators: OLS and the *Splus* function `lmsreg`. Figures 4a and 4b show that a response transformation is needed while 4c and 4d both suggest that $\log(y)$ is the appropriate response transformation. Using OLS and a resistant estimator as in Figure 4 may be very useful if outliers are present.

4. Improving 1D Estimators

Assume that the 1D regression model (1.2) holds but both g and $\boldsymbol{\beta}$ are unknown. If a good estimator $\hat{\boldsymbol{b}} \approx c\boldsymbol{\beta}$ can be found where $c \neq 0$, then the ESSP can be used to visualize g. The next step might be to fit a tentative parametric or nonparametric model in the ESP.

Since methods for estimating the sufficient predictor such as OLS, PHD, SAVE and SIR can fail if strong nonlinearities such as outliers are present in the predictors or if $\hat{\boldsymbol{b}} \approx 0\ \boldsymbol{\beta}$, techniques for improving these methods are needed. The basic tool will be to use $(\hat{\boldsymbol{b}}, T, \boldsymbol{C})$ trimmed views where $\hat{\boldsymbol{b}}$ is an estimator of $c\boldsymbol{\beta}$ and $(T, \boldsymbol{C})$ is an estimator of multivariate location and dispersion. Two good choices for $(T, \boldsymbol{C})$ are the classical estimator (Cook 1998a, p. 152) or a robust estimator

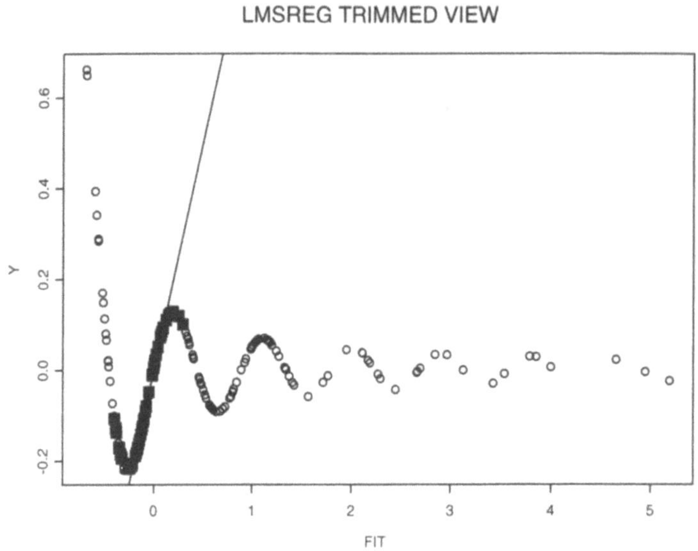

FIGURE 6. The Weighted `lmsreg` Fitted Values vs. Y.

such as `cov.mcd`. Then, for example, SIR trimmed views generalize the SIR ESSP in the same way that OLS trimmed views generalize the OLS view: use ellipsoidal trimming to delete the $K\%$ of the cases with the largest Mahalanobis distances. Next, plot $\hat{\boldsymbol{b}}_K^T \boldsymbol{x}_i$ versus y_i where $\hat{\boldsymbol{b}}_K$ is the first SIR direction computed from the untrimmed cases. Again use 10 values of K where $K = 0$ corresponds to the usual SIR ESSP, and the best SIR trimmed view is the trimmed view with a smooth mean function and the smallest variance function.

Using trimmed views seems to work for several reasons. The ellipsoidal trimming divides the data into two groups: the *trimmed cases* and the *untrimmed cases*. Trimming often removes strong nonlinearities from the predictors, and the untrimmed predictor distribution is often more nearly elliptical contoured than the predictor distribution of the entire data set (recall Winsor's principle: "data are roughly Gaussian in the middle"). Secondly, under heavy trimming, the mean function of the untrimmed cases may be more linear than the mean function of the entire data set. Thirdly, if $|c|$ is very large, then the bias vector $\boldsymbol{u}_g$ may be small relative to $c\boldsymbol{\beta}$. From Remarks 1 and 2, any of these three reasons could produce a better estimated sufficient predictor. Also notice that trimmed views are resistant to y-outliers since the y values are plotted, and trimmed views are resistant to $\boldsymbol{x}$-outliers if $(T, \boldsymbol{C})$ is a resistant estimator.

Example 5. To illustrate the above discussion, an artificial data set with 200 trivariate vectors $\boldsymbol{x}_i$ was generated. The marginal distributions of $x_{i,j}$ are iid lognormal for $j = 1, 2$, and 3. Since the response $y_i = \sin(\boldsymbol{\beta}^T \boldsymbol{x}_i)/\boldsymbol{\beta}^T \boldsymbol{x}_i$ where $\boldsymbol{\beta} = (1, 2, 3)^T$, the random vector $\boldsymbol{x}_i$ is not elliptically contoured and the function g is strongly

TABLE 1. Estimated Sufficient Predictors Estimating $c(1,2,3)^T$.

method	$\hat{b}_1$	$\hat{b}_2$	$\hat{b}_3$
OLS View	0.0032	0.0011	0.0047
SIR	-0.394	-0.361	-0.845
PHD	-0.072	-0.029	-0.0097
SAVE	-1.09	0.870	-0.480
OLS 90% Trimmed View	0.086	0.182	0.338
LMSREG 70% Trimmed View	0.143	0.287	0.428

nonlinear. The `cov.mcd` estimator was used for trimming, and Figure 5 shows the estimated sufficient summary plots for SIR, PHD, SAVE (using 8 slices), and the OLS 90% trimmed view. The OLS trimmed view is the best, while SAVE completely fails. Figure 6 shows that for this data, the `lmsreg` trimmed view is very useful for visualizing g. The `lmsreg` estimator attempts to make the median squared MLR residual small and implements the PROGRESS algorithm described in Rousseeuw and Leroy (1987, pp. 197–204). Table 1 shows the estimated sufficient predictor coefficients $\hat{\boldsymbol{b}}$ when the sufficient predictor coefficients are $c(1,2,3)^T$. Only the OLS and `lmsreg` trimmed views produce estimated sufficient predictors that are highly correlated with the sufficient predictor. (The SAVE 40% trimmed view was also very good.)

Figure 6 helps illustrate why the best `lmsreg` trimmed view worked. This view used 70% trimming, and the open circles denote cases that were trimmed while the highlighted squares are the untrimmed cases. Note that the highlighted cases are far more linear than the data set as a whole. Also `lmsreg` will give about half of the highlighted cases zero weight, further linearizing the function. In Figure 6 the weighted `lmsreg` constant $\hat{\alpha}_{70}$ is included, and the plot is simply the forward response plot of the weighted `lmsreg` fitted values versus y. The vertical deviations from the identity line are the "MLR residuals" $y_i - \hat{\alpha}_{70} - \hat{\boldsymbol{\beta}}_{70}^T \boldsymbol{x}_i$ and at least half of the highlighted cases have small MLR residuals. There exist data sets where OLS is better than `lmsreg` for showing curvature (see Cook et al., 1992), but, as illustrated by Example 5, `lmsreg` often performed better for single index models when m was smooth.

A great deal of work remains to be done in the area of resistant dimension reduction. When a 1D regression model holds, the trimmed views work at least as well as the untrimmed view since the untrimmed view corresponds to 0% trimming. The webpage (http://www.math.siu.edu/olive/) contains programs and a good introduction to 1D regression models.

References

[1] G.E.P. Box and D.R. Cox, *An Analysis of Transformations.* J. Roy. Statist. Soc. Ser. B Stat. Methodol. **26** (1964), 211–246.

[2] D.R. Brillinger, *The Identification of a Particular Nonlinear Time Series.* Biometrika **64** (1977), 509–515.

[3] D.R. Brillinger, *A Generalized Linear Model with "Gaussian" Regressor Variables.* In P.J. Bickel, K.A. Doksum, and J.L. Hodges, editors, A Festschrift for Erich L. Lehmann, pages 97–114. Wadsworth, Pacific Grove, CA, 1983.

[4] J.M. Chambers, W.S. Cleveland, B. Kleiner, and P. Tukey. *Graphical Methods for Data Analysis.* Duxbury Press, Boston, 1983.

[5] R.D. Cook, *Deletion of Influential Observations in Linear Regression.* Technometrics **19** (1977), 15–18.

[6] R.D. Cook, *Regression Graphics: Ideas for Studying Regression Through Graphics.* John Wiley and Sons, New York, 1998.

[7] R.D. Cook, *Principal Hessian Directions Revisited.* J. Amer. Statist. Assoc. **93** (1998), 84–100.

[8] R.D. Cook, *SAVE: A Method for Dimension Reduction and Graphics in Regression.* Comm. Statist. Theory Methods **29** (2000), 2109–2121.

[9] R.D. Cook, D.M. Hawkins, and S. Weisberg, *Comparison of Model Misspecification Diagnostics Using Residuals from Least Mean of Squares and Least Median of Squares.* J. Amer. Statist. Assoc. **87** (1992), 419–424.

[10] R.D. Cook and B. Li, *Dimension Reduction for Conditional Mean in Regression.* Ann. Statist. **30** (2002), 455–474.

[11] R.D. Cook and C.J. Nachtsheim, *Reweighting to Achieve Elliptically Contoured Covariates in Regression.* J. Amer. Statist. Assoc. **89** (1994), 592–599.

[12] R.D. Cook and D.J. Olive, *A Note on Visualizing Response Transformations in Regression.* Technometrics **43** (2001), 443–449.

[13] R.D. Cook and S. Weisberg, *Transforming a Response Variable for Linearity.* Biometrika **81** (1994), 731–737.

[14] R.D. Cook and S. Weisberg, *Applied Regression Including Computing and Graphics.* John Wiley and Sons, New York, 1999.

[15] R.D. Cook and S. Weisberg, *Graphs in Statistical Analysis: Is the Medium the Message?* Amer. Statist. **53** (1999), 29–37.

[16] W.K. Fung, X. He, L. Liu, and P.D. Shi, *Dimension Reduction Based on Canonical Correlation.* Statist. Sinica **12** (2002), 1093–1114.

[17] U. Gather, T. Hilker, and C. Becker, *A Note on Outlier Sensitivity of Sliced Inverse Regression.* Statistics **36** (2002), 271–281.

[18] D.M. Hawkins, D. Bradu, and G.V. Kass, *Location of Several Outliers in Multiple Regression Data Using Elemental Sets.* Technometrics **26** (1984), 197–208.

[19] K.C. Li, *Sliced Inverse Regression for Dimension Reduction.* J. Amer. Statist. Assoc. **86** (1991), 316–342.

[20] K.C. Li and N. Duan, *Regression Analysis Under Link Violation.* Ann. Statist. **17** (1989), 1009–1052.

[21] D.J. Olive, *Applications of Robust Distances for Regression.* Technometrics **44** (2002), 64–71.

[22] P.J. Rousseeuw and A.M. Leroy, *Robust Regression and Outlier Detection.* John Wiley and Sons, New York, 1987.

[23] P.J. Rousseeuw and K. Van Driessen, *A Fast Algorithm for the Minimum Covariance Determinant Estimator.* Technometrics **41** (1999), 212–223.

[24] T.M. Stoker, *Consistent Estimation of Scaled Coefficients.* Econometrica **54** (1986), 1461–1481.

D.J. Olive
Department of Mathematics
Mailcode 4408
Southern Illinois University
Carbondale, IL 62901-4408
USA
e-mail: `dolive@math.siu.edu`

Statistics for Industry and Technology, 235–246

Robust Redundancy Analysis by Alternating Regression

M.R. Oliveira, J.A. Branco, C. Croux and P. Filzmoser

Abstract. Given two groups of variables redundancy analysis searches for linear combinations of variables in one group that maximize the variance of the other group that is explained by each one of the linear combination. The method is important as an alternative to canonical correlation analysis, and can be seen as an alternative to multivariate regression when there are collinearity problems in the dependent set of variables. Principal component analysis is itself a special case of redundancy analysis.

In this work we propose a new robust method to estimate the redundancy analysis parameters based on alternating regressions. These estimators are compared with the classical estimator as well as other robust estimators based on robust covariance matrices. The behavior of the proposed estimators is also investigated under large contamination by the analysis of the empirical breakdown point.

Mathematics Subject Classification (2000). 62F35; 62J99.

Keywords. Redundancy analysis, alternating regression, robust regression.

1. Introduction

The problem of finding relationships between groups of variables is central in multivariate analysis. A number of methods have been suggested to pursue this objective but canonical correlation analysis is by far the most commonly used. Given two sets of variables the goal of canonical correlation analysis is to construct pairs of canonical variates (linear combinations of the original variables, one in each set) such that they have maximum correlation. The correlations between canonical variates of the same pair are important to the study of the correlations between the two sets but they cannot be interpreted as the degree of relation between the sets of variables. In particular the squared canonical correlations represent the variance shared by the two canonical variates of the same pair but not the variance shared by the two sets of observed random variables.

To overcome the difficulty in using the squared canonical correlations as a measure of the shared variance between the two sets, a redundancy index was proposed by Stewart and Love (1968). This index is a measure of the proportion of the variance in one set $\boldsymbol{y} = (y_1, \ldots, y_q)^t$ (dependent or criterion set) that is accounted for by the other set $\boldsymbol{x} = (x_1, \ldots, x_p)^t$ (independent or predictor set). The redundancy analysis model, proposed by van den Wollenberg (1977), searches for the linear combination, $u_1 = \boldsymbol{\alpha}_1^t \boldsymbol{x}$ (the first redundancy variate), of the independent set that maximizes the redundancy index, R_y, defined as

$$R_y = \sum_{j=1}^{q} Corr(\boldsymbol{\alpha}^t \boldsymbol{x}, y_j)^2 / q, \tag{1.1}$$

under the restriction $Var(\boldsymbol{\alpha}_1^t \boldsymbol{x}) = 1$. The second redundancy variate, $u_2 = \boldsymbol{\alpha}_2^t \boldsymbol{x}$, is defined as the linear combination of the independent set, non-correlated with u_1, that maximizes R_y under the restriction $Var(\boldsymbol{\alpha}_2^t \boldsymbol{x}) = 1$. A maximum of $r = \min(p, q)$ redundancy variates can be sequentially defined, following the scheme as above. The objective of this paper is to provide robust estimators for $\boldsymbol{\alpha}$ and R_y. Robust estimation is particularly useful in the analysis of multivariate data where the presence of outlying observations is common.

In Section 2, the relationship between redundancy and canonical correlation analysis is highlighted. This will help to address the problem of robust estimation in redundancy analysis and in particular to introduce the robust method based on alternating regressions. The algorithm to estimate the redundancy variates by alternating regressions is also presented in Section 2, and in Section 3 a simulation study is developed in order to compare the performance of the proposed estimators with others based on robust covariance matrices. In the last section a brief discussion about the results obtained is made and links with canonical correlation analysis and suggestions for future work are discussed.

2. Robust Estimation

The classical solution to (1.1) comes down to an eigenvector/eigenvalue problem, as was shown by van den Wollenberg (1977). The coefficients $\boldsymbol{\alpha}$ are the eigenvectors of the matrix $R_{11}^{-1} R_{12} R_{21}$, where R_{ij} $(i, j = 1, 2)$ are the usual partition matrices of the correlation matrix associated with the two sets of variables. A simple approach to robust estimation is to robustify the correlation matrix and then apply the traditional methods of estimation. So given a robust estimate of the correlation matrix the eigenvectors of $R_{11}^{-1} R_{12} R_{21}$ are calculated in order to estimate the coefficients $\boldsymbol{\alpha}$. This approach will be considered in Section 3. For the estimation of the robust correlation matrix we will use the M-estimator (M) as outlined by Maronna (1976). As an alternative we will consider the minimum covariance determinant (MCD) estimator of Rousseeuw (1985).

The above approaches were studied for the first latent variate (redundancy variate) in Oliveira and Branco (2003). Additionally, a method based on robust alternating regression was considered. This technique, originally suggested by Wold (1966), has recently been used in the context of robust factor analysis (Croux et al., 2003) and robust canonical correlation analysis (Branco et al., 2005). Like Tenenhaus (1998) has pointed out, the algorithm is quite similar to the one proposed for canonical correlation analysis, initially discussed by Wold (1966), Lyttkens (1972) and later on mentioned by Tenenhaus (1998). For higher-order redundancy variates, the algorithm is based on repeating the idea underlying the construction of the first redundancy variate in successive residual spaces. Since these spaces have reduced rank, problems occur in the regression procedure.

In this paper we will focus on estimating higher order redundancy variates by robust alternating regressions. This procedure has a main advantage over the approach based on robust correlation matrix estimation: While the latter method discards an outlying observation completely, robust alternating regression is still using the information of the non-outlying cells for parameter estimation. For this reason, the method based on robust alternating regressions can also deal with missing values (Croux et al., 2003).

The general idea of the algorithm takes advantage of the links between redundancy analysis and canonical correlation analysis. In the context of canonical correlation analysis, the redundancy index (1.1) can be written as in Rencher (1998) by:

$$R_y = \rho^2 \sum_{j=1}^{q} Corr(\boldsymbol{\beta}^t \boldsymbol{y}, y_j)^2/q, \tag{2.1}$$

where ρ is the canonical correlation coefficient and $\boldsymbol{\beta}$ the canonical coefficient associated with the $\boldsymbol{y}'s$. Looking at the redundancy coefficient from these two different perspectives, an alternating procedure based on regression models can be built. To clarify this idea, let us consider that an initial value $\boldsymbol{\beta}$ is given. If the redundancy index is written in the form (2.1), to maximize R_y we need to determine the vector $\boldsymbol{\alpha}$ that maximizes $\rho^2 = Corr(\boldsymbol{\alpha}^t\boldsymbol{x}, \boldsymbol{\beta}^t\boldsymbol{y})^2$. From standard results on multiple regression, it follows that $\boldsymbol{\alpha}$ is proportional to the regression coefficient $\boldsymbol{a}$ in the model

$$\boldsymbol{\beta}^t\boldsymbol{y} = \boldsymbol{a}^t\boldsymbol{x} + \gamma_1 + \varepsilon_1. \tag{2.2}$$

This step is analogous to the procedure followed in canonical correlation analysis (Branco et al., 2005).

Once $\boldsymbol{\alpha}$ has been obtained, the value $\boldsymbol{\beta}$ has to be updated. Taking into account that R_y is a measure of the proportion of the variance in set $\boldsymbol{y}$ explained by $\boldsymbol{\alpha}^t\boldsymbol{x}$, Tenenhaus (1998) suggested that $\boldsymbol{\beta}$ can be chosen proportional to $\boldsymbol{b} = (b_1, \ldots, b_q)^t$, where its components are the coefficients of the regression equations

$$y_j = b_j\boldsymbol{\alpha}^t\boldsymbol{x} + \gamma_{2j} + \varepsilon_{2j}, \tag{2.3}$$

that maximizes the variance of y_j explained by $\boldsymbol{\alpha}^t\boldsymbol{x}$, i.e., $\hat{b}_j$ is such that $Corr\big(y_j, \hat{b}_j\boldsymbol{\alpha}^t\boldsymbol{x}\big)^2$ is maximum. From regression standard results $Corr(y_j, \hat{y}_j)^2 =$

$Corr(y_j, \hat{b}_j \boldsymbol{\alpha}^t \boldsymbol{x})^2 = Corr(y_j, \boldsymbol{\alpha}^t \boldsymbol{x})^2$. Choosing $\hat{b}_j$ in this way leads to an updated value of β, that maximizes the variance of y_j explained by $\boldsymbol{\alpha}^t \boldsymbol{x}$. By (1.1)

$$\sum\nolimits_{j=1}^{q} Corr(\hat{y}_j, y_j)^2/q = R_y,$$

where R_y is the value of the redundancy index obtained in the previous step.

This scheme can be implemented using least squares regression, but it leads to non robust estimates. In order to robustify the parameters in redundancy analysis we can simply use robust regression estimators. In the case of least squares estimators convergence of the algorithm is fast. However, using a robust estimator yields more local minima, and less smooth objective functions, so higher risks of lack of convergence appear. One benefits from choosing weighted L_1 regression estimators since they have bounded influence functions and the algorithm is fast to compute, once the weights are properly calculated. These weights are defined smoothly according to an appropriate distance measure associated with each observation. Although, weighted L_1 regression is a good choice other robust regression estimators could be used.

In the next subsection we give a general description of the algorithm to estimate the redundancy variates by alternating regressions.

2.1. Algorithm

It is convenient to center the observations, so that the intercept terms, γ_1 and γ_{2j}, can be eliminated from the equations (2.2) and (2.3). The observations are centered using robust estimators of location. In the present work, the median of each variable, ($\tilde{\boldsymbol{x}}$, $\tilde{\boldsymbol{y}}$), has been chosen for robust estimators of location. So, if $\boldsymbol{X}$ and $\boldsymbol{Y}$ represent the data matrices of size $(n \times p)$ and $(n \times q)$, respectively, the centered data are

$$\boldsymbol{X}_0 = \boldsymbol{X} - \mathbf{1}_p \tilde{\boldsymbol{x}}^t \quad \text{and} \quad \boldsymbol{Y}_0 = \boldsymbol{Y} - \mathbf{1}_q \tilde{\boldsymbol{y}}^t,$$

where $\mathbf{1}_k$ is a k-vector of ones.

2.1.1. First Redundancy Variate. The algorithm starts by choosing an initial value, $\boldsymbol{\beta}^{(0)}$, followed by the estimation of the regression coefficients in (2.2).

Starting values: In the context of robust estimation, the starting values can be of crucial importance. For the classical version of the algorithm Tenenhaus (1998) suggested to use the vector $\boldsymbol{\beta}^{(0)} = (1, 0, \ldots, 0)^t$. However, we have built the starting value using the first robust principal component of $\boldsymbol{X}_0$, $\boldsymbol{z}_1$ (see e.g. Croux and Ruiz-Gazen, 1996). Let $\hat{b}_j^{(0)}$ $(j = 1, \ldots, q)$ be the estimated regression coefficient associated with the model

$$\boldsymbol{y}_{0j} = b_j \boldsymbol{z}_1 + \boldsymbol{\varepsilon}_{3j}, \tag{2.4}$$

where $\boldsymbol{y}_{0j}$ is the jth column of $\boldsymbol{Y}_0$. The starting value is defined as

$$\boldsymbol{\beta}^{(0)} = \hat{\boldsymbol{b}}^{(0)} / ||\hat{\boldsymbol{b}}^{(0)}||, \tag{2.5}$$

where $\hat{\boldsymbol{b}}^{(0)} = (\hat{b}_1^{(0)}, \ldots, \hat{b}_q^{(0)})^t$, and the corresponding latent variate is $\boldsymbol{v}^{(0)} = \boldsymbol{Y}_0 \boldsymbol{\beta}^{(0)}$.

Step s: Given the values $\boldsymbol{\beta}^{(s-1)}$ and $\boldsymbol{v}^{(s-1)}$ (for $s > 1$), the $\hat{\boldsymbol{a}}^{(s)}$ is the estimated regression coefficient of the model

$$\boldsymbol{v}^{(s-1)} = \boldsymbol{X}_0 \boldsymbol{a}^{(s)} + \boldsymbol{\varepsilon}_4. \tag{2.6}$$

The estimated vector of redundancy coefficients is

$$\boldsymbol{\alpha}^{(s)} = \hat{\boldsymbol{a}}^{(s)} / ||\hat{\boldsymbol{a}}^{(s)}||, \tag{2.7}$$

and the associated redundancy variate is $\boldsymbol{u}^{(s)} = \boldsymbol{X}_0 \boldsymbol{\alpha}^{(s)}$.

Given these new estimated values, $\boldsymbol{\beta}^{(s-1)}$ and $\boldsymbol{v}^{(s-1)}$ have to be updated. Let $\hat{b}_j^{(s)}$ $(j = 1, \ldots, q)$ be the estimate of the regression coefficient associated with the model

$$\boldsymbol{y}_{0j} = b_j^{(s)} \boldsymbol{u}^{(s)} + \boldsymbol{\varepsilon}_{5j}. \tag{2.8}$$

So, the updated vector is defined as

$$\boldsymbol{\beta}^{(s)} = \hat{\boldsymbol{b}}^{(s)} / ||\hat{\boldsymbol{b}}^{(s)}||, \tag{2.9}$$

where $\hat{\boldsymbol{b}}^{(s)} = (\hat{b}_1^{(s)}, \ldots, \hat{b}_q^{(s)})^t$, and the corresponding latent variate is $\boldsymbol{v}^{(s)} = \boldsymbol{Y}_0 \boldsymbol{\beta}^{(s)}$.

Repeat this procedure until convergence of the algorithm, and let $\boldsymbol{\alpha}_1$ and $\boldsymbol{u}_1$ be proportional to the last values of $\boldsymbol{\alpha}^{(s)}$ and $\boldsymbol{u}^{(s)}$, respectively, where the proportional constant is such that $\boldsymbol{u}_1$ has norm equal to one.

The first redundancy coefficient, R_{y1}, is estimated using a robust estimator of the correlation coefficient (in the present case, reweighed MCD estimator, RMCD, Rousseeuw and Van Driessen, 1999).

$$R_{y1} = \sum_{j=1}^{q} Corr(\boldsymbol{u}_1, \boldsymbol{y}_{0j})^2 / q. \tag{2.10}$$

We recall that RMCD is the empirical covariance matrix computed from the subset of size $h \approx 0.75n$ with the smallest determinant, where n is the sample size.

2.1.2. Redundancy Variates of Higher Order. The next redundancy variate, $\boldsymbol{u}_2$, has to be uncorrelated with the previous one. In other words, we can say that $\boldsymbol{u}_1$ and $\boldsymbol{u}_2$ have to be orthogonal. This restriction can be fulfilled if the data matrix, $\boldsymbol{X}_0$, is projected into the space orthogonal to $\boldsymbol{u}_1$. Let us consider the following regression model

$$\boldsymbol{X}_0 = \boldsymbol{u}_1 \boldsymbol{c}^t + \boldsymbol{\varepsilon}_6 \tag{2.11}$$

where $\boldsymbol{X}_1 = \boldsymbol{X}_0 - \hat{\boldsymbol{X}}_0$ are the corresponding estimated residuals. Due to standard results from multiple linear regression, the residuals, $\boldsymbol{X}_1$, are orthogonal to $\boldsymbol{u}_1$. Hence, repeating the procedure to obtain the first redundancy variate using the residuals, $\boldsymbol{X}_1$, instead of the original data, $\boldsymbol{X}_0$, we can guarantee that the next redundancy variate (a linear combination of the columns of $\boldsymbol{X}_1$) is, in fact, orthogonal to $\boldsymbol{u}_1$.

However, this idea raises some difficulties, since $\boldsymbol{X}_1$ is orthogonal to $\boldsymbol{u}_1$, and therefore, it has rank $(p-1)$, because $rank\,(\boldsymbol{X}_0) = p$. In order to overcome this collinearity problem we consider, as in Branco et al. (2005), the singular value decomposition

$$X_1 = UDV^t = U^*D^*V^{*t}, \tag{2.12}$$

where D is a diagonal matrix of the p singular values, D^* is diagonal matrix reduced by one row and column associated with the null singular value. A similar idea is used to define the matrices U^* and V^*. Using (2.12), we can define $X_1^* = U^{*t}X_1 = D^*V^{*t}$, and this matrix of size $(n \times (p-1))$ has full rank. Finally, the procedure proposed to estimate the first redundancy variate can be applied to X_1^* and Y_0. As opposed to the estimation method developed for canonical correlation analysis (Branco et al., 2005), in the redundancy analysis there is no need to transform Y_0.

Let u^* and α^* be the results, after convergence, of the iterative procedure developed to estimate the second redundancy variate. These quantities are defined in residual spaces. Hence, a transformation to the original space has to be done, under the orthogonality condition, that is, $u_2 = X_0\alpha_2$ has to be orthogonal to u_1. Let us consider the regression model

$$u^* = eu_1 + \varepsilon_7, \tag{2.13}$$

and let $\tilde{u}$ be the estimated residuals. This vector, $\tilde{u}$, is orthogonal to the previous estimated redundancy variate, u_1. Moreover, in order to express the redundancy variate as a linear combination of X_0 with coefficients α_2 we need to regress $\tilde{u}$ on X_0

$$\tilde{u} = X_0 f + \varepsilon_8. \tag{2.14}$$

Then $\alpha_2 = k\hat{f}$, where k is such that $u_2 = X_0\alpha_2$ has norm equal to one.

To obtain redundancy variates of order $l \in \{3, \ldots, r\}$, with $r = \min\{p, q\}$, a similar procedure has to be developed, but to transform the estimates into the original space, instead of using (2.13), a slightly different regression model has to be considered

$$u^* = Ue + \varepsilon_9, \tag{2.15}$$

where $U = [u_1, \ldots, u_{l-1}]$. This model guarantees that u_l is orthogonal to $u_1, \ldots, u_{l-1}$.

2.2. Robust Alternating Regressions

As suggested in Croux et al. (2003) and Branco et al. (2005), the regression models considered in the iterative procedure are estimated using weighted L_1 regression, with weights $w_i(X_{l-1}^*)$ defined by

$$w_i(X_{l-1}^*) = \min\left(1, \frac{\chi^2_{p^*,0.95}}{D_i^2(X_{l-1}^*)}\right), \quad i = 1, \ldots, n, \tag{2.16}$$

where $\chi^2_{p^*,0.95}$ is the upper 5% critical value of a chi-squared distribution with $p^* = p - l + 1$ degrees of freedom (the number of columns of X_{l-1}^*), and

$$D_i(X_{l-1}^*) = \sqrt{(x_i^{*(l-1)} - T(X_{l-1}^*))^t C(X_{l-1}^*)^{-1}(x_i^{*(l-1)} - T(X_{l-1}^*))}, i=1,\ldots,n,$$

where $\boldsymbol{x}_i^{*(l-1)}$ is the ith column of $\boldsymbol{X}_{l-1}^*$ and D_i is a robust distance. By the same reasons pointed out in Croux et al. (2003) and Branco et al. (2005) T and C were chosen to be MVE estimators of location and scatter, respectively. The regressions used to rewrite $\boldsymbol{u}^*$ in the original space were estimated using LTS regression (Rousseeuw, 1984). However, other robust estimators of regression can also be used.

3. Simulation Study

It is convenient to keep the simulation conditions as in Branco et al. (2005). By doing so, it is possible to compare the performance of the robust estimators based on alternating regressions not only with other robust estimators but also with the robust estimators of canonical correlation analysis based on alternating regression. This makes sense since the correlation matrices chosen are such that the redundancy analysis and the canonical correlation analysis lead to the same linear combination of the observed variables, i.e. the true values for the canonical coefficients are equal to the true values for the redundancy coefficients.

The performance of the robust estimators based on alternating regressions (RAR) are compared with the performance of the estimators based on robust correlation matrices (RMCD - $h \approx 0.75n$ and M-Estimator (M), using Huber psi-function: $w(s) = \min\left(1, \chi^2_{(p+q),0.95}/|s|\right)$) as well as with the classical estimator (Class). To do so, samples of size 500 were generated from four different distributions: normal distribution (NOR), with zero mean and covariance matrix Σ ($N(\mathbf{0}, \Sigma)$); symmetric contaminated normal (SCN) where each vector of observations is generated from a normal distribution ($N(\mathbf{0}, \Sigma)$) with probability 0.9 and from $N(\mathbf{0}, 9\Sigma)$ with probability 0.1; t-distribution with 3 degrees of freedom (T3) and asymmetric contamination (ACN) where 90% of the data are generated from a $N(\mathbf{0}, \Sigma)$ and the other 10% of the observations are equal to the vector $tr(\Sigma)\mathbf{1}^t_{(p+q)}$.

For each type of distribution and each estimation method $m = 300$ samples of 500 observations were produced. To assess the performance of each estimation method the following measures of MSE, were defined

$$\begin{aligned}
\text{MSE}(\hat{\boldsymbol{\alpha}}_l) &= \frac{1}{m}\sum_{j=1}^{m} \cos^{-1}\left(\frac{|\boldsymbol{\alpha}^t \hat{\boldsymbol{\alpha}}_l^j|}{||\hat{\boldsymbol{\alpha}}_l^j|| \cdot ||\boldsymbol{\alpha}_l||}\right), \\
\text{MSE}(\hat{R}_{yl}) &= \frac{1}{m}\sum_{j=1}^{m} \left(\hat{R}_{yl}^j - R_{yl}^j\right)^2,
\end{aligned}$$

where $\hat{\boldsymbol{\alpha}}_l^j$ is the jth estimate of the lth redundancy coefficient, $\boldsymbol{\alpha}_l$, and $\hat{R}_{yl}^j$ is the jth estimate of the lth redundancy index, R_{yl}.

Table 1 shows the values of p and q that were considered, together with the correlation matrix between the two groups. The correlation matrix of each group was taken as the identity matrix.

TABLE 1. Simulation setup. $R_{11} = \boldsymbol{I}_p$ and $R_{22} = \boldsymbol{I}_q$.

n	m	p	q	R_{12}
500	300	2	2	$\begin{bmatrix} 0.9 & 0 \\ 0 & 1/2 \end{bmatrix}$
500	300	2	4	$\begin{bmatrix} 0.9 & 0 & 0 & 0 \\ 0 & 1/2 & 0 & 0 \end{bmatrix}$
500	300	4	4	$\begin{bmatrix} 0.9 & 0 & 0 & 0 \\ 0 & 1/2 & 0 & 0 \\ 0 & 0 & 1/3 & 0 \\ 0 & 0 & 0 & 1/4 \end{bmatrix}$

The results for the case $p = q = 2$ are summarized in Figure 1 and 2 (see Figure 5(a) to identify each contamination scheme). In this case, it can be said that the contamination influences the vectors of redundancy coefficients equally. ACN is the contamination scheme that produces more damages in the estimated values, and only the estimates of redundancy coefficients based on RMCD and RAR are robust for this kind of contamination. ACN has also a serious effect on the first redundancy index when the classical method is used. In the case $p = 2$ and $q = 4$ something quite similar happens, except that RAR reveals to be even better to estimate the first redundancy coefficient. Because of lack of space the figures associated with this case are omitted.

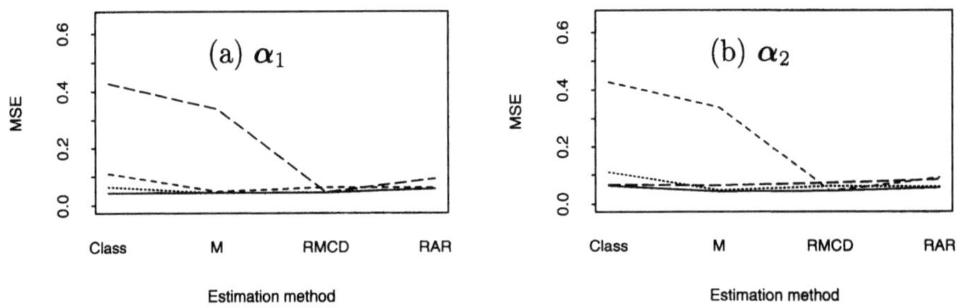

FIGURE 1. Mean square error of the redundancy coefficients, $p = 2$, $q = 2$.

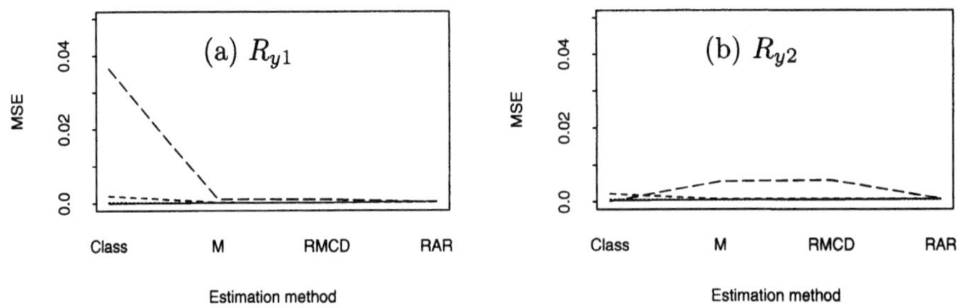

FIGURE 2. Mean square error of the redundancy indexes, $p = 2$, $q = 2$.

For $p = q = 4$ we obtain very similar results for the first redundancy coefficient and index as for $p = q = 2$. However, the effect of the contamination increases with the order of the redundancy coefficients (see Figures 3, 4, and Figure 5(a) to identify each contamination scheme). The method based on alternating regression performs slightly worse than RMCD for the two last redundancy coefficients ($\boldsymbol{\alpha}_3$ and $\boldsymbol{\alpha}_4$), and for the two last redundancy indexes (R_{y3} and R_{y4}).

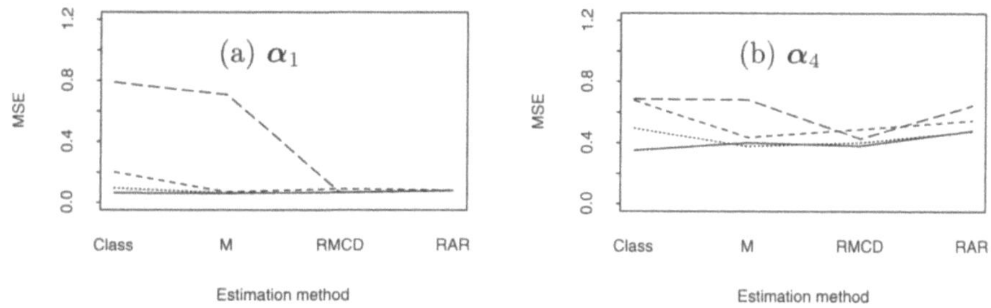

FIGURE 3. Mean square error of the first and last redundancy coefficients, $p = 4$, $q = 4$.

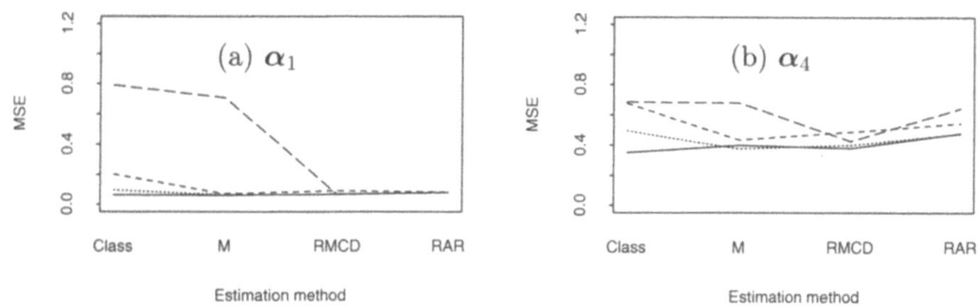

FIGURE 4. Mean square error of the first and last redundancy indexes, $p = 4$, $q = 4$.

FIGURE 5. (a) On the left – legend of Figures 1, 2, 3 and 4. (b) On the right – legend of the figures summarizing the performance and the empirical breakdown point of the various estimators.

3.1. Empirical Breakdown Point

Another criteria to compare the performance of the estimators is based on the empirical breakdown point. To do so, a simulation study was carried out, with the same setup that was used in Branco et al. (2005), where each group has 3 variables ($p = q = 3$). For each sample, $(100 - \epsilon)\%$ of the points were generated from a normal distribution with zero mean and correlation matrix, R, with $R_{11} = \boldsymbol{I}_3$ and $R_{22} = \boldsymbol{I}_3$ and

$$R_{12} = \begin{bmatrix} 0.9 & 0 & 0 \\ 0 & 1/2 & 0 \\ 0 & 0 & 1/3 \end{bmatrix},$$

and the other $\epsilon\%$ of the observations are equal $tr(\Sigma)\mathbf{1}^t$, (in the present case, $tr(\Sigma) = 6$). The values of ϵ were chosen from zero (no contamination) to 25 (25% of contamination), i.e., $\epsilon \in \{0, 1, \ldots, 25\}$. As before we chose $n = 500$. The procedure was repeated 200 times for each estimation method. The results, for the first and last redundancy coefficients and indexes, are summarized in Figures 6 and 7 (see Figure 5(b) to identify each estimator considered). In Figure 7(a), the MSE associated with the M, RMCD and RAR estimators seems to be equally low due to the apparent large magnitude of the MSE of the Class estimator. Nevertheless, the empirical breakdown point for R_{y2} and R_{y3} (see Figure 7(b)) show that the RAR has lower empirical breakdown point than RMCD but higher than the estimators based on the M-estimator (M). Similar conclusions apply to $\boldsymbol{\alpha}_1$, $\boldsymbol{\alpha}_2$ and $\boldsymbol{\alpha}_3$, as showed in Figure 6.

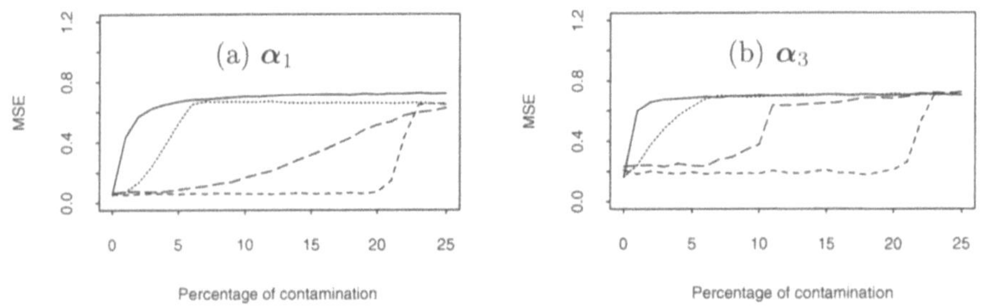

FIGURE 6. Empirical breakdown point, redundancy coefficients.

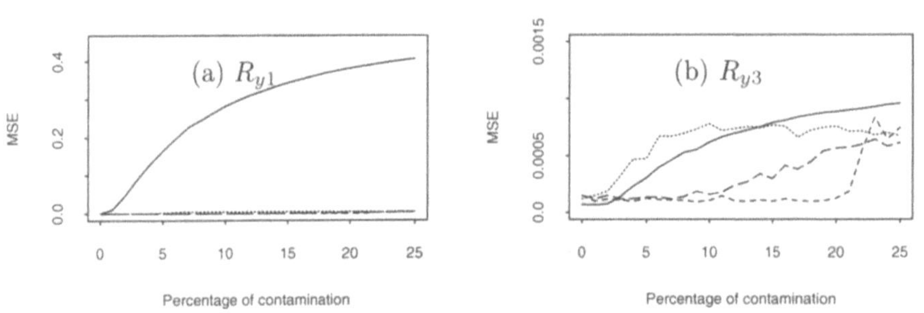

FIGURE 7. Empirical breakdown point, redundancy indexes.

4. Discussion and Future Work

The RAR procedure produced useful results and it has the advantage that it can be carried out in the case of more variables than observations. It is also capable of dealing with missing values and it can cope with outlying cells (Croux et al., 2003). Moreover, RAR aims at directly maximizing the redundancy index, and is therefore in a way intuitively more appealing than a covariance matrix based procedure. However, the search for other robust estimators should be pursued. In Oliveira and Branco (2003) various robust estimators have been studied for the first redundancy variate, where the estimators based on projection pursuit revealed promising results. This estimator has to be developed for higher order variates.

The simulation developed in this study was designed to facilitate the comparison with the study in Branco et al. (2005). So far, it can be said that the behavior of RAR is similar in both canonical correlation analysis and redundancy analysis. Having resolved the problem of robustifying (by alternating regression) the canonical correlation analysis and redundancy analysis we are compelled to consider the generalization of these two methods proposed by DeSarbo (1981).

As it was pointed out by van den Wollenberg (1977), principal components analysis is a special case of redundancy analysis. If we choose the dependent variables, $\boldsymbol{y}$, equal to the independent variables $\boldsymbol{x}$, the principal components differ from the redundancy variates by a constant term. In principal components analysis, we search for a linear combination of the $\boldsymbol{x}$, $\boldsymbol{a}$, that maximizes $\boldsymbol{a}^t R\boldsymbol{a}$ and $\boldsymbol{a}^t\boldsymbol{a} = 1$ ($R = Corr(\boldsymbol{x})$). In redundancy analysis we seek for a linear combination of $\boldsymbol{x}$, $\boldsymbol{\alpha}$, that maximizes $\boldsymbol{\alpha}^t R^2\boldsymbol{\alpha}$, where $\boldsymbol{\alpha}^t R\boldsymbol{\alpha} = 1$. Both the solutions of the two problems have the directions of the eigenvectors of R. Therefore, the algorithm based on alternating regressions can be used to estimate principal components. It would be interesting to compare this approach with other suggestions made in the literature, like the procedure based on alternating regressions to estimate principal components proposed in Wold (1966), and other robust estimators, e.g., based on projection pursuit.

References

[1] J.A. Branco, C. Croux, P. Filzmoser, and M.R. Oliveira, *Robust Canonical Correlations: A Comparative Study.* Computational Statistics, **2** (2005). To appear, available at *http://www.statistik.tuwien.ac.at/public/filz/publications/*.

[2] C. Croux and A. Ruiz-Gazen, *A fast Algorithm for Robust Principal Components Based on Projection Pursuit.* In A. Prat, editor, COMPSTAT: Proceedings in Computational Statistics, pages 211–216. Physica-Verlag, Heidelberg, 1996.

[3] C. Croux, P. Filzmoser, G. Pison, and P.J. Rousseeuw, *Fitting Multiplicative Models by Robust Alternating Regressions.* Statist. Comput. **13** (2003), 23–36.

[4] W.S. DeSarbo, *Canonical/Redundancy Factoring Analysis.* Psychometrika **46** (1981), 307–329.

[5] E. Lyttkens, *Regression Aspects of Canonical Correlation.* J. Multivariate Anal. **2** (1972), 418–439.

[6] R.A. Maronna, *Robust M-Estimators of Multivariate Location and Scatter.* Ann. Statist. **4** (1976), 51–67.

[7] M.R. Oliveira and J.A. Branco, *Comparison of Three Methods for Robust Redundancy Analysis.* In R. Dutter, P. Filzmoser, U. Gather, and P. J. Rousseeuw, editors, Developments in Robust Statistics, pages 287–295. Physica-Verlag, Heidelberg, 2003.

[8] A.C. Rencher, *Multivariate statistical inference and applications.* Wiley Series in Probability and Statistics, New York, 1998.

[9] P.J. Rousseeuw, *Multivariate Estimation With High Breakdown Point.* In W. Grossmann, G. Pflug, I. Vincze, and W. Wertz, editors, Mathematical Statistics and Applications, Vol. B, pages 283–297. Reidel, Dordrecht, The Netherlands, 1985.

[10] P.J. Rousseeuw and K. Van Driessen, *A Fast Algorithm For the Minimum Covariance Determinant Estimator.* Technometrics **41** (1999), 212–223.

[11] P.J. Rousseeuw, *Least Median of Squares Regression.* J. Amer. Statist. Assoc. **79** (1984), 871–881.

[12] D.K. Stewart and W.A. Love, *A General Canonical Correlation Index.* Psychol. Bullet. **70** (1968), 160–163.

[13] M. Tenenhaus, *La Régression PLS. Théorie et pratique.* Éditions Technip, Paris, 1998.

[14] H. Wold, *Nonlinear Estimation by Iterative Least Squares Procedures.* In F.N. David, editor, A Festschrift for J. Neyman, pages 411–444. John Wiley and Sons, New York, 1966.

[15] A.L. van den Wollenberg, *Redundancy Analysis: An Alternative for Canonical Correlation Analysis.* Psychometrika **42** (1977), 207–219.

M.R. Oliveira and J.A. Branco
Instituto Superior Técnico
Av. Rovisco Pais
1049-001 Lisboa
Portugal
e-mail: `{rosario.oliveira, jbranco}@math.ist.utl.pt`

C. Croux
Katholieke Universiteit Leuven
Naamsestraat 69
B-3000 Leuven
Belgium
e-mail: `Christophe.Croux@econ.kuleuven.ac.be`

P. Filzmoser
Vienna University of Technology
Wiedner Hauptstr. 8–1
A-1040 Vienna
Austria
e-mail: `P.Filzmoser@tuwien.ac.at`

Statistics for Industry and Technology, 247–258

Robust ML-estimation of the Transmitter Location

E. Ollila and V. Koivunen

Abstract. Location of a radio transmitter may be estimated using the triangulation principle and angular measurements from at least two sensor arrays. For example, two base-stations equipped with antenna arrays observe the waveforms emitted by a transmitter such as a mobile phone. Based on the estimates of angle-of-arrival (AoA) of waveforms impinging the antenna array at base-stations, an estimate of the location of the transmitter can be formed using elemental geometry. In this paper, we discuss the ML-estimation of the location of the transmitter by modelling the distribution of the AoAs by two well-known angular distributions, namely the von Mises and the wrapped Cauchy distribution. These distributions are well justified by the physical measurements on radio wave propagation. Since the received signals at the base-stations are corrupted by noise which may be impulsive in nature, robust estimation of the signal parameters is an important issue.

Mathematics Subject Classification (2000). Primary 62F35; Secondary 62F25.

Keywords. Angular data, signal processing, ML-estimation, robustness.

1. Introduction

Location estimation using triangulation principle has numerous applications, for example in wireless communications, electronic warfare and navigation. In such problems, the angle of arrival (AoA) of the wavefield emitted by a transmitter may be estimated using observations acquired by a sensor array. Using the knowledge on the locations of the sensor arrays and the estimates of the angles of arrivals of the waveforms impinging the arrays, the location of the emitter may be estimated. See, e.g., Krim and Viberg (1996) for a nice tutorial on estimating AoA's using sensor arrays.

As a motivating practical example, envisage that $\mathbf{z}_1$ and $\mathbf{z}_2$ represent the locations of two base-stations of cellular phone system such as GSM and the unknown $\mathbf{z}$ is the location of the mobile phone of a subscriber. An accurate esti-

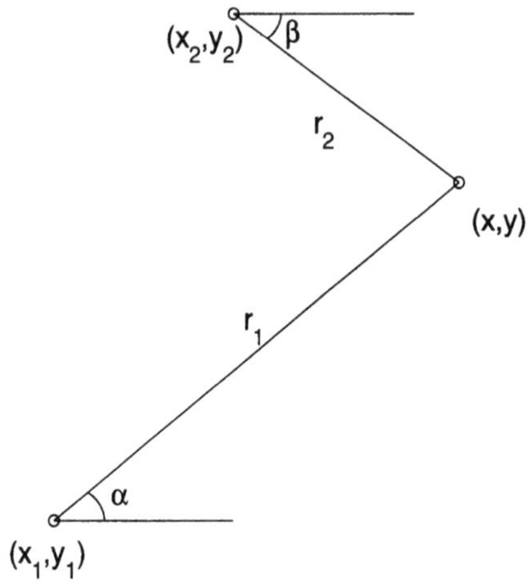

FIGURE 1. The location of the transmitter $\mathbf{z} = (x, y)^T$ is estimated using the angles of arrival α and β and the known locations of the sensor arrays $\mathbf{z}_1 = (x_1, y_1)^T$ and $\mathbf{z}_2 = (x_2, y_2)^T$.

mate of the location of the phone may be needed in case of emergency calls or just to provide subscribers information about different services available in the neighborhood of the current location. The base-stations equipped with antenna array are programmed to produce estimates of the (azimuth) AoA of the signals the phone is transmitting. It is often sensible to model the AoA measurements $\Theta_n = \{\theta_1, \ldots, \theta_n\}$ and $\Omega_n = \{\omega_1, \ldots, \omega_n\}$ obtained from the base-stations during n time steps as i.i.d. random deviates symmetric about the true angles α and β, respectively (given the transmitter at $\mathbf{z}$ has been static during this measurement interval). The received signals at base-stations are corrupted by environmental noise plus receiver noise. Especially man-made noise is often impulsive in nature (see Middleton, 1973) and therefore robust estimation of the signal parameters (such as AoA's of the signals or the location of the transmitter) based on the sensor array output vectors (snapshots in engineering jargon) is of major importance. Hence, robust estimation of the transmitter location is an important issue.

If the location of the points $\mathbf{z}_1 = (x_1, y_1)^T$ and $\mathbf{z}_2 = (x_2, y_2)^T$ and the angle α between $\mathbf{z}_1$ and a point $\mathbf{z}$ and angle β between $\mathbf{z}_2$ and a point $\mathbf{z}$ are known, then the "unknown" $\mathbf{z} = (x, y)^T$ can be written as the crossing point of the lines with angles α and β originating from $\mathbf{z}_1$ and $\mathbf{z}_2$, i.e., $\mathbf{z} := [x(\alpha, \beta), y(\alpha, \beta)]^T$. The above scenario becomes an estimation problem of $\mathbf{z}$ if the angle α is unknown and measurements, say $\theta_1, \ldots, \theta_n$, of α are provided. Similarly β is unknown and measurements, say $\omega_1, \ldots, \omega_n$, are provided. Figure 1 illustrates the geometry of the measurement system.

In this paper, we model the given data sets $\Theta_n = \{\theta_1, \ldots, \theta_n\}$ and $\Omega_n = \{\omega_1, \ldots, \omega_n\}$ as i.i.d. (and also mutually independent) random deviates from angular distribution symmetric about α and β, respectively. We use two well-known angular distributions, namely the von Mises (VM-)distribution and the wrapped Cauchy (WrC-)distribution as our models. The role of the former distribution for angular measurements is similar to that of the normal distribution for measurements on the line. The latter distribution is an alternative to the classical VM-distribution and its MLE of the symmetry center is a robust alternative for the "mean direction", i.e., the MLE of the VM-distribution. These model distributions are briefly reviewed in Section 3. We refer to monographs by Mardia (1972) and Jammalamadaka and SenGupta (2001) for a detailed account on angular models. A practical application oriented motivation for this study and the used model distributions is given in Section 2, where the statistics of wireless communication channels are briefly described.

Let $\hat{\alpha}$ and $\hat{\beta}$ denote the ML-estimates of the angles α and β calculated from the samples Θ_n and Ω_n, respectively. The MLE of the location is then $\hat{\mathbf{z}} = [x(\hat{\alpha}, \hat{\beta}), y(\hat{\alpha}, \hat{\beta})]^T$. Throughout the paper we assume that $\beta \neq \alpha$ and $\beta \neq (\alpha + \pi)\mathrm{mod}(2\pi)$. This restriction is due to estimation technique since otherwise the lines originated from $\mathbf{z}_1$ and $\mathbf{z}_2$ are parallel and do not cross. Note also that the estimation technique leads to the following ambiguity: the estimated angle pairs $(\hat{\alpha}, \hat{\beta})$ and $((\hat{\alpha} + \pi)\mathrm{mod}(2\pi), \hat{\beta})$ will lead to the same location estimate. See Krim and Viberg (1996) for a brief description on different antenna array configurations and ambiguities associated with them.

In this paper we investigate the statistical properties of $\hat{\mathbf{z}}$: large sample properties and an estimator of standard errors are derived using ML-theory. Moreover, accompanying the estimate and its standard errors, also the $100p\%$ likelihood contours are easily provided. This is discussed in Section 4. We will demonstrate the properties (robustness, variability) of the ML-estimators of the location based on VM- and WrC-distribution using Monte-Carlo studies. This is the subject of Section 5. Finally, Section 6 concludes.

2. Statistics of Wireless Communication Channels

During the recent years a significant research interest in the area of wireless telecommunications has focused on modelling the statistics of indoor and outdoor mobile radio channels (Middleton, 1973; Pedersen et al., 2000; Spencer et al., 2000). The modelling then includes, for example, the pdf of the azimuth AoA and time delay of the impinging wavefields as well as their expected power. Several experimental measurements are available to which we can compare the validity of the angular model distributions used in this paper. For example, Pedersen et al. (2000) have published a number of *outdoor* measurement results collected in a macrocellular typical urban environment. Their measurements were conducted at 1.8 GHz in Stockholm, Sweden and Arhus, Denmark, using an eight element Uniform Linear Antenna (ULA) array at an elevated base-station (the base-station antenna is located above the surrounding buildings). The signal parameters (e.g., the AoA's, delays, etc) of the impinging wavefields are estimated using the array snapshots. In Fig. 10 of (Pedersen et al., 2000), the empirical pdf of the AoAs is depicted. The authors then suggest to model the pdf by the Gaussian distribution. They however note that the Gaussian function does not provide a good match for the empirical pdf at the tails of the distribution which are heavier than those of the Gaussian distribution. The empirical pdf indicates that the VM-distribution may provide a better fit as its shape is close to Gaussian at the center but its tails are heavier on the ends of the angular interval. Indeed, the Gaussian function fails since it is a valid pdf for measurements on the line; thus assigning probability mass outside the angular interval $(0°, 360°]$.

Spencer et al. (2000) have published a number of *indoor* measurements for two different buildings. Their measurements were conducted at 7 GHz and scanning was done mechanically with antenna beamwidth 6°. In Fig. 12 of Spencer et al. (2000) the empirical pdf of the AoA's is depicted. The authors note that the Laplacian function gives a good match except at the tails of the distribution which are heavier than those of the Laplace distribution. The empirical AoA pdf in this case indicate that the WrC-distribution may provide a better fit than the Laplace distribution.

The outdoor measurements in Pedersen et al. (2000) and the indoor measurements in Spencer et al. (2000) thus suggest a different angular model for the AoA distribution (i.e., the VM- and WrC-distribution, respectively). One may speculate whether these distributions have their origin in characterizing the true distribution of the AoA's for indoor and outdoor mobile radio channels respectively, or whether they are artifacts of different measurement systems and of the algorithms used to extract the signal parameters. Nonetheless, these physical real-world measurements validate that the angular distributions considered in this paper are reasonable models for the AoA's of the indoor and outdoor mobile radio channels.

3. Angular Model Distributions

3.1. Von Mises Distribution

VM-distribution plays a crucial role in statistical inference for angular measurements and its role is in many ways the same as that of the normal distribution for measurements on the line. VM-distribution is also referred as "Circular Normal distribution", e.g, in Jammalamadaka and SenGupta (2001), to highlight its importance and similarities with the normal distribution on the line.

An angular random variable $\theta \in (0, 2\pi]$ is said to have von Mises distribution, denoted by $\mathrm{VM}(\alpha, \kappa)$, if its pdf is given by

$$f(\theta; \alpha, \kappa) = \frac{1}{2\pi i_0(\kappa)} \exp\{\kappa \cos(\theta - \alpha)\}, \tag{3.1}$$

where $i_0(\kappa)$ is the modified Bessel function of the first kind and order zero, i.e., $i_0(\kappa) = \sum_{r=0}^{\infty} \frac{1}{r!^2}(\kappa/2)^{2r}$. The parameter α is the symmetry center and the parameter $\kappa > 0$ is the concentration parameter: the larger the κ, the greater is the concentration about the center α. In the limit $\kappa \to \infty$, (3.1) approaches point mass distribution at α and conversely, in the limit $\kappa \to 0$, (3.1) approaches the uniform distribution on the circle. Computation of the ML-estimates of the VM-distribution is discussed in Jammalamadaka and SenGupta (2001).

3.2. Wrapped Cauchy Distribution

The Cauchy distribution is a well-known alternative to the classical normal distribution and its MLE of the symmetry center serves as a robust alternative for the mean. The wrapped Cauchy distribution is an alternative to the classical von Mises distribution for modelling symmetric angular data and its MLE of the symmetry center serves as a robust alternative for the mean direction, i.e., the MLE of the center of symmetry of the VM-distribution. The WrC-distribution is simply the distribution generated by wrapping the Cauchy distribution on the line around the unit circle, or in other words, expressing the latter mod(2π). An angular random variable $\theta \in (0, 2\pi]$ has wrapped Cauchy distribution, denoted by WrC(α, ρ), if its pdf is of the form

$$f(\theta; \alpha, \rho) = \frac{1}{2\pi} \frac{1-\rho^2}{1+\rho^2 - 2\rho\cos(\theta-\alpha)}, \tag{3.2}$$

where the parameter α represents the symmetry center and ρ represents a shape parameter with $0 \le \rho < 1$. When $\rho \to 1$, (3.2) approaches point mass distribution at α and when $\rho = 0$, (3.2) is simply the uniform distribution on the circle. A novel iterative algorithm to calculate the ML-estimates of WrC-distribution was proposed by Kent and Tyler (1988).

4. ML-estimation of the Transmitter Location

As noted earlier, the location $\mathbf{z} = (x, y)^T$ of the transmitter can be written as the crossing point of the lines with angles α and β originating from the location of the base-stations $\mathbf{z}_1 = (x_1, y_2)^T$ and $\mathbf{z}_2 = (x_2, y_2)^T$, mathematically,

$$x = x(\alpha, \beta) = \frac{y_2 - y_1 + \tan(\alpha)x_1 - \tan(\beta)x_2}{\tan(\alpha) - \tan(\beta)}, \tag{4.1}$$

$$y = y(\alpha, \beta) = \frac{\tan(\alpha)y_2 - \tan(\beta)y_1 - \tan(\alpha)\tan(\beta)(x_2 - x_1)}{\tan(\alpha) - \tan(\beta)} \tag{4.2}$$

where, by definition,

$$\alpha = \alpha(x, y) = \tan^{-1}\{(y - y_1)/(x - x_1)\}, \tag{4.3}$$

$$\beta = \beta(x, y) = \tan^{-1}\{(y - y_2)/(x - x_2)\}. \tag{4.4}$$

Recall that we model the given data sets $\Theta_n = \{\theta_1, \ldots, \theta_n\}$ and $\Omega_n = \{\omega_1, \ldots, \omega_n\}$ as i.i.d. (and also mutually independent) random deviates from angular distributions symmetric about α and β, respectively. Let $\hat{\alpha}$ and $\hat{\beta}$ denote the MLEs of the angles α and β calculated from the samples Θ_n and Ω_n, respectively. Then the **MLE of the location z** of the transmitter is simply

$$\hat{\mathbf{z}} = \begin{bmatrix} \hat{x} \\ \hat{y} \end{bmatrix} = \begin{bmatrix} x(\hat{\alpha}, \hat{\beta}) \\ y(\hat{\alpha}, \hat{\beta}) \end{bmatrix}.$$

We denote the symmetric angular distribution of θ_i's by F_1, which is parameterized by α, the symmetry center of the distribution, and τ_1 (say), the shape parameter of the distribution. Similarly, denote the angular distribution of ω_i's by F_2, which is parameterized by β, the symmetry center of the distribution, and τ_2 (say), the shape parameter of the distribution. F_1 and F_2 need not to be same angular distributions.

Next consider the asymptotic distribution of the MLE $\hat{\mathbf{z}} = (\hat{x}, \hat{y})^T$ in the case that the shape parameters τ_1 and τ_2 of the angular distributions are known. Then under sufficient regularity conditions,

$$\sqrt{n} \begin{bmatrix} \hat{\alpha} - \alpha \\ \hat{\beta} - \beta \end{bmatrix} \xrightarrow{D} N_2 \left(\begin{bmatrix} 0 \\ 0 \end{bmatrix}, \begin{bmatrix} I_1(\alpha)^{-1} & 0 \\ 0 & I_2(\beta)^{-1} \end{bmatrix} \right)$$

where $I_1(\alpha)$ and $I_2(\beta)$ denote the expected information function for a single observation from F_1 and F_2. If the above holds, then the MLE $\hat{\mathbf{z}}$ has a limiting bivariate normal distribution with the variance-covariance matrix $I(x,y)^{-1}$, where $I(x,y) = Q^T I(\alpha,\beta) Q$ is the expected information function for (x,y), $I(\alpha,\beta) = \operatorname{diag}\{I_1(\alpha), I_2(\beta)\}$ and

$$Q = \begin{bmatrix} \partial\alpha(x,y)/\partial x & \partial\alpha(x,y)/\partial y \\ \partial\beta(x,y)/\partial x & \partial\beta(x,y)/\partial y \end{bmatrix} = \begin{bmatrix} -\sin(\alpha)/r_1 & \cos(\alpha)/r_1 \\ -\sin(\beta)/r_2 & \cos(\beta)/r_2 \end{bmatrix}, \tag{4.5}$$

where r_1,r_2 denote the Euklidean distances between $\mathbf{z}$ and $\mathbf{z}_1$,$\mathbf{z}_2$, respectively. Now, denote

$$I(x,y)^{-1} = \begin{bmatrix} \sigma_x^2 & \sigma_{xy} \\ \sigma_{xy} & \sigma_y^2 \end{bmatrix}. \tag{4.6}$$

Then we obtain the following expressions for its components:

$$\sigma_x^2 = \frac{1}{(\tan(\alpha) - \tan(\beta))^2} \left\{ r_1^2 \frac{I_1(\alpha)^{-1}}{\cos^2(\alpha)} + r_2^2 \frac{I_2(\beta)^{-1}}{\cos^2(\beta)} \right\},$$
$$\sigma_y^2 = \frac{\tan^2(\alpha)\tan^2(\beta)}{(\tan(\alpha) - \tan(\beta))^2} \left\{ r_1^2 \frac{I_1(\alpha)^{-1}}{\sin^2(\alpha)} + r_2^2 \frac{I_2(\beta)^{-1}}{\sin^2(\beta)} \right\},$$
$$\sigma_{xy} = \frac{2\tan(\alpha)\tan(\beta)}{(\tan(\alpha) - \tan(\beta))^2} \left\{ r_1^2 \frac{I_1(\alpha)^{-1}}{\sin(2\alpha)} + r_2^2 \frac{I_2(\beta)^{-1}}{\sin(2\beta)} \right\}.$$

Note that for the existence of the covariance matrix, we also need to assume that $\alpha, \beta \neq \{0, \pi/2, \pi, 3\pi/2, 2\pi\}$.

From the expression of $I(x,y)^{-1}$ we may conclude the following: (A) the accuracy of the location estimate decreases when the distance r_1 (and/or r_2) between the transmitter $\mathbf{z}$ and the base-station $\mathbf{z}_1$ (and/or $\mathbf{z}_2$) increases; (B) the better the accuracy of the angle estimates $\hat{\alpha}$ and $\hat{\beta}$ (i.e., the smaller their asymptotic variances $I_1(\alpha)^{-1}$ and $I_2(\beta)^{-1}$ are), the better is the accuracy of the location estimate.

As an estimator of the covariance matrix of $\hat{\mathbf{z}} = (\hat{x}, \hat{y})^T$ we may use the inverse of the observed information function $J(\hat{x}, \hat{y})^{-1}$, where

$$J(\hat{x}, \hat{y}) = \hat{Q}^T \mathrm{diag}\{J_1(\hat{\alpha}), J_2(\hat{\beta})\}\hat{Q}. \tag{4.7}$$

Here we evaluate Q at the MLE to obtain $\hat{Q}$, and $J_1(\hat{\alpha})$ is the observed information function for the sample Θ_n from F_1 evaluated at the MLE $\hat{\alpha}$. Similarly, $J_2(\hat{\beta})$ is the observed information function for sample Ω_n from F_2 evaluated at the MLE $\hat{\beta}$.

Next consider the case that the shape parameters τ_1 and τ_2 are unknown. Under sufficient regularity conditions, $\sqrt{n}(\hat{\alpha}-\alpha, \hat{\tau}_1-\tau_1)^T$ and $\sqrt{n}(\hat{\beta}-\beta, \hat{\tau}_2-\tau_2)^T$ have a limiting bivariate normal distribution. If $I_{12}(\alpha, \tau_1) = E[-\partial^2 \log f_1/\partial\alpha\partial\tau_1] = 0$ and $I_{12}(\beta, \tau_2) = E[-\partial^2 \log f_2/\partial\beta\partial\tau_2] = 0$, then the ML-estimates $\hat{\alpha}$ and $\hat{\tau}_1$ and the ML-estimates $\hat{\beta}$ and $\hat{\tau}_2$ are asymptotically independent and consequently asymptotic behavior of $\hat{\mathbf{z}}$ remains unchanged if τ_1 and τ_2 are known parameters or to be estimated from the data. This holds for the model distributions considered in this paper. Thereby, if τ_1 and τ_2 are unknown, the covariance matrix of $\hat{\mathbf{z}}$ is estimated by $J(\hat{x}, \hat{y})^{-1}$ given in (4.7) with the exception that τ_1 and τ_2 in the resulting formulae are replaced by their ML-estimates.

Finally, we note that accompanying the MLE $\hat{\mathbf{z}}$ and its standard error estimates, also the related $100p\%$ likelihood contours (i.e., the approximate $100(1-p)\%$ confidence regions; see Kalbfleisch (1985, Sec. 11.5) can easily be calculated. In the case that τ_1 and τ_2 are known parameters, the log relative likelihood function $\mathrm{R}(x, y)$ may be calculated due to invariance of the likelihood functions under reparameterization as follows: $\mathrm{R}(x_0, y_0)$ of any pair of values (x_0, y_0) is found by computing the corresponding values $\alpha_0 = \alpha(x_0, y_0)$ and $\beta_0 = \beta(x_0, y_0)$ and then equating $\mathrm{R}(x_0, y_0)$ with $\mathrm{R}_1(\alpha_0) + \mathrm{R}_2(\beta_0)$, where $\mathrm{R}_1(\alpha)$ and $\mathrm{R}_2(\beta)$ are the log relative likelihood functions for a sample Θ_n from F_1 and a sample Ω_n from F_2. Then the desired $100p\%$ likelihood contours may be plotted on the $x-y$ plane. Often, a computationally more feasible solution is to calculate the normal approximation of the likelihood contours. In the case that τ_1 and τ_2 are unknown, the related $100p\%$ maximum likelihood contours (Kalbfleisch, 1985), i.e., the profile likelihoods using the terminology of Lindsey (1996), can be calculated in a similar fashion.

5. Simulation Studies

5.1. Simulation Arrangements

We now explain the simulation setup with an example. The transmitter is located at $\mathbf{z} = (200, 300)^T$ and the base-stations are at $\mathbf{z}_1 = (20, 100)^T$ and $\mathbf{z}_2 = (100, 400)^T$. This yields $\alpha = 0.838$ and $\beta = 5.498$ for the angles (measured in radians). We generated samples $\theta_1, \ldots, \theta_n$ and $\omega_1, \ldots, \omega_n$ from VM-distributions $\mathrm{VM}(\alpha, \kappa_1)$ and $\mathrm{VM}(\beta, \kappa_2)$, respectively, with shape parameters $\kappa_1 = 200$ and $\kappa_2 = 50$. The sample size is $n = 15$. From the simulated samples we calculated the MLE's of the angles $\hat{\alpha}$ and $\hat{\beta}$ and the location $\hat{\mathbf{z}}$. In Figure 2, the configuration of the transmitter and the base-stations are shown together with the MLE of the transmitter. The estimates for the covariance matrix of $\hat{\mathbf{z}}$, calculated by $J(\hat{x}, \hat{y})^{-1}$ were

$$\begin{bmatrix} 26.195 & 1.971 \\ 1.971 & 27.216 \end{bmatrix} \quad \text{and} \quad \begin{bmatrix} 27.492 & 0.711 \\ 0.711 & 28.410 \end{bmatrix},$$

where the last matrix correspond to case that κ_1 and κ_2 were estimated from the data.

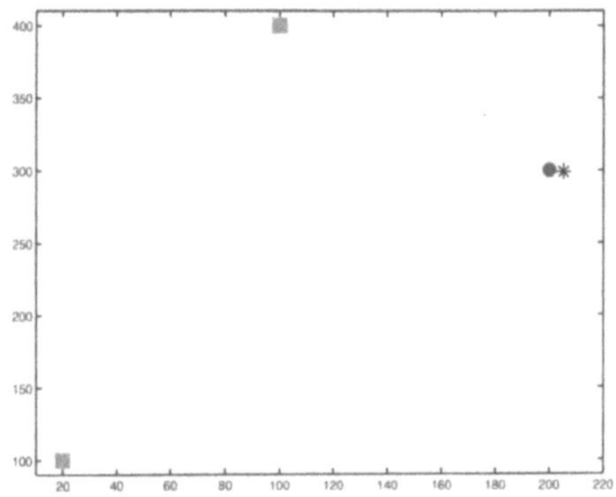

FIGURE 2. The location of the transmitter is indicated by a filled circle and its MLE by an asterisk. The location of the base-stations are indicated by filled squares.

Figure 3a depicts the 1%, 10% and 50% likelihood contours and the normal approximations (κ_1 and κ_2 are taken to be known). Note that the normal approximation provides a good approximation despite of the small sample size ($n = 15$). In the case that κ_1 and κ_2 are not known, the relevant summary about (x, y) is obtained by the $100p\%$ maximum likelihood (profile likelihood) contours which are depicted in Figure 3b.

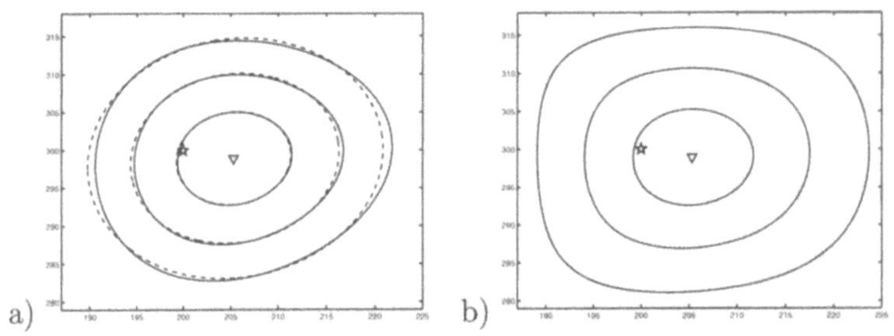

FIGURE 3. Figure a) depicts the 1%, 10% and 50% likelihood contours and their normal approximations (dotted lines). Figure b) depicts the 1%, 10% and 50% maximum likelihood (profile likelihood) contours. The location of the transmitter is indicated by the star and its MLE by the triangle.

5.2. Study 1: Estimation of the Accuracy of the Location Estimator

As described in Section 4, the accuracy (the variance-covariance matrix) of the VM-estimate or WrC-estimate of the location can be conveniently estimated by $J(\hat{x},\hat{y})^{-1}$. In the present simulation study we investigate the performance of this estimator of accuracy for small sample sizes and also compare its performance to non-parametric estimators of the variance-covariance matrix based on the principles of Jackknife and Bootstrap. We refer to Efron and Tibshirani (1993), Shao and Wu (1989) and Shao (1989) for a comprehensive treatment on the Bootstrap and the Jackknife.

Let $\hat{\alpha}$ and $\hat{\beta}$ be the angle estimates based on the samples $\theta_1,\ldots,\theta_n$ and $\omega_1,\ldots,\omega_n$ from symmetric angular distributions F_1 and F_2, and let $\hat{\mathbf{z}} = (\hat{x},\hat{y})^T$ be the corresponding location estimate. Then calculate, say M bootstrap samples from each of the data sets. From the M bootstrap samples calculate the estimates $\hat{\alpha}_i$ and $\hat{\beta}_i$ and the location estimate $\hat{\mathbf{z}}_i = (x(\hat{\alpha}_i,\hat{\beta}_i), y(\hat{\alpha}_i,\hat{\beta}_i))^T$. The estimator of the variance-covariance matrix of $\hat{\mathbf{z}} = (\hat{x},\hat{y})^T$ based on the bootstrap is then defined as

$$\mathrm{BS}(\hat{x},\hat{y}) = \frac{1}{M}\sum_{i=1}^{M}(\hat{\mathbf{z}}_i - \hat{\mathbf{z}}^*)(\hat{\mathbf{z}}_i - \hat{\mathbf{z}}^*)^T,$$

where $\hat{\mathbf{z}}^* = M^{-1}\sum_{i=1}^{M}\hat{\mathbf{z}}_i$. Let $\hat{\alpha}_{(i)}$ and $\hat{\beta}_{(i)}$ be the estimates of α and β after deleting observations θ_i and ω_i from the samples, and let $\hat{\mathbf{z}}_{(i)} = (x(\hat{\alpha}_{(i)},\hat{\beta}_{(i)}), y(\hat{\alpha}_{(i)},\hat{\beta}_{(i)}))^T$ be the corresponding location estimate. The estimator of the variance-covariance matrix based on the Jackknife is then

$$\mathrm{JK}(\hat{x},\hat{y}) = \frac{n-1}{n}\sum_{i=1}^{n}(\hat{\mathbf{z}}_{(i)} - \hat{\mathbf{z}}_{(\cdot)})(\hat{\mathbf{z}}_{(i)} - \hat{\mathbf{z}}_{(\cdot)})^T,$$

where $\hat{\mathbf{z}}_{(\cdot)} = n^{-1}\sum_{i=1}^{n}\hat{\mathbf{z}}_{(i)}$.

The means of the standard error estimates of the VM-estimates and WrC-estimates of the location $\hat{\mathbf{z}} = (\hat{x},\hat{y})^T$ for 2000 simulated samples of size $n = 15$ and $n = 100$ from VM- and WrC-distribution are tabulated in Table 1. Standard deviations are reported in parentheses. Also the Monte Carlo standard error estimates are reported, i.e., the sample standard deviations of $\hat{x}$ and $\hat{y}$ of the 2000 simulated data sets. The parameters of the sample distributions were $\alpha = 0.838$ and $\beta = 5.498$, $\kappa_1 = \kappa_2 = 200$ and $\rho_1 = \rho_2 = 0.8$. We assume that the shape parameters of the distributions are unknown and thus they are also estimated from the data sets. A reasonable choice of estimator to use in practice is then the one that is consistently close to the Monte Carlo estimate. We see that the $J(\hat{x},\hat{y})^{-1}$ estimator has the best performance (i.e., smaller bias, smaller variability) when the samples are from the true model distribution, but is clearly biased when this is not the case. The BS-estimator is also performing well except in the case that the samples are from WrC-distribution and the sample size is small ($n = 15$). This is due to the poor robustness properties of the Bootstrap (Stromberg, 1997). The JK-estimator however does not seem to have this problem. The simulation thus

TABLE 1. Monte Carlo standard errors of VM- and WrC-estimates of $(\hat{x},\hat{y})$ together with means of the various standard error estimators for the 2000 simulated samples of sample size $n = 15$ and $n = 100$. Standard deviations are reported in parentheses.

Esti-mate	sample size	sample distr.	Monte Carlo	Estimator of Standard Error $J(\hat{x},\hat{y})^{-1}$	$\mathrm{BS}(\hat{x},\hat{y})$	$\mathrm{JK}(\hat{x},\hat{y})$
VM	15	VM	3.806	3.732 (0.60)	3.727 (0.64)	3.985 (0.77)
			3.962	3.822 (0.60)	3.812 (0.64)	4.057 (0.79)
		WrC	32.48	38.14 (14.10)	84.51 (847)	32.90 (12.65)
			31.75	37.25 (11.50)	83.79 (777)	32.24 (10.05)
	100	VM	1.493	1.495 (0.09)	1.493 (0.12)	1.508 (0.11)
			1.534	1.529 (0.09)	1.524 (0.12)	1.543 (0.11)
		WrC	11.55	14.08 (1.67)	11.47 (1.50)	11.39 (1.42)
			11.78	14.34 (1.38)	11.63 (1.27)	11.64 (1.21)
WrC	15	VM	4.838	3.628 (1.09)	5.340 (1.56)	5.329 (2.54)
			5.133	3.720 (1.08)	5.469 (1.55)	5.442 (2.56)
		WrC	20.43	18.80 (7.42)	144.8 (2168)	21.48 (10.67
			20.33	18.86 (6.74)	137.8 (2102)	21.58 (9.76)
	100	VM	1.914	1.491 (0.16)	1.962 (0.31)	1.961 (0.33)
			1.963	1.525 (0.16)	2.003 (0.31)	2.013 (0.33)
		WrC	6.796	6.790 (0.89)	7.025 (1.13)	6.906 (1.12)
			6.957	6.948 (0.84)	7.180 (1.08)	7.079 (1.10)

suggest to use the JK-estimator instead of $J(\hat{x},\hat{y})^{-1}$ if one suspects violations in the model assumptions. Since the BS-estimator is clearly biased for samples from heavy tailed distributions (such as WrC) with small sample sizes, we favor the JK-estimator over the BS-estimator.

5.3. Study 2: Robustness of the Location Estimator

The geometry of the transmitter and the base-stations are as in the earlier example of Section 5.1. We generated a pair of samples $\theta_1,\dots,\theta_n$ and $\omega_1,\dots,\omega_n$ from the ε-contaminated VM-distributions $\mathrm{CVM}(\varepsilon,\alpha,\kappa_1)$ and $\mathrm{CVM}(\varepsilon,\beta,\kappa_2)$, respectively, using $\kappa_1 = 200$ and $\kappa_2 = 50$. The $\mathrm{CVM}(\varepsilon,\alpha,\kappa)$ is defined so that $100(1-\varepsilon)\%$ of the observations are good observations from $\mathrm{VM}(\alpha,\kappa)$ and $100\varepsilon\%$ are bad observations from $\mathrm{VM}(2\pi-\alpha,\kappa)$. Note that this kind of situation may arise in real-world. For example, in the case that the base-stations receive desired signal at $100(1-\varepsilon)\%$ of the time but "listen" to a different transmitter (interferer) $100\varepsilon\%$ of the time. The sample size was $n = 50$ and the number of sample pairs was $m = 500$. We chose $\varepsilon = 0.2$. From the ith simulated sample pair we calculated the WrC- and VM-estimates $(\hat{\alpha}_i,\hat{\beta}_i)$ of the angles (α,β) and the corresponding estimates $\hat{\mathbf{z}}_i$ of the location $\mathbf{z}$. This procedure was repeated for all $i = 1,\dots,m$ pairs of samples.

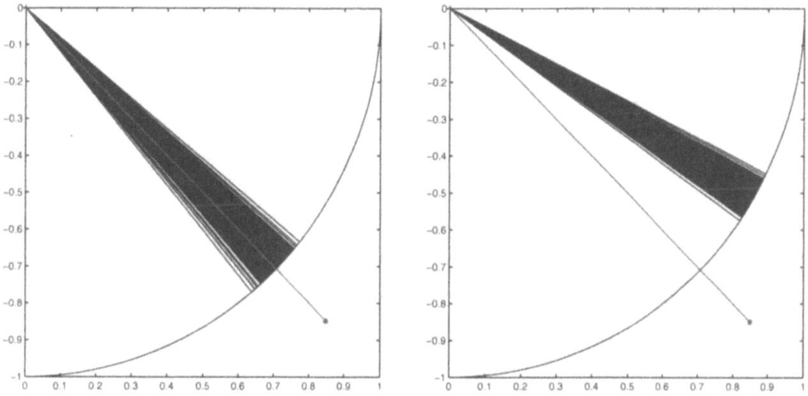

FIGURE 4. Plots of the WrC- (left) and VM-estimates (right) of the angle β calculated from 500 simulated samples from CVM-distribution. Also the true angle value is plotted (the line extending over the unit circle).

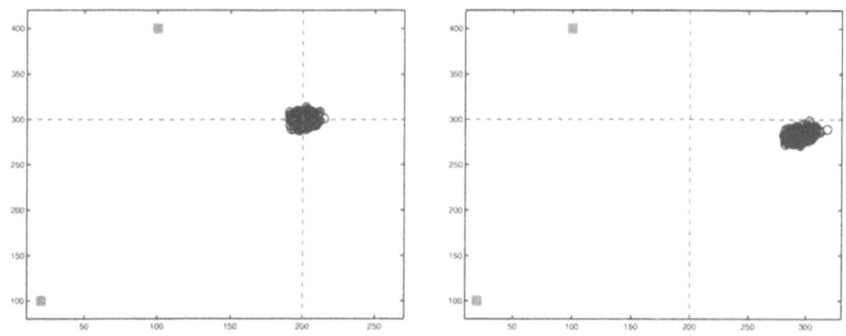

FIGURE 5. Plots of the WrC- (left) and VM-estimates (right) of the location $\mathbf{z}$ calculated from the 500 simulated samples from CVM-distribution. The location of the base-stations $\mathbf{z}_1$ and $\mathbf{z}_2$ are indicated by filled squares. The crossing point of the dashed line is at the true location of the transmitter $\mathbf{z}$.

We visualized the bias and the variability of the different estimators by plotting the calculated angle estimates $\hat{\alpha}_i$ and $\hat{\beta}_i$, $i = 1, \ldots, m$, as radii on the unit circle. Figure 4 demonstrates that the VM-estimates of the angle are greatly affected and drawn towards the contaminated observations. The WrC-estimates are very robust: their variability is small and they are distributed around the true angle. Plots of the corresponding location estimates are shown in Figure 5. This confirms the findings for the angle estimates: the WrC-estimates of transmitter location are robust whereas the VM-estimates are highly sensitive in the face of outlying observations.

6. Final Remarks

In this paper we considered the ML-estimation of the location of a transmitter based on the Angle-of-Arrival data provided by the base-stations and modelled by angular distributions. We used von Mises and wrapped Cauchy distributions which can be motivated by real-world measurements as discussed in Section 2. The transmitter location was then estimated using the MLE of the angles and elemental geometry. We demonstrated that accompanying the MLE, also its standard errors and likelihood regions (approximate confidence regions) can easily be calculated. Our simulation demonstrated that the WrC-estimate of location is robust.

References

[1] B. Efron and R.J. Tibshirani, *An Introduction to the Bootstrap.* Chapman & Hall, New York, 1993.

[2] S.R. Jammalamadaka and A. SenGupta, *Topics in Circular Statistics.* World Scientific Press, Singapore, 2001.

[3] J.G. Kalbfleisch, *Probability and Statistical Inference Volume* 2: *Statistical Inference.* Springer-Verlag, New York, 1985.

[4] J.T. Kent and D.E. Tyler, *Maximum Likelihood Estimation for the Wrapped Cauchy Distribution.* J. Appl. Stat. **15** (1988), 247–254.

[5] H. Krim and M. Viberg, *Two Decades of Array Signal Processing: The Parametric Approach.* IEEE Signal Proc. Mag. **13** (1996), 67–94.

[6] J.K. Lindsey, *Parametric Statistical Inference.* Clarendon Press, Oxford, 1996.

[7] K.V. Mardia, *Statistics of Directional Data.* Academic Press, London, 1972.

[8] D. Middleton, *Man-made Noise in Urban Environments and Transportation Systems: Models and Measurements.* IEEE Trans. Commun. **21** (1973), 1232–1241.

[9] K.I. Pedersen, P.E. Mogensen, and B.H. Fleury, *A Stochastic Model of the Temporal and Azimuthal Dispersion Seen at the Base Station in Outdoor Propagation Environments.* IEEE Trans. Veh. Tech. **49** (2000), 437–447.

[10] J. Shao, *The Efficiency and Consistency of Approximations to the Jackknife Variance Estimators.* J. Amer. Statist. Assoc. **84** (1989), 114–119.

[11] J. Shao and C.F.J. Wu, *A General Theory For Jackknife Variance Estimation.* Ann. Statist. **17** (1989), 1176–1197.

[12] Q.H. Spencer, B.D. Jeffs, M.A. Jensen, and A.L. Swindlehurst, *Modeling the Statistical Time and Angle of Arrival Characteristics of an Indoor Multipath Channel.* IEEE J. Sel. Areas Comm. **18** (2000), 347–360.

[13] A.J. Stromberg, *Robust Covariance Estimates Based on Resampling.* J. Statist. Plann. Inference **57** (1997), 321–334.

E. Ollila and V. Koivunen
SMARAD CoE, Helsinki University of Technology, P.O. Box 3000
Fin-02015 HUT, Finland
e-mail: {esa.ollila, visa.koivunen}@hut.fi

Statistics for Industry and Technology, 259–269

A Family of Scale Estimators by Means of Trimming

J.F. Ortega

Abstract. A very common practice in literature is that of building scale estimators by means of a location estimator. The most common scale estimator used is the standard deviation, which is obtained by using the mean; its use is not the most suitable one due to its weakness under the presence of outliers. We can find other scale estimators, based on location estimators, which are more resistant under the presence of outliers, as for instance, the so called *Mad*, which uses for its construction the median.

Since the mean and the median can be considered extreme elements of a location estimators family known as *trimmed means*, in this paper we propose a scale estimator family called $\alpha\beta$*_Trimmed*. For its definition, we will use the above mentioned *trimmed means* family and different parameters α and β, which are called *trimming levels*. We will demonstrate the good *robustness* behavior of the elements of such a family. It will be proved that they are *affine equivariant*, and (depending on the *trimming levels*) have a high *exact fit point*, a high *breakdown point* and a bounded *sensitivity curve*.

Mathematics Subject Classification (2000). Primary 62G35; Secondary 62G05.

Keywords. Robust scale estimators, robustness, efficiency.

1. Introduction

In certain cases, scale estimators are defined by using a location estimator as the basis for their construction. Generally speaking, the behavior of the scale estimators under the presence of outliers is inherited by the behavior of the location estimators used.

The most commonly used scale estimator is the square root of the variance, which is built using the mean. Regarding the mean, we know that just one observation of the sample is able to influence its calculation, in such a way that it can "lead" it to any desirable value. Therefore, other scale estimators have been defined by using other location estimators, which are more resistant in the presence of outliers (for instance, Rousseeuw and Croux, 1993).

The median (*Med*) is an estimator with good robustness properties (it is *equivariant*, with a maximum *breakdown point* and bounded *influence function*, that is, *B-Robust*), so that it has been used in different papers for the construction of scale estimators. The most well known is the *median absolute deviation* (*Mad*) given by Hampel et al. (1986) whose definition, considering a sample $x = \{x_1, x_2, \ldots, x_n\}$, is:

$$Mad(x) = C(Mad)\ Med(\{|x_i - Med(x)|\}_i)$$

where $C(Mad)$ is a *consistency coefficient* under the assumption of normality which has value 1.4826.

In the controversy between *robustness* and *efficiency*, the behavior of the Mad in contaminated samples is very good; with the best possible *asymptotic breakdown point* (1/2) and bounded *influence function*. Its weakest point is its *efficiency*, achieving 37% under the assumption of normality. On the other hand, there are scale estimators with high *breakdown point* and high *efficiency*. For instance, Yohai and Zamar (1988) obtained a class of regression estimates based on scale estimators which simultaneously have high *breakdown point* and high *efficiency* under normality, and Croux (1994) provided M-estimators of scale which combine high *breakdown point* and high *efficiency*.

The mean and the median can be regarded as extreme elements of a family known as *trimmed means*. The elements of this family use for their definition a parameter α, that will be called *trimming level*, where $\alpha \in [0, 50)$, the *trimmed mean at trimming level* α (α*_Trimmed*) is the mean of the observations of the sample, eliminating the $\alpha\%$ of the highest observations and the $\alpha\%$ of the lowest. That is, for a sample $x = \{x_1, x_2, \ldots, x_n\}$, where we will call $x_{[i]}$ the observation in the ith position of a sequence ranging from the smallest to the largest of the elements of x, and considering $a = Int(\alpha n/100)$, the α*_Trimmed* of the sample x, denoted $\alpha\_Trim(x)$ is defined in the following way:

$$\alpha\_Trim(x) = \frac{1}{n-2a} \sum_{i=a+1}^{n-a} x_{[i]} \tag{1.1}$$

In this way, the 0*_Trimmed* is the mean and if α tends towards 50 then the α*_Trimmed* tends to the median. So, if we define the 50*_Trimmed* to be the median, we can define a trimming for every α in the interval [0,50].

The properties of these estimators were studied in different papers (for instance, Hoaglin et al., 1983) proving that: for a variable which follows a symmetric distribution of parameter μ, the α*_Trimmed* is a class of *unbiased* estimators of μ; which are *affine equivariant*, have *breakdown point* equal to $(a+1)/n$, *asymptotic breakdown point* $\alpha/100$, which depends on the *trimming level*, and they are *B-Robust* iff $a \geq 1$. On the other hand, its *efficiency* (considering the *relative efficiency* with respect to the mean) under normality can be approximated by means of $(100 - 0.723\alpha)\%$, so the efficiency of the elements in this family varies between the 64% of the median and the maximum 100% of the mean.

As a consequence, the mean (*0_Trimmed*) has the highest possible *efficiency* for normal distributions (100%), together with the worst behavior in the presence of outliers (*breakdown point* of value 0, unbounded *influence function*, that is, it is not a *B-Robust* estimator). On the other hand, the median (*50_Trimmed*) has a lower *efficiency* under normality (64%), its *asymptotic breakdown point* is the maximum possible, with a bounded *influence function*, and definitely, more suitable behavior for the treatment of the outliers.

The possibility of being able to choose an arrangement option amongst the generic properties of *robustness* and *efficiency*, and considering the structure of the two scale estimators defined by means of extreme estimators of the *trimmed means* family, leads us to propose some scale estimators by using the elements of such a family.

Our paper is organized as follows: in Section 2, we propose the new scale estimators family, by using trimmed means. In Section 3 we study the main properties of the elements of this family in the presence of outliers. In Section 4 the behavior of this family under normality is analyzed. Finally, in Section 5, we present some conclusions.

2. A New Scale Estimators Family

Starting from the variance and replacing the means by elements of the *trimmed means* family, it is possible to build a spread estimators family which, as we will see later, has suitable behavior in the presence of outliers.

Since the variance is defined as the mean of the squares of the distances between the observations to the mean, we propose to substitute the two used means for other two *trimmed means*. Therefore, in order to define the new spread estimators we will need two parameters; one of them used to calculate a location estimator, which acts as the mean; and another one, which provides us with a mean of the distances of the observations of this location estimator. In this way, we define the following concept:

Definition 2.1. Given a sample $x = \{x_1, x_2, \ldots, x_n\}$, the parameters $\alpha, \beta \in [0, 50]$ and the trimmed means family being known, we define the elements $\alpha\beta$*_Trimmed* over the sample x, termed $\alpha\beta\_Trim(x)$, as:

$$\alpha\beta\_Trim(x) = C(\alpha, \beta)\ \beta\_Trim(\{(x_i - \alpha\_Trim(x))^2\}_i)$$

where $C(\alpha, \beta)$ is a *consistency coefficient.*

As in the definition of the Mad where the *consistency coefficient* is supposed to be invariable under changes of location and/or scale of the sample x, the coefficients $C(\alpha, \beta)$ will also be independent of these changes.

On the other hand, by changing α and β in the interval [0,50] a family of estimators is obtained, so that each of its elements provides a *trimmed mean* (at level β) of the distances between the collection of x and a "representative" of these ones, calculated by means of another *trimmed mean* (now at level α). Thus, the extremes of such a family are constituted by the variance of x, in the case of $\alpha = \beta = 0$, and by an estimator that we will call *median of deviations to the median* (MDM), in case that $\alpha = \beta = 50$, defined in the following way:

$$MDM(x) = C(MDM)\ Med(\{(x_i - Med(x))^2\}_i)$$

The square root of the MDM is very similar to the Mad, with slight differences between them because of the parity of the sample size. However, MDM respects, in a stricter way, the structure of the variance, when considering the squares of the distances and not absolute values, which, in practice, can be useful at the moment of handling such a concept.

The square root of the elements of the family given in Definition 2.1, that is $\sqrt{\alpha\beta\text{\_Trimmed}}$, based on samples of random variables, can be regarded as scale estimators, since they provide some information about the "deviation" or "variability" of the elements of the considered sample.

3. Properties of the New Estimators

The properties that we are going to analyze in this section are the usual ones in order to prove the *robustness* of a scale estimator (see Rousseeuw and Leroy, 1987; or Barnett and Lewis, 1994).

Efficiency and Consistency

In the definition of the elements of the new estimators family *consistency coefficients* appear, which depend on the distribution of the sample variable over which the estimator applies, and also depend on the parameters α and β. These coefficients need to be carefully studied, and this will be done in Section 4 in the case of samples from normal populations.

Equivariance

The *equivariance* property deals with the behavior of the estimators when there are changes in position and/or scale in the sample data. For the proposed scale estimators the following proposition is true:

Proposition 3.1. *For a sample* $x = \{x_1, x_2, \dots, x_n\}$ *and* $r, s \in I\!R$ *with* $rx + s = \{rx_1 + s, rx_2 + s, \dots, rx_n + s\}$, *then it is true that for every* $\alpha, \beta \in [0, 50]$ *which:*

$$\alpha\beta\text{\_Trim}(rx + s) = r^2\ \alpha\beta\text{\_Trim}(x)$$

Proof. For the construction of the elements of the *αβ_Trimmed* family (Definition 2.1), based on the definition of the *α_Trimmed*, and using the invariance of the *consistency coefficients* under changes in location and in scale, the result is immediate. □

Therefore, the changes of position do not affect the elements of the family, whereas the behavior of such elements under changes of scale are the most usual ones in scale estimators. To sum up, the elements of the $\sqrt{\alpha\beta\_Trimmed}$ family are *affine equivariant* scale estimators.

Exact Fit

For the study of the *exact fit* property, the concept *exact fit point* is used as an indicator of the grade of fulfillment. The *exact fit point* δ_n^* of a scale estimator for a sample $x = \{x_1, x_2, \ldots, x_n\}$ of size n where $x_i = x_0 \; \forall i$ measures the minimal percentage of contamination in x that is necessary before the estimator yields any value other than the desirable one, which is 0.

Let us consider the following proposition:

Proposition 3.2. *Consider a sample* $x = \{x_1, x_2, \ldots, x_n\}$, *then the* exact fit point *of* $\alpha\beta$_Trimmed *for* x *is:*

$$\delta_n^*(\alpha\beta\text{\_Trimmed}, x) = Min\{\frac{a+1}{n}, \frac{b+1}{n}\}$$

where $a = Int(\alpha n/100)$ *and* $b = Int(\beta n/100)$ *for every* $\alpha, \beta \in [0, 50)$.

In the particular case of MDM, *we have that* $\delta_n^*(MDM, x) = Int(\frac{n+1}{2})/n$.

Proof. Let us distinguish between the MDM (case a) and the rest of the family elements (case b). Later, in (c) we will show their connection.

Let us assume, in all the cases, that $x = \{x_1, x_2, \ldots, x_n\}$ is a sample where $x_i = x_0 \; \forall i$, and $xc_m = \{x_i'\}_{i=1,2,\ldots,n}$ is a contaminated sample with m observations of x replaced by arbitrary points. Denote J the set of indices corresponding to contaminated data points in xc_m.

(a): In the case of MDM, we find that the *exact fit point* is defined as:

$$\delta_n^*(MDM, x) = Min\{\frac{m}{n} \; / \; \exists \; xc_m \text{ such that } MDM(xc_m) \neq 0\}$$

For the median, we know that:

- If $m < Int(\frac{n+1}{2})$ then $Med(xc_m) = x_0$, and therefore:

$$(x_i' - Med(xc_m))^2 = \begin{cases} (x_i' - x_0) \neq 0 & \text{if } i \in J \\ 0 & \text{otherwise} \end{cases}$$

  so that the $MDM(xc_m) = 0$.
- If $m \geq Int(\frac{n+1}{2})$ then we can find a contaminated sample xc_m such that $Med(xc_m) \neq x_0$, and therefore $MDM(xc_m) \neq 0$.

Thus, we find:

$$\delta_n^*(MDM, x) = Int(\frac{n+1}{2})/n$$

(b): In the case of the $\alpha\beta$_*Trimmed*, the *exact fit point* is defined:

$$\delta_n^*(\alpha\beta\text{\_Trimmed}, x) = Min\{\frac{m}{n} \; / \; \exists \; xc_m \text{ such that } \alpha\beta\text{\_Trim}(xc_m) \neq 0\}$$

Note that the *exact fit point* for the α_*Trimmed* is $(a+1)/n$ which yields:

- If $m > a$ then we can find a xc_m such that $\alpha\_Trim(xc_m) \neq x_0$. So, $(x'_i - \alpha\_Trim(xc_m))^2$ is higher than 0 for every i, and hence $\alpha\beta\_Trim(xc_m)$ is different from 0 for every β.
- If $m \leq a$ then the number of elements with $(x'_i - \alpha\_Trim(xc_m))^2$ different from 0 is exactly m. Therefore, if $m > b$ then $\alpha\beta\_Trim(xc_m) \neq 0$.

To sum up, $\alpha\beta\_Trim(xc_m) \neq 0$ always that $m > a$ or $m > b$ and therefore:

$$\delta_n^*(\alpha\beta\_Trimmed, x) = Min\{\frac{a+1}{n}, \frac{b+1}{n}\}$$

(c): It is easy to prove that, if α and β tend towards 50 then $\delta_n^*(\alpha\beta\_Trimmed, x)$ tend toward $\delta_n^*(MDM, x)$, since in this case $a = b$ and then:
- if n is even, $n = 2k$ with $k \in \mathbb{N}$, then $a = Int(\alpha n/100) = k - 1$ and $Int(\frac{n+1}{2}) = k$; whereas,
- if n is odd, $n = 2k + 1$ with $k \in \mathbb{N}$, then $a = Int(\alpha n/100) = k$ and $Int(\frac{n+1}{2}) = k + 1$.

Therefore, for every n we have $Int(\frac{n+1}{2}) = a + 1 = b + 1$.

□

Thus, the grade of fulfillment of the *exact fit* property depends on the considered *trimming levels*; this grade being the highest one in the case of MDM, extreme element of the $\alpha\beta\_$*Trimmed* family.

Sensitivity Curve

The *sensitivity curve* investigates the changes of the estimator that occur when we introduce a particular contamination. The *sensitivity curve* must be bounded, which means that the estimator produces controlled values for every contamination.

Let us look at the behavior of the *sensitivity curve* for the new estimators, by means of the following proposition:

Proposition 3.3. *Given the trimming levels* $\alpha, \beta \in [0, 50]$, *where* $a = Int(\alpha n/100)$ *and* $b = Int(\beta n/100)$, *the* sensitivity curve *of the* $\alpha\beta$_Trimmed *estimator is bounded iff* $Min\{a, b\} \geq 1$.

Proof. For a sample of n elements $x = \{x_1, x_2, \ldots, x_n\}$ and for any value $x_0 \in \mathbb{R}$, the *sensitivity curve* for every $\alpha\beta\_$*Trimmed* is:

$$SC_n(x_0; \alpha\beta\_Trimmed, x) = (n+1)\ [\alpha\beta\_Trim(x') - \alpha\beta\_Trim(x)]$$

where $x' = x \bigcup \{x_0\}$ and

$$\alpha\beta\_Trim(x') = \\ C(\alpha, \beta)\ \beta\_Trim(\{(x_i - \alpha\_Trim(x'))^2\}_i, (x_0 - \alpha\_Trim(x'))^2)$$

Under these conditions, since x' is a contamination of x in an observation, if $a \geq 1$ then $\alpha\_Trim(x')$ is between the minimum and the maximum of the x. Moreover, it will be necessary that $b \geq 1$ in order to ensure that $(x_0 - \alpha\_Trim(x'))^2$ is eliminated, if required, so that the $\alpha\beta\_Trim(x')$ is "controlled".

On the other hand, if $a = 0$ or $b = 0$ then $\alpha\beta\_Trim(x')$ is "non-controlled", since the value $(x_0 - \alpha\_Trim(x'))^2$ can be any value.

In short, if $Min\{a, b\} \geq 1$ we can state that the $SC_n(x_0; \alpha\beta\_Trimmed, x)$ is bounded. □

In this way, the *sensitivity curve* of the $\alpha\beta$*_Trimmed* estimator is bounded if and only if a and b are greater or equal than 1, that is, if some observation of the sample is eliminated. Remember that if $\alpha = \beta = 0$, then we are considering the variance which has a unbounded *sensitivity curve.* However, if $\alpha = \beta = 50$ then a and b will be greater than 1, so 50,50*_Trimmed*=MDM will also have bounded *sensitivity curve.*

Breakdown Point

The *breakdown point* $\epsilon_n^*(S, x)$ of a scale estimator S for a sample x of size n is defined in Rousseeuw and Croux (1993) by:

$$\epsilon_n^*(S, x) = Min\{\epsilon_n^+(S, x), \epsilon_n^-(S, x)\}$$

where

$$\epsilon_n^+(S, x) = Min\{\frac{m}{n} \ / \ Sup_{xc_m}\{S(xc_m)\} = \infty\}$$

is called explosion breakdown point, and

$$\epsilon_n^-(S, x) = Min\{\frac{m}{n} \ / \ Inf_{xc_m}\{S(xc_m)\} = 0\}$$

is called implosion breakdown point, and where xc_m is a contamination of m elements of x then

Let us calculate the *breakdown point* of the new estimators.

Proposition 3.4. *For a sample* $x = \{x_1, x_2, \ldots, x_n\}$ *we have that the* breakdown point *of the* $\alpha\beta$_Trimmed *over* x *is:*

$$\epsilon_n^*(\alpha\beta\text{\_Trimmed}, x) = Min\{\frac{a+1}{n}, \frac{b+1}{n}\}$$

for every $\alpha, \beta \in [0, 50]$, *with* $a = Int(\alpha n/100)$ *and* $b = Int(\beta n/100)$.

Proof. Assuming that xc_m is a contamination of m elements of x, and in the case of $\epsilon_n^+(\alpha\beta\_Trimmed, x)$, if $m \geq (a + 1)$ then we can find a contaminated sample xc_m such that $\alpha\_Trim(xc_m)$ is not "controlled" if only the information of the sample x is known; so, also $\alpha\beta\_Trim(xc_m)$ is not "controlled". On the other hand, if $m < (a + 1)$, so that $\alpha\_Trim(xc_m)$ is between the highest element and the lowest one of the sample x, it is necessary that $m \geq (b + 1)$ in order to ensure that we can find a contaminated sample xc_m such that $\alpha\beta\_Trim(xc_m) = C(\alpha, \beta)\beta\_Trim(\{(x_i' - \alpha\_Trim(xc_m))^2\}_i)$, where $x_i' \in xc_m$, is not "controlled" if only the information of the sample x is known.

So, it is necessary that m is lower than $(a + 1)$ and also lower than $(b + 1)$ in order to ensure that, for every contamination the estimator $\alpha\beta$*_Trimmed* is "controlled" if only the information of the sample x, that is, $Sup_{xc_m}\{\alpha\beta\_Trim(xc_m)\}$

is finite. Therefore:

$$\epsilon_n^+(\alpha\beta\_Trimmed, x) = Min\{\frac{a+1}{n}, \frac{b+1}{n}\}$$

Let us concentrate on the case of $\epsilon_n^-(\alpha\beta\_Trimmed, x)$. It is easy to see that at least $n-b$ observations of the set $\{(x_i' - \alpha\_Trim(xc_m))^2\}_i$ being 0 are needed in order to ensure $\alpha\beta\_Trim(xc_m) = 0$. Thus, we need a contaminated sample xc_m with $m \geq n-b$. Therefore:

$$\epsilon_n^-(\alpha\beta\_Trimmed, x) = \frac{n-b}{n}$$

It follows that $\epsilon_n^+ \leq \epsilon_n^-$ because:

- if n is even, $n = 2k$ with $k \in \mathbb{N}$, then $a+1 \leq k$ and $b+1 \leq k$. So, since $k < k+1 = 2k-k+1 = n-(k-1) \leq n-b$, then $a+1 \leq n-b$ and $b+1 \leq n-b$;
- if n is odd, $n = 2k+1$ with $k \in \mathbb{N}$, then $a \leq k$ and $b \leq k$. So, since $k+1 = 2k+1-k = n-k \leq n-b$, then $a+1 \leq n-b$ and $b+1 \leq n-b$.

To sum up, the breakdown point of the *αβ_Trimmed* is:

$$\epsilon_n^*(\alpha\beta\_Trimmed, x) = Min\{\frac{a+1}{n}, \frac{b+1}{n}\}$$

□

Regarding this last proposition, we can claim that the *asymptotic breakdown point* of *αβ_Trimmed* is:

$$\epsilon^*(\alpha\beta\_Trimmed) = Min\{\frac{\alpha}{100}, \frac{\beta}{100}\}$$

and as a consequence, for the extreme elements of the *αβ_Trimmed* family it follows that ϵ^*(variance)=0 and $\epsilon^*(MDM) = 1/2$.

4. The New Estimators Under Normality

In order to have the elements of the *αβ_Trimmed* family defined, it is necessary to start by determining the *consistency coefficient*, which is used in their definition. We will then determine the *efficiency* of the elements of the *αβ_Trimmed* family.

The analysis will be carried out under the normality assumption, since it is the most common one.

Consistency Coefficients

Assuming that x is an i.i.d. sample of the variable $X \equiv N(\mu, \sigma)$, the Definition 2.1 provides an estimator family where the coefficient $C(\alpha, \beta)$ is defined. Therefore, the corresponding element of such a family, for the α and β *trimming levels*, is consistent (in mean square) for σ^2.

TABLE 1. Consistency coefficients under normality.

β	$C(\alpha,\beta)$
0	1.0000000
5	1.2485537
10	1.4280902
15	1.5884787
20	1.7344188
25	1.8653508
30	1.9788504
35	2.0717771
40	2.1409476
45	2.1836602
50	2.1981094

Under this assumption and considering that every element of the *trimmed means* family is *affine equivariant* and that the *α_Trimmed* is an *unbiased* estimator of μ, it is true that:

$$\begin{aligned} \alpha\beta\_Trim(X) &= C(\alpha,\beta)\ \beta\_Trim((X-\alpha\_Trim(X))^2) \\ &= C(\alpha,\beta)\ \sigma^2\ \beta\_Trim(X') \end{aligned}$$

where $X' = (\frac{X-\mu}{\sigma})^2$. Assuming that the values $C(\alpha,\beta)$ must ensure the *consistency* of $\alpha\beta\_Trim(X)$, it must be true that $\alpha\beta\_Trim(X) = \sigma^2$. Applying this condition we obtain:

$$C(\alpha,\beta) = [\beta\_Trim(X')]^{-1}$$

where X' follows a distribution χ^2_1.

If the distribution function of an χ^2_1 is denoted F then for $\beta \in [0,50)$, and $a_1, a_2 \in \mathbb{R}$ where $F(a_1) = \beta/100$ and $F(a_2) = 1-(\beta/100)$ we find that

$$\beta\_Trim(X') = \frac{\frac{1}{\sqrt{2\pi}} \int_{a_1}^{a_2} x^{1/2} e^{-x/2} dx}{1-(2\beta/100)}$$

The *consistency coefficients* for some values of β in the interval [0,50] are given in Table 1, where for $\beta = 50$ we find that $\beta\_Trim(X') = F^{-1}(1/2)$, so that the *consistency coefficient* for MDM coincides with the square of the one calculated for the Mad estimator.

Efficiency

Once the values of the *consistency coefficients* are calculated, it is possible to determine the *efficiency* of the elements of the *αβ_Trimmed* family, where the considered *efficiency* is the *relative efficiency* of the elements of the *αβ_Trimmed* family with regard to the corrected variance.

The *efficiency* of the elements of the *αβ_Trimmed* family will be estimated by means of a simulation study. For that reason, 10000 samples of size 1000 have been

TABLE 2. Efficiency of the *αβ_Trimmed* under normality with respect to the corrected variance.

α β	0	5	10	15	20	25	30	35	40	45	50
0	100	87.17	79.04	72.31	65.42	58.72	53.33	49.77	44.78	43.06	40.02
5	99.97	87.08	78.69	71.84	64.98	58.53	53.11	48.25	44.76	42.98	39.66
10	99.95	87.06	78.61	71.64	64.81	58.50	53.01	48.23	44.76	42.81	39.64
15	99.94	86.97	78.55	71.50	64.75	58.46	52.96	48.23	44.74	42.63	39.62
20	99.91	86.85	78.47	71.50	64.69	58.44	52.91	48.23	44.73	42.63	39.61
25	99.86	86.77	78.41	71.49	64.66	58.35	52.80	48.22	44.72	42.49	39.35
30	99.78	86.69	78.35	71.44	64.65	58.30	52.66	48.22	44.70	42.48	39.31
35	99.82	86.66	78.14	71.27	64.62	58.26	52.59	48.16	44.68	42.48	39.26
40	99.76	86.48	77.86	71.08	64.58	58.06	52.44	47.95	44.63	42.07	39.11
45	99.51	86.38	77.56	70.90	64.40	57.91	52.18	47.54	44.39	42.01	38.85
50	99.20	86.31	77.44	70.88	64.19	57.82	51.84	47.51	44.27	41.68	38.35

generated from $N(0,1)$. For each sample the elements of the *αβ_Trimmed* family have been calculated for values of α and β which are within the interval [0,50] in 5 by 5. Afterwards we have calculated the variance of the resulting estimates and compared it to the variance of the classical variance estimator, which is also calculated for these samples.

As a result of this simulation we have obtained Table 2, where it can be noticed that the *αβ_Trimmed* family varies in a wide range of efficiencies.

5. Final Comments

We have defined the scale estimators family $\sqrt{\alpha\beta\_Trimmed}$ and analyzed the *robustness* and *efficiency* properties of the elements in this family.

With regard to the *robustness*, for a sample of size n with $a = Int(\alpha n/100)$ and $b = Int(\beta n/100)$ for $\alpha, \beta \in [0,50]$, it has been proved that the estimators in this class are *affine equivariant* with *exact fit point* of value $Min\{(a+1)/n, (b+1)/n\}$, bounded *sensitivity curve* when $Min\{a,b\} \geq 1$, *breakdown point* equal to $Min\{(a+1)/n, (b+1)/n\}$ and corresponding *asymptotic breakdown point* $Min\{\alpha/100, \beta/100\}$.

On the other hand, the *efficiency* of the elements of this family (assuming *relative efficiency* with regard to the corrected variance) has been obtained. Depending on the *trimming levels*, the *efficiency* of the *αβ_Trimmed* is between 38.35% and 100%.

To sum up, the *αβ_Trimmed* provides a family of spread estimators where it is possible to choose certain *efficiency* levels or *robust* qualities. This is the main advantage of this family compared to other scale estimators, such as the Mad, the *Shorth* or the ones defined in Rousseeuw and Croux (1993), called S^* and Q^*. Thus, depending on the case, or even on the researcher's preferences, it is possible to choose an element of the *αβ_Trimmed* family which provides an adequate scale estimator.

References

[1] C. Croux, *Efficiency High-Breakdown M-Estimators of Scale.* Statist. Probab. Lett. **19** (1994), 371–379.

[2] V. Barnett and T. Lewis, *Outliers in Statistical Data.* 3rd Edition, Wiley and Sons, New York, 1994.

[3] F.R. Hampel, E.M. Ronchetti, P.J. Rousseeuw, and W.A. Stahel, *Robust Statistics: The Approach Based on Influence Functions.* Wiley and Sons, New York, 1986.

[4] D.C. Hoaglin, F. Mosteller, and J.W. Tukey, *Understanding Robust and Exploratory Data Analysis.* Wiley and Sons, New York, 2000.

[5] P.J. Rousseeuw and C. Croux, *Alternatives To the Median Absolute Deviation.* J. Amer. Statist. Assoc. **88** (1993), 1273–1283.

[6] P.J. Rousseeuw and A.M. Leroy, *Robust Regression and Outlier Detection.* Wiley and Sons, New York, 1987.

[7] V.J. Yohai and R.H. Zamar, *High Breakdown-Point Estimates of Regression By Means of the Minimization of an Efficient Scale.* J. Amer. Statist. Assoc. **83** (1988), 406–413.

Acknowledgements

The author is grateful to two referees for their helpful comments and suggestions.

J.F. Ortega
Universidad de Castilla-La Mancha
Facultad de CC. Económicas y Empresariales de Albacete
Plaza de la Universidad, 1
02071 Albacete
Spain
e-mail: JuanFco.Ortega@uclm.es

Statistics for Industry and Technology, 271–282

Robust Efficient Method of Moments Estimation

C. Ortelli and F. Trojani

Abstract. This paper focuses on the robust Efficient Method of Moments (EMM) estimation of a general parametric stationary process and proposes a broad framework for constructing robust EMM statistics in this context. We characterize the local robustness properties of EMM estimators for time series by computing the corresponding influence functions and propose two versions of a robust EMM (REMM) estimator with bounded influence function. We then show by Monte Carlo simulation that our REMM estimators are very successful in controlling for the asymptotic bias under model misspecification while maintaining a high efficiency under the ideal structural model.

Mathematics Subject Classification (2000). 62F10, 62F35, 62G05, 62G35, 62M09.

Keywords. Efficient Method of Moments, indirect inference, influence function, robust estimation, robust statistics.

1. Introduction

This paper analyzes the local robustness properties of estimators derived from the Efficient Method of Moments (EMM, Gallant and Tauchen, 1996) and develops a new class of robust statistics for the statistical analysis of parametric models that are estimated within a general EMM framework. Specifically, we focus on the robust EMM estimation of a general parametric stationary process where the implied stationary density may be not computed analytically and propose a broad framework for constructing robust EMM statistics in this context.

Some authors already addressed some important issues related to a robust inference on time series models, starting from different perspectives. For instance, some first definitions of an influence function (IF) for times series (extending the basic definitions in Hampel's (1974) seminal work) have been developed in Künsch

Both authors gratefully acknowledge the financial support of the Swiss National Science Foundation (grant 12-65196.01 and NCCR FINRISK).

(1984) and Martin and Yohai (1986). Künsch (1984) also derived the optimality of a robust M-estimator for the parameters of a linear autoregressive process with normally distributed error terms. In a similar vein, de Luna and Genton (2001) proposed more recently a robust estimator for the parameters of a linear ARMA process. In the nonlinear time series context Ronchetti and Trojani (2001) introduced a robust version of a general Generalized Method of Moments (GMM) statistic for situations where a reference model for the underlying data distribution can be assumed. Finally, Genton and Ronchetti (2003) have addressed in some simple model setting (using an indirect inference approach à la Gourieroux et al., 1993) the issue of robust estimation for models where the stationary density of the structural model cannot be computed explicitly.

We address the problem of a robust inference on general parametric models for time series from a broad perspective that allows us to extend the application field of robust procedures for time series essentially to all models that can be estimated by a classical (nonrobust) EMM estimator. We compute the time series influence function (Künsch, 1984) of a general EMM estimator and propose two robust EMM estimators with bounded IF that ensure a bounded asymptotic bias in neighborhoods of the given structural model. We then present some Monte Carlo experiments attempting to quantify the trade-off between robustness and efficiency in the REMM estimation of a highly nonlinear model. To this end we estimate a structural ARMA(1,1)-ARCH(2) model by means of an auxiliary AR(3)-ARCH(2) model. Both our REMM estimators yield very satisfactory results in these experiments. Indeed, in our simulations the efficiency loss implied by a REMM estimation under a perfectly specified structural model is virtually negligible, when compared with the results of the classical EMM procedure. Further, we observe that even a quite moderate model contamination can induce a very important bias and a strong loss in efficiency of a classical EMM estimator while both our REMM estimators are very successful in bounding the induced asymptotic bias and efficiency loss. Finally, we find the outlier identification procedure implied by the robust weights of our REMM procedures to be very efficient in all our experiments.

The paper is organized as follows. Section 2 introduces the standard (nonrobust) EMM setting. Section 3 computes the time series IF of an EMM estimator and introduces two REMM estimators that bound the implied IF in an appropriate metric. Section 4 presents some Monte Carlo experiments where the performance of our REMM estimators is evaluated in a nontrivial EMM setting. Section 5 summarizes and concludes.

2. Basic EMM Setting

Let $\mathbb{X} = \{X_t : \Omega \longrightarrow \mathbb{R}^v, v \in \mathbb{N}, t = 1, 2, \dots\}$ be a strictly stationary and ergodic stochastic process on a complete probability space $(\Omega, \Im, Q_0)$. The goal in EMM estimation is to produce statistical inference on the probability $P_0 := Q_0\mathbb{X}^{-1}$ based on a *structural model* $\mathcal{P} = \{P_\rho : \rho \in \mathcal{R} \subseteq \mathbb{R}^l, l \in \mathbb{N}\}$ that defines for any *structural parameter* $\rho \in \mathcal{R}$ a probability measure P_ρ on the measurable space $(\mathbb{R}^{v\infty} := \times_{t=1}^{\infty}\mathbb{R}^v, \mathcal{B}(\mathbb{R}^{v\infty}))$. In classical EMM estimation $\mathcal{P}$ is assumed to be correctly specified for P_0.

We denote by $x^n := (x_1', x_2', \ldots, x_n')'$ the observed finite history of $X^n := (X_1', X_2', \ldots, X_n')'$ and by X_{n-m+1}^n the subvector consisting of the last m components of X^n, $0 < m \leq n$. In EMM estimation the focus is on model settings where the (conditional) density of X^n $(X_n \mid X^{n-1})$ cannot be expressed analytically, so that direct estimation of P_{ρ_0} is not feasible. Instead, an auxiliary parametric model and an auxiliary parameter $\theta \in \Theta \subset \mathbb{R}^q$, $q \geq l$ are introduced, which can be estimated for instance by pseudo maximum likelihood. This defines an auxiliary estimation procedure that estimates in a first step of the EMM the underlying probability P_{ρ_0} in an approximate way. When a well-defined binding function (see Gourieroux et al., 1993) between the structural and auxiliary parameters exists, an estimate of ρ can be recovered from the estimate of θ by minimizing a quadratic form.

Often the auxiliary model is defined using a parametric family of pseudo-likelihood functions that induce a corresponding set of estimating equations, and we could do the same to develop our robust EMM (REMM) methodology. However, it is convenient for our robust analysis and straightforward from a statistical perspective to work with a larger class of auxiliary models, which can be estimated by some M-estimator (Huber, 1964) defined by a general score function

$$\psi : \mathbb{R}^{vL} \times \Theta \longrightarrow \mathbb{R}^q, q \geq l, L \geq 1 \quad . \tag{2.1}$$

To characterize the robustness of EMM statistics we need to write EMM estimators and tests as functionals on a suitable space of distributions. For our purposes, focusing on the finite dimensional distributions induced by the strictly stationary distributions on $(\mathbb{R}^{v\infty}, \mathcal{B}(\mathbb{R}^{v\infty}))$, as in Künsch (1984), will be enough. Thus, define for $L \geq 1$ the following set of finite-dimensional distributions

$$\mathcal{M}_{stat}^L = \{(L)\text{-finite-dimensional marginals of strictly stationary processes}\}.$$

Moreover, let for $L < n$ the empirical (L)-dimensional marginal distribution of x^n be defined by

$$P_n^L = n^{-1} \sum_{i=1}^{n} \delta_{(x_i', \ldots, x_{i+L-1}')'} \quad ; x_i = x_{i-n} \text{ for } i > n \,, \tag{2.2}$$

where δ_{x^L} is the Dirac mass at $x^L \in \mathbb{R}^{vL}$. By construction $P_n^L \in \mathcal{M}_{stat}^L$.

To analyze the local robustness of statistical functionals derived from an EMM estimator, the first step is to define a statistical functional $\widehat{\theta}(\cdot)$ for the estimator of the auxiliary parameter θ. By definition, $\widehat{\theta}(\cdot)$ is the functional solution of the asymptotic estimating equations implied by the score function (2.1), that is

$$\widehat{\theta} : dom(\widehat{\theta}) \subset \mathcal{M}_{stat}^L \to \Theta, \; P^L \longmapsto \widehat{\theta}\left(P^L\right) := \widehat{\theta}_{P^L} \quad ,$$

is the functional solution of the implicit equation

$$E_{P^L}\left(\psi(X^L; \theta)\right) = 0 \quad , \tag{2.3}$$

in θ. Notice, that when restricting the auxiliary functional $\widehat{\theta}(\cdot)$ on the set $\mathcal{P}^L := \{P_\rho^L : \rho \in \mathcal{R} \subseteq \mathbb{R}^l\}$ it follows

$$E_{P_\rho^L}\left(\psi\left(X^L;\widehat{\theta}_{P_\rho^L}\right)\right) = 0 \quad , \tag{2.4}$$

so that to any $\rho \in \mathcal{R}$ a unique $\theta_{P_\rho^L}$ can be associated, provided that (2.3) can be uniquely solved for any $P_\rho^L \in dom(\widehat{\theta})$. Thus, in this case a well-defined *binding function* mapping ρ to $\widehat{\theta}_{P_\rho^L}$ is obtained.

Axiom 2.1. *There exists a smooth injective binding function*

$$b_\psi : \mathcal{R} \longrightarrow \Theta \quad , \quad \rho \longmapsto b_\psi(\rho) := \widehat{\theta}(\rho) := \widehat{\theta}_{P_\rho^L} \quad .$$

In the new notation we thus have $\theta_0 = \widehat{\theta}(\rho_0) = b_\psi(\rho_0)$. Furthermore, notice that different auxiliary score functions ψ, ψ', of the form (2.1) inducing possibly different functionals $\widehat{\theta}$, $\widehat{\theta}'$, will produce different binding functions b_ψ, $b_{\psi'}$, if and only if $\widehat{\theta}\Big|_{P_\rho^L} \neq \widehat{\theta}'\Big|_{P_\rho^L}$.

The pseudo-true value θ_0 is estimated by $\widehat{\theta}_n := \widehat{\theta}\left(P_n^L\right)$ which solves (2.3) with respect to the empirical measure P_n^L. Under regularity conditions the auxiliary estimator is consistent and asymptotically normal at the model.

Proposition 2.2. *Under regularity conditions on* $\left(\widehat{\theta}_n\right)_{n\in\mathbb{N}}$ *(see, e.g., Gallant and Tauchen, 1996) it follows*

$$\begin{cases} \lim_{n\to\infty} \| \widehat{\theta}_n - \theta_0 \| = 0 \quad , \quad a.s. - P_0 \\ \sqrt{n}(\widehat{\theta}_n - \theta_0) \xrightarrow[n\to\infty]{\mathcal{D}} N(0, \mathcal{V}_0) \end{cases} , \tag{2.5}$$

where

$$\mathcal{V}_0 = A_0^{-1} B_0 A_0^{-1} \tag{2.6}$$

and

$$A_0 = E_{P_0^L}\left(\frac{\partial}{\partial\theta'}\psi(X^L;\theta)|_{\theta=\theta_0}\right) \quad , \tag{2.7}$$

$$B_0 = B_{0,0} + \sum_{\tau=1}^{\infty} B_{0,\tau} + \left(\sum_{\tau=1}^{\infty} B_{0,\tau}\right)' \quad , \tag{2.8}$$

$$B_{0,\tau} := E_{P_0}(\psi(X_{1+\tau}^{L+\tau};\widehat{\theta}(P_0^L))\psi(X^L;\widehat{\theta}(P_0^L))') \, . \tag{2.9}$$

The statistical functional $\widehat{\rho}(\cdot)$ for the EMM estimator of the structural model is defined as the minimizer of a quadratic form in (2.4), given the value $\widehat{\theta}_{P_L}$ of the auxiliary estimator, i.e.,

$$\widehat{\rho}(P_L) = \arg\min_{\rho\in\mathcal{R}} E_{P_\rho^L}\left(\psi(X^L;\widehat{\theta}_{P_L})\right)' S E_{P_\rho^L}\left(\psi(X^L;\widehat{\theta}_{P_L})\right) ,$$

where $\mathcal{S}$ is some positive definite deterministic matrix. Thus, when denoting by $m : \mathcal{R} \times \Theta \longrightarrow \mathbb{R}^q$ the function given by

$$m(\rho, \theta) = E_{P_\rho^L}\left(\psi(X^L;\theta)\right) \quad ,$$

the functional

$$\widehat{\rho} : dom(\widehat{\rho}) \subset dom(\widehat{\theta}) \to \mathcal{R} \quad , \quad P^L \longmapsto \widehat{\rho}\left(P^L\right) := \widehat{\rho}_{P^L} \quad ,$$

is the functional solution of the implicit equation

$$m(\rho, \widehat{\theta}_{P^L})' \, \mathcal{S} \, \frac{\partial}{\partial \rho'} m(\rho, \widehat{\theta}_{P^L}) = 0 \quad , \tag{2.10}$$

in ρ. Under the given assumptions we have $\widehat{\rho}\left(P_0^L\right) = \rho_0$. Remind that we focus on structural models where P_ρ^L is not expressible in an analytical form; therefore the expectation in (2.10) will have to be computed by Monte Carlo simulation, using some simulated series $x^k(\rho)$, for k sufficiently large.

The structural parameter is estimated by $\widehat{\rho}_n := \widehat{\rho}\left(P_n^L\right)$, which solves (2.10) with respect to the empirical measure P_n^L, and for some positive definite sequence $\mathcal{S}_n$ converging $a.s. - P_0$ to $\mathcal{S}$. Specifically, $\widehat{\rho}_n$ is such that

$$m(\widehat{\rho}_n, \widehat{\theta}_n)' \, \mathcal{S}_n \, \frac{\partial}{\partial \rho'} m(\widehat{\rho}_n, \widehat{\theta}_n) = 0 \quad .$$

Weak convergence $a.s. - P_0$ of P_n^L to P_0^L implies with Proposition 2.2 and some further regularity conditions on m (see Gallant and Tauchen, 1996), that also the sequence $(\widehat{\rho}_n)_{n \in \mathbb{N}}$ is consistent and asymptotically normal at the model P_0.

Proposition 2.3. *Under regularity conditions on* $(\widehat{\rho}_n)_{n \in \mathbb{N}}$ *it follows*

$$\begin{cases} \lim_{n \to \infty} \| \widehat{\rho}_n - \rho_0 \| = 0 \quad , \quad a.s. - P_0 \\ \sqrt{n}(\widehat{\rho}_n - \rho_0) \xrightarrow[n \to \infty]{\mathcal{D}} N\left(0, \Sigma_{\mathcal{S}}\right) \end{cases} ,$$

where

$$\Sigma_{\mathcal{S}} = \left(M_\rho' \mathcal{S} M_\rho\right)^{-1} M_\rho' \mathcal{S} M_\theta \mathcal{V}_0 M_\theta' \mathcal{S} M_\rho \left(M_\rho' \mathcal{S} M_\rho\right)^{-1} , \tag{2.11}$$

with

$$\underset{(q \times l)}{M_\rho} := \frac{\partial}{\partial \rho'} m(\rho, \theta)|_{\rho=\rho_0, \theta=\theta_0} \quad , \quad \underset{(q \times q)}{M_\theta} := \frac{\partial}{\partial \theta'} m(\rho, \theta)|_{\rho=\rho_0, \theta=\theta_0} \quad . \tag{2.12}$$

Standard regularity conditions allowing the interchange of integration and differentiation in (2.12) *imply* $M_\theta = A_0$ *(see 2.7), so that finally* (2.11) *reads*

$$\Sigma_{\mathcal{S}} = \left(M_\rho' \mathcal{S} M_\rho\right)^{-1} M_\rho' \mathcal{S} B_0 \mathcal{S} M_\rho \left(M_\rho' \mathcal{S} M_\rho\right)^{-1} . \tag{2.13}$$

The next section analyzes the local robustness properties of EMM estimators and statistics derived from EMM estimators.

3. Robust EMM Estimators

We denote by *dom* (T) the domain of a statistical functional T. For an M-estimator defined by a score function ψ the IF is given by the standard expression (see also Hampel et al., 1986, p. 230)

$$IF(x^L;\widehat{\theta},P_0^L) = -M_\theta^{-1}\,\psi(x^L;\theta_0). \tag{3.1}$$

The IF of the EMM functional $\widehat{\rho}$ is obtained by implicitly differentiating the EMM first-order condition

$$m\left(\widehat{\rho}(P^L),\widehat{\theta}(P^L)\right)'\,S\,\frac{\partial}{\partial\rho'}m\left(\widehat{\rho}(P^L),\widehat{\theta}(P^L)\right) = 0 \quad , \tag{3.2}$$

in direction δ_{x^L}. We then have

$$IF(x^L;\widehat{\rho},P_0^L) = -\left(M_\rho' S M_\rho\right)^{-1} M_\rho' S M_\theta\, IF(x^L;\widehat{\theta},P_0^L) \quad . \tag{3.3}$$

Thus, the IF of the EMM estimator depends linearly on the IF of the estimator of the auxiliary parameters, implying that the IF of the EMM estimator is bounded if and only if the estimator of the auxiliary parameter has a bounded IF. Finally, replacing (3.1) in (3.3) it follows

$$IF(x^L;\widehat{\rho},P_0^L) = \left(M_\rho' S M_\rho\right)^{-1} M_\rho' S\,\psi(x^L;\theta_0) \quad . \tag{3.4}$$

Hence, for a $\xi \in \mathbb{R}^{vL}$ such that $\psi(\xi;\theta_0)$ is large, ξ may have no influence on the structural parameters if the vector $\psi(\xi;\theta_0)$ belongs to the kernel of $M_\rho' S$. Note that the $l \times q$ matrix M_ρ' has full row rank l so that the dimension of its kernel is equal to $q-l$.

We consider two robust EMM estimators with bounded IF that correct simultaneously for two sources of asymptotic bias arising in the REMM setting: (i) the standard bias arising in the EMM estimation of an auxiliary model and (ii) the usual bias induced by a robust M-estimation of the auxiliary model via the truncation of an unbounded auxiliary score function.

3.1. Bounding the IF

Following Ronchetti and Trojani (2001), the construction of a REMM estimator is performed by bounding the IF of $\widehat{\rho}$ with respect to the metric induced by its variance-covariance matrix. Formally, we want the IF of $\widehat{\rho}$ to satisfy

$$\| IF(x^L;\widehat{\rho},P_0^L) \|_{\Sigma_S^{-1}} := \| \Sigma_S^{-1/2}\, IF(x^L;\widehat{\rho},P_0^L) \| \leq c, \tag{3.5}$$

where c is an a priori positive bound on the self-standardized sensitivity of $\widehat{\rho}$ (see also Hampel et al., 1986, Chapter 4). By (3.5) and (2.13) this gives

$$\| IF(x^L;\widehat{\rho},P_0^L) \|^2_{\Sigma_S^{-1}} = \psi(x^L;\theta_0)'\,S M_\rho \left(M_\rho' S B_0 S M_\rho\right)^{-1} M_\rho' S\,\psi(x^L;\theta_0). \tag{3.6}$$

As a simple alternative, the properties of orthogonal projections for the matrix

$$K := B_0^{1/2} S M_\rho \left(M_\rho' S B_0 S M_\rho\right)^{-1} M_\rho' S B_0^{1/2}$$

imply

$$\| IF(x^L;\widehat{\rho},P_0^L) \|^2_{\Sigma_S^{-1}} \leq \psi(x^L;\theta_0)' B_0^{-1} \psi(x^L;\theta_0) = \| IF(x;\widehat{\theta},P_0^L) \|^2_{V_0^{-1}} . \quad (3.7)$$

The implications of this last result are very useful. Firstly, bounding the self-standardized norm of the score of the auxiliary model is sufficient in order to bound the self-standardized IF of the EMM estimator of the structural parameters. Secondly, in contrast to $\| IF(x^L;\widehat{\rho},P_0^L) \|_{\Sigma_S^{-1}}$ in order to bound ψ in the metric defined by B_0^{-1} it is not necessary to compute numerically the matrix M_ρ.

3.2. Definition of Robust EMM Estimators

Construction of robust EMM estimators having a self-standardized IF bounded by c can be performed by truncating the function $\psi(x^L;\theta_0)$ with an appropriate algorithm; see also Hampel et al. (1986), pp. 238 ff. We now present two procedures by which this can be achieved. More details on the algorithms for computing our REMM estimators are available in Ortelli and Trojani (2002). In both procedures we replace the (unbounded) function ψ of a classical (nonrobust) EMM estimator by a new bounded one, denoted by ψ_c^A and $\psi_c^{\widetilde{A}}$, respectively. Computation of M_ρ is required for $\psi_c^{\widetilde{A}}$ but not for ψ_c^A as the latter makes use of inequality (3.7). In the sequel we refer to REMM1 for the robust estimator based on ψ_c^A and to REMM2 for that based on $\psi_c^{\widetilde{A}}$. Because of its simplicity, we start with the construction of ψ_c^A.

For a non-singular matrix $A \in \mathbb{R}^q \times \mathbb{R}^q$ we define the truncated auxiliary score function

$$\psi_c^A(x^L;\theta) := A\psi(x^L;\theta)\, w_c(A\psi(x^L;\theta))\,, \quad (3.8)$$

where $w_c(x) = 1$ if $x = 0$, $w_c(x) = \min(1, c/\| x \|)$ otherwise, and $\| \cdot \|$ denotes the Euclidean norm. The new binding function $b_{\psi_c^A} : \mathcal{R} \longrightarrow \Theta$, $\rho \mapsto \theta = b_{\psi_c^A}(\rho)$ and the functional estimator $\widehat{\theta}$ are now implicitly defined as the solution of the nonlinear system of equations

$$E_{P_\rho^L}\left[\psi_c^A(X^L;\theta)\right] = 0 \text{ and } E_{P^L}\left[\psi_c^A(X^L;\theta)\right] = 0\,, \quad (3.9)$$

respectively. The binding function $b_{\psi_c^A}$ is generally different from that implied by the score function ψ. However, Fisher consistency of the structural parameters is naturally maintained in the EMM framework because of the second step. The non-singular matrix A is determined by solving

$$\begin{aligned} I &= B_0 = E_{P_0^L}\left(\psi_c^A(X^L;\theta_0)\psi_c^A(X^L;\theta_0)'\right) \\ &\quad + \sum_{\tau=1}^{\infty} E_{P_0}(\psi_c^A(X^L;\theta_0)\psi_c^A(X_{\tau+1}^{L+\tau};\theta_0)') \\ &\quad + \sum_{\tau=1}^{\infty} E_{P_0}(\psi_c^A(X_{\tau+1}^{L+\tau};\theta_0)\psi_c^A(X^L;\theta_0)'). \end{aligned} \quad (3.10)$$

Equation (3.10) ensures $B_0 = I$ for the robust EMM estimator implied by the score ψ_c^A so that the self standardized IF of $\widehat{\rho}$ is automatically bounded by c.

Construction of the second version of a robust EMM estimator is similar to the first one. The truncated auxiliary score is defined as

$$\psi_c^{\widetilde{A}}(x^L;\theta) := \psi(x^L;\theta)\, w_c(\widetilde{A}M_\rho'\mathcal{S}\,\psi(x^L;\theta))\,, \tag{3.11}$$

where $\widetilde{A} \in \mathbb{R}^l \times \mathbb{R}^l$ is non-singular. Here, the norm of ψ is measured with respect to the semi-metric induced by the matrix $\mathcal{S}M_\rho\widetilde{A}'\widetilde{A}M_\rho'\mathcal{S}$ which takes into account the fact that we are interested in the robustness of the structural estimator and not necessarily in the one of the auxiliary estimator. Assume that $l < q$. Because $rank\,(M_\rho) = l$ and $rank\,(\mathcal{S}) = q$, observations with large influence on the auxiliary parameters may have small or even zero influence on the structural parameters and, therefore, do not have to be downweighted. Furthermore, (3.5) is true whenever

$$\widetilde{A}'\widetilde{A} = \left(M_\rho'\mathcal{S}B_0\mathcal{S}M_\rho\right)^{-1} \tag{3.12}$$

is satisfied. In fact, substituting (3.12) in (3.6) we obtain

$$\|\, IF(x^L;\widehat{\rho},P_0^L)\,\|^2_{\Sigma_S^{-1}} = \psi_c^{\widetilde{A}}(x^L;\theta_0)'\,\mathcal{S}M_\rho\widetilde{A}'\widetilde{A}M_\rho'\mathcal{S}\,\psi_c^{\widetilde{A}}(x^L;\theta_0) \leq c^2$$

4. Monte Carlo Simulations

We extend the simple MA(1) model analyzed in Martin and Yohai (1986) to an ARMA(1,1)-ARCH(2) process given by

$$\begin{aligned} Y_t &= 0.9Y_{t-1} + \epsilon_t - 0.5\epsilon_{t-1}, \quad \epsilon_t = \sqrt{h_t}\nu_t, \quad \nu_t \text{ iid } N(0,1) \\ h_t &= 0.04 + 0.6\epsilon_{t-1}^2 + 0.3\epsilon_{t-2}^2\,. \end{aligned} \tag{4.1}$$

We estimate the structural model (4.1) by means of an AR(3)-ARCH(2) auxiliary model. Since we include a constant term in the estimation of the structural and the auxiliary model the total number of auxiliary parameters q is equal to 7 while l is equal to 6. For both robust estimators REMM1 and REMM2 the robustness tuning constant c is set equal to 10. Other levels of c gave similar findings.

The upper panel of Figure 1 shows a simulated path from model (4.1) where the observations have been contaminated with the replacement outlier model defined by

$$X_t = (1 - H_t^\varepsilon)\,Y_t + H_t^\varepsilon\xi\,. \tag{4.2}$$

$\mathbb{H}^\varepsilon$ represents an iid 0-1 process independent of $\mathbb{Y}$ and $P\,(H_t^\varepsilon = 1) = \varepsilon$. We fixed $\xi = 3$ and $\epsilon = 1\%$. For this path only 4 observations have been replaced (observations 180, 548, 660 and 797). Note there the infrequent large positive and negative movements which typically occur in periods of higher volatility. At first sight they may easily be identified as outliers in a naive model misspecification analysis. Even when knowing the form of the outliers generating process, their exact identification in Figure 1 could be a very difficult task.

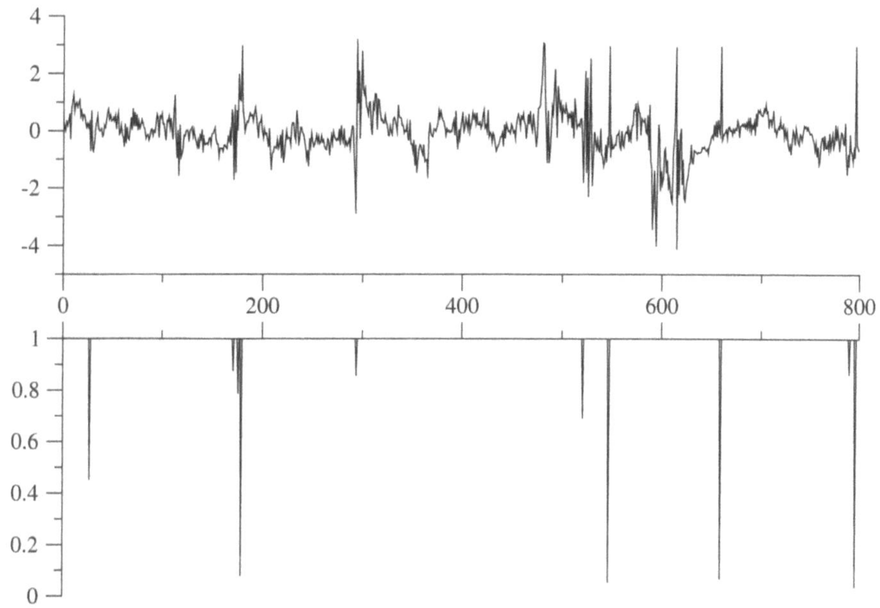

FIGURE 1. Contaminated series (top panel) and weights implied by the REMM.

Table 1 summarizes the results obtained from 1000 uncontaminated simulations of model (4.1) by presenting summary statistics and mean squared errors. For each $\widehat{\rho}_i$ in Table 1 the first row contains the summary statistics for the parameter estimates under a classical EMM estimation. The summary statistics for the REMM1 and REMM2 estimators are given, respectively, in the second and the third row for each $\widehat{\rho}_i$ in Table 1. The sample size behind all simulations is 800.

From Table 1 it is evident that the efficiency loss at the model when using REMM is virtually negligible. Indeed, the mean squared errors of all parameters estimates are very similar across the different EMM and REMM estimators. The effects of the contamination on the resulting parameter estimates are summarized in Table 2 with corresponding summary statistics. The results for the classical EMM estimator in Table 2 highlight some large biases and mean squared errors of a standard EMM model estimation. Indeed, we observe that the mean squared error of all EMM parameter estimates in Table 2 is highly inflated by the presence of contamination (compare for instance the mean squared errors for the estimates of ρ_1 and ρ_4 given in Table 1 and 2) and that some large biases are obtained especially in the parameter estimates of the conditional variance equation. The REMM procedures, on the other side, are very successful in controlling both for bias and efficiency in the presence of contamination. Indeed, when compared with

TABLE 1. Summary statistics under the uncontaminated model (4.1).

True	Method	Mean	Median	q_{25}	q_{75}	Stdv	$q_{75}-q_{25}$	MSE
0	EMM	0.0005	0.0005	-0.0025	0.0036	0.0048	0.0061	2.33E-5
	REMM1	0.0008	0.0010	-0.0022	0.0039	0.0049	0.0061	2.43E-5
	REMM2	0.0008	0.0009	-0.0023	0.0037	0.0048	0.0060	2.38E-5
0.9	EMM	0.8963	0.8978	0.8842	0.9098	0.0199	0.0255	4.10E-4
	REMM1	0.8954	0.8974	0.8837	0.9083	0.0196	0.0246	4.06E-4
	REMM2	0.8964	0.8985	0.8847	0.9088	0.0197	0.0241	4.02E-4
-0.5	EMM	-0.5036	-0.5033	-0.5408	-0.4653	0.0563	0.0755	3.19E-3
	REMM1	-0.5014	-0.5000	-0.5360	-0.4644	0.0556	0.0716	3.09E-3
	REMM2	-0.5054	-0.5034	-0.5401	-0.4697	0.0556	0.0704	3.11E-3
0.04	EMM	0.0397	0.0393	0.0363	0.0429	0.0048	0.0066	2.33E-5
	REMM1	0.0400	0.0396	0.0365	0.0431	0.0048	0.0066	2.30E-5
	REMM2	0.0403	0.0399	0.0369	0.0434	0.0048	0.0065	2.33E-5
0.6	EMM	0.5846	0.5844	0.5339	0.6389	0.0745	0.1050	5.78E-3
	REMM1	0.5900	0.5923	0.5379	0.6436	0.0761	0.1057	5.90E-3
	REMM2	0.5744	0.5746	0.5256	0.6257	0.0730	0.1001	5.98E-3
0.3	EMM	0.2922	0.2923	0.2513	0.3332	0.0607	0.0818	3.75E-3
	REMM1	0.2959	0.2950	0.2539	0.3375	0.0617	0.0836	3.82E-3
	REMM2	0.2986	0.2992	0.2575	0.3414	0.0606	0.0839	3.67E-3

EMM, all mean squared errors in Table 2 are much smaller and the estimates in the conditional variance equation present a quite reasonable bias.

Finally, in all our simulations we have observed REMM to identify correctly outliers when they were generated by the contaminating model used. To illustrate this point Figure 1 presents again the contaminated path with the estimated REMM weights (in the bottom panel). Weights clearly below one indicate an influential observation and can be used to identify outlying observations or more general model misspecifications. As shown in the bottom panel of Figure 1 all generated outliers were clearly indicated with the lowest REMM weights. On the other hand, even some model-generated very erratic movements, as for instance the ones at the observation 294, 480 and 614, were correctly identified as observations generated by the underlying structural model.

5. Conclusions

We characterized the local robustness properties of EMM estimators for time series and proposed two versions of a REMM estimator with bounded IF. We verified in a Monte Carlo simulation study that REMM estimators are successful in controlling for the asymptotic bias under model misspecification while maintaining a high efficiency under the ideal structural model. REMM extends the application field

TABLE 2. Summary statistics under model (4.1) contaminated by the replacement model (4.2).

True	Method	Mean	Median	q_{25}	q_{75}	Stdv	$q_{75} - q_{25}$	MSE
0	EMM	0.0100	0.0091	0.0018	0.0172	0.0127	0.0154	2.59E-4
	REMM1	0.0009	0.0010	-0.0025	0.0045	0.0056	0.0071	3.17E-5
	REMM2	0.0010	0.0011	-0.0024	0.0046	0.0056	0.0070	3.28E-5
0.9	EMM	0.8793	0.8852	0.8585	0.9096	0.0464	0.0511	2.58E-3
	REMM1	0.8861	0.8889	0.8730	0.9023	0.0225	0.0293	7.02E-4
	REMM2	0.8865	0.8895	0.8732	0.9031	0.0235	0.0299	7.33E-4
-0.5	EMM	-0.4913	-0.4877	-0.5484	-0.4266	0.0979	0.1218	9.65E-3
	REMM1	-0.4854	-0.4857	-0.5219	-0.4488	0.0541	0.0731	3.14E-3
	REMM2	-0.4855	-0.4856	-0.5216	-0.4495	0.0570	0.0721	3.46E-3
0.04	EMM	0.1095	0.1023	0.0703	0.1422	0.0474	0.0719	7.08E-3
	REMM1	0.0482	0.0472	0.0424	0.0531	0.0088	0.0107	1.45E-4
	REMM2	0.0488	0.0480	0.0430	0.0537	0.0082	0.0106	1.44E-4
0.6	EMM	0.5499	0.5473	0.4216	0.6672	0.1722	0.2456	3.21E-2
	REMM1	0.5996	0.5995	0.5375	0.6671	0.0925	0.1296	8.55E-3
	REMM2	0.5795	0.5825	0.5180	0.6490	0.0973	0.1310	9.88E-3
0.3	EMM	0.2415	0.2187	0.1277	0.3254	0.1499	0.1977	2.59E-2
	REMM1	0.2554	0.2567	0.2003	0.3053	0.0748	0.1050	7.59E-3
	REMM2	0.2651	0.2648	0.2078	0.3174	0.0797	0.1096	7.57E-3

of robust statistics to very general times series models and permits in a natural way a robust estimation of non Markovian processes.

References

[1] X. de Luna and M.G. Genton, *Robust Simulation-Based Estimation of ARMA Models*. J. Comput. Graph. Statist. **10** (2001), 370–387.

[2] A.R. Gallant and G. Tauchen, *Which moments to match?* Econometric Theory **12** (1996), 657–681.

[3] M.G. Genton and E. Ronchetti, *Robust Indirect Inference*. J. Amer. Statist. Assoc. **98** (2003), 67–76.

[4] C. Gourieroux, A. Monfort, and E. Renault, *Indirect Inference*. J. Appl. Econom. **8** (1993), 85–118.

[5] F.R. Hampel, *The Influence Curve and its Role in Robust Estimation*. J. Amer. Statist. Assoc. **68** (1974), 383–393.

[6] F.R. Hampel, E.M. Ronchetti, P.J. Rousseeuw, and W.A. Stahel, *Robust Statistics: The Approach Based on Influence functions*. Wiley, New York, 1986.

[7] P.J. Huber, *Robust Estimation of a Location Parameter*. Ann. Math. Statist. **35** (1964), 73–101.

[8] H. Künsch, *Infinitesimal Robustness for Autoregressive Processes.* Ann. Statist. **12** (1984), 843–863.

[9] R.D. Martin and V.J. Yohai, *Influence Functionals for Time Series.* Ann. Statist. **14** (1986), 836–837.

[10] C. Ortelli and F. Trojani, *Robust Efficient Method of Moments.* Working paper, University of Southern Switzerland, 2002, available at *http://papers.ssrn.com/sol3/papers.cfm?abstract_id=335700.*

[11] E. Ronchetti and F. Trojani, *Robust Inference With GMM Estimators.* J. Econometrics **101** (2001), 37–69.

C. Ortelli
Institute of Finance
University of Lugano
Switzerland
e-mail: `claudio.ortelli@lu.unisi.ch`

F. Trojani
Institute of Finance
University of Lugano

and

Department of Economics
University of St. Gallen
Switzerland
e-mail: `fabio.trojani@lu.unisi.ch`

Statistics for Industry and Technology, 283–295

Computational Geometry and Statistical Depth Measures

E. Rafalin and D.L. Souvaine

Abstract. The computational geometry community has long recognized that there are many important and challenging problems that lie at the interface of geometry and statistics (e.g., Shamos, 1976; Bentley and Shamos, 1977). The relatively new notion of *data depth* for non-parametric multivariate data analysis is inherently geometric in nature, and therefore provides a fertile ground for expanded collaboration between the two communities. New developments and increased emphasis in the area of multivariate analysis heighten the need for new and efficient computational tools and for an enhanced partnership between statisticians and computational geometers.

Over a decade ago point-line duality and combinatorial and computational results on arrangements of lines contributed to the development of an efficient algorithm for two-dimensional computation of the LMS regression line (Souvaine and Steele, 1987; Edelsbrunner and Souvaine, 1990). The same principles and refinements of them are being used today for more efficient computation of data depth measures. These principles will be reviewed and their application to statistical problems such as the LMS regression line and the computation of the half-space depth contours will be presented. In addition, results of collaborations between computational geometers and statisticians on data-depth measures (such as half-space depth and simplicial depth) will be surveyed.

Mathematics Subject Classification (2000). 68U05, 62-07.

Keywords. Computational geometry, data depth, LMS regression, duality, half-space depth, simplicial depth.

1. Introduction

The field of *Computational Geometry* deals with the systematic study of algorithms and data structures for geometric objects (de Berg et al., 1997). Computational geometry usually focuses at the outset on exact algorithms that are asymptotically fast. Yet once exact algorithms have been obtained, refined, and are still slow,

Partially supported by NSF grant EIA-99-96237.

approximation algorithms of provable performance are sought. The field emerged in the 1970's with the work of Michael Shamos (1978). Even in these early days the computational geometry community has recognized that there are many important and challenging problems that lie at the interface of geometry and statistics (e.g., Shamos, 1976; Bentley and Shamos, 1977).

The field is commonly related to problems in *robotics, CAD/CAM* and *geographic information systems.* However, any problem that can be represented using geometric objects and operators can be viewed as a computational geometry problem, including the relatively new notion of *data depth* for non-parametric multivariate data analysis. Data depth is inherently geometric in nature, and therefore provides a fertile ground for expanded collaboration between the two communities.

A *data depth* measures how deep (or central) a given point $x \in \mathbb{R}^d$ is relative to F, a probability distribution in $\mathbb{R}^d$ or relative to a given data cloud. Some examples of data depth functions are *Half-space Depth* (Hodges, 1955; Tukey, 1975), *Majority Depth* (Singh, 1993), *Simplicial Depth* (Liu, 1990), *Oja Depth* (Oja, 1983), *Convex Hull Peeling Depth* (Barnett, 1976; Eddy, 1982) and *Regression Depth* (Rousseeuw and Hubert, 1999). The data depth concept provides *center-outward* orderings of points in Euclidean space of any dimension and leads to a new *non-parametric* multivariate statistical analysis in which *no* distributional assumptions are needed. Most depth functions are defined in respect to a probability distribution F, considering $\{X_1, \ldots, X_n\}$ random observations from F. The *finite sample version* of the depth function is obtained by replacing F by F_n, the empirical distribution of the sample $\{X_1, \ldots, X_n\}$. In general, computational geometers study the finite sample case in which sets of points are investigated.

Over a decade ago, point-line duality and combinatorial and computational results on arrangements of lines contributed to the development of efficient algorithms for two-dimensional computation of the LMS regression line (Souvaine and Steele, 1987; Edelsbrunner and Souvaine, 1990). The same principles and refinements of them are being used today for more efficient computation of data depth measures.

The technique of point-line duality, its application to LMS, and various sweep techniques will be sketched in Section 2. More recent results for data-depth related problems will be reviewed in Section 3 (including half-space depth and simplicial depth). Sections 4 and 5 will include suggestions for future collaboration and a summary.

2. History

2.1. The Duality Transform (Brown, 1980; Dobkin and Souvaine, 1987)

The structure of a collection of lines is more readily observed than the structure implied by a set of points. This structure can be exploited to create efficient algorithms. A set of points can be transformed into an arrangement of lines using a duality transform that preserves key properties (Brown, 1980). Many problems

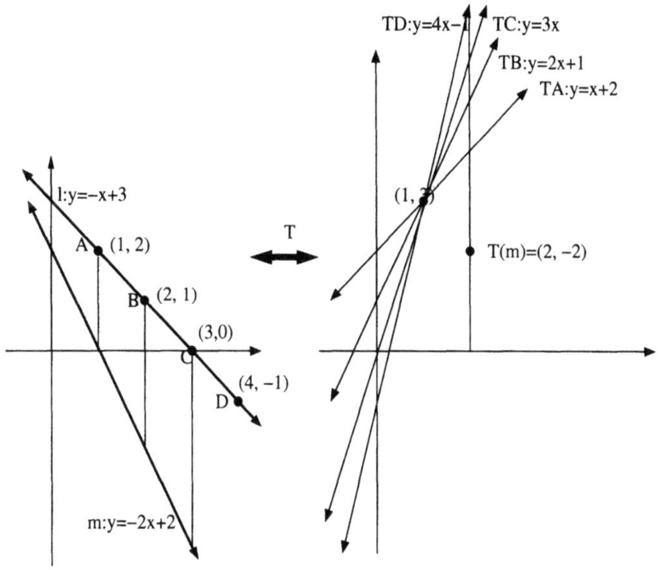

FIGURE 1. The duality transform T: The set of points and lines on the left is transformed into the set on the right. The points $A = (1,2), B = (2,1), C = (3,0)$ and $D = (4,-1)$ lying on the line $l : y = x + 3$ are mapped to the lines $T(A) : y = x + 2$, $T(B) : y = 2x + 1$, $T(C) : y = 3x$ and $T(D) : y = 4x - 1$, all passing through the point $(1,3)$, dual to the line l. The points A, B, C, D are located above $m : y = 2x + 2$, as are their dual lines, above point $T(m) : (2,-2)$. In addition, the vertical distance between every point A, B, C, D to the line m is identical to the vertical distance between their dual lines and the dual point $T(m)$.

based on sets of points have been solved efficiently by first transforming the points into a set of lines, solving a related problem on the set of lines, and converting the answer to a solution for the original problem. The transformation itself is conducted in linear time, and consequently does not affect the total computational complexity of the solution. Different transformations apply to different problems. Here, we will focus on a particular transform suitable for the statistical problems we treat. For more details see Dobkin and Souvaine (1987).

The *duality transform* T (see Figure 1) maps a point $P = (a, b)$ to a line $T(P) : y = ax + b$. We need to define the image of a line $l : y = cx + d$ under transform T consistently. Let us pick points $Q = (q, cq + d)$, $R = (r, cr + d)$ lying on line l. $T(Q) : y = qx + (cq + d)$ and $T(R) : y = rx + (cr + d)$. Both of these lines pass through the point $(-c, d)$. Every other point on l will also be mapped to a line passing through $(-c, d)$. Consequently we say that $T(l) : (-c, d)$.

Note that the slope of l is preserved in the x-coordinate of the image point. If the slope of line l exceeds the slope of line k, then $T(l)$ lies to the left of $T(k)$. The vertical distance of a point $P = (a, b)$ to a line $l : y = cx + d$ equals $b - (ca + d)$.

In the dual, the vertical distance from the line $T(p) : y = ax + b$ to the point $T(l) = (-c, d)$ is $(-ac + b) - d$, the same value. Hence, the transformation T preserves slope, vertical distance and the above/below relationship.

For a vertical line $l : x = m$, T maps it to a pencil of parallel lines of slope m. We assume that these parallel lines intersect at a point ∞_m (this point is called an *improper point*). For every direction m, there is an improper point ∞_m, associated with it. Conversely, each line $y = mx + a$ which passes through ∞_m gets mapped to a point $(-m, a)$ which lies on the line $x = -m$. Consequently, the image of the improper point ∞_m under T is the vertical line $x = -m$.

The duality transform T can be extended to higher dimensions: in $\mathbb{R}^3$ the dual of the point $P = (a, b, c)$ is the plane $z = ax + by + c$.

2.2. Least Median of Squares Regression

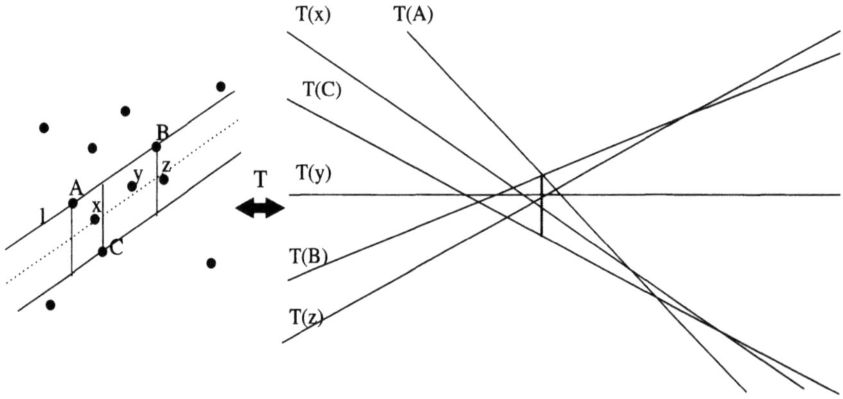

FIGURE 2. LMS regression: A set of points and its dual arrangement (only some of the lines in the arrangement are drawn), and the LMS slab and regression line. Line l passes through points A and B, and its dual point $T(l)$ is the intersection of lines $T(A)$ and $T(B)$. The slab created from line l and the line parallel to it passing through point C contains 6 points. In the dual, the vertical lines between $T(l)$ and the line $T(C)$ crosses 6 lines.

Consider the problem of fitting a line to a set of data points. The familiar *ordinary least squares (OLS)* method minimizes the sum of the squares of the y-distance between the fitted line and the data points (the *residual*). This method has the disadvantage that a single corrupt point (*outlier*) can significantly perturb the fitted line. The *Least Median of Squares (LMS) Regression line* (Rousseeuw, 1984) is the line that minimizes the median of the squares of the residuals. This method has a high breakdown point as up to 50% of the data points can be outliers, without perturbing the fitted line.

It is easy to prove that the problem of finding the line l that minimizes the median residual is equivalent to the problem of finding the slab bounded by a pair of parallel lines of minimum vertical separation that contains $\lceil \frac{n}{2} \rceil$ of the data points. Furthermore, one bounding line L_1 must pass through two points and the other line L_2 through a point whose x-coordinate lies between those of L_1's points. To find the slab, one can check each pair of points (p, q) and find a line l parallel to them, passing through some other point r, such that precisely $\lceil \frac{n}{2} \rceil$ points lie in the closed slab defined by $\overline{pq}$ and l. This can be solved naively using an $O(n^3)$ time algorithm.

Mapping the set of points using the transform T, the above problem becomes one of looking for the intersection point of two lines $T(p)$ and $T(q)$ such that the y-distance between this point and the line $T(r)$, which is $\lceil \frac{n}{2} \rceil$ above or below, is the shortest among all such distances. By sweeping (see Section 2.3) a line through the arrangement dual to the set of points, one can check all intersection points and lines such that the closed vertical segment from the point to the line intersects $\lceil \frac{n}{2} \rceil$ lines and is as short as possible.

Souvaine and Steele (1987) proposed an $O(n^2 \log n)$ time algorithm for computing the LMS line in $\mathbb{R}^2$ using the duality concept described above. Their result was improved by Edelsbrunner and Souvaine (1990) to an $O(n^2)$ time algorithm, by using a topological line sweep (see Section 2.3). A practical approximation algorithm for 2 and higher dimensions, using additional computational geometry concepts, was recently presented by Mount et al. (1997).

2.3. Line Sweep, Topological Line Sweep and Levels in an Arrangement

Vertical line sweep (Bentley and Ottmann, 1979) is a classical technique in computational geometry. The algorithm sweeps a vertical line across an arrangement of objects (points, lines, segments, ...) from left to right, reporting all intersection points, in a series of elementary steps. For an arrangement of n line segments consisting of k intersection points this technique achieves a time complexity of $O((n+k)\log n)$ and requires $O(n)$ space (Brown, 1981). For an arrangement of n infinite lines that contains $k = \frac{n(n-1)}{2} = O(n^2)$ intersections, vertical line sweep could report all intersection pairs sorted in order of x-coordinate in $\Theta(n^2 \log n)$ time and $O(n)$ space. If one needs to report the intersection points of the lines according only to a partial order related to the *levels* in the arrangement greater efficiency is possible using *topological sweep* (Edelsbrunner and Guibas, 1989). In topological line sweep, to report all the intersection points of the lines, a topological line, which is monotonic in the y-direction but not necessarily straight, and which intersects each of the n lines in the arrangement exactly once, sweeps the arrangement, in a series of elementary steps, in $O(n^2)$ time and $O(n)$ space.

The kth *level* of an arrangement of lines is the set of points that lie on lines and have at most $k-1$ lines above them and at most $n-k$ lines below them (see Figure 3). Consider an arrangement that was created from a set of points, using the duality transform T. If a point $T(p)$, the image of line p, that lies on line $l = T(L)$, is in the kth level, then line p has at most $k-1$ points above it and at most $n-k$ points below it. This property will be used in Section 3.1.1, to compute half-space depth contours.

3. Current Applications

3.1. Half-space Depth and Half-space Depth Contours

The *half-space depth* (in the literature this is sometimes called *location depth* or *Tukey depth*) of a point x relative to a set of points $\mathcal{S} = \{X_1, \ldots, X_n\}$ is the minimum number of points of $\mathcal{S}$ lying in any closed half-space determined by a line through x (Hodges, 1955; Tukey, 1975).

Using the duality transform T, the set of points $\mathcal{S}$ maps to an arrangement of lines $T(\mathcal{S})$. The number k of points of $\mathcal{S}$ above (resp. below) a line L through x, equals the number of lines of $T(\mathcal{S})$ above (resp. below) the dual point $T(L)$. The depth of a point x relative to $\mathcal{S}$ is the minimum number of points of $\mathcal{S}$ lying in any closed half-space determined by a line through x (above or below the line through x). Consequently, the depth of a line $T(x)$, dual to the point x, relative to $T(\mathcal{S})$, is the minimal number of lines of $T(\mathcal{S})$ above or below $T(x)$.

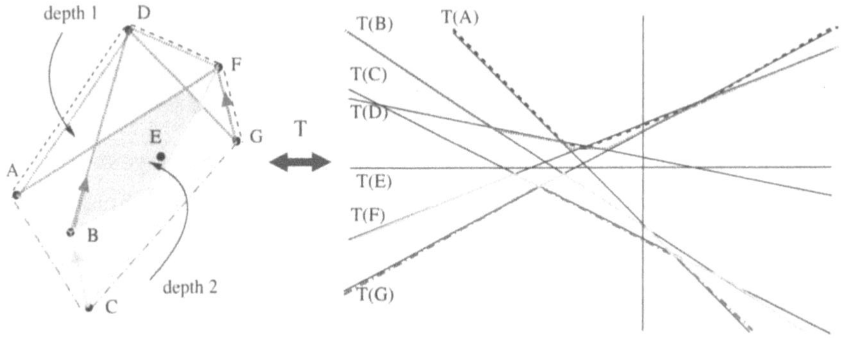

FIGURE 3. Levels in an arrangement and half-space depth computation: The set of 7 points $A-G$ on the left is transformed using the duality transform T into the arrangement of lines $T(A)-T(G)$ on the right. The levels $1, 2, 6, 7$ are drawn in blue, green, yellow and red, respectively. Any vertical line cutting the kth level will pass through exactly $k-1$ lines of the arrangement above it. The intersection points on the 1st and 7th level are candidate points for the first half-space depth contour (as are the 2nd and 6th levels for the second depth contour). Therefore transforming the intersection points on each level back to the primal plane (as lines) and computing their intersection will result in that depth contour. The 1st and 2nd contours are drawn as grayed areas.

3.1.1. Efficient computation of Depth Contours using Duality. The kth *depth contour* for a set of points $\mathcal{S}$ in $\mathbb{R}^d$ is the boundary of the points of $\mathbb{R}^d$ with depth $\geq k$, according to the chosen depth function.

The kth depth contours for half-space depth in $\mathbb{R}^2$ can be constructed by taking the intersection of all half-planes containing k points of $\mathcal{S}$. Specifically, the significant half-planes are the ones bounded by lines connecting pairs of points of $\mathcal{S}$. Every line L passing through a pair of points l, m is dualized to an intersection point I of two lines $T(l), T(m)$. The vertical line through I in the dual intersects every line of the arrangement, some above I and some below. The number of lines intersected above (resp. below) I equals the number of points lying above (resp. below) $T^{-1}(I) = L$ in the primal. The contour to which L contributes is the minimum of these two numbers (this is also said to be the *level* of L, see Section 2.3). Examining all intersection points in the arrangement of dual lines will produce a list of half-planes/intersections for each potential contour.

Miller et al. (2003) presented an $O(n^2)$ time algorithm for computing all depth contours for a planar data set, using the idea sketched above. The authors used *topological sweep* (see Section 2.3) to examine all intersection points and to report the level of each intersection point in optimal time and space.

Note that the duality concept and the algorithmic idea can be easily extended to higher dimensions. For example, in $\mathbb{R}^3$ the half-space depth of a point x is the minimum number of points of a given set $\mathcal{S}$ lying in any closed half-space bounded by a hyperplane (instead of a line, as in $\mathbb{R}^2$) through x.

3.1.2. Overview of Related Results. The half-space depth of a single point can be computed in $O(n \log n)$ time as shown by Rousseeuw and Ruts (1996). This matches the lower bound that was recently proved by Aloupis et al. (2002).

Algorithms for computing the deepest point relative to the data set (also known as the *Tukey median*) were introduced by Rousseeuw and Ruts. In Ruts and Rousseeuw (1996), they give an $O(n^2 \log n)$ time implementation to compute a single contour. In Rousseeuw and Ruts (1998), the same technique is applied repeatedly to give an $O(n^2 \log^2 n)$ time implementation to compute the two-dimensional Tukey median. A fast implementation to approximate the deepest location in high dimensions is given in (Rousseeuw and Struyf, 1998; Struyf and Rousseeuw, 2000; Verbarg, 1997). Theoretical complexity results on two-dimensional *depth contours*, or k-hulls, have been known for some time (Cole et al., 1987). The best known theoretical result for computing the $2D$ Tukey median is an $O(n \log^5 n)$ algorithm by Matoušek (1991), but its complex structure makes it an impractical candidate for implementation. The improved $O(n \log^3 n)$ algorithm by Langerman and Steiger (2003) uses parametric search, also difficult to implement.

Ruts and Rousseeuw (1996) developed a program called ISODEPTH, that computes the kth contour in $O(n^2 \log n)$ time and all contours in $O(n^3 \log n)$. Johnson et al. (1998) gave a program called FDC to compute the k outermost depth contours which outperforms ISODEPTH for small ks. The $\Theta(n^2)$ time implementation by Miller et al. (2003) that was described above computes all the depth contours, the depth of all the data points, and the Tukey median. The implementation was expanded by Rafalin et al. (2002) to handle degenerate data sets, that contain 3 or more points on a line or points that share the same x-coordinate.

A different approach based on parallel arrangement construction by Fukuda and Rosta (2003) allows to compute high dimensional depth contours. In addition, half-space depth contours can be computed for display in 2D using hardware assisted computation, as suggested by Krishnan et al. (2002).

Centerpoints, which are points of depth $\geq n/(d+1)$ (by Helly's theorem (Edelsbrunner, 1987), such points are guaranteed to exist), have been widely studied in the computational geometry literature (Clarkson et al., 1996; Jadhav and Mukhopadhyay, 1994; Matoušek, 1991; Naor and Sharir, 1990).There are efficient algorithms for centerpoints in $\mathbb{R}^2$ by Matoušek (1991) and in linear time by Jadhav and Mukhopadhyay (1994). Naor and Sharir (1990) give an algorithm in $\mathbb{R}^3$. The best known time in high dimensions is $O(n^{d+1})$, but a faster approximation algorithm is given by Clarkson et al. (1996). The algorithms are difficult to implement.

3.2. Simplicial Depth

Simplicial depth was introduced by Liu (1990): The *simplicial depth* of a point x with respect to a data set $\mathcal{S} = \{X_1, \ldots, X_n\}$ is the fraction of the closed simplices formed by $d+1$ points of $\mathcal{S}$ containing x, where I is the indicator function: $SD_{Liu}(\mathcal{S}; x) = \binom{n}{d+1}^{-1} \sum I_{(x \in S[X_{i_1}, \ldots, X_{i_{d+1}}])}$ (see Figure 4(a)).

A simplex in $\mathbb{R}^d$ is the boundary of the region defined by $d+1$ vertices in general position, and represents the simplest possible polytope in any dimension. In one dimension, a simplex is simply a line segment; in 2 dimensions, it is a triangle; and in 3 dimensions, a tetrahedron. Simplicial depth has been studied by computational geometers since it was first introduced (Khuller and Mitchell, 1990; Gil et al, 1992). The simplicial depth is computationally more difficult in the finite sample case than some other depth measures. Recently new results and concepts were discovered by computational geometers, some discussed below.

3.2.1. A Revised Definition. Several problems arise in the finite sample case of simplicial depth under Liu's definition. As suggested by Zuo and Serfling (2000), a depth function should attain maximum value at the center (*the maximality property*) and as a point x moves away from the *deepest point* along any fixed ray through the center, the depth at x should decrease monotonically (*monotonicity property*). However, the simplicial depth function for the finite sample case fails to satisfy these properties (Zuo and Serfling, 2000) (see, e.g., Figure 4(b)). In addition, the depth of points on facets causes discontinuities in the depth function: The depth of all points on the boundary of a cell is at least the depth of a point on the interior of the cell. In most cases the depth values on the boundaries can be higher than the depth in each of the adjacent cells (see, e.g., Figure 4(b)).

Burr et al. (2003) proposed a modification to the definition that corrects the irregularity at boundaries of simplices by making the depth of a point on the boundary between two cells the *average* of the depth of the two cells and fixes the counterexamples provided by Zuo and Serfling (2000):

Given a data set $\mathcal{S} = \{X_1, \ldots, X_n\}$ in $\mathbb{R}^d$, the *simplicial depth* of a point x is the average of the fraction of closed simplices containing x and the fraction of open

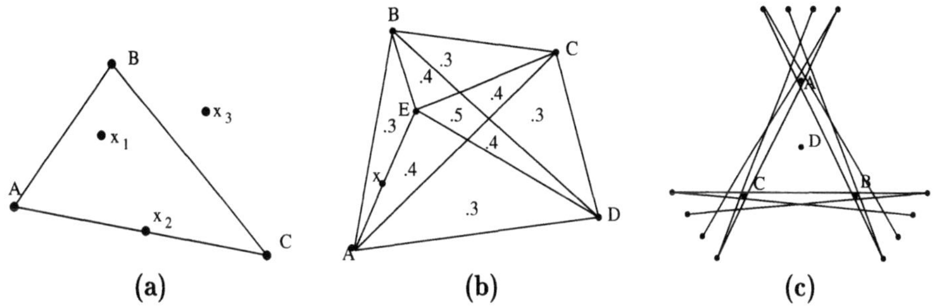

FIGURE 4. (a) Computation of simplicial depth, according to Liu and according to Burr et al. $SD_{Liu}(x_1) = SD_{BRS}(x_1) = 1$, $SD_{Liu}(x_2) = 1$, $SD_{BRS}(x_2) = .5$. $SD_{Liu}(x_3) = SD_{BRS}(x_3) = 0$. (b) Problems with simplicial depth: The total number of simplices is $\binom{5}{3} = 10$. The depth of the open regions is drawn on the figure. $SD_{Liu}(p) = \binom{4}{2}/10 = .6$, for $p = \{A, B, C, D\}$, while $SD_{Liu}(E) = .8$. For a point x on $\overline{AE}$ $SD_{Liu}(x) = .5$, violating the monotonicity property and causing discontinuities in the depth function at edges. According to the revised definition $SD_{BRS}(p) = .3$, for $p = \{A, B, C, D\}$, $SD_{BRS}(E) = .5$ and $SD_{BRS}(x) = .35$. (c) A problem with the revised definition. The data points A,B, and C all have depth $(241 + \frac{1}{2}\binom{15}{2})/\binom{16}{3} = \frac{587}{1120}$ and the data point D, which is at the unique center of the data set has depth $(5^3 + \frac{1}{2}\binom{15}{2})/\binom{16}{3} = \frac{355}{1120}$. For clarity reasons not all cells are drawn.

simplices containing x:

$$SD_{BRS}(\mathcal{S}; x) = \frac{1}{2}\binom{n}{d+1}^{-1}\left(\sum I_{(x \in S[X_{i_1},\ldots X_{i_{d+1}}])} + I_{(x \in int(S[X_{i_1},\ldots X_{i_{d+1}}]))}\right)$$

where *int* refers to the open relative interior (see Edelsbrunner, 1987, page 401) of $S[X_{i_1}, \ldots, X_{i_{d+1}}]$. Equivalently, this could be formulated as: $SD_{BRS}(\mathcal{S}; x) = \rho(\mathcal{S}, x) + \frac{1}{2}\sigma(\mathcal{S}, x)$ where $\rho(\mathcal{S}, x)$ is the number of simplices with data points as vertices which contain x in their open interior, and $\sigma(\mathcal{S}, x)$ is the number of simplices with data points as vertices which contain x in their boundary.

The revised definition reduces to the original definition for continuous distributions and for points lying in the interior of cells, and it maintains the ranking order of data points. In addition, it can be calculated using the existing algorithms (see Section 3.2.2), with slight modifications. However, it does not achieve all desired properties in the sample case. Figure 4(c) shows an example where the data set has a unique center, D, but it neither attains maximality at the center, nor does it have monotonicity relative to the deepest point.

In addition, data points are still over-counted. For example data points in $\mathbb{R}^2$ are a vertex of $\binom{n-1}{2}$ simplicies, whereas edges are counted only $n-2$ times. This implies that the weight λ of a data point should be $\frac{1}{2}(n-2) = \lambda\binom{n-1}{2} \Rightarrow \lambda = \frac{1}{n-1}$.

However, this factor isn't enough: consider a data set of n points, where $n-1$ points are evenly distributed angularly around one point. Then the depth of the center point should be at least as large as the depth of the n cells which use it as a vertex; this is not guaranteed by the λ factor. Thus the depth of a data point in $\mathbb{R}^d$ should be based both on the λ factor and the geometry of the $n-1$ other data points.

3.2.2. Overview of Related Results. The simplicial depth of a point in $\mathbb{R}^2$ can be computed in $O(n \log n)$ time, as proposed independently by Gil et al. (1992), Khuller and Mitchell (1990), and Rousseeuw and Ruts (1996). This bound matches the lower bound, as proved by Gil et al. (1992) and Aloupis et al. (2002). In 3 dimensions, Gil et al. (1992) gave an $O(n^2)$ algorithm for computing the simplicial depth of x relative to $\mathcal{S}$ that was recently improved by Cheng and Ouyang (2001). Cheng and Ouyang also give a generalization of this algorithm to $\mathbb{R}^4$, with a time complexity of $O(n^4)$. Rousseeuw and Ruts (1996) proposed an $O(n^2)$ algorithm for $\mathbb{R}^3$, but some missing details may not be resolvable (Cheng and Ouyang, 2001). For space higher than 4 dimensional, there are no known algorithms faster than the straightforward $O(n^{d+1})$ method (generate all simplices and count the number of containments). The depth of all n points in $\mathbb{R}^2$ can be computed in $O(n^2)$ using the duality transform (Gil et al., 1992; Khuller and Mitchell, 1990). The *simplicial median* is the point with the highest simplicial depth. Aloupis et al. (2003) showed that in 2 dimensions it can be computed in $O(n^4)$. Their method was slightly improved by Burr et al. (2003).

No known algorithm exists for the computation of simplicial depth contours, apart from the straightforward one. Recently an approximation method, using local information about the depth function (using a discretized version of the gradient, the vector where simplicial depth of positions is increasing most rapidly) was proposed by Burr et al. (2003).

4. Future Work

Collaborations between computational geometers and statisticians on data-depth measures have produced more efficient algorithms and implementations with significant speedup. Some problems are solved in theory by algorithms with efficient asymptotic running times (see Sections 3.1.2, 3.2.2) where, in practice, the hidden constants make implementations infeasible, prompting the need for continued research and implementable solutions. In addition, most of the algorithms and their implementations work for the low dimensional case. High dimensional data sets pose more interesting research questions. Exact computation of the depth functions will probably never be efficient enough, as the complexity of the algorithm is in most cases exponential in the dimension of the data. This calls for more approximation algorithms, that can compute depth functions and depth contours efficiently, with provable error bounds.

5. Summary

The computational geometry community is thirsty for new and challenging open problems with real applications; the statistical community clearly needs improved computation speed in order to handle the large data sets of today. Much is to be gained by increased collaboration and solutions that are not only provably efficient but also effective in practice.

References

[1] G. Aloupis, C. Cortes, F. Gomez, M. Soss, and G. Toussaint, *Lower Bounds for Computing Statistical Depth.* Computat. Statist. Data Anal. **40** (2002), 223–229.

[2] G. Aloupis, S. Langerman, M. Soss, and G. Toussaint, *Algorithms for Bivariate Medians and a Fermat-Torricelli Problem for Lines.* Comput. Geom. **26** (2003), 69–79.

[3] V. Barnett, *The Ordering of Multivariate Data.* J. Roy. Statist. Soc. Ser. A **139** (1976), 318–355.

[4] J.L. Bentley and T.A. Ottmann, *Algorithms for Reporting and Counting Geometric Intersections.* IEEE Trans. Comput. **C-28** (1979), 643–647.

[5] J.L. Bentley and M.I. Shamos, *A Problem in Multivariate Statistics: Algorithm, Data Structure, and Applications.* In Proc. 15th Allerton Conf. Commun. Control Comput., pages 193–201, 1977.

[6] K.Q. Brown, *Geometric Transforms for Fast Geometric Algorithms.* Ph.D. thesis, Dept. Comput. Sci., Carnegie-Mellon Univ., Pittsburgh, PA, Report CMU-CS-80-101, 1980.

[7] K.Q. Brown, *Comments on "Algorithms for Reporting and Counting Geometric Intersections".* IEEE Trans. Comput. **C-30** (1981), 147–148.

[8] M. Burr, E. Rafalin, and D.L. Souvaine, *Simplicial Depth: An Improved Definition, Analysis, and Efficiency for the Discrete Case.* DIMACS technical report 2003-28, Rutgers University–New Brunswick, 2003, available at *http://dimacs.rutgers.edu/TechnicalReports/2003.html.*

[9] A.Y. Cheng and M. Ouyang, *On Algorithms for Simplicial Depth.* In Proc. 13th Canad. Conf. Comput. Geom., pages 53–56, 2001.

[10] K.L. Clarkson, D. Eppstein, G.L. Miller, C. Sturtivant, and S.-H. Teng, *Approximating Center Points with Iterated Radon Points.* Int. J. Comput. Geom. Appl. **6** (1996), 357–377.

[11] R. Cole, M. Sharir, and C.K. Yap, *On k-hulls and Related Problems.* SIAM J. Comput. **15** (1987), 61–77.

[12] M. de Berg, M. van Kreveld, M. Overmars, and O. Schwarzkopf, *Computational Geometry: Algorithms and Applications.* Springer-Verlag, Berlin, 1997.

[13] D.P Dobkin and D.L. Souvaine, *Advances in Robotics 1: Algorithmic and Geometric Aspects of Robotics*, chapter Computational Geometry – A User's Guide, pages 43–93. Lawrence Erlbaum Associates, 1987.

[14] W. Eddy, *Convex Hull Peeling.* In H. Caussinus, editor, COMPSTAT, pages 42–47. Physica-Verlag, Wien, 1982.

[15] H. Edelsbrunner, *Algorithms in Combinatorial Geometry.* Springer-Verlag, Berlin, 1987.

[16] H. Edelsbrunner and L.J. Guibas, *Topologically Sweeping an Arrangement.* J. Comput. System Sci. **38** (1989), 165–194.

[17] H. Edelsbrunner and D.L. Souvaine, *Computing Median-of-Squares Regression Lines and Guided Topological Sweep.* J. Amer. Statist. Assoc. **85** (1990), 115–119.

[18] K. Fukuda and V. Rosta, *Exact Parallel Algorithms for the Location Depth and the Maximum Feasible Subsystem Problems.* In C.A. Floudas and P.M. Pardalos, editors, Frontiers in Global Optimization. Kluwer Academic Publishers, 2003.

[19] J. Gil, W. Steiger, and A. Wigderson, *Geometric Medians.* Discrete Math. (Topological, algebraical and combinatorial structures. Frolík's memorial volume.) **108** (1992), 37–51.

[20] J. Hodges, *A Bivariate Sign Test.* Ann. Math. Statist. **26** (1955), 523–527.

[21] S. Jadhav and A. Mukhopadhyay, *Computing a Centerpoint of a Finite Planar Set of Points in Linear Time.* Discrete Comput. Geom. **12** (1994), 291–312.

[22] T. Johnson, I. Kwok, and R. Ng, *Fast Computation of 2-Dimensional Depth Contours.* In Proc. 4th Int. Conf. on Knowledge Discovery and Data Mining, pages 224–228, 1998.

[23] S. Khuller and J.S.B. Mitchell, *On a Triangle Counting Problem.* Inform. Process. Lett. **33** (1990), 319–321.

[24] S. Krishnan, N.H. Mustafa, and S. Venkatasubramanian, *Hardware-Assisted Computation of Depth Contours.* In 13th ACM-SIAM Symposium on Discrete Algorithms, 2002.

[25] S. Langerman and W. Steiger, *Optimization in Arrangements.* Proceedings of the 20th International Symposium on Theoretical Aspects of Computer Science (STACS 2003), 2003.

[26] R. Liu, *On a Notion of Data Depth Based on Random Simplices.* Ann. Statist. **18** (1990), 405–414.

[27] J. Matoušek, *Computing the Center of Planar Point Sets.* DIMACS Ser. Discrete Math. Theoret. Comput. Sci. **6** (1991), 221–230.

[28] K. Miller, S. Ramaswami, P. Rousseeuw, T. Sellarés, D. Souvaine, I. Streinu, and A. Struyf, *Efficient Computation of Location Depth Contours by Methods of Combinatorial Geometry.* Statist. Comput. **13** (2003), 153–162.

[29] D.M. Mount, N.S. Netanyahu, K.R. Romanik, R. Silverman, and A.Y. Yu, *A Practical Approximation Algorithm for the LMS Line Estimator.* In Proc. 8th ACM-SIAM Sympos. Discrete Algorithms, pages 473–482, January 1997.

[30] N. Naor and M. Sharir, *Computing a Point in the Center of a Point Set in Three Dimensions.* In Proc. 2nd Canad. Conf. on Computational Geometry, pages 10–13, 1990.

[31] H. Oja, *Descriptive Statistics for Multivariate Distributions.* Statist. Probab. Lett. **1** (1983), 327–332.

[32] E. Rafalin, D. Souvaine, and I. Streinu, *Topological Sweep in Degenerate Cases.* In Proc. 4th Workshop on Algorithm Engineering and Experiments (ALENEX), volume 2409 of Lect. Notes Comput. Sci., pages 577–588. Springer-Verlag, Berlin, 2002.

[33] P.J. Rousseeuw, *Least Median of Squares Regression.* J. Amer. Statist. Assoc. **79** (1984), 871–880.

[34] P.J. Rousseeuw and M. Hubert, *Regression Depth* (with discussion). J. Amer. Statist. Assoc. **94** (1999), 388–433.

[35] P.J. Rousseeuw and I. Ruts, *Bivariate Location Depth.* J. Roy. Statist. Soc. Ser. C **45** (1996), 516–526.

[36] P.J. Rousseeuw and I. Ruts, *Constructing the Bivariate Tukey Median.* Statist. Sinica **8** (1998), 827–839.

[37] P.J. Rousseeuw and A. Struyf, *Computing Location Depth and Regression Depth in Higher Dimensions.* Statist. Comput. **8** (1998), 193–203.

[38] I. Ruts and P.J. Rousseeuw, *Computing Depth Contours of Bivariate Point Clouds.* Comput. Statist. Data Anal. **23** (1996), 153–168.

[39] M.I. Shamos, *Geometry and Statistics: Problems at the Interface.* In J.F. Traub, editor, Algorithms and Complexity: New Directions and Recent Results, pages 251–280. Academic Press, New York, NY, 1976.

[40] M.I. Shamos, *Computational Geometry.* Ph.D. thesis, Dept. Comput. Sci., Yale Univ., New Haven, CT, 1978.

[41] K. Singh, *On the Majority Data Depth.* Technical report, Department of Statistics, Rutgers University, 1993.

[42] D.L. Souvaine and J.M. Steele, *Time- and Space-Efficient Algorithms for Least Median of Squares Regression.* J. Amer. Statist. Assoc. **82** (1987) 794–801.

[43] A. Struyf and P. Rousseeuw, *High-Dimensional Computation of the Deepest Location.* Comput. Statist. Data Anal. **34** (2000), 415–426.

[44] J.W. Tukey, *Mathematics and the Picturing of Data.* In Proc. Int. Congress of Mathematicians, Vancouver, Vol. 2, pages 523–531. Canad. Math. Congress, Montreal, Que., 1975.

[45] K. Verbarg, *Approximate Center Points in Dense Point Sets.* Inform. Process. Lett. **61** (1997), 271–278.

[46] Y. Zuo and R. Serfling, *General Notions of Statistical Depth Function.* Ann. Statist. **28** (2000), 461–482.

E. Rafalin and D.L. Souvaine
Department of Computer Science
Tufts University
Medford, MA 02155
USA
e-mail: {erafalin, dls}@cs.tufts.edu

Statistics for Industry and Technology, 297–306

Partial Mixture Estimation and Outlier Detection in Data and Regression

D.W. Scott

Abstract. The covariance matrix is a key component of many multivariate robust procedures, whether or not the data are assumed to be Gaussian. We examine the idea of robustly fitting a mixture of multivariate Gaussian densities in the situation when the number of components estimated is intentionally too few. Using a minimum distance criterion, we show how useful results may be obtained in practice. Application areas are numerous, and examples will be provided.

Mathematics Subject Classification (2000). Primary 62F35; Secondary 62H12.

Keywords. Mixture models, covariance estimation, minimum distance estimation.

1. Introduction

The problem of identifying and handling outliers is an important problem in data analysis; see Barnett and Lewis (1994) and Huber (1981), for example. Finding outliers through interactive graphical exploration of small multivariate sets can be accomplished using Swayne, Cook, and Buja's XGobi system (1998); however, with very large medium-dimensional data sets nonparametric density algorithms such as Scott's ASH (1992) can be recommended. Of course, purely graphical and nonparametric approaches to identifying outliers are prone to errors since many smooth densities give rise to data that may appear to have outliers, but do not. The Cauchy distribution is one example of such a density.

Thus the identification of outliers without an explicit probability model should always be viewed as preliminary and exploratory. If a probability model is known, then the tasks of parameter estimation and outlier identification can be more rigorously defined. However, even probability models are usually known only

This work was supported in part by NSF grant number DMS 02-04723 and NSF Digital Government contract number EIA-9983459.

approximately at best, and hence outliers so identified are still subject to certain biases.

The assumption of a multivariate normal model is most common. Often data may be transformed so that normality holds approximately. Successful robust estimation of the mean and covariance allows one to tag outliers more than three or four standard units from the mean. Finding the correct shape of the covariance matrix is the more challenging and interesting task. This shape can be sought without assuming normality. The minimum volume ellipse (MVE) has been investigated (Rousseeuw, 1985; Rousseeuw and Leroy, 1987; Poston et al., 1997; Cook et al., 1993; Woodruff and Rocke, 1994). Finding the MVE exactly is a challenging combinatorial problem but reasonably good approximate solutions exist. The MVE approach is especially appealing when the number of outliers is nontrivial and/or the outliers themselves form clusters.

Another approach that can be quite successful in this setting was proposed by Aitkin and Wilson (1992), who investigated fitting a gaussian mixture model for $\mathbf{x} \in \mathbb{R}^p$

$$f(\mathbf{x}) = \sum_{k=1}^{K} w_k \phi(\mathbf{x}|\mu_k, \Sigma_k) \tag{1.1}$$

using the expectation-maximization (EM) algorithm (Dempster et al., 1980). This model is often used for other purposes such as cluster analysis (McLachan and Peel, 2001). Here we treat the outliers as members of "nuisance" clusters. Once successful estimation is achieved, we re-order the labels so that the cluster weights are in decreasing magnitude, $w_1 > w_2 > \cdots > w_K$. Then in many situations, we may view the smallest $K-1$ clusters as representing various kinds of outliers and outlying clusters, with (w_1, μ_1, Σ_1) being the parameters of interest. For example, the fraction of outliers would be estimated as being $1 - w_1$.

Getting the EM algorithm to work requires a number of steps. Choosing K is especially tricky since many of the outliers may be singletons. Even more difficult is getting good initial guesses for the mixture parameters, especially when K is much larger or smaller than the "correct" model. Finally, small clusters cannot support estimation of a full covariance matrix, so that special structure may be assumed, such as a diagonal form or pooling of covariance information. Nevertheless, as a conceptual framework for handling large numbers of outliers in a range of challenging situations, the mixture model has great intuitive appeal.

A practical concern with the Aitkin and Wilson approach is that while a normality assumption may make good sense for the "good" data, knowledge of the distribution of the outliers and outlier clusters is more suspect. Since the primary parameters of interest are (w_1, μ_1, Σ_1), a procedure which estimates only that component would be of great interest. In one sense, the MVE approach is capable of this task. In this paper, we propose an alternative estimation approach to EM that can also estimate only a subfraction of the components of the mixture model. Applications are numerous.

2. Mixture Estimation by Minimum Distance

Practical algorithms for estimating parameters in a model $f(x|\theta)$ by minimum distance have been discussed by Beran (1977), for example, by minimizing Hellinger distance between the model and a kernel density estimate. Such an indirect approach is not feasible for a parameter-rich model such as the mixture model (Scott, 1999). One distance criterion, integrated squared error (ISE), affords better numerical properties. ISE has been the criterion of choice for nonparametric function estimation (Scott, 1992), and Rudemo (1982) showed how to formulate a cross-validated estimate of ISE for histograms using a leave-one-out approach. Terrell (1990) first proposed using the ISE as an alternative to maximum likelihood, with extensions by his student Kim (1995). The use of ISE for parametric estimation has been described (Hjort, 1994; Scott, 1999; Scott and Szewczyk, 2001; Wojciechowski and Scott, 1999; Scott, 2001). A more general divergence criterion (which includes ISE) has been described by Basu et al. (1998). Mixture estimation by ISE has been discussed in Scott (1999).

In the usual presentation, we seek to find the true parameter, θ_0, from the parametric model, $f(x|\theta)$. In our case, our estimate will be of the form $f(x|\hat{\theta})$, but the true density, $g(x)$, will not (necessarily) be from the parametric family, $f(x|\theta)$. Nevertheless, we seek to find the value of θ such that $f(x|\theta)$ is closest to $g(x)$ in the sense of integrated squared error.

$$\begin{aligned}\hat{\theta} &= \arg\min_\theta \left[\int [f(x|\theta) - g(x)]^2 \, dx \right] \\ &= \arg\min_\theta \left[\int f(x|\theta)^2 dx - 2 \int f(x|\theta) g(x) dx + \int g(x)^2 dx \right] \\ &= \arg\min_\theta \left[\int f(x|\theta)^2 dx - 2E\left[f(X|\theta)\right] \right],\end{aligned}$$

since $\int g(x)^2 dx$ is a constant in the second line with respect to choice of θ, and where X is a random variable from the true density, $g(x)$. Now for many models (including the normal mixture model), the first integral can be computed in closed form for any value of the parameter vector, θ. Given a random sample, an unbiased estimate of $E[f(X|\theta)]$ is the mean. Thus a completely data-based version of the ISE criterion is given by

$$\hat{\theta} = \arg\min_\theta \left[\int f(x|\theta)^2 dx - \frac{2}{n} \sum_{i=1}^{n} f(x_i|\theta) \right] . \tag{2.1}$$

In practice, this nonlinear optimization problem falls in the well-studied class of M-estimators (Hjort, 1994; Basu et al., 1998; Scott, 2001), and $\hat{\theta}$ is often asymptotically normal. Scott (2001) calls the value of $\hat{\theta}$ which minimizes (2.1) the *L2E* estimator, since integrated squared error and the L_2 distance are equivalent.

A careful examination of the argument leading up to (2.1) reveals that the density model, $f(x|\theta)$, need not be a density function, whereas the fact that $g(x)$ is a density was critical in estimating $E[f(X|\theta)]$. Thus we may consider an incomplete mixture model for $f(x|\theta)$. The simplest such model is the multivariate partial density component (MPDC) given by

$$f(\mathbf{x}|\theta) = w\,\phi(\mathbf{x}|\mu, \Sigma)\,, \tag{2.2}$$

where $\theta = (w, \mu, \Sigma)$. For the MPDC (2.1) is given explicitly by

$$\hat{\theta} = \arg\min_\theta \left[w^2\,\phi(0|0, 2\Sigma) - \frac{2w}{n} \sum_{i=1}^{n} \phi(\mathbf{x}_i|\mu, \Sigma) \right]. \tag{2.3}$$

Splus code that finds θ is available from the author.

3. Bivariate Examples

A thorough simulation study of robust properties often requires a book; see the Princeton robustness study (Andrews et al., 1972), for example. Here we show a sample of bivariate examples of normal mixtures where the number of components, K, ranges from two to five. We try to avoid any hidden assumptions, such as symmetry of outliers. There are no singleton outliers. In fact, most outliers are in clusters with at least 50 or 100 points. The main central cluster has 500 or 1000 points.

In the first five and eighth examples, the "true" mixture component is located at the origin, $\mu = (0,0)^T$, with $\Sigma = I_2$, the identity matrix. In the bivariate case, the parameter estimated is six-dimensional:

$$\theta = (w, \mu_x, \mu_y, \sigma_{xx}, \sigma_{xy}, \sigma_{yy})^T\,.$$

The iteration was started with the maximum likelihood estimates of the parameters

$$\theta^{(0)} = (1, \bar{x}, \bar{y}, s_x^2, r s_x s_y, s_y^2)^T\,.$$

Both initial and final estimates of θ are displayed graphically in each figure as one-sigma ellipses.

Here are the particulars of the mixture samples for the four examples displayed in Figure 1:

(i) $K = 2$, $n_1 = 500$, $n_2 = 100$, $\mu_2 = (5,5)^T$, $\Sigma_2 = I_2$;
(ii) $K = 3$, $n_1 = 500$, $n_2 = n_3 = 100$, $\mu_2 = (5,5)^T$, $\mu_3 = -\mu_2$, $\Sigma_2 = \Sigma_3 = I_2$;
(iii) $K = 2$, $n_1 = 500$, $n_2 = 100$, $\mu_2 = (0,0)^T$, $\Sigma_2 = 25\,I_2$;
(iv) $K = 2$, $n_1 = 500$, $n_2 = 100$, $\mu_2 = (7,7)^T$, $\Sigma_2 = 9\,I_2$.

In the first frame of Figure 1, 100 points are centered at $(5,5)^T$. The estimated value $\hat{w} = 81.5\%$, which is very close to 500/600. Notice the lack of any visual correlation in the final estimate $\hat{\Sigma}_1$.

In the second frame of Figure 1, two groups of 100 outlying points are centered at $(5,5)^T$ and $(-5,-5)^T$. The estimated value $\hat{w} = 68.3\%$, which is very close to 500/700. Again, the apparent correlation in the initial maximum likelihood estimate of Σ is correctly missing in the $L2E$ estimate.

In the third frame of Figure 1, the 100 outliers share the same center but their scale is 5 times larger. Given the overlap of "outliers" and "inliers," $\hat{w}$ is biased upwards. However, observe how $\hat{\Sigma}_1$ is much closer to I_2.

In the fourth frame of Figure 1, which is a hybrid of frames one and three, the 100 outlying points are centered at $(7,7)^T$ and their scale is $9\,I_2$. The estimated value $\hat{w} = 84.1\%$, which is very close to $500/600$. Notice the lack of any correlation in the final estimate of $\hat{\Sigma}_1$.

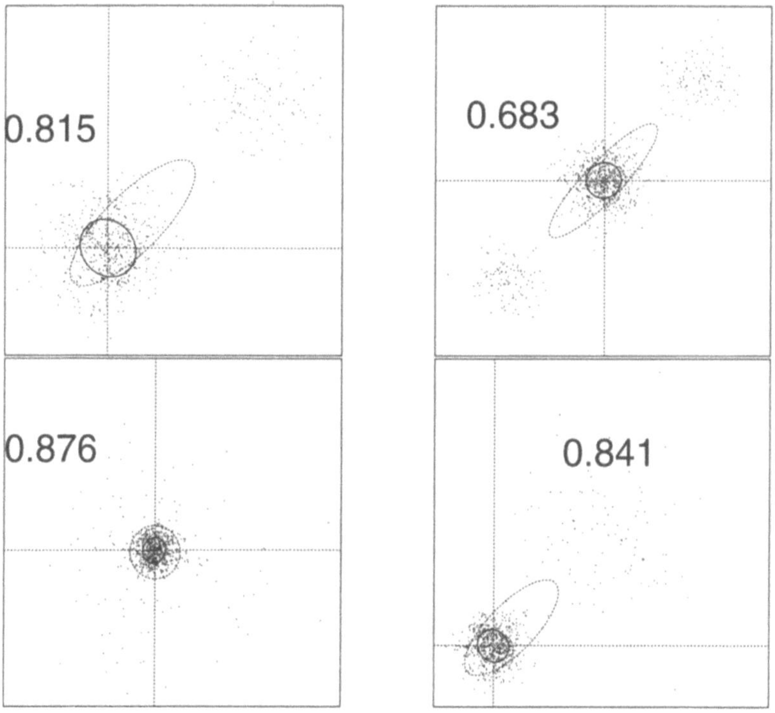

FIGURE 1. Multivariate partial density component estimation examples. The maximum likelihood one-sigma ellipse is shown as a dotted line, while the $L2E$ ellipse is shown as a thick solid line. $\hat{w}$ is also given. The origin is at the intersection of the two axes. See text for true parameter values.

Our second set of examples are displayed in Figure 2. In two of these, the true correlation coefficient of the main cluster is $\rho = 0.7$. Here are the particulars of the mixture samples for the four examples displayed in Figure 2:

(v) $K = 4$, $n_1 = 1000$, $n_2 = n_3 = n_4 = 100$, $\mu_2 = (5,5)^T$, $\mu_3 = -\mu_2$, $\mu_4 = (5,-5)^T$, $\Sigma_2 = \Sigma_3 = \Sigma_4 = I_2$;

(vi) $K = 2$, $n_1 = 500$, $\rho_1 = 0.7$, $n_2 = 100$, $\mu_2 = (5,5)^T$, $\Sigma_2 = I_2$;

(vii) $K = 2$, $n_1 = 500$, $\rho_1 = 0.7$, $n_2 = 100$, $\mu_2 = (5, -5)^T$, $\Sigma_2 = I_2$;

(viii) Multiple starting values with $K = 5$, $n_1 = 1000$, $n_2 = 50$, $n_3 = 100$, $n_4 = 150$, $n_5 = 200$, $\mu_2 = (5, 5)^T$, $\mu_3 = (-5, -5)^T$, $\mu_4 = (5, -5)^T$, $\mu_5 = (-5, 5)^T$, $\Sigma_2 = \Sigma_3 = \Sigma_4 = \Sigma_5 = I_2$.

In the first frame of Figure 2, three outlying clusters of 100 points each surround 1000 points at the origin. The estimated value $\hat{w} = 74.9\%$, which is very close to 1000/1300.

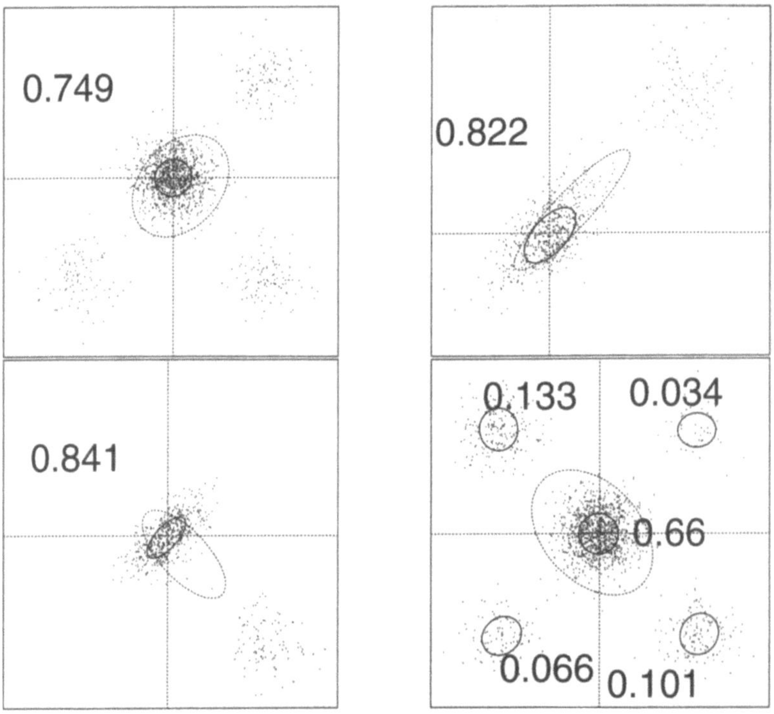

FIGURE 2. Second set of multivariate partial density component estimate examples. In the final frame, solutions for five different initial guesses of θ are displayed. See text for true θ values.

In the second frame of Figure 2, the 500 center points have a correlation of $\rho = 0.7$. The group of 100 outlying points are centered at $(5, 5)^T$, which is in line with the main axis of the covariance matrix. Here, the apparent correlation in the initial maximum likelihood estimate of Σ is correctly retained in the $L2E$ estimate, and the center is correctly estimated, too.

In the third frame of Figure 2, the main cluster is identical, but now the 100 outliers are centered at $(5, -5)^T$. Thus the initial covariance estimate shows a strong, but negative, correlation, when the true correlation is strong, but positive. The $L2E$ estimates properly reorient the covariance matrix.

The fourth frame of Figure 2 is similar to the first, but with four outlying clusters (of different size 50, 100, 150, and 200). Starting from the maximum likelihood estimates, a $L2E$ MPDC component of size $\hat{w} = 0.66$ is found (actual value is 1000/1500).

Of special interest are the four other $L2E$ MPDC estimates displayed. These were found using different starting values for θ that were closer to the true parameters of each outlying cluster. This confirms the earlier suggestion that the "size" of a component is not as important as how separated the component is from the remaining data. The smallest cluster found (in one of five runs) has $\hat{w} = 0.034$, which is very close to 50/1500.

We have run a number of limited experiments in higher dimensions. The algorithm initially failed to converge on the Fisher Iris data. Closer examination revealed the reason to be that the data only had one decimal point. Adding a small amount of random noise (blurring) fixed the convergence problem.

Examples similar to those in Figure 1 were run in dimensions up to seven, with full covariance estimation. In seven dimensions, that amounts to $1 + 7 + 28 = 36$ quantities in the parameter vector, θ. An unfortunate aspect of the choice of integrated squared error is that ISE is not dimensionless (as is Hellinger distance, L_1 distance, maximum likelihood). The practical implication of this fact is that numerical optimization does not behave well due to scaling issues in these dimensions. Gaining a better understanding of this phenomenon and extending the range of dimensions where $L2E$ may be applied should be a valuable area of research.

4. Regression

The extension of $L2E$ to regression

$$y_i = \mathbf{x}_i^T \beta + \epsilon_i$$

is described and illustrated in Scott (2001). The algorithm is driven by a normal assumption on the residuals, $\epsilon_i \sim N(0, \sigma_\epsilon^2)$. In our experience, we often see clusters of outliers in the regression plot, or even multiple regression curves mixed together. Here we briefly describe the extension of the MPDC approach to regression. The beauty of the regression setting is that the error distribution is univariate even with p predictor variables. Thus the MPDC is given by this assumption for the error variable:

$$\epsilon_i \sim w \cdot N(0, \sigma_\epsilon^2)\,.$$

The parameter vector estimated is $\theta = (w, \beta, \sigma_\epsilon)^T$, which is of length $p + 3$, assuming the parameter vector β includes an intercept term, β_0.

We illustrate an application to a well-known set of data first described by Harrison and Rubinfeld (1978) on the median house value in census tracts in Boston, following the usual transformation of the $p = 13$ variables. The residuals from the least-squares fit were smoothed using a kernel estimate, as shown in the first frame of Figure 3. Also shown is a $N(0, \sigma_\epsilon^2)$ curve, with residual variance estimated in the usual manner. The residual distributions are not too close, but only a few outliers are apparent.

We next fit the $L2E$ model with noise distribution $w \cdot N(0, \sigma_\epsilon^2)$. The kernel smooth of the residuals and the estimated MPDC are both drawn in the right frame of Figure 3. Here, $\hat{w} = 84.5\%$. Interpreting and diagnosing the fit is much easier with the $L2E$ fit. For example, while there are some outliers on the low end, almost 14% of the outliers appear on the high side. In a more complete analysis where the outlying census tracts are interactively displayed on maps, we can learn that these tracts fall in certain regions (some along the Charles River) and not at random; see Scott and Christian (2003).

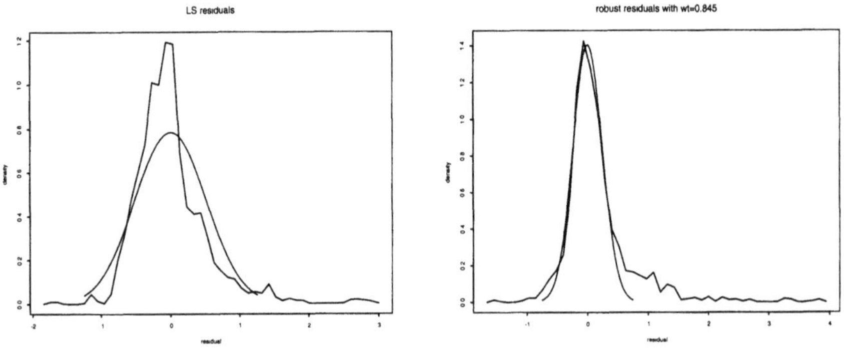

FIGURE 3. Kernel estimate of estimated residuals and fitted normal error density for the least squares fit (left frame) and the $L2E$ fit (right frame), for the Boston housing data set.

5. Discussion

The PMDC approach is useful for outlier detection in many situations, and for clustering in particular; see Scott and Szewczyk (2004). The concept of breakdown is more complicated in this setting, as the algorithm is local. That is, the algorithm converges to local normal clusters for certain ranges of initial parameter settings. We have carefully examined the attractiveness of a single univariate component as a function of the parameters of an adjacent component. Generally, if the components overlap to a significant amount, the algorithm may not converge to either of the separate components, but rather to a single large component.

We have found a way to overcome this limitation by optimizing over a subset of the MPDC parameters. In Scott and Szewczyk (2004), numerous random initial guesses are used and the collection of parameter estimates, θ, is clustered to find the most common solutions. We take these to be cluster locations.

Models other than normal may be chosen, however, the closed form expression of the $L2E$ criterion is very convenient for optimization. If the main cluster is not approximately normal, the $L2E$ solution may find the cluster, but the estimated parameters will vary depending upon the degree of non-normality.

There are numerous other extensions which we only touch upon here. For example, the MPDC may be formulated with more than one mixture component, or sequentially with some fixed mixture components, but then getting good initial estimates for θ becomes more difficult. Other regression applications may be formulated, including image processing tasks. We describe these separately.

References

[1] M. Aitkin and G.T. Wilson, *Mixture Models, Outliers, and the EM Algorithm.* Technometrics **22** (1980), 325–331.

[2] D.F. Andrews, P.J. Bickel, F.R. Hampel, P.J. Huber, W.H. Rogers, and J.W. Tukey, *Robust Estimates of Location: Survey and Advances.* Princeton University Press, Princeton, 1972.

[3] V. Barnett and T. Lewis, *Outliers in Statistical Data.* John Wiley & Sons, New York, 1994.

[4] A. Basu, I.R. Harris, H.L. Hjort, and M.C. Jones, *Robust and Efficient Estimation by Minimising a Density Power Divergence.* Biometrika **85** (1998), 549–560.

[5] R. Beran, *Robust Location Estimates,* Ann. Statist. **5** (1977), 431–444.

[6] R.D. Cook, D.M. Hawkins, and S. Weisberg, *Exact Iterative Computation of the Robust Multivariate Minimum Volume Ellipsoid Estimator.* Statist. Probab. Lett. **16** (1993), 213–218.

[7] A.P. Dempster, N.M. Laird, and D.B. Rubin, *Maximum Likelihood from Incomplete Data via the EM Algorithm.* J. Roy. Statist. Soc. Ser. B **39** (1977), 1–22.

[8] D. Harrison and D.L. Rubinfeld, *Hedonic Housing Prices and the Demand for Clean Air.* J. Environ. Econ. Manag. **5** (1978), 81–102.

[9] H.L. Hjort, *Minimum L2 and Robust Kullback-Leibler Estimation.* In P. Lachout and J.Á. Víšek, editors, Proceedings of the 12th Prague Conference on Information Theory, Statistical Decision Functions and Random Processes, pages 102–105. Prague Academy of Sciences of the Czech Republic, 1994.

[10] P.J. Huber, *Robust Statistics.* John Wiley & Sons, New York, 1981.

[11] D. Kim, *Least Squares Mixture Decomposition Estimation.* Unpublished doctoral dissertation, Department of Statistics, Virginia Tech, Blacksburg, VA, 1995.

[12] G.J. McLachlan and D. Peel, *Finite Mixture Models.* John Wiley & Sons, New York, 2001.

[13] W.L. Poston, E.J. Wegman, C.E. Priebe, and J.L. Solka, *A Deterministic Method for Robust Estimation of Multivariate Location and Shape,* J. Comput. Graph. Statist. **6** (1997), 300–313.

[14] P.J. Rousseeuw, *Multivariate Estimation with High Breakdown Point.* In W. Grossmann, G. Pflug, I. Vincze, and W. Wertz, editors, Mathematical Statistics and Applications, Vol B, pages 283–297. Reidel Publishing Company, Dordrecht, 1985.

[15] P.J. Rousseeuw and A.M. Leroy, *Robust Regression and Outlier Detection,* John Wiley & Sons, New York, 1987.

[16] M. Rudemo, *Empirical Choice of Histogram and Kernel Density Estimators.* Scand. J. Statist. **9** (1982), 65–78.

[17] D.W. Scott, *Multivariate Density Estimation: Theory, Practice, and Visualization.* John Wiley & Sons, New York, 1992.

[18] D.W. Scott, *Remarks on Fitting and Interpreting Mixture Models.* In K. Berk and M. Pourahmadi, editors, Computing Science and Statistics **31**, pages 104–109, 1999.

[19] D.W. Scott, *Parametric Statistical Modeling by Minimum Integrated Square Error.* Technometrics **43** (2001), 274–285.

[20] D.W. Scott and J.B. Christian, *Finding Outliers in Models of Spatial Data.* In E. Hovy, editor, Proceedings of the Third National Conference on Digital Government. Digital Government Research Center, Boston, 2003, available at *http://www.dgrc.org/dgo2003/start.html.*

[21] D.W. Scott and W.F. Szewczyk, *From Kernels to Mixtures.* Technometrics **43** (2001), 323–335.

[22] D.W. Scott and W.F. Szewczyk, *The Stochastic Mode Tree and Clustering.* J. Comput. Graph. Statist. **12** (2004), in press.

[23] D.F. Swayne, D. Cook, and A. Buja, *XGobi: Interactive Dynamic Data Visualization in the X Window System.* J. Comput. Graph. Statist. **7** (1998), 113–130.

[24] G.R. Terrell, *Linear Density Estimates.* In Proceedings of the Statistical Computing Section of the Americ. Statist. Assoc., pages 297–302, 1990.

[25] W.C. Wojciechowski and D.W. Scott, *Robust Location Estimation with L2 Distance.* In K. Berk and M. Pourahmadi, editors, Computing Science and Statistics **31**, pages 292–295, 1999.

[26] D.L. Woodruff and D.M. Rocke, *Computable Robust Estimation of Multivariate Location and Shape in High Dimension using Compound Estimators.* J. Amer. Statist. Assoc. **89** (1994), 888–896.

Acknowledgment

The author would also like to thank William Szewczyk and the Department of Defense for support and collaboration.

D.W. Scott
Department of Statistics, MS-138
Rice University, P.O. Box 1892
Houston, Texas 77251-1892
USA
e-mail: `scottdw@rice.edu` `http://www.stat.rice.edu/~scottdw`

Statistics for Industry and Technology, 307–318

Robust Fitting Using Mean Shift: Applications in Computer Vision

D. Suter and H. Wang

Abstract. Much of computer vision and image analysis involves the extraction of "meaningful" information from images using concepts akin to regression and model fitting. Applications include: robot vision, automated surveillance (civil and military) and inspection, biomedical image analysis, video coding, human-machine interface, visualization, historical film restoration etc.

However, problems in computer vision often have characteristics that are distinct from those usually addressed by the statistical community. These include *pseudo-outliers*: in a given image, there are usually several populations of data. Some parts may correspond to one object in a scene and other parts will correspond to other, rather unrelated, objects. When attempting to fit a model to this data, one must consider all populations as outliers to other populations – the term pseudo-outlier has been coined for this situation. Thus it will rarely happen that a given population achieves the critical size of 50% of the total population and, therefore, techniques that have been touted for their high breakdown point (e.g., Least Median of Squares) are no longer reliable candidates, being limited to a 50% breakdown point.

Computer vision researchers have developed their own techniques that perform in a robust fashion. These include RANSAC, ALKS, RESC and MUSE. In this paper new robust procedures are introduced and applied to two problems in computer vision: range image fitting and segmentation, and image motion estimation. The performance is shown, empirically, to be superior to existing techniques and effective even when as little as 5–10% of the data actually belongs to any one structure.

Mathematics Subject Classification (2000). Primary 68T45; Secondary 62F35.
Keywords. Robust Statistics, model fitting, computer vision, breakdown-point.

This work was completed with the support of the Australian Research Council: grant A10017082. in particular, the second author was supported by that grant. This support is gratefully acknowledged.

1. Introduction

Computer Vision involves the extraction of "meaningful" information from images. Subareas include: robot vision, automated surveillance (civil and military) and inspection. Moreover, the techniques involved in computer vision also find their way into a diverse range of other areas, such as biomedical image analysis, video coding, human-machine interface, visualization, historical film restoration etc.

Many of the tasks to be carried out, for the purposes of computer vision, can be cast as forms of statistical estimation and fitting. This has lead to a strong community interest in statistical methods - particularly robust statistical methods. Indeed, robust statistical methods in computer vision have a history, at least two decades old, using or adapting standard methods from the statistical community: for example, M-Estimators (Black and Rangarajan, 1996) and Least Median of Squares (Meer et al., 1991; Bab-Hadiashar and Suter, 1998). At least one novel technique, RANSAC (Fischler and Bolles, 1981) was developed by vision researchers in the early days (and is still a widely used technique).

However, problems in computer vision often have characteristics that are distinct from those addressed by the statistical community. These include:

- **Pseudo-outliers.** In a given image, there are usually several populations of data. Some parts may correspond to one object in a scene and other parts will correspond to other, rather unrelated, objects. When attempting to fit a model to this data, one must consider all populations as outliers to other populations - the term pseudo-outlier has been coined (Stewart, 1995). Thus it will rarely happen that a given population achieves the critical size of 50% of the total population and, therefore, techniques that have been touted for their high breakdown point (e.g., Least Median of Squares) are no longer reliable candidates from this point of view.

- **Unknown sizes of populations – and unknown location.** Computer vision requires fully automated analysis in, generally, rather unstructured environments. Thus, the sizes and locations of the populations involved, will fluctuate greatly. Moreover, there is no "human in the loop" to select regions of the image dominated by a single population, or to adjust various thresholds. In contrast, statistical problems studied in most other areas usually have a single dominant population plus some percentage of outliers (typically mis-recordings - not the pseudo-outliers mentioned above). Typically a human expert is there to assess the results (and, if necessary, crop the data, adjust thresholds, try another technique etc.).

- **Large data sizes.** Modern digital cameras exist with around 4 million pixels per image. Image sequences, typically at up to 50 frames per second, contain many images. Thus, computer vision researchers typically work with data sets in the tens of thousands of elements, *at least*, and data sets in the 10^6 and 10^9 range are not uncommon.

- **Emphasis on fast calculation.** Most tasks in computer vision must be performed "on-the-fly". Offline analysis that takes seconds, let alone minutes or hours, is usually a luxury afforded by relatively few applications.

These rather peculiar circumstances have lead computer vision researchers to develop their own techniques that perform in a robust fashion (perhaps "empirically-robust" should be used, as few have formal proved robust properties, though many trace their heritage to techniques that do have such proved properties). These include ALKS (Lee et al., 1998), RESC (Yu et al., 1994) and MUSE (Miller and Stewart, 1996). Results presented in those papers *suggest* that ALKS can be tolerant to at least 65% outliers, RESC to about 80% outliers, and MUSE to 55% outliers. However, it has to be admitted that a complete solution, addressing all of the above problems, is far from being achieved. Indeed, none of the techniques, with present hardware limitations, are really "real-time" when applied to the most demanding tasks. None have been *proved* to reliably tolerate high percentages of outliers and, indeed, we have found with our experiments that RESC and ALKS, although clearly better than Least Median of Squares, in this respect, are not always reliable. For recent surveys, see Meer et al. (2000) and Stewart (1999).

The purpose of this paper is to highlight recent research carried out by the authors, in developing techniques with robust behavior, to address various problems in computer vision – specifically, range image fitting and segmentation, and image motion estimation.

2. Basic Idea and Results

Because of the presence of multiple structures in the image, we need approaches that are robust to (pseudo)-outliers in the sense of having a high breakdown point. Established techniques, in use, that meet this criteria are based upon random sampling: e.g., Least Median of Squares, Least Trimmed Squares, and RANSAC. Random sampling techniques aim to explore the search space of possible solutions well enough to have at least one candidate which is determined solely by inliers (to a single structure in the data). However, since one does not have an "oracle" to tell us which of the candidates are unpolluted by outliers; we require some form of model/fit scoring. In Least Median of Squares, this is obviously the median of the residuals. In RANSAC, it is the number of data residuals inside a certain bound. Of course, each form of model scoring has potential weaknesses. In Least Median of Squares, the median of the residuals of the entire data set (with respect to the candidate model) will obviously not be a good measure if the inliers to that model contain less than 50% of the total data. Generalizing to the kth other statistic (rather than the Median) is one way out of this dilemma but now one either has to know in advance what value of k to use, or one has to attempt some form adaptation e.g., ALKS (which will perhaps be costly and limited in reliability). Even still, it is overly optimistic to expect a single statistic (the kth order residual, for Least Median of Squares and ALKS; or the number of inliers within a certain bound, for RANSAC) to be an entirely reliable/informative measure of the quality of a solution.

This observation has lead the authors to seek alternative ways of scoring candidate models, so that greater robustness may be achieved. An early attempt (Wang and Suter, 2002) employed possible symmetry in the data set as one such statistic: though somewhat limited in versatility, such an approach definitely restores robustness in situations where standard Least Median of Squares will break down. Our more recent estimators seek to use more information from the residual distribution. In particular, we have used Kernel Density estimation and Mean Shift Techniques (Fukunaga and Hostetler, 1975) to formulate model/fit scores that lead to empirically observed higher breakdown points than all existing methods.

2.1. Mean Shift and Kernel Density Estimation in Model Scoring

Consider a candidate model fit (as part of a random selection and trial fitting as in the usual Least Median of Squares algorithm). We seek to approximate the pdf of the residuals and find the mode of this distribution.

Although other kernels could be employed, we use the Epanechnikov kernel in its 1-D form:

$$K(x) = \begin{cases} \frac{3}{4}(1-x^2) & x^2 < 1 \\ 0 & \text{otherwise} \end{cases} \tag{2.1}$$

The kernel density estimator is then:

$$\hat{f}(x) = \frac{1}{nh}\sum_{i=1}^{n} K(\frac{x - X_i}{h}) \tag{2.2}$$

for a data set of n residuals X_i and using "bandwidth" h.

The mean shift is calculated by (Fukunaga and Hostetler, 1975):

$$\hat{f}(x) = M_h(x) = \frac{1}{n_x}\sum_{X_i \in S_h} (X_i - x) \tag{2.3}$$

where S_h is an interval of half-width h centered on x – the "mean shift window".

Simple calculations show that iterating the above will converge to a local maximum of the estimated pdf. Because of the limited extent of the mean shift window (in the particular case of the kernel we choose, there is an inherent limit to the spatial influence of a data point, coming from the finite support of the Kernel - although a window could be imposed on other kernels not having finite support), this process is reasonably insensitive to outlier residuals. Comaniciu and co-workers (Comaniciu et al., 2002) have recently popularized this mean-shift technique for various applications in computer vision (e.g., clustering). Here we use the method as a way of scoring candidate fits.

The basic notion is that the robust estimate should produce a strong peak in the pdf of the residuals for that fit, and that the value of the residual corresponding to that peak should be small (ideally zero, of course). We have experimented with several ways to encapsulate such notions and have found maximization of the following objective function performs well:

$$MPDE = \frac{\left(\sum_{X_i \in W_c} \hat{f}(X_i)\right)^{\alpha}}{\exp(|X_c|)} \tag{2.4}$$

where the mean shift procedure is used to find the mode of the residual density $\hat{f}$ and to limit mean estimation to using only "inliers" X_i (with center value X_c) within the mean-shift window W_c centered on the mode of the pdf. This method is similar to the Residual Concensus (RESC) method (Yu et al., 1994). Essentially that method estimates the pdf by using a histogram (whose bin size is chosen by compressing a large histogram of residuals using a heuristic procedure so that 12% of the residuals are found in the first bin). The RESC criterion is then:

$$RESC = \sum_{i}^{m} \frac{{h_i}^{\alpha}}{|r_i|^{\beta}} \tag{2.5}$$

where h_i is the histogram value and r_i the residual of the ith bin. The value m is chosen by another heuristic (4.4% of h_{max}), so as to exclude outliers. (Note: in Yu et al., 1994, $\alpha = 1.3$ and $\beta = 1$.) In essence, the scores given in equations 2.4 and 2.5 differ in how one restricts attention to likely inliers (we use the mean shift window) and how one models the pdf (we use kernel density estimation *and* mean shift mode seeking). These differences lead to significant improvements in robustness.

2.2. Examples

In this section we show that approaches based upon these procedures can tolerate up to 90% or so of outliers (including pseudo-outliers) and outperform previous computer vision methods (MUSE, RESC, ALKS, RANSAC) and more widely known methods (Least Median of Squares and Least Trimmed Squares) in that regard. We demonstrate this with experiments using synthetic data and with "real life" data in the areas of: line and circle finding, range image segmentation and image motion estimation.

2.2.1. Lines and Circle Finding. Various edge detection routines can be employed to try to find edges in a scene. Typically, these routines produce very noisy output (gaps in edges, isolated points etc. – see Figures 1 and 2) If one is seeking structures that have particular parametric forms (e.g., lines and circles) one can use regression type frameworks to find the structures of interest amongst the edge detector output.

Thus this group of problems is similar to the standard regression problem: although we acknowledge that one should consider using geometric distance, rather than residuals, in the minimization (Kanatani, 1996). However, here, we avoid the non-linear theory and resultant approximations of geometric fitting, and apply our procedure to the residuals produced by substituting the data points into the defining parametric form of an circle or line. Note: as opposed to standard settings for regression, we have *multiple* structures. To find *every line* we could find *any*

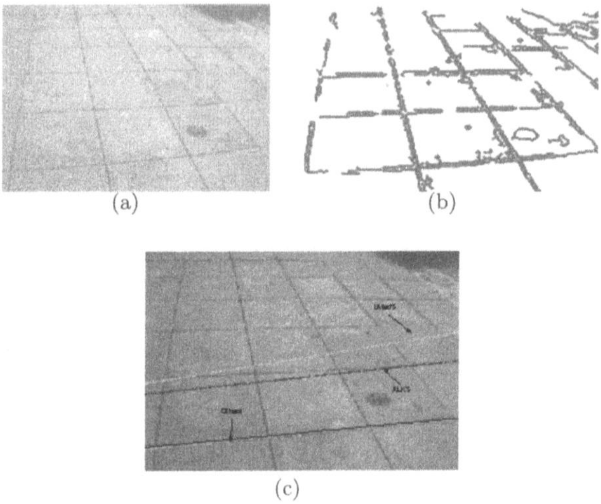

FIGURE 1. Line finding.

line, first, then find the inliers to that line, remove them from the pool of data and re-apply the procedure to find the next line. Here we just demonstrate the first stage – finding any valid structure (though in segmentation, section 3, we do find all of the structures sequentially.)

In computer vision, probably the most commonly used technique, for such problems, is the Hough Transform. Essentially, this transform discretizes the parameter space and each data point "votes" for the parameters it is consistent with. Therefore, it produces an estimate of the parameters with a finite precision which is determined by the "bin" size in parameter space. In this respect, it is unlike the other methods we consider: all of the other methods produce estimates with potentially infinite precision (limited by round-off errors, and, of course, in accuracy rather than precision). Moreover, since the parameter space has to be discretized, the Hough transform is simple to apply when the parameter space is compact or known limits apply (e.g., as imposed by the finite boundaries of an image) but is less simple when the parameter space has infinite extent or the practical bounds on likely encountered parameters are less clear. For these reasons, despite the early and continuing successes of such a technique, we prefer to seek improved methods. Nonetheless, it is instructive to compare the results with the Hough transform.

In addition to the Hough transform, we experiment with various established robust fitting methods Least Median of Squares, and several techniques not generally known outside of the computer vision community: RANSAC, ALKS, and RESC. An example can be found in Figure 1. The pavement image 1(a) is processed with an edge detector 1(b). After various forms of robust line fitting are applied, we obtain the results typically as shown in 1(c). Least Median of Squares (light upper line) fails because it is not robust to some many outliers. ALKS (mid-

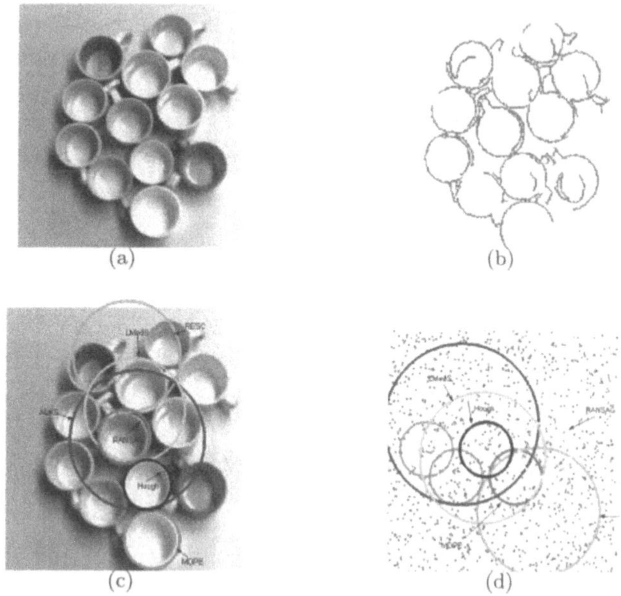

FIGURE 2. Cirle finding.

dle dark line) also fails. The remaining techniques (MDPE, RESC and Hough – lower dark line) all produce similar results on this example – as the discretization of the parameter space for the Hough transform is adequate, its limit of precision is not apparent.

Similarly, examples of circle fitting can be found in Figure 2. The cups image 2(a) is processed with an edge detector 2(b). After various forms of robust line fitting are applied, we obtain the results shown in 2(c). MDPE (finds bottom cup), Hough (finds cup above it) and RANSAC (finds next cup above) all produce robust fits. Least Median of Squares, ALKS and RESC all breakdown (larger circles in the figure). As a second example, a synthetic data set was created to resemble "olympic rings" and then artificially corrupted with uniformly distributed noise samples – see Figure 2(d). In this example, the data that are "inliers" to any one ring constitute only about 5% of the data – that is, for a given fit, there are around 95% outliers to that fit. MDPE (bottom right ring), RANSAC (top right ring) and the Hough transform (top middle ring) can all find a ring. However, Least Median of Squares, ALKS and RESC again breakdown. Note: it is not significant *which structure is found.*

These and other similar experiments we have performed, show that algorithms based upon our MDPE criterion outperform other robust techniques (Least Median of Squares, techniques based upon least kth order statistics such as ALKS, and the RESC approach). The technique is challenged for robustness only by RANSAC (which requires a priori knowledge of the expected sizes of the majority of the inlier residuals) and the non-regression (limited precision) Hough transform (which also requires a well-chosen bin size for the parameter discretization). We must acknowledge that almost no method has an implementation that is free of certain parameters that have to be set. In our experiments, where there are such parameters, we have tried first to use the settings advocated by the authors of those methods (if, indeed, they are available to us). We have also tried to vary those parameters and to select the best performance delivered by a method over the range of parameters investigated. We believe that this is as fair as we can be in our comparisons. The essential parameter required for our technique is the bandwidth of the kernel density estimator (although there are the inevitable minor bounds and tolerance parameters that plague code in order to guard against certain numerical limits, or to decide when to cease iteration). In the experiments reported in this section, we empirically investigated the behavior of our approach with varying bandwidth and found that overall performance, at least over a reasonable range of bandwidth choices, was reasonably stable. In the next section we present results where we have automatically chosen the bandwidth.

2.2.2. Optic Flow. Optic flow ((u, v)) is the apparent motion on the image plane caused by relative motion between an observer and the imaged points. If one assumes that the imaged point at position (x, y) and time t maintains a constant brightness $I(x, y, t)$, as it moves under the optic flow, then a simple differentiation reveals to first order:

$$\frac{\partial I}{\partial x}u + \frac{\partial I}{\partial y}v + \frac{\partial I}{\partial t} = 0 \tag{2.6}$$

– the so-called "Optic Flow Constraint" (see, e.g., Bab-Hadiashar and Suter, 1998). Since this constraint provides only one constraint in two unknowns, it has been solved by assuming, within a small image patch, that the flow is of a simple parametric form (e.g., locally constant, or locally affine). Such low order local polynomial approximations can be justified in the usual manner - providing the flow is locally smooth. In fact, it can be shown that if the scene contains rigid planar patches, then to a good approximation, the flow will be an 8 parameter quadratic

$$u = ax^2 + bxy + cx + dy + e; \quad v = by^2 + axy + fx + gy + h \tag{2.7}$$

(i.e., having less degrees of freedom than the full quadratic defined over a 2-D domain.)

Within each patch, we can measure the quantities $\frac{\partial I}{\partial x}$, $\frac{\partial I}{\partial y}$, and $\frac{\partial I}{\partial t}$ so that we can solve for the parameters of the assumed flow model. However, since the measurements can contain very large errors, and since the patches may straddle different flows (moving according to different parameters), we need methods of solution that are robust to large numbers of outliers (including pseudo-outliers). Robust statistical methods such as Least Median of Squares have been demonstrated as superior to Least Squares and other competing methods in this respect (Bab-Hadiashar and Suter, 1998).

TABLE 1. Yosemite Valley Sequence – Optic Flow.

Technique	Average Error
Bab-Hadiashar and Suter (1998)	1.94
QMDPE ($\sigma = 2.0$, 25×25, m=30)	1.34

We have applied our method, as described above but with some modifications, to the problem of optic flow estimation (Wang and Suter, 2003a). Firstly, we have implemented a form of automatic selection of the kernel bandwidth. Using standard techniques (Wand and Jones, 1995):

$$\hat{h} = \left[\frac{243R(K)}{35u_2(K)^2 n} \right]^{\frac{1}{5}} s \tag{2.8}$$

where $R(K) = \int_{-1}^{1} K(\xi)^2 \, d\xi$, $u_2(K) = \int_{-1}^{1} \xi^2 K(\xi) \, d\xi$ and s is the sample standard deviation. This still requires the estimation of the sample scale in a robust fashion and, since the above is the recommended upper bound on $\hat{h}$, we also have to employ a multiplicative adjustment $c\hat{h}$ for $0 < c < 1$. Space does not permit us to go into details here, but we have investigated, with some success, both the use of order statistics for scale estimation and a rather more novel scheme of using a mean-shift like procedure to also find the pdf valley (thus providing a useful classification of inliers as those between the peak and the valley in the pdf). We have also, for speed reasons, modified the MDPE criterion to use what we call QMDPE:

$$QMDPE = \frac{\left(\hat{f}\left(X_c\right)\right)^{\alpha}}{\exp\left(|X_c|\right)} \tag{2.9}$$

One sample of our results is provided in Table 1. To estimate the optic flow at a point, we center a rectangular patch on that point and solve using all optic flow constraints from that patch. The table reports an error measure (see Bab-Hadiashar and Suter, 1998, for a definition) averaged over every result. We compare against the closest (and to this point best) competing robust method, and to other methods available in the literature. Clearly QMDPE performs well.

3. Segmentation

Segmentation is the process of dividing the image (or a sequence of images such as a movie) into spatial and/or temporal structures of interest. In principle, one can use a parametric fitting method to sequentially find structures in the data, find the inliers to that fitted model, remove the inliers and repeat by fitting another model. Such approaches, using variants of Least kth order fitting have been shown to be reasonably promising (Bab-Hadiashar and Suter, 1999; Bab-Hadiashar and Suter, 2000). However, the implementation of such a relatively simple scheme becomes complicated by a number of practical issues that are beyond the scope of this paper and possibly of little interest to the intended audience – suffice to say that one can devise methods, based upon robust fitting schemes, that outperform traditional and other competing methods.

Consider for example, a range image, such as one collected via a laser rangefinder, or stereo camera, or structured lighting technique: ultimately, though these devices differ in their principles, they all produce an "image" which is a set of 3D points sampled from some real world scene. The challenge is then to make sense of this mass of data points – to extract meaningful structures and surfaces. We have shown that one can segment a range image, into smooth parametric surfaces (Wang and Suter, 2003b).

Likewise, using variants of Least kth Squares, one can segment a movie sequence purely on the basis of the motion (Bab-Hadiashar et al., 2002). We are working on using MDPE and QMDPE on this task. It is worthwhile noting that to completely solve the problem of segmentation, in situations where the observed objects can have varying shapes and motions, one must not only use a robust fitting technique (and solve the attendant implementation issues), but one also needs to solve the *model selection problem*. Though the cited papers show some promising investigations into that issue, robust model selection in this context remains the major (and daunting) problem.

4. Conclusion

In this paper we have adopted the philosophy that a single statistic, such as the kth order statistic, or the number of inliers within a certain bound; is unlikely to be a sophisticated enough measure to reliably discriminate between candidate model fits in a robust procedure that has to cope with a wide range of possible outlier/pseudo-outlier populations. Instead, we have looked at characterizing the quality of a model fit by more complex measures of the residual distribution: capturing information such as how peaked around zero the residual probability density function is. To this end, we have devised procedures that, at their core, use kernel density estimation of the pdf and a mean-shift approach to located the peak of that pdf. Experiments have shown that such procedures can considerably outperform existing robust techniques in terms of apparent breakdown point behavior.

In concluding, we must remark on the shortcomings of the approaches we are hereby promoting. From a practical point of view, the methods are somewhat costly. Though we haven't extensively studied the issue of just how computationally efficient the approaches can become: it would be optimistic to expect that they can compete with methods relying on much more simple measures of candidate solution quality. From a theoretical point of view, a lot remains to be studied. The reader would be justified in concluding that some aspects of our proposal are ad-hoc and that many variants can be easily dreamt up – without this paper providing any justification for the particular variants we have examined. Secondly, though we promote our schemes in terms of "breakdown point", we acknowledge a number of issues in respect of this. We have not formally defined "breakdown point"; nor, consequently, have we in any way attempted to prove attainment of a high breakdown point. In these respects, our approach is intuitive and empirical. However, we trust, despite these shortcomings, the techniques we have described will be of use to the computer vision community (and wider) as the basis of proven practical methods which can be refined, and whose theoretical underpinnings can be explored. Moreover, we must point out that, despite impressions that may be obtained by reading much of the literature, particularly that aimed more at the practitioner, more traditionally accepted techniques still have their shortcomings in similar ways. For example, though it is often cited that Least Median of Squares has a proven breakdown point of 50%, it is often overlooked that all practical implementations of Least Median of Squares are an approximate form of Least Median of Squares (and thus only have a weaker guarantee of robustness). Indeed, the robustness of practical versions of Least Median of Squares hinges on the robustness of two components (and in two different ways): the robustness of the median residual as a measure of quality of fit and the robustness of the random sampling procedure to find at least one residual distribution whose median is not greatly affected by outliers. Our procedures, like many other procedures, share the second vulnerability as we too rely on random sampling techniques. The first vulnerability is sometimes disregarded for practical versions of Least Median of Squares, because robustness is viewed as being guaranteed by virtue of the proof of robustness for the ideal Least Median of Squares. However, two comments should be made in this respect. Firstly, that proof relies on assumptions regarding the outlier distribution and it can easily be shown that clustered outliers will invalidate that proof. Secondly, there is an inherent "gap" between a proof for an ideal procedure and what one can say about an approximation to that procedure. We believe that our method of scoring the fits better protects against the vulnerabilities that structure in the outliers expose. We have presented empirical evidence to support that. The challenge is to not only continue to amass empirical evidence, but to also explore the theoretical properties of these (and similar) schemes.

References

[1] A. Bab-Hadiashar and D. Suter, *Robust Optic Flow Computation.* Int. J. Comput. Vision **29** (1998), 59–77.

[2] A. Bab-Hadiashar and D. Suter, *Robust Segmentation of Visual Data Using Ranked Unbiased Scale Estimate.* ROBOTICA, Int. J. Information, Education and Research in Robotics and Artificial Intelligence **17** (1999), 649–660.

[3] A. Bab-Hadiashar and D. Suter, *Range and Motion Segmentation.* In A. Bab-Hadiashar and D. Suter, editors, Data Segmentation and Model Selection for Computer Vision, pages 119–142. Springer-Verlag, 2000.

[4] A. Bab-Hadiashar, N. Gheissari, and D. Suter, *Robust Model Based Visual Data Using Segmentation.* In R. Kasturi, D. Laurendeau, and G. Suen, editors, Proceedings of ICPR2002, Vol. 2, pages 753–757, 2002.

[5] M. Black and A. Rangarajan, *On the Unification of Line Processes, Outlier Rejection, and Robust Statistics With Applications in Early Vision.* Int. J. Comput. Vision **19** (1996), 57–92.

[6] D. Comaniciu, V. Ramesh, and A. Del Bue, *Multivariate Saddle Point Detection for Statistical Clustering.* In A. Heyden, G. Sparr, M. Nielsen, and P. Johansen, editors, Proceedings of European Conf. Computer Vision, Vol 3 (ECCV'02), pages 561–576. Copenhagen, Denmark, 2002.

[7] M.A. Fischler and R.C. Bolles, *Random Consensus: A Paradigm for Model Fitting with Applications to Image Analysis and Automated Cartography.* Comm. ACM **24** (1981), 381–395.

[8] K. Fukunaga and L.D. Hostetler, *The Estimation of the Gradient of a Density Function, with Applications in Pattern Recognition.* IEEE Trans. Info. Theory **IT-21** (1975), 32–40.

[9] K. Kanatani, *Statistical Optimization for Geometric Computation: Theory and Practice.* Elsevier Science, Amsterdam, 1996.

[10] K.-M. Lee, P. Meer, and R.-H. Park, *Robust Adaptive Segmentation of Range Images.* IEEE Trans. Pattern Anal. Mach. Intell. **20** (1998), 200–205.

[11] P. Meer, D. Mintz, D.Y. Kim, and A. Rosenfeld, *Robust Regression Methods in Computer Vision: A Review.* Int. J. Comput. Vision **6** (1991), 59–70.

[12] P. Meer, C.V. Stewart, and D. Tyler, *Robust Computer Vision: an Interdisciplinary Challenge.* Comput. Vis. Image Und. **78** (2000), 1–7.

[13] J.V. Miller and C.V. Stewart, *MUSE: Robust Surface Fitting Using Unbiased Scale Estimates.* In Proc. Computer Vision and Pattern Recognition'96, pages 300–306, 1996.

[14] C.V. Stewart, *MINPRAN: A New Robust Estimator For Computer Vision.* IEEE Trans. Pattern Anal. Mach. Intell. **17** (1995), 925–938.

[15] C.V. Stewart, *Robust Parameter Estimation in Computer Vision.* SIAM Rev. **41** (1999), 513–537.

[16] M.P. Wand and M. Jones, *Kernel Smoothing.* Chapman and Hall, 1995.

[17] H. Wang and D. Suter, *LTSD: A Highly Efficient Symmetry-Based Robust Estimator.* In Proceedings ICARCV2002, pages 332–337, 2002.

[18] H. Wang and D. Suter, *Variable Bandwidth QMDPE and its Application in Robust Optic Flow Estimation.* In Proceedings ICCV03, International Conference on Computer Vision, to appear, 2003a, available at *http://www.ds.eng.monash.edu.au/suter_publications.*

[19] H. Wang and D. Suter, *MDPE: A Very Robust Estimator for Model Fitting and Range Image Segmentation.* Int. J. Comput. Vision (2003b), to appear, available at *http://www.ds.eng.monash.edu.au/suter_publications.*

[20] X. Yu, T.D. Bui, and A. Krzyzak, *Robust Estimation for Range Image Segmentation and Reconstruction.* IEEE Trans. Pattern Anal. Mach. Intell. **16** (1994), 530–538.

D. Suter and H. Wang
Dept. Elect. & Comp. Syst. Eng.
P.O. Box 35
Monash University
Clayton, Victoria, 3800, Australia
e-mail: `{d.suter, hanzi.wang}@eng.monash.edu.au`

Statistics for Industry and Technology, 319–328

Testing the Equality of Location Parameters for Skewed Distributions Using S_1 with High Breakdown Robust Scale Estimators

S.S. Syed Yahaya, A.R. Othman and H.J. Keselman

Abstract. A simulation study had been carried out to compare the Type I error and power of S_1, a statistic recommended by Babu et al. (1999) for testing the equality of location parameters for skewed distributions. Othman et al. (in press) showed that this statistic is robust to the underlying populations and is also powerful. In our work, we modified this statistic by replacing the standard errors of the sample medians with four alternative robust scale estimators; the median absolute deviation (MAD) and three of the scale estimators proposed by Rousseeuw and Croux (1993); Q_n, S_n, and T_n. These estimators were chosen based on their high breakdown value and bounded influence function, and in addition, they are simple and easy to compute. Even though MAD is more appropriate for symmetric distributions (Rousseeuw and Croux, 1993), due to its popularity and for the purpose of comparison, we decided to include it in our study. The comparison of these methods was based on their Type I error and the power for J = 4 groups in an unbalanced design having heterogeneous variances. Data from the Chi-square distribution with 3 degrees of freedom were considered. Since the null distribution of S_1 is intractable, and its asymptotic null distribution may not be of much use for practical sample sizes, bootstrap methods were used to give a better approximation. The S_1 statistic combined with each of the scale estimators was shown to have good control of Type I errors.

Mathematics Subject Classification (2000). 62G10.

Keywords. Type I error, power, bootstrap, skewed distributions, breakdown value.

1. Introduction

Progress has been made in terms of finding better methods for controlling Type I error and power to detect treatment effects in the one-way independent group designs. Through a combination of impressive theoretical developments, more flexible statistical methods, and faster computers, serious practical problems that seemed insurmountable only a few years ago can now be addressed. These developments are important to applied researchers because they greatly enhance the ability to discover true differences between groups and improved their effort in trying to guard against seeing benefits that do not exist.

Analysis of variance (ANOVA) is one of the most commonly used statistical methods for locating treatment effects in the one-way independent groups design. Generally, violating the assumptions associated with the standard ANOVA method can seriously hamper its ability to detect true differences. Non-normality and heteroscedasticity are the two general problems in ANOVA. In particular, when these problems occur at the same time, rates of Type I error usually are inflated or depressed, resulting in spurious rejections of null hypotheses. They can also substantially reduce the power of a test, resulting in treatment effects going undetected. Reductions in the power to detect differences between groups occur because the usual standard deviation (σ) is very sensitive to outliers and will be greatly influenced by their presence. Consequently, the standard error of the mean (σ^2/n) can become seriously inflated when the underlying distribution has heavy tails (Wilcox et al., 1998). Therefore, the standard error of the F statistics in ANOVA is larger than it should be and power accordingly will be depressed.

To achieve a good test, one needs to be able to control Type I errors and to increase the power. We do not want to lose power, and at the same time we do not want to inflate the Type I error. In recent years, numerous methods for locating treatment effects simultaneously controlling Type I error and power to detect treatment effects have been studied. The classical least squares estimators can be highly inefficient in non-normal models. In their effort to control the Type I error and power rate, investigators were looking into numerous robust methods. Robust methods generally are insensitive to assumptions about the overall nature of the data. Robust measures of location such as trimmed means, medians or M-estimators were considered as the alternatives for the usual least squares estimator, that is, the usual mean. These measures of location tendency have been shown to have better control over Type I error and power to detect treatment effects (Othman, Keselman, Padmanabhan, Wilcox, & Fradette, in press). Using trimmed means and variances based on Winsorized sum of squares will enable one to obtain test statistics which do not suffer losses in power due to non-normality (Wilcox, Keselman, & Kowalchuk, 1998).

Babu, Padmanabhan, and Puri (1999) proposed a more flexible statistical method that can deal with asymmetric distributions and heteroscedastic settings. Known as the S_1 statistics, this method is one of the latest procedures in assessing the effects of a treatment variable across groups. Othman et al. (in press) replaced the standard errors of the sample medians in S_1 with asymptotic variances but this modification did not result in better Type I error control compared to the former.

Unlike methods using trimmed means, when using S_1 one can work with the original data without having to transform or to trim the data in achieving symmetry. Simple transformations may fail to deal effectively with outliers and heavy tailed distributions. Even the popular strategy of taking logarithms of all the observations does not necessarily reduce problems due to outliers (Wilcox, 1997).

2. Methods

This paper focuses on the S_1 method and the modified S_1 methods. The modified S_1 methods are the S_1 statistics combined with each of the four scale estimators proposed by Rousseeuw and Croux (1993). These estimators, MAD_n, S_n, Q_n, and T_n were chosen for their high breakdown value and bounded influence function, the basic tools for judging robustness (Wilcox, 1997). The S_1 methods were compared in terms of Type I error and power under conditions of normality and non-normality. Non-normality will be represented by skewed distributions since S_1 is more appropriate for skewed data. S_1 methods use sample medians as the central tendency. Being simple and having the highest possible breakdown value, the sample median is still a popular robust estimator of location.

2.1. S_1 Method

When dealing with a skewed distribution, the parameter of interest is therefore the population median. For this particular case, the S_1 statistics which are based on sample medians will be more appropriate to compare distributions (Babu et al., 1999). S_1 is a solution to the problem when the assumption of symmetry is suspect. To understand S_1, consider the problem of comparing location parameters for skewed distributions. Let $Y_j = (Y_{1j}, Y_{2j}, \dots, Y_{n_j j})$ be a sample from an unknown skewed distribution F_j and let M_i be the population median of $F_j : j = 1, 2, \dots, J$. For testing $H_0 : M_1 = M_2 = \cdots = M_J$ versus $H_1 : M_i \neq M_j$ for at least one pair (i, j), the S_1 statistic is defined as

$$S_1 = \sum_{1 \le i < j \le J} |s_{ij}| \qquad \text{where } s_{ij} = \frac{(\hat{M}_i - \hat{M}_j)}{\sqrt{(\hat{w}_i + \hat{w}_j)}}$$

$$w_j = \left(\frac{1}{n_j}\sum \left|Y_{ij} - \hat{M}_j\right|\right)^2 \qquad \hat{w}_j = \frac{w_j}{n_j}, \text{ and}$$

$\hat{M}_j$ is the sample median from the jth group,
ω_j is the squared mean absolute deviation from sample median $M \cap j$, and
n_j is the sample size for group j.

S_1 is the sum of all possible differences of sample medians from the J distributions divided by their respective sample standard errors, $\hat{\omega}$. Therefore, if there are J distributions, then the number of possible differences equals $J(J-1)/2$. Since the sampling distribution of S_1 is unknown, Babu et al. (1999), followed by Othman et al. (in press), used the bootstrap percentile method for obtaining p-values. According to Babu et al. (1999), the bootstrap method is known to give a better approximation than the one based on the normal approximation theory and this method is attractive, especially when the samples are of moderate size. Taking into consideration the intractability of the sampling distribution of S_1 and the reliability of the bootstrap method, the p-values in our study were obtained by using the percentile bootstrap method (see, e.g., Efron and Tibshirani, 1993). To obtain the p-value, the percentile bootstrap method is used as follows:

1. Calculate S_1 based on the available data.
2. Generate bootstrap samples by randomly sampling with replacement n_j observations from the jth group yielding $Y_{1j}^*, Y_{2j}^*, \ldots, Y_{n_j j}^*$
3. Each of the sample points in the bootstrapped groups must be centered at their respective estimated medians.
4. Use the bootstrap sample to compute the S_1 statistic, denoted by $S_1{}^*$.
5. Repeat Step 2 to Step 4 B times yielding $S_{11}^*, S_{12}^*, \ldots, S_{1B}^*$. B = 599 appears sufficient in most situations when $n \geq 12$ (Wilcox, 1997).
6. Calculate the p-value as ($\sharp$ of $S_{1b}^* > S_1)/B$

The amount of computer time depends mainly on how long it takes to evaluate the bootstrap replications and increases linearly with B. The number B varies according to approximations. For estimating the standard error, $B = 50$ is often enough to give a good estimate, while larger B is needed for estimating the percentiles. Efron and Tibshirani (1993) suggested that B should be at least 500 or 1000 in order to make the variability of the estimated percentile acceptably low. Hypothesis testing will adopt the same range of B as the percentile to achieve acceptable accuracy.

Type I error and power of the test corresponding to each method will be determined and compared.

2.2. Scale Estimators

When searching for measures of scale, the breakdown value turns out to have considerable practical importance (Wilcox, 1997). The four scale estimators proposed by Rousseeuw and Croux (1993) have the optimum breakdown value of 0.5. These scale estimators have explicit formulas, which guarantee uniqueness of the estimates. They also have bounded influence functions, which is one of the most important properties for robust estimators. Another advantage of using these estimators is their simplicity, which make them easy to compute.

Let $X = (x_1, x_2, \ldots, x_n)$ be a random sample from any distribution and let the sample median be denoted by $\text{med}_i\, x_i$.

2.2.1. MAD$_n$. A very robust scale estimator is the median absolute deviation about the median, given by

$$\text{MAD}_n = b\,\text{med}_i \left| x_i - \text{med}_j\, x_j \right|.$$

The constant b is needed to make the estimator consistent for the parameter of interest.

The MAD$_n$ has the best possible breakdown value, and its influence function is bounded, with the sharpest possible bound among all scale estimators

(Rousseeuw and Croux, 1993). Huber (1981) identified MAD_n as the single most useful ancillary estimate of scale due to its high breakdown property.

Despite all these advantages, MAD_n has some drawbacks. It has very low efficiency (only 37%) with Gaussian distributions. MAD_n takes a symmetric view on dispersion, because one first estimates a central value (the median) and then attaches equal importance to positive and negative deviations from it, which does not seem to be a natural approach for asymmetric distributions.

2.2.2. S_n. Rousseeuw and Croux (1993) suggested alternatives to MAD_n that can be used as initial or ancillary scale estimates in the same way but that are more efficient and not slanted towards symmetric distributions.

One such estimator is S_n, defined as

$$S_n = c\,\text{med}_i\,\{\text{med}_j\,|x_i - x_j|.\}$$

S_n is very similar to MAD_n. The only difference being that the med_j operation is moved outside the absolute value. This makes S_n a location free estimator. Instead of measuring the deviation of observations from a central value, S_n looks at a typical distance between observations. Another advantage is its explicit formula which means that this estimator is always uniquely defined. A modest simulation study by Rousseeuw and Croux found that the correction factor $c = 1.1926$ succeeded in making S_n unbiased for finite samples. They also proved that S_n has the highest possible breakdown value. In terms of efficiency, S_n was proven to be more efficient (58.23 %) than MAD_n

2.2.3. Q_n. Even though the influence functions for MAD_n and S_n are bounded, they have discontinuities. For a smooth influence function, Rousseeuw and Croux proposed an estimator Q_n defined as

$$Q_n = d\,\{|x_i - x_j|\,; i < j\}_{(k)}$$

where d is a constant factor,

$$k = \binom{h}{2} \approx \binom{n}{2}\Big/4$$

and $h = [n/2] + 1$.

The estimator Q_n shares the attractive properties of S_n; a simple and explicit formula, suitable for asymmetric distributions, and attains the optimal value for its breakdown value (50 %). Other added advantages are the smooth influence function and the high efficiency (82 %) with Gaussian distributions. However, with small samples, S_n performs better than Q_n.

2.2.4. T_n. Another promising scale estimator proposed by Rousseeuw and Croux (1993) which possesses the attractive properties of the robust scale estimator is T_n defined as

$$T_n = 1.3800\frac{1}{h}\sum_{k=1}^{h}\left\{\underset{j\neq i}{med}\,|x_i - x_j|\right\}_{(k)}.$$

It was proven that T_n has a 50 % breakdown value, a continuous influence function, and an efficiency of 52 %, which makes it more efficient than MAD_n. Like S_n and Q_n, this estimator has a simple and explicit formula which guarantees uniqueness and it is suitable for asymmetric distributions.

Taking into consideration all the attractive properties attached to the scale estimators, such as the breakdown value, continuous influence function, and their efficiency, we substituted the standard errors derived from them in place of $\hat{\omega}$ in S_1.

3. Procedures

The procedures investigated were:

1. S_1 with MAD_n
2. S_1 with Q_n
3. S_1 with S_n
4. S_1 with T_n
5. S_1 with $\hat{\omega}$

Each of these five methods was tested for treatment group equality under two types of distributions, the normal and skewed distributions. Note that for the rest of this paper, each of these methods will be referred to by its scale estimator, MAD_n, Q_n, S_n, T_n, and $\hat{\omega}$. We compared MAD_n, Q_n, S_n, and T_n with the existing procedure, $\hat{\omega}$ in terms of their Type I error and power rate.

4. Empirical Investigation

For comparison with the work done by Othman et al. (in press), this paper focused on an unbalanced completely randomized design containing four groups with small samples. Since S_1 is appropriate for skewed distributions, we chose the χ_3^2 distribution for simulating the non-normality condition. The skewness and kurtosis values for the χ_3^2 distribution are 1.63 and 4.00 respectively. This distributional shape was chosen for reasons of comparability to the work. Type I error rates had been found to be distorted when the underlying distribution is skewed, e.g., the case of the two sample t-test in Sawilowsky and Blair (1992). Other conditions which are known to highlight the strengths and weaknesses of test for equality of location are heteroscedasticity, and the pairing of variances and group sizes.

For this reason, only unbalanced designs and unequal variance of 36:1 ratio will be considered (see Table 1). Variances and group sizes are both positively and negatively paired. For positive pairings, the group having the fewest number of observations was associated with the population having the smallest variance, while the group having the greatest number of observations was associated with the population having the largest variance, whereas for the negative pairings, the group with largest observations was paired with smallest variance and the group with smallest observations was paired with population having largest variance. These conditions were chosen since they typically produce conservative results for

TABLE 1. Design Specification for the Four Groups.

PAIRING	*GROUP SIZES*				*GROUP VAR*			
	1	2	3	4	1	2	3	4
POSITIVE	10	15	25	30	1	1	1	36
NEGATIVE	10	15	25	30	36	1	1	1

TABLE 2. Type I Error rates.

Distribution	*Pairing*	*S1* with corresponding scale estimators				
		MAD_n	Q_n	S_n	T_n	$\hat{\omega}$
χ_3^2	pos	0.027	0.017	0.029	0.032	0.004
	neg	0.034	0.019	0.029	0.037	0.007
Normal	pos	0.025	0.012	0.021	0.023	0.017
	neg	0.029	0.023	0.028	0.030	0.013
	Average	0.029	0.018	0.027	0.031	0.010

the positive pairings and liberal results for the negative pairings (Othman et al., in press). We set the samples at $n_1 = 10, n_2 = 15, n_3 = 25$ and $n_4 = 30$ and heterogeneous variances at 1,1,1, and 36 respectively for positive pairings and 36, 1,1,1 respectively for the negative pairings.

Our choices of these extreme conditions (skewness, heteroscedasticity, and unbalanced designs) were based on the premise that if a procedure works under extreme conditions, it is likely to work under most conditions to be encountered by researchers.

The random samples were drawn using SAS generator RANNOR (SAS institute, 1989). The variates were standardized, and then transformed to χ_3^2 variates having mean μ_j and variance σ_j^2. The design specification for the four groups is shown in Table 1.

For Type I error, the group means were (0, 0, 0, 0). For power, one of the group means will be non-zero. Cohen (1977) stated that for the effect size to be uniquely determined, the pattern separation of the means should be specified. Three patterns were identified, the minimum, intermediate, and maximum variability. Our study focused on the intermediate variability, where the J means were equally spaced over the range. In this case, the group means were (−1, −0.5, 0.5, 1).

For each of the designs, 1000 datasets were simulated, and 599 bootstrap samples were generated.

5. Results

The results for Type I error and power rates for the methods investigated were outlined in Table 2 and Table 3, respectively.

TABLE 3. Power rates.

Distribution	*Pairing*	*S1* with corresponding scale estimators				
		MAD_n	Q_n	S_n	T_n	$\hat{\omega}$
χ_3^2	pos	0.078	0.059	0.075	0.091	0.100
	neg	0.088	0.064	0.082	0.102	0.131
Normal	pos	0.403	0.365	0.396	0.408	0.588
	neg	0.227	0.260	0.221	0.278	0.715
	Average	0.199	0.187	0.194	0.220	0.384

Based on the liberal criterion of robustness (Bradley, 1978), a test can be considered robust if its empirical rate of Type I error, $\hat{\alpha}$, is within the interval $0.5\alpha \leq \hat{\alpha} \leq 1.5\alpha$. For the nominal level $\alpha = 0.05$, the Type I error rate should be between 0.025 and 0.075. The empirical Type I error rates in Table 2 indicate robustness in three of the methods investigated. MAD_n, S_n and T_n produced average values ranging from 0.028 to 0.031, all within the Bradley's liberal criterion. These methods produced higher Type I error rates for skewed distribution compared to the normal distributions. Even though the method using Q_n estimator did not satisfy Bradley's liberal criterion, the average error rate for both distributions were higher than the default S_1 method (with $\hat{\omega}$). The average value for Q_n was 0.018 while for $\hat{\omega}$ the average value was only 0.010. However, both methods produced average Type I error rates which were considered to be too conservative, meaning that the estimated rates of Type I error were below 0.025. $\hat{\omega}$ produced a more conservative Type I error rate for skewed distribution.

For both distributions, the empirical Type I error rates for the positive pairings were smaller than for negative pairings, except for the pairings for S_n which showed no variability when the data were skewed.

Our new methods, combining the Babu et al. (1999) S_1 and Rousseeuw and Croux (1993) scale estimators were able to show some improvement over the default S_1 using $\hat{\omega}$ in terms of Type I error rate. The T_n method resulted in the best average error rate of 0.031, which was nearest to the nominal level. All the methods studied (excluding $\hat{\omega}$) produced better average error rate for the skewed distribution compared to normal distribution. The average error rate across the three methods, MAD_n, S_n, and T_n, exhibited small variability for both distributions.

The average power values outlined in Table 3 show two sets of results. The low values belong to the four methods when data were skewed, whereas the larger values were obtained when data were normal. Ranging from 0.075 to 0.131, the average power rate for the new methods under skewed distribution were low. The average values when data were normal ranged from a low of 0.221 and to a high of 0.408. The default S_1 under normal distribution resulted in an average value rate of 0.7. Even though the default S_1 under skewed distributions produced the highest average power rate compared to the rest of the methods, the value of

0.131 was still very low. The mean values across the four methods showed very little variability for both distributions.

The negative and positive pairings for the skewed distributions did not show much variability for each method, but for the normal distribution, the variability was obvious for different pairings.

6. Conclusion

This paper focused on the situation, common in psychological and educational research, where the observations are from skewed distributions. One requires statistics which are robust especially in locating treatment effects. Realizing the need of a good statistic in addressing this problem, we integrate the S_1 statistic by Babu et al. (1999) with the high breakdown scale estimators of Rousseeuw and Croux (1993). This paper has shown some improvement in the statistical solution of locating treatment effects. In controlling the Type I error rate, the study reported in this paper leads us to formulate the following conclusions and recommendations. When symmetry is suspect, we can avoid trimming or transforming the observations by using one of the methods in our paper. These new methods produced better Type I error rates than the default S_1 using $\hat{\omega}$. Three of the investigated methods, MAD_n, S_n, and T_n, reasonably controlled Type I errors; the remaining methods, Q_n and $\hat{\omega}$, were conservative at a significance level of 0.05.

The methods are considered robust when they meet the criteria for robustness with values in between 0.025 and 0.075 for 0.05 level.

The findings on power rate did not show any improvement from the previous research done by Othman et al. (in press). Babu et al. (1999) in their investigation on exponential and log-normal distributions with S_1 also produced low power rates which were less than 0.10.

References

[1] G.J. Babu, A.R. Padmanabhan, and M.L. Puri, *Robust One-Way ANOVA Under Possibly Non-Regular Conditions.* Biom. J. **41** (1999), 321–339.

[2] J.V. Bradley, *Robustness?* British J. Math. Statist. Psych. **31** (1978), 144–152.

[3] J. Cohen, *Statistical Power Analysis For the Behavioral Sciences.* Academic Press, New York, 1977.

[4] B. Efron, *Bootstrap Methods: Another Look at the Jackknife.* Ann. Statist. **7** (1979), 1–26.

[5] B. Efron and R.J. Tibshirani, *An Introduction to the Bootstrap.* Chapman and Hall Inc, New York, 1993.

[6] A.R. Othman, H.J. Keselman, A.R. Padmanabhan, R.R. Wilcox, and K. Fradette, *Comparing Measures of the "Typical" Score Across Treatment Groups.* British J. Math. Statist. Psych. (in press).

[7] P.J. Rousseeuw and C. Croux, *Alternatives to the Median Absolute Deviation.* J. Amer. Statist. Assoc. **88** (1993), 1273–1283.

[8] P.J. Rousseeuw and S. Verboven, *Robust Estimation in Very Small Samples.* Comput. Statist. Data Anal. **40** (2002), 741–758.
[9] SAS Institute Inc., *IML Software: Usage and Reference, Version 6.* 1st Edition, Cary, New York, 1989.
[10] S.S. Sawilowsky and R.C. Blair, *A More Realistic Look at the Robustness and Type II Error Probabilities of the t Test to Departures From Population Normality.* Psychol. Bullet. **111** (1992), 352–360.
[11] R.G. Staudte and S.J. Sheather, *Robust Estimation and Testing.* John Wiley and Sons, Inc., New York, 1990.
[12] R.R. Wilcox, *Introduction to Robust Estimation and Hypothesis Testing.* Academic Press, New York, 1997.
[13] R.R. Wilcox and H.J. Keselman, *Power Analyses When Comparing Trimmed Means.* J. Modern Appl. Statist. Methods **1** (2002), 24–31.
[14] R.R. Wilcox, H.J. Keselman, and R.K. Kowalchuk, *Can Tests For Treatment Group Equality Be Improved?: The Bootstrap and Trimmed Means Conjecture.* British J. Math. and Statist. Psych. **51** (1998), 123–134.

S.S. Syed Yahaya
School of Quantitative Sciences, Universiti Utara Malaysia
06010 Sintok, Kedah, Malaysia
e-mail: `syedys8@yahoo.com`

A.R. Othman
School of Distance Education, Universiti Sains Malaysia
11800 Minden, Penang, Malaysia
e-mail: `oarahman@usm.my`

H.J. Keselman
Department of Psychology, University of Manitoba, 190 Dysart Road
Winnepeg, Manitoba R3T 2N2, Canada
e-mail: `kesel@cc.umanitoba.ca`

Statistics for Industry and Technology, 329–341

Rank Scores Tests of Multivariate Independence

S. Taskinen, A. Kankainen and H. Oja

Abstract. New rank scores test statistics are proposed for testing whether two random vectors are independent. The tests are asymptotically distribution-free for elliptically symmetric marginal distributions. Recently, Gieser and Randles (1997), Taskinen et al. (2003a) and Taskinen et al. (2003b) introduced and discussed different multivariate extensions of the quadrant test, Kendall's tau and Spearman's rho statistics. In this paper, standardized multivariate spatial signs and the (univariate) ranks of the Mahalanobis-type distances of the observations from the origin are combined to construct rank scores tests of independence. The limiting distributions of the test statistics are derived under the null hypothesis as well as under contiguous sequences of alternatives. Three different choices of the score functions, namely the sign scores, the Wilcoxon scores and the van der Waerden scores, are discussed in greater detail. The small sample and limiting efficiencies of the test procedures are compared and the robustness properties are illustrated by an example. It is remarkable that, in the multinormal case, the limiting Pitman efficiency of the van der Waerden scores test equals to that of the classical parametric Wilks' test.

Mathematics Subject Classification (2000). Primary 62G10; Secondary 62H15.
Keywords. Affine equivariant signs, efficiency, Kendall's tau, robustness, Spearman's rho, Wilks' test.

1. Introduction

Let $\boldsymbol{x}_i^T = (\boldsymbol{x}_i^{(1)^T}, \boldsymbol{x}_i^{(2)^T})$ for $i = 1, \dots, n$ be a random sample of vector pairs, where $\boldsymbol{x}_i^{(1)}$ and $\boldsymbol{x}_i^{(2)}$ are p- and q-dimensional continuous random vectors. We wish to test the null hypothesis

$$H_0: \quad \boldsymbol{x}_i^{(1)} \text{ and } \boldsymbol{x}_i^{(2)} \text{ are independent.}$$

The research was partially supported by the Academy of Finland.

The classical parametric test due to Wilks (1935) is based on the partitioned sample covariance matrix S and is defined as

$$W_n = \frac{|S|}{|S_{11}||S_{22}|}.$$

Puri and Sen (1971) introduced a nonparametric analogue to Wilks' test where the data vectors are replaced by the vectors of their componentwise ranks. Gieser and Randles (1997) and Taskinen et al. (2003a) proposed invariant extensions of the univariate quadrant test of Blomqvist (1950). The former test procedure is based on interdirection counts and the latter on standardized spatial signs. If the marginal distributions of $\boldsymbol{x}_i^{(1)}$ and $\boldsymbol{x}_i^{(2)}$ are elliptic, these two tests are asymptotically equivalent. Later Taskinen et al. (2003b) proposed multivariate invariant extensions of Kendall's tau and Spearman's rho.

Our plan is as follows. In Section 2, we explain the test constructions starting with standardized spatial signs and ranks of the lengths of the standardized vectors. The test statistics for multivariate dependence are then introduced. Special choices of the score functions then yield the sign test, the Wilcoxon scores test and the van der Waerden scores test. In Section 3, the limiting distribution of the test statistic is derived under the null hypothesis and under interesting sequences of contiguous alternatives. The finite-sample and limiting efficiencies of the new procedures are then compared to that of the classical Wilks' test in Section 4, and the robustness properties are illustrated by an example in the final Section 5. The proofs can be found in Taskinen et al. (2003c) on the web site http://www.maths.jyu.fi/~ojahannu.

2. The Rank Scores Test Statistics

2.1. Spatial Signs and Ranks of the Distances From the Origin

Consider a random sample $\boldsymbol{x}_1, \ldots \boldsymbol{x}_n$ from a k-variate distribution. The spatial sign of vector $\boldsymbol{x}$ is defined as

$$\boldsymbol{S}(\boldsymbol{x}) = \begin{cases} \|\boldsymbol{x}\|^{-1}\boldsymbol{x}, & \boldsymbol{x} \neq \boldsymbol{0} \\ \boldsymbol{0}, & \boldsymbol{x} = \boldsymbol{0}, \end{cases}$$

where $\|\boldsymbol{x}\| = (\boldsymbol{x}^T\boldsymbol{x})^{1/2}$ is the (Euclidean) length of the vector $\boldsymbol{x}$. The spatial signs $\boldsymbol{S}(\boldsymbol{x}_i)$ and ranks rank($||\boldsymbol{x}_i||$) of the distances from the origin are not invariant under affine transformations to the data vectors, however. In order to construct invariant test statistics, the data points have to be standardized before spatial signs and ranks are formed. For the standardization we need affine equivariant $\sqrt{n}$-consistent location vector and scatter matrix estimates, $\widehat{\boldsymbol{\mu}}$ and $\widehat{C}$. The transformed data points are then given as $\boldsymbol{z}_i = \widehat{C}^{-1/2}(\boldsymbol{x}_i - \widehat{\boldsymbol{\mu}})$, $i = 1, \ldots, n$.

The vectors $\widehat{\boldsymbol{u}}_i = \boldsymbol{S}(\boldsymbol{z}_i)$, $i = 1, \ldots, n$, are called **standardized spatial sign vectors**. Standardized sign vectors are affine invariant in the sense that if $\widehat{\boldsymbol{u}}_i^*$ are calculated from $\boldsymbol{x}_i^* = A\boldsymbol{x}_i + \boldsymbol{b}$, $i = 1, \ldots, n$, with a nonsingular $k \times k$ matrix A and k-vector $\boldsymbol{b}$, then $\widehat{\boldsymbol{u}}_i^* = P\widehat{\boldsymbol{u}}_i$, $i = 1, \ldots, n$, for some orthogonal P. See, e.g.,

Taskinen et al. (2003b). The ranks $\widehat{R}_i = \text{rank}(||\boldsymbol{z}_i||)$ are naturally affine invariant (in the usual sense). Note that, in the standardization, the scatter matrix estimate $\widehat{C}$ may be replaced by a $\sqrt{n}$-consistent affine equivariant shape matrix estimate $\widehat{V}$ as only the directions and ranks of distances are used in the analysis. For the shape matrices, see Ollila et al. (2003). Note also that, if the standardization is done using such location vector and scatter (or shape) matrix estimates that do not require any moment assumptions of the underlying data (e.g., Tyler's shape matrix and the transformation retransformation spatial median in Hettmansperger and Randles, 2002), then the resulting test procedures are valid without any moment assumptions.

2.2. New Test Statistics

Our test statistic for testing the null hypothesis of independence is obtained as follows. For $i = 1, \ldots, n$, write $\widehat{\boldsymbol{u}}_i^{(1)} = ||\boldsymbol{z}_i^{(1)}||^{-1}\boldsymbol{z}_i^{(1)}$ for p-dimensional standardized sign vectors based on the first components $\boldsymbol{x}_i^{(1)}$ and let $\widehat{R}_i^{(1)}$ denote the rank of $\widehat{r}_i^{(1)} = ||\boldsymbol{z}_i^{(1)}||$ among $\widehat{r}_1^{(1)}, \ldots, \widehat{r}_n^{(1)}$. For the second random vector, write similarly $\widehat{\boldsymbol{u}}_i^{(2)}$ for q-dimensional standardized sign vectors based on $\boldsymbol{x}_i^{(2)}$ and let $\widehat{r}_i^{(2)}$ and $\widehat{R}_i^{(2)}$ be constructed as before. The test statistic is then as follows.

Definition 2.1. Let $a : (0,1) \to \mathbb{R}$ and $b : (0,1) \to \mathbb{R}$ be continuous, monotone and square integrable score function and write

$$\widehat{H} = \text{ave}_i \left\{ a\left(\frac{\widehat{R}_i^{(1)}}{n+1}\right) b\left(\frac{\widehat{R}_i^{(2)}}{n+1}\right) \widehat{\boldsymbol{u}}_i^{(1)} \widehat{\boldsymbol{u}}_i^{(2)^T} \right\}.$$

The rank test statistic for testing H_0 is then

$$T_n = \frac{npq}{\sigma_a^2 \sigma_b^2} \, ||\widehat{H}||^2,$$

where $||\widehat{H}||^2 = Tr(\widehat{H}^T \widehat{H})$, $\sigma_a^2 = E[a^2(U)]$ and $\sigma_b^2 = E[b^2(U)]$ with U uniformly distributed on $(0,1)$.

Note that since standardized sign vectors and ranks are invariant with respect to the group of affine transformations, the invariance of T_n easily follows. As score functions, one may use optimal location score functions. See Hallin and Paindaveine (2002), for that. In the following, some choices of the score functions and resulting test statistics are given.

Definition 2.2. For $a(u) = 1$ and $b(u) = 1$, the **sign test** of independence (Taskinen et al., 2003a) with test statistic

$$T_{0n} = npq \, ||\text{ave}_i \{ \widehat{\boldsymbol{u}}_i^{(1)} \widehat{\boldsymbol{u}}_i^{(2)^T} \}||^2$$

is obtained. For $a(u) = u$ and $b(u) = u$, one gets the **Wilcoxon (scores) test** of independence with test statistic

$$T_{1n} = \frac{9npq}{(n+1)^4} \, ||\text{ave}_i \{ \widehat{R}_i^{(1)} \widehat{R}_i^{(2)} \widehat{\boldsymbol{u}}_i^{(1)} \widehat{\boldsymbol{u}}_i^{(2)^T} \}||^2.$$

Finally the choices $a(u) = [\Psi_p^{-1}(u)]^{1/2}$ and $b(u) = [\Psi_q^{-1}(u)]^{1/2}$, where Ψ_k is a cdf of chi-square distribution with k degrees of freedom, yield the **van der Waerden (scores) test** of independence with test statistic

$$T_{2n} = n \left\| \text{ave}_i \left\{ \left[\Psi_p^{-1} \left(\frac{\widehat{R}_i^{(1)}}{n+1} \right) \right]^{1/2} \left[\Psi_q^{-1} \left(\frac{\widehat{R}_i^{(2)}}{n+1} \right) \right]^{1/2} \widehat{\boldsymbol{u}}_i^{(1)} \widehat{\boldsymbol{u}}_i^{(2)^T} \right\} \right\|^2 .$$

3. Limiting Distributions

In order to derive the limiting distribution of T_n, we assume that the marginal distributions of $\boldsymbol{x}^{(1)}$ and $\boldsymbol{x}^{(2)}$ are elliptically symmetric. The marginal density functions are then of the form

$$f(\boldsymbol{x}) = |\Sigma|^{-1/2} f_0(\Sigma^{-1/2}(\boldsymbol{x} - \boldsymbol{\mu})),$$

where Σ is a positive definite symmetric matrix and $f_0(\boldsymbol{z}) = \exp\{-\rho(||\boldsymbol{z}||)\}$ with $\boldsymbol{z} = \Sigma^{-1/2}(\boldsymbol{x}-\boldsymbol{\mu})$. Note that if $r = ||\boldsymbol{z}||$ and $\boldsymbol{u} = \boldsymbol{z}/r$, then r and $\boldsymbol{u}$ are independent. In the following we denote the cdf of $r^{(1)}$ as G_1 and the cdf of $r^{(2)}$ as G_2.

To establish a limiting distribution of our test statistic under the null hypothesis, we need the following lemma.

Lemma 3.1. *Let*

$$H = \text{ave}_i\{a(G_1(r_i^{(1)}))b(G_2(r_i^{(2)}))\boldsymbol{u}_i^{(1)}\boldsymbol{u}_i^{(2)^T}\},$$

then $\sqrt{n}(\widehat{H} - H) \to_p 0$.

Now the limiting distribution can be found easily.

Theorem 3.2. *Under H_0 and for elliptically distributed $\boldsymbol{x}^{(1)}$ and $\boldsymbol{x}^{(2)}$, the limiting distribution of T_n is a chi-square distribution with pq degrees of freedom.*

Next we derive the limiting distribution of T_n under alternative sequences similar to those used in Gieser and Randles (1997). As T_n is affine invariant, we restrict to the spherical case only. See Appendix, for a discussion on the alternative sequences. Let thus $\boldsymbol{x}_i^{(1)}$ and $\boldsymbol{x}_i^{(2)}$ be independent with spherical marginal densities $\exp\{-\rho_1(||\boldsymbol{x}^{(1)}||)\}$ and $\exp\{-\rho_2(||\boldsymbol{x}^{(2)}||)\}$, respectively, and write

$$\begin{pmatrix} \boldsymbol{y}_i^{(1)} \\ \boldsymbol{y}_i^{(2)} \end{pmatrix} = \begin{pmatrix} (1-\Delta)I_p & \Delta M_1 \\ \Delta M_2 & (1-\Delta)I_q \end{pmatrix} \begin{pmatrix} \boldsymbol{x}_i^{(1)} \\ \boldsymbol{x}_i^{(2)} \end{pmatrix}, \tag{3.1}$$

where $\Delta = \delta/\sqrt{n}$. If T_n^* is calculated from transformed observations in (3.1), we get

Theorem 3.3. *Under general assumptions, the limiting distribution of T_n^* is a noncentral chi-square distribution with pq degrees of freedom and noncentrality parameter*

$$\frac{\delta^2}{pq\sigma_a^2\sigma_b^2}||c_1M_1 + c_2M_2^T||^2,$$

where

$$c_1 = E[a(G_1(r_i^{(1)}))\psi_1(r_i^{(1)})]E[b(G_2(r_i^{(2)}))r_i^{(2)}]$$

and

$$c_2 = E[b(G_2(r_i^{(2)}))\psi_2(r_i^{(2)})]E[a(G_1(r_i^{(1)}))r_i^{(1)}],$$

with optimal location score functions $\psi_1(r_i^{(1)}) = \rho_1'(r_i^{(1)})$ *and* $\psi_2(r_i^{(2)}) = \rho_2'(r_i^{(2)})$.

4. Limiting and Finite-Sample Efficiencies

4.1. Limiting Pitman Efficiencies

In this section we consider the sign, Wilcoxon and van der Waerden tests of independence: We compare the limiting and finite-sample efficiencies of the new tests to those of the Wilks' likelihood ratio test W_n. The comparisons are made in the multivariate normal distribution, t distribution and contaminated normal distribution cases. Since $-n \log W_n$ has, under the alternative sequences, a limiting noncentral chi-squared distribution with pq degrees of freedom and noncentrality parameter $\delta^2||M_1 + M_2^T||^2$, the asymptotic efficiencies are simply

$$\text{ARE}(T_n, W_n) = \frac{||c_1 M_1 + c_2 M_2^T||^2}{pq\sigma_a^2\sigma_b^2||M_1 + M_2^T||^2},$$

where c_1 and c_2 are given in Theorem 3.3. Note that for multivariate normal distribution, $\psi(r) = r$, for k-variate t distribution with ν degrees of freedom, $\psi(r) = (k+\nu)r/(\nu + r^2)$ and for k-variate contaminated normal distribution with cdf $F(\boldsymbol{x}) = (1-\epsilon)\Phi(\boldsymbol{x}) + \epsilon\Phi(c^{-1}\boldsymbol{x})$, where $c > 0$ and Φ is the cdf of $N_k(\mathbf{0}, I_k)$,

$$\psi(r) = \frac{(1-\epsilon)\exp(-r^2/2) + \epsilon c^{-k-2}\exp(-r^2/2c^2)}{(1-\epsilon)\exp(-r^2/2) + \epsilon c^{-k}\exp(-r^2/2c^2)}\, r.$$

Assume now for simplicity that $M_1 = M_2^T$. For the limiting efficiency of the sign test of independence, we refer to Taskinen et al. (2003a). The limiting efficiency of the Wilcoxon test T_{1n} with respect to the Wilks' test W_n is

$$\text{ARE}(T_{1n}, W_n) = \frac{9(c_1 + c_2)^2}{4pq},$$

where

$$c_1 = E[G_1(r_i^{(1)})\psi_1(r_i^{(1)})]E[G_2(r_i^{(2)})r_i^{(2)}]$$

and

$$c_2 = E[G_2(r_i^{(2)})\psi_2(r_i^{(2)})]E[G_1(r_i^{(1)})r_i^{(1)}].$$

The resulting efficiencies for t distributions with selected degrees of freedom and dimensions are listed in Table 1 and for contaminated normal distributions with $\epsilon = 0.1$ and for selected values of c in Table 2. The efficiencies were derived using numerical integration.

TABLE 1. $ARE(T_{1n}, W_n)$ at different p- and q-variate t distributions for selected $\nu = \nu_1 = \nu_2$.

	q	p = 2	3	5	8	10
	2	1.089	1.064	1.023	0.986	0.970
	3		1.039	0.998	0.961	0.946
$\nu = 5$	5			0.958	0.922	0.907
	8				0.886	0.871
	10					0.857
	2	0.970	0.960	0.934	0.907	0.893
	3		0.950	0.925	0.898	0.884
$\nu = \infty$	5			0.901	0.874	0.861
	8				0.848	0.835
	10					0.823

TABLE 2. $ARE(T_{1n}, W_n)$ at different p- and q-variate contaminated normal distributions for $\epsilon = 0.1$ and for selected values of c.

	q	p = 2	3	5	8	10
	2	1.216	1.204	1.172	1.137	1.121
	3		1.192	1.161	1.126	1.109
$c = 3$	5			1.130	1.096	1.080
	8				1.063	1.048
	10					1.034
	2	1.833	1.815	1.767	1.714	1.689
	3		1.797	1.749	1.697	1.672
$c = 6$	5			1.703	1.652	1.628
	8				1.603	1.579
	10					1.556

Further, the limiting efficiency of the van der Waerden test T_{2n} as compared to the W_n is

$$\text{ARE}(T_{2n}, W_n) = \frac{(c_1 + c_2)^2}{4p^2q^2},$$

where

$$c_1 = E\{[\Psi_p^{-1}(G_1(r_i^{(1)}))]^{1/2}\psi_1(r_i^{(1)})\}E\{[\Psi_q^{-1}(G_2(r_i^{(2)}))]^{1/2}r_i^{(2)}\}$$

and

$$c_2 = E\{[\Psi_q^{-1}(G_2(r_i^{(2)}))]^{1/2}\psi_2(r_i^{(2)})\}E\{[\Psi_p^{-1}(G_1(r_i^{(1)}))]^{1/2}r_i^{(1)}\}.$$

Now for the multivariate normal distribution, $\text{ARE}(T_{2n}, W_n) = 1$, and for the contaminated normal distribution, $\text{ARE}(T_{2n}, W_n) = (1 - \epsilon + \epsilon/c)^2(1 - \epsilon + \epsilon c)^2$.

These efficiencies do not depend on the dimensions at all. For the efficiencies at certain contaminated normal distributions, see Figure 1. The efficiencies for t distribution with 5 degrees of freedom were derived using numerical integration and are listed in Table 3.

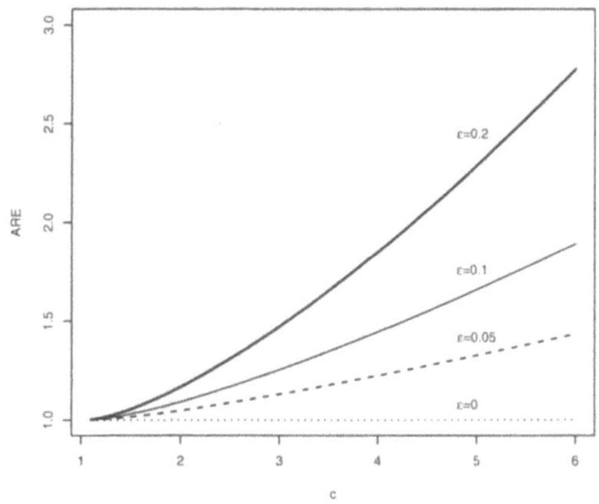

FIGURE 1. $ARE(T_{2n}, W_n)$ as a function of c at the contaminated normal model with $\epsilon = 0, 0.05, 0.10, 0.20$.

TABLE 3. $ARE(T_{2n}, W_n)$ at different p- and q-variate t distributions with $\nu_1 = \nu_2 = 5$.

			p		
q	2	3	5	8	10
2	1.125	1.132	1.144	1.155	1.160
3		1.140	1.151	1.162	1.168
5			1.162	1.174	1.179
8				1.185	1.190
10					1.195

Now some comments follow. First of all, the limiting efficiencies of the Wilcoxon test T_{1n} decrease with increasing dimension while the efficiencies of sign test T_{0n} and van der Waerden test T_{2n} increase or stay constant. Due to this property, for low dimensions, the efficiencies of T_{1n} are higher than those of T_{0n}, but for high dimensions, T_{0n} outperforms T_{1n}. The van der Waerden scores test is the most efficient one in all considered cases. When the underlying distribution is multivariate normal, it is as efficient as the Wilks' test. When the distribution becomes heavy-tailed, the efficiencies are higher than those of T_{0n} and T_{1n} (for the contaminated normal distribution with $\epsilon = 0.1$ and $c = 3$ and $c = 6$, the efficiencies of T_{2n} are 1.254 and 1.891). For comparisons of limiting efficiencies, see also Figures 2 and 3.

4.2. A Simulation Study

A simple simulation study was used to compare the finite sample efficiencies of W_n, T_{0n}, T_{1n} and T_{2n}. 1500 independent $\boldsymbol{x}^{(1)}$- and $\boldsymbol{x}^{(2)}$-samples of sizes $n = 50$ and 200 were generated from a multivariate standard normal distribution, from a t distribution with 5 degrees of freedom and from a contaminated normal distribution with $\epsilon = 0.1$ and $c = 6$. The transformation in (3.1) with $M_1 = M_2^T = I$ was applied for chosen values of $\Delta = \delta/\sqrt{n}$ to introduce dependence into the model. The tests were applied using the location and shape estimates chosen to satisfy

$$\text{ave}\{\widehat{\boldsymbol{S}}_i^{(1)}\} = \mathbf{0} \quad \text{and} \quad p\,\text{ave}\{\widehat{\boldsymbol{S}}_i^{(1)}\widehat{\boldsymbol{S}}_i^{(1)^T}\} = I_p$$

and

$$\text{ave}\{\widehat{\boldsymbol{S}}_i^{(2)}\} = \mathbf{0} \quad \text{and} \quad q\,\text{ave}\{\widehat{\boldsymbol{S}}_i^{(2)}\widehat{\boldsymbol{S}}_i^{(2)^T}\} = I_q,$$

that is, the transformation retransformation spatial median and the Tyler's M-estimate (Tyler, 1987; Hettmansperger and Randles, 2002). For the transformation retransformation technique, see also Chakraborty et al. (1998). The critical values used in test constructions were based on the chi-square approximations to the null distributions.

In Figure 2, the empirical powers as well as exact limiting powers ($n = \infty$) computed using Theorem 3.3 are given for $p = q = 3$. In the multivariate normal case W_n is slightly better than T_{1n} and T_{2n} and much better than T_{0n}. In the t distribution case no big differences can be seen between tests and in the contaminated normal case T_{1n} and T_{2n} outperform W_n and T_{0n}. In Figure 3, the empirical powers are illustrated for $p = q = 8$. In the multivariate normal case T_{0n} and T_{2n} are slightly more powerful than T_{1n}. In the considered t distribution case T_{1n} performs poorly, but as the underlying distribution is contaminated normal, T_{1n} performs very well. As $p = q = 8$, the sizes of T_{0n} and T_{2n} are often slightly below 0.05. The size of T_{1n} is very close to 0.05 in all cases and for heavy-tailed distributions, the size of W_n often exceeds 0.05.

5. A Robustness Study and Final Comments

Finally, a simple simulation study was used to illustrate the robustness of test statistics proposed above. Independent $\boldsymbol{x}^{(1)}$- and $\boldsymbol{x}^{(2)}$-samples of size $n = 30$ were generated from a bivariate standard normal distribution and the transformation in (3.1) with $M_1 = M_2^T = I_2$ was applied for chosen values of Δ to introduce "positive" dependence into the model. By positive dependence we mean that each $\boldsymbol{x}^{(1)}$-coordinate is positively dependent on each $\boldsymbol{x}^{(2)}$-coordinate. Finally, the first observation vectors in each sample were replaced by contaminated vectors $\boldsymbol{x}_1^{(1)} =$

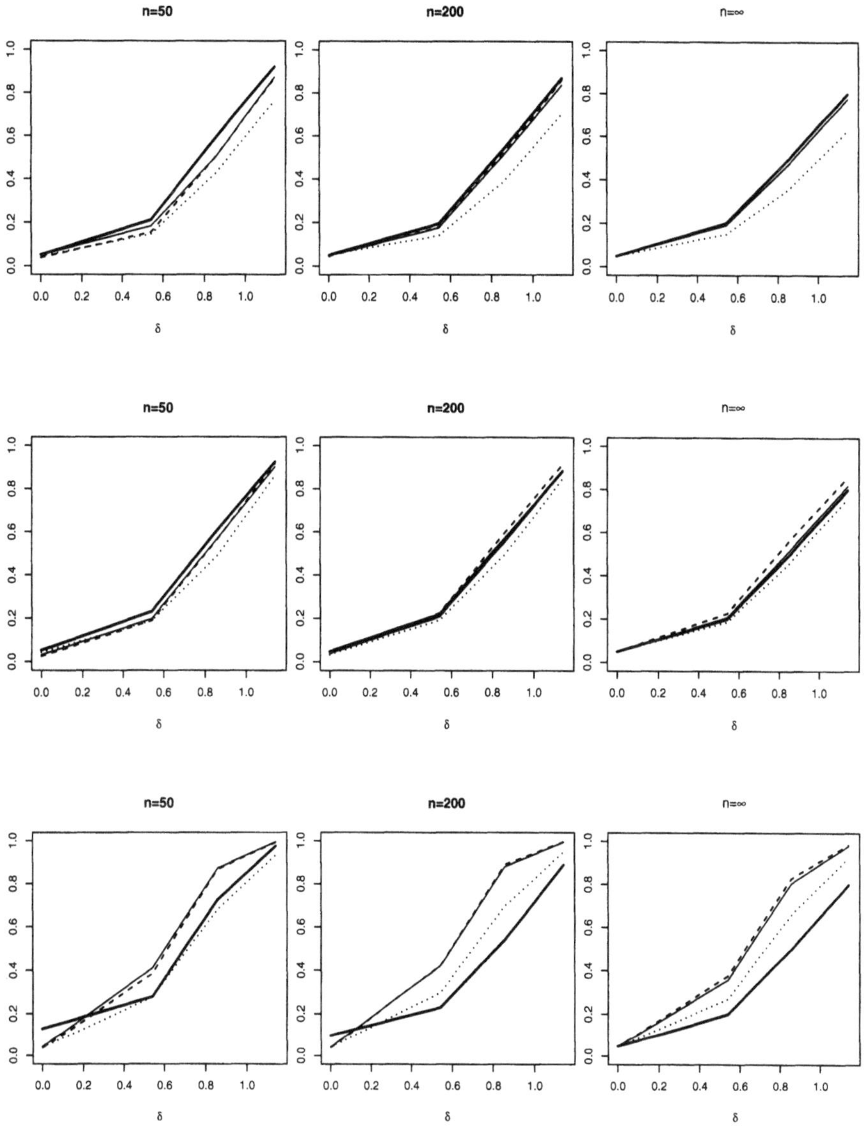

FIGURE 2. Empirical powers for $p = q = 3$ using the multivariate normal distribution (first row), multivariate t distribution with $\nu = 5$ (second row) and contaminated normal distribution with $\epsilon = 0.1$ and $c = 6$ (third row). The thick solid line denotes W_n, the thin solid line T_{1n}, the thick dotted line T_{2n} and the thin dotted line T_{0n}.

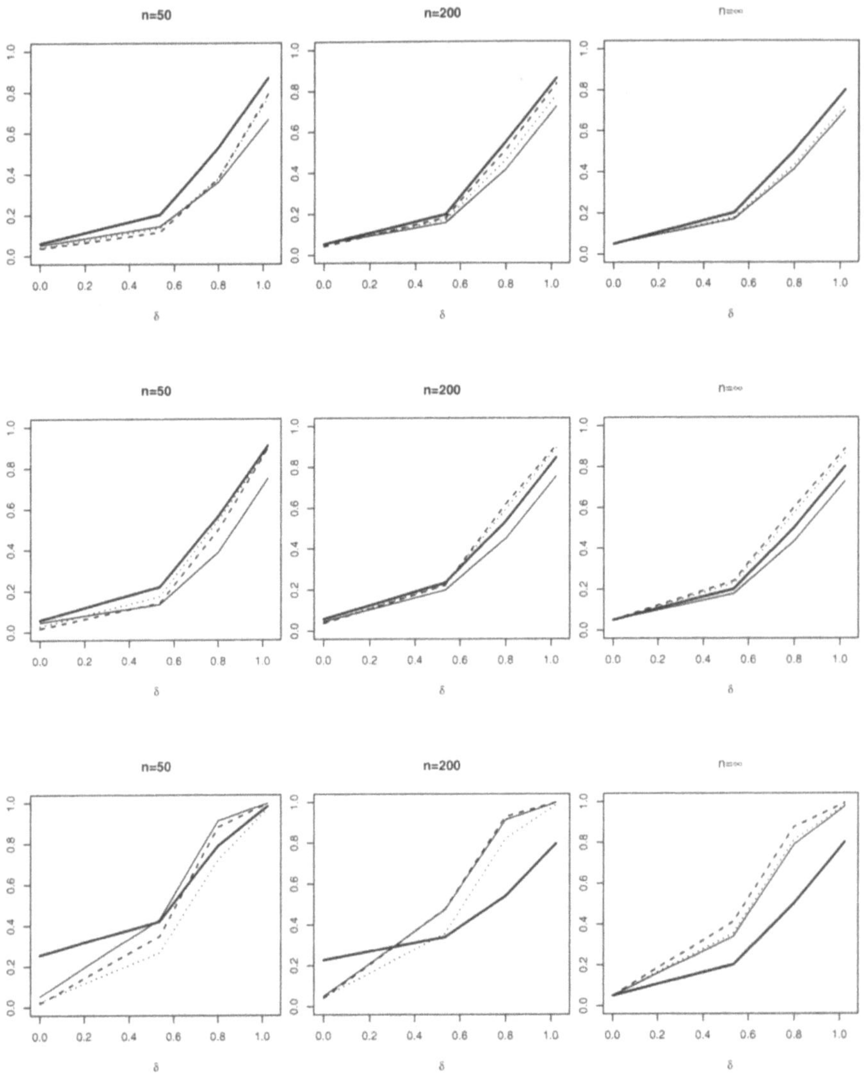

FIGURE 3. Empirical powers for $p = q = 8$ using the multivariate normal distribution (first row), multivariate t distribution with $\nu = 5$ (second row) and contaminated normal distribution with $\epsilon = 0.1$ and $c = 6$ (third row). The thick solid line denotes W_n, the thin solid line T_{1n}, the thick dotted line T_{2n} and the thin dotted line T_{0n}.

$(c,c)^T$ and $\boldsymbol{x}_1^{(2)} = (-c,-c)^T$ with "negative" dependence. The procedure was repeated 1000 times and mean p-values were computed.

In Figure 4, the mean p-values are illustrated as a function of contamination value c for $\Delta = 0$ and for $\Delta = 0.2$. In the null hypothesis case ($\Delta = 0$), all tests give p-values close to 0.5, as the contamination value is near zero. Note also that T_{0n} and T_{1n} give practically the same p-values as Wilks' test. When the contamination value is high, p-values given by Wilks' test decrease considerably and some decrease is also seen in the p-values of T_{1n} and T_{2n}. In the considered case, the sign test T_{0n} seems to be the most robust one, since the mean p-value is constant as a function of c. As $\Delta = 0.2$, the contamination slightly increases the mean values of rank scores tests. In the case of Wilks' test the p-values first increase and then decrease to zero with the contamination value. The careful analysis shows that the small p-values for large contamination values erroneously indicate "negative" dependence, however.

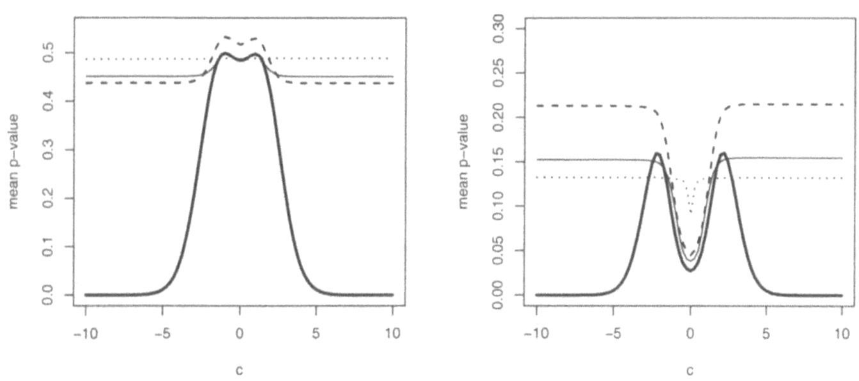

FIGURE 4. Mean p-values for the true null hypothesis $H_0 : \Delta = 0$ (left figure) and for the alternative hypothesis $H_1 : \Delta = 0.2$ (right figure) as a function of contamination value as described in the text. The thick solid line refers to W_n, the thin solid line to T_{1n}, the thick dotted line to T_{2n} and the thin dotted line to T_{0n}.

In the paper, new affine invariant rank scores procedures were proposed for testing whether two random vectors are independent. The test statistics were constructed using standardized spatial signs and ranks of the lengths of the standardized vectors. It is remarkable that, the proposed tests are valid without any moment assumptions on the underlying data as far as the standardization is done using such location vector and scatter (or shape) matrix estimates that do not require any moment assumptions. In the paper, three different score functions, namely the sign scores, the Wilcoxon scores and the van der Waerden scores were considered in more detail. The tests have good limiting and finite-sample efficiencies and as illustrated by an example, the tests are resistant to outliers.

Appendix: Some Notions on Alternative Sequences

For all elliptic cases, it is enough to consider the alternative sequences

$$\begin{pmatrix} \boldsymbol{y}_i^{(1)} \\ \boldsymbol{y}_i^{(2)} \end{pmatrix} = \begin{pmatrix} (1-\Delta)I_p & \Delta M_1 \\ \Delta M_2 & (1-\Delta)I_q \end{pmatrix} \begin{pmatrix} \boldsymbol{x}_i^{(1)} \\ \boldsymbol{x}_i^{(2)} \end{pmatrix},$$

where $\Delta = \delta/\sqrt{n}$ and $\boldsymbol{x}_i^{(1)}$ and $\boldsymbol{x}_i^{(2)}$ are independent with **spherical marginal distributions**. This is because, for the weighted sum of elliptical marginals,

$$\begin{aligned} \begin{pmatrix} \boldsymbol{y}_i^{(1)} \\ \boldsymbol{y}_i^{(2)} \end{pmatrix} &= \begin{pmatrix} (1-\Delta)I_p & \Delta M_1 \\ \Delta M_2 & (1-\Delta)I_q \end{pmatrix} \begin{pmatrix} A\boldsymbol{x}_i^{(1)} \\ B\boldsymbol{x}_i^{(2)} \end{pmatrix} \\ &= \begin{pmatrix} A & 0 \\ 0 & B \end{pmatrix} \begin{pmatrix} (1-\Delta)I_p & \Delta A^{-1}M_1B \\ \Delta B^{-1}M_2A & (1-\Delta)I_q \end{pmatrix} \begin{pmatrix} \boldsymbol{x}_i^{(1)} \\ \boldsymbol{x}_i^{(2)} \end{pmatrix} \\ &= \begin{pmatrix} A & 0 \\ 0 & B \end{pmatrix} \begin{pmatrix} (1-\Delta)I_p & \Delta M_1' \\ \Delta M_2' & (1-\Delta)I_q \end{pmatrix} \begin{pmatrix} \boldsymbol{x}_i^{(1)} \\ \boldsymbol{x}_i^{(2)} \end{pmatrix}. \end{aligned}$$

Hence (due to affine invariance) one can as well consider the sequence

$$\begin{pmatrix} (1-\Delta)I_p & \Delta M_1' \\ \Delta M_2' & (1-\Delta)I_q \end{pmatrix} \begin{pmatrix} \boldsymbol{x}_i^{(1)} \\ \boldsymbol{x}_i^{(2)} \end{pmatrix},$$

for different choices of M_1' and M_2'. In all cases, the efficiencies are then of the same type $||c_1M_1 + c_2M_2^T||$, where c_1 and c_2 depend on the marginal spherical distributions and the test used. If the marginal distributions are of the same type (that is $c_1 = c_2$ for all tests to be compared), then the efficiencies do not depend on M_1 and M_2. Note also that the tests are not "unbiased" (noncentrality parameter > 0) for all alternative sequences. They are all unbiased for normal marginals. If the marginals are nonnormal but of the same type, then they all fail under $M_1 = -M_2^T$.

References

[1] N. Blomqvist, *On a Measure of Dependence Between Two Random Variables.* Ann. Math. Statist. **21** (1950), 593–600.

[2] B. Chakraborty, P. Chaudhuri, and H. Oja, *Operating Transformation Retransformation On Spatial Median and Angle Test.* Statist. Sinica **8** (1998), 767–784.

[3] P.W. Gieser and R.H. Randles, *A Nonparametric Test of Independence Between Two Vectors.* J. Amer. Statist. Assoc. **92** (1997), 561–567.

[4] M. Hallin and D. Paindaveine, *Optimal Tests For Multivariate Location Based On Interdirections and Pseudo-Mahalanobis Ranks.* Ann. Statist. **30** (2002), 1103–1133.

[5] T.P. Hettmansperger and R.H. Randles, *A Practical Affine Equivariant Multivariate Median.* Biometrika **89** (2002), 851–860.

[6] E. Ollila, T.P. Hettmansperger and H. Oja, *Affine Equivariant Multivariate Sign Methods.* Under revision, 2003, available at *http://www.maths.jyu.fi/~ojahannu/manu.html.*

[7] M.L. Puri and P.K. Sen, *Nonparametric Methods in Multivariate Analysis.* Wiley, New York, 1971.

[8] S. Taskinen, A. Kankainen, and H. Oja, *Sign Test of Independence Between Two Random Vectors.* Statist. Probab. Lett. **62** (2003a), 9–21.

[9] S. Taskinen, H. Oja, and R.H. Randles, *Multivariate Nonparametric Tests of Independence.* Under revision, 2003b, available at *http://www.maths.jyu.fi/~ojahannu/manu.html.*

[10] S. Taskinen, A. Kankainen, and H. Oja, *Rank Scores Tests of Multivariate Independence* (with proofs). Technical report, 2003c, available at *http://www.maths.jyu.fi/~ojahannu/manu.html.*

[11] D.E. Tyler, *A Distribution-Free M-Estimator of Multivariate Scatter.* Ann. Statist. **15** (1987), 234–251.

[12] S.S. Wilks, *On the Independence of k Sets of Normally Distributed Statistical Variables.* Econometrica **3** (1935), 309–326.

Acknowledgment

The authors wish to thank the referees for several helpful comments and suggestions.

S. Taskinen, A. Kankainen and H. Oja
Department of Mathematics and Statistics
University of Jyväskylä
P.O.Box 35
FIN-40014 Jyväskylä, Finland
e-mail: {slahola, kankaine, Hannu.Oja}@maths.jyu.fi

Statistics for Industry and Technology, 343–353

The Influence of a Stochastic Interest Rate on the n-fold Compound Option

L. Thomassen and M. Van Wouwe

Abstract. We reintroduced the idea of an n-fold compound option as a generalization of Geske's (2-fold) compound option in the same framework of constant interest rates. For the valuation of long-term financial agreements (life insurance products) this assumption is not always realistic so that the stochastic modelling of the interest rates might be a better approach.
According to Miltersen et al. (1997), we will use the requirement of simple interest rates over a fixed finite period to be log-normal distributed. With these assumptions, closed-form solutions are determined for the n-fold compound call options written on zero-coupon bonds.
A numerical illustration of the application of robust methods to interest rates is discussed.

Mathematics Subject Classification (2000). Primary 91B28; Secondary 62P05.

Keywords. Financial, n-fold compound options, log-normal interest rates.

1. Introduction

In constructing an n-fold compound call option model, Thomassen and Van Wouwe (2002) used the same economic assumptions as Geske (1979) did when introducing his (2-fold) compound option. One of the economic hypotheses is the assumption of a deterministic risk-free interest rate both for the original compound call option and the generalized compound call option.

If however one looks at possible application fields of the n-fold compound call option, it appears that long-term commitments are involved. Indeed, the introduction of for instance new life insurance contracts are long-term processes. So one can start questioning the validity of a constant interest rate. The life insurance practice shows that the currently used approaches to represent interest rates in actuarial models may be broadly categorized as deterministic and stochastic.

The most familiar deterministic approach is still a single interest rate with a slight generalization to a series of interest rates for future years. In recent years one also sees the development of a tremendous profusion of models for valuation of interest rate sensitive financial instruments. Wall street dealers routinely use a multiplicity of models based on widely varying assumptions in different markets. For example, an options desk most likely uses a version of the Black formula to value interest rate caps and floors, implying an approximately log-normal distribution of the interest rates.

A critical issue in selecting an interest rate model is ease of application. It is indeed difficult or impossible to provide efficient valuation algorithms for some models. For that reason we focused on the literature to find closed-form solutions for term structure derivatives with log-normal interest rates. We were inspired by the reasoning developed by Miltersen et al. (1997).

They derived a unified model that gives closed-form solutions for caps and floors written on interest rates as well as for puts and calls written on zero-coupon bonds. Their crucial assumption is that the simple interest rates over a fixed finite period are log-normal distributed, an assumption that is shown to be consistent with the Heath-Jarrow-Morton model for a specific choice of volatility. As they pointed out, log-normal distributed interest rate models avoid the problem of negative interest rates but the rates explode with positive probability, implying zero prices for bonds and hence allowing arbitrage opportunities.

The problem of exploding interest rates disappears as shown in Sandmann and Sondermann (1994) if instead of assuming the continuously compounded interest rates to be log-normal distributed, one assumes the simple interest rates over a fixed finite period to be log-normal distributed. The main result of their article is a unified model that provides closed-form solutions for interest rate caps and floors as well as for puts and calls written on zero-coupon bonds within the context of a log-normal interest rate model.

The closed-form valuation is a Black-Scholes alike formula which we could generalize for n-fold compound call options written on zero-coupon bonds in the same framework as by Miltersen et al. (1997).

2. A Model for Simple Forward Rates over a Fixed Period

Let $f(t,T,\alpha)$ denote the simple forward rate at time t over a fixed period of length α prevailing at date t for the future time interval $[T,T+\alpha]$. According to Miltersen et al. (1997) the simple forward rates will be modelled as log-normal diffusions:

$$df(\cdot,T,\alpha)_t = \mu(t,T,\alpha)\cdot f(t,T,\alpha)\,dt + \gamma(t,T,\alpha)\cdot f(t,T,\alpha)\,dW_t$$

Let $P(t,T)$ denote the price at time t of a default-free zero-coupon bond that pays 1 at maturity date T. It then follows that

$$P(t,T+\alpha) = P(t,T)\,\frac{1}{1+\alpha f(t,T,\alpha)}$$

If the forward price $F(t,T,\alpha)$ of the contract is defined as the fixed price which the buyer agrees to pay at date T for the bond with maturity $T+\alpha$, such that the value of the forward contract at date t is zero, by no-arbitrage assumptions the forward price is given by

$$F(t,T,\alpha) = \frac{P(t,T+\alpha)}{P(t,T)}.$$

3. A Closed-form Solution for the n-fold Compound Call Option

We reintroduce the notion of an n-fold compound call option:

Definition 3.1. By induction:
A compound call option (of order 2) is a call option on a call option. This can be generalized to a compound call of order n (with exercise date and price given by t_1 and K_1) with an underlying call of order $n-1$.

Notations:

V : current market value of the firm,
t_i : maturity date of investment for the compound call option C_i,
K_i : exercise price for the compound call option C_i,
C_i : current value of the compound call option on the option C_{i+1},
r : risk-free rate of interest,
σ^2 : instantaneous variance of the return on the assets of the firm,
$N_n(a_1, a_2, \ldots, a_n; F)$: n-variate cumulative normal distribution function with a_i as upper limits and F as the correlation matrix.

Theorem 3.2. *Suppose for $s = n, n-1, \ldots, 2$ the calls C_s of order $n+1-s$ are known and given by:*

$$\begin{aligned} C_s &= V\, N_{n+1-s}\left(a_s, a_{s+1}, \ldots, a_n; F_s^{n+1-s}\right) \\ &\quad - \sum_{m=s}^{n} K_m\, e^{-r(t_m-t)}\, N_{m+1-s}\left(b_s, b_{s+1}, \ldots, b_m; F_s^{m+1-s}\right), \end{aligned}$$

with the notations

$$\begin{aligned} a_\ell &= b_\ell + \sigma\sqrt{t_\ell - t} && \ell = 1,2,\ldots,n \\ b_\ell &= \frac{\ln\frac{V}{\overline{V}_\ell} + (r - \frac{\sigma^2}{2})(t_\ell - t)}{\sigma\sqrt{t_\ell - t}} && \ell = 1,2,\ldots,n \\ \overline{V}_\ell &= \textit{solution of the equation} \\ &\quad\; C_{\ell+1}(V, t_\ell) = K_\ell && \ell = 1,2,\ldots,n-1 \\ \overline{V}_n &= K_n \\ \rho_{ij} &= \sqrt{\frac{t_i - t}{t_j - t}} && i < j \\ F_s^\ell &= \left(f_{ij}^\ell\right)_{i,j=1,2,\ldots,\ell} \quad \textit{where} \begin{cases} f_{ii} = 1 \\ f_{ij} = \rho_{i+s-1,j+s-1} & i<j \end{cases}. \end{aligned}$$

Then the n-fold compound call option can be found:

$$\begin{aligned} C_1 &= V\,N_n\left(a_1, a_2, \ldots, a_n; F_1^n\right) \\ &\quad - \sum_{m=1}^{n} K_m\, e^{-r(t_m-t)}\, N_m\left(b_1, b_2, \ldots, b_m; F_1^m\right). \end{aligned}$$

For a proof the reader is referred to the article by Thomassen and Van Wouwe (2002).

We now make a particular choice for the underlying by taking a forward F instead of the current market value of the firm V. This implies that we will use the forward price $\hat{C}$ for a call option C, defined by

$$\hat{C}(t, F(t,T,\alpha)) \equiv \frac{C}{P(t,T)}.$$

With the adapted definition for an n-fold compound call option we are in the possibility to value the forward price of an n-fold compound call option. The valuation is proved in the following theorem by induction:

Theorem 3.3. *The following notations are introduced:*

$$\begin{array}{lcll} \alpha_i &=& t_{i+1} - t_i & \forall i = 1, \ldots, n \\ s_i &=& \displaystyle\int_t^{t_i} \gamma^2(r, t_n, \alpha_n)\, dr & \forall i = 1, \ldots, n \\ F_i && \text{is solution of } P(t_i, t_{i+1}) \cdot \hat{C}_{i+1}(t_i, F) = K_i & \forall i = 1, \ldots, n-1 \\ F_n &=& K_n & \\ a_i &=& b_i + \sqrt{s_i} & \forall i = 1, \ldots, n \\ b_i &=& \dfrac{1}{\sqrt{s_i}}\left(\ln\left|\dfrac{F(1-F_i)}{F_i(1-F)}\right| - \dfrac{s_i}{2}\right) & \forall i = 1, \ldots, n \end{array}$$

where F is used as the abbreviated notation for the forward rate $F(t, t_n, \alpha_n)$. Suppose $\forall v = n, n-1, \ldots, 3, 2$ the value of the (n-v+1)-fold is known and given by

$$\begin{aligned} \hat{C}_v &= \prod_{i=v}^{n-1} P(t_i, t_{i+1}) \cdot F \cdot N_{n-v+1}(a_v, a_{v+1}, \ldots, a_n; G_n^v) \\ &\quad - \sum_{i=v}^{n} K_i \cdot F \cdot N_{i-v+1}(a_v, a_{v+1}, \ldots, a_i; G_i^v) \cdot \prod_{j=v}^{i-1} P(t_j, t_{j+1}) \\ &\quad - \sum_{i=v}^{n} K_i \cdot (1-F) \cdot N_{i-v+1}(b_v, b_{v+1}, \ldots, b_i; G_i^v) \cdot \prod_{j=v}^{i-1} P(t_j, t_{j+1}) \end{aligned}$$

then the value of the n-fold compound call is determined by the following closed-form solution:

$$\begin{aligned}\hat{C}_1 &= \prod_{i=1}^{n-1} P(t_i,t_{i+1}) \cdot F \cdot N_n(a_1,a_2,\ldots,a_n;G_n^1)\\ &\quad -\sum_{i=1}^{n} K_i \cdot F \cdot N_i(a_1,a_2,\ldots,a_i;G_i^1) \cdot \prod_{j=1}^{i-1} P(t_j,t_{j+1})\\ &\quad -\sum_{i=1}^{n} K_i \cdot (1-F) \cdot N_i(b_1,b_2,\ldots,b_i;G_i^1) \cdot \prod_{j=1}^{i-1} P(t_j,t_{j+1})\end{aligned}$$

where the correlation matrices are defined by the structure

$$G_y^x = (g_{ab})_{a,b=1,2,\ldots,y-x+1} \quad \text{with} \left\{ \begin{array}{l} g_{aa} = 1 \\ g_{ab} = g_{ba} = \sqrt{\dfrac{s_{a+x-1}}{s_{b+x-1}}} \quad \text{if } a < b \end{array} \right.$$

(where $\prod$ equals 1 if the upper limit is smaller than the lower limit and where $\sum$ equals 0 if the upper limit is smaller than the lower limit).

Proof. $n=1$: This result was obtained by Miltersen et al. (1997).

$n>1$: Suppose $\forall v = n, n-1, \ldots, 3, 2$ the value of the $(n-v+1)$-fold compound call $\hat{C}_v$ is given by

$$\begin{aligned}\hat{C}_v &= \prod_{i=v}^{n-1} P(t_i,t_{i+1}) \cdot F \cdot N_{n-v+1}(a_v,a_{v+1},\ldots,a_n;G_n^v)\\ &\quad -\sum_{i=v}^{n} K_i \cdot F \cdot N_{i-v+1}(a_v,a_{v+1},\ldots,a_i;G_i^v) \cdot \prod_{j=v}^{i-1} P(t_j,t_{j+1})\\ &\quad -\sum_{i=v}^{n} K_i \cdot (1-F) \cdot N_{i-v+1}(b_v,b_{v+1},\ldots,b_i;G_i^v) \cdot \prod_{j=v}^{i-1} P(t_j,t_{j+1}).\end{aligned}$$

For the n-fold compound call C_1 we construct the following hedge:

$$C_1 = \psi_t^1 \cdot P(t,t_1) \cdot F(t,t_n,\alpha_n) + \psi_t^2 \cdot P(t,t_1)$$

so that the forward value fulfills

$$\hat{C}_1 = \psi_t^1 \cdot F(t,t_n,\alpha_n) + \psi_t^2.$$

Using Itô calculus on the one hand and the fact of a self-financing portfolio on the other, we obtain the PDE for the forward price of the compound call option:

$$\frac{\partial \hat{C}_1}{\partial t} + \frac{1}{2}\gamma^2(t,t_n,\alpha_n) \cdot (1-F(t,t_n,\alpha_n))^2 \cdot F^2(t,t_n,\alpha_n) \cdot \frac{\partial^2 \hat{C}_1}{\partial F^2} = 0. \tag{3.1}$$

By analogous substitutions as in Miltersen et al. (1997):

$$\begin{aligned} t &\rightarrow s_1 = \int_t^{t_1} \gamma^2(r, t_n, \alpha_n)\, dr \\ F &\rightarrow z = \ln\left|\frac{F}{1-F}\right| \quad \left(\text{suppose } F < 1 \text{ so } F = \frac{1}{1+\exp(-z)}\right) \\ g(s_1, z) &= \hat{C}_1(t, F) \\ a(z) &= \frac{1}{\exp(z/2) + \exp(-z/2)} \\ b(s_1) &= \exp(-s_1/8) \\ h(s_1, z) &= \frac{g(s_1, z)}{a(z) \cdot b(s_1)} \end{aligned}$$

the above equation (3.1) is rewritten as a diffusion equation $\frac{1}{2}\, h_{zz} - h_{s_1} = 0$ and solved using a Green's function of the form

$$G(s_1, z - z') = \frac{1}{\sqrt{2\pi s_1}} \exp\left(-\frac{(z-z')^2}{2s_1}\right).$$

After some calculations, the solution of the diffusion equation is given by:

$$\begin{aligned} h(s_1, z) &= \prod_{i=1}^{n-1} P(t_i, t_{i+1}) \exp\left[\frac{s_1}{8} + \frac{z}{2}\right] \\ &\int_{\left\{\ln\left|\frac{F_1}{1-F_1}\right| - z - \frac{s_1}{2}\right\}/\sqrt{s_1}}^{+\infty} N_{n-1}(a_2^*, \cdots, a_n^*; G_n^{2*})\, \phi(u)\, du \\ &- \sum_{i=1}^{n} K_i \prod_{j=1}^{i-1} P(t_j, t_{j+1}) \exp\left[\frac{s_1}{8} + \frac{z}{2}\right] \\ &\int_{\left\{\ln\left|\frac{F_1}{1-F_1}\right| - z - \frac{s_1}{2}\right\}/\sqrt{s_1}}^{+\infty} N_{i-1}(a_2^*, \cdots, a_i^*; G_i^{2*})\, \phi(u)\, du \\ &- \sum_{i=1}^{n} K_i \prod_{j=1}^{i-1} P(t_j, t_{j+1}) \exp\left[\frac{s_1}{8} - \frac{z}{2}\right] \\ &\int_{\left\{\ln\left|\frac{F_1}{1-F_1}\right| - z + \frac{s_1}{8}\right\}/\sqrt{s_1}}^{+\infty} N_{i-1}(b_2^*, \cdots, b_i^*; G_i^{2*})\, \phi(w)\, dw \end{aligned}$$

with the notation

$$\phi(u) = \frac{1}{\sqrt{2\pi}} \exp\left[-\frac{u^2}{2}\right]$$

and with

$$G_j^{2*} = (g_{ab})_{a,b=1,2,\cdots,j-1} \quad \text{with} \quad \begin{cases} g_{aa} = 1 \\ g_{ab} = g_{ba} = \sqrt{\dfrac{s_{a+1}^*}{s_{b+1}^*}} \quad \text{if } a < b. \end{cases}$$

So the final boundaries and correlations are given by the following expressions:

$$s_j^* = s_j(t_1) = \int_{t_1}^{t_j} \gamma^2(r, t_n, \alpha_n)\, dr.$$

Observing that

$$s_1 + s_j^* = s_j,$$

we obtain

$$\begin{aligned} b_j^* &= \frac{1}{\sqrt{s_j^*}} \left\{ \sqrt{s_1} w + \ln \left| \frac{F[1-F_j]}{F_j[1-F]} \right| - \frac{s_j}{2} \right\} \\ a_j^* &= \frac{1}{\sqrt{s_j^*}} \left\{ \sqrt{s_1} u + \ln \left| \frac{F[1-F_j]}{F_j[1-F]} \right| + \frac{s_j}{2} \right\}. \end{aligned}$$

Using the appendix the final equation writes:

$$\begin{aligned} h(s_1, z) &= \prod_{i=1}^{n-1} P(t_i, t_{i+1}) e^{s_1/8} e^{z/2} N_n(a_1, a_2, \ldots, a_n; M_n) \\ &\quad - \sum_{i=1}^{n} K_i \prod_{j=1}^{i-1} P(t_j, t_{j+1}) e^{s_1/8} e^{z/2} N_i(a_1, a_2, \ldots, a_i; M_i) \\ &\quad - \sum_{i=1}^{n} K_i \prod_{j=1}^{i-1} P(t_j, t_{j+1}) e^{s_1/8} e^{-z/2} N_i(b_1, b_2, \ldots, b_i; M_i) \end{aligned}$$

with correlation matrices

$$M_j = (m_{uv})_{u,v=1,2,\ldots,j} \quad \text{with} \quad \begin{cases} m_{uu} = 1 \\ m_{uv} = m_{vu} = \sqrt{\dfrac{s_1 + s_u^*}{s_1 + s_v^*}} = \sqrt{\dfrac{s_u}{s_v}} & \text{if } u < v \end{cases}$$

or in other words: $M_j \equiv G_j^1$.
Returning to the value of the forward n-fold compound call option we conclude:

$$\begin{aligned} \hat{C}_1(t, F) &= \prod_{i=1}^{n-1} P(t_i, t_{i+1}) \cdot F \cdot N_n(a_1, \ldots, a_n; G_n^1) \\ &\quad - \sum_{i=1}^{n} K_i \cdot \prod_{j=1}^{i-1} P(t_j, t_{j+1}) \cdot F \cdot N_i(a_1, \ldots, a_i; G_i^1) \\ &\quad - \sum_{i=1}^{n} K_i \cdot \prod_{j=1}^{i-1} P(t_j, t_{j+1}) \cdot [1 - F] \cdot N_i(b_1, \ldots, b_i; G_i^1) \end{aligned}$$

For a detailed proof, the reader is referred to Thomassen and Van Wouwe (2003).
□

4. Robust Parameter Estimate: the Influence on the Numerical Calculation of the n-fold Compound Option

For the numerical estimate of the n-fold compound option we return to the situation of a constant interest rate model, because no data on forwards are available. Under this constraint it was proved in Thomassen and Van Wouwe (2002) that the value of the n-fold compound option is given by Theorem 3.2.

The problem for the numerical evaluation essentially reduces to carefully choosing the estimates for the risk-free interest rate and the volatility of the underlying asset. In Thomassen and Van Wouwe (2002) the sensitivity of the n-fold compound option towards these two variables was analyzed and they found the following expressions

$$\begin{aligned}\frac{\partial C_1}{\partial \sigma} &= \sum_{j=0}^{n-1} \frac{1}{\sqrt{2\pi}} \exp\left[-\frac{1}{2} b_{j+1}^2\right] \overline{V}_{j+1} \sqrt{t_{j+1}-t} \exp\left[-r(t_{j+1}-t)\right] \\ &\quad \cdot N_{n-1}\left(\frac{b_1-\rho_{1,j+1}\, b_{j+1}}{\sqrt{1-\rho_{1,j+1}^2}}, \ldots, \frac{b_j-\rho_{j,j+1}\, b_{j+1}}{\sqrt{1-\rho_{j,j+1}^2}}, a_{j+2}^*, \ldots, a_n^*; G^{nj}\right)\end{aligned}$$

with the boundaries $a_{j+2}^*, \cdots, a_n^*$ taken at time $t = t_{j+1}$ and in the value $V = V_{j+1}$,

$$\frac{\partial C_1}{\partial r} = \sum_{m=1}^{n} K_m\, e^{-r(t_m-t)} (t_m - t)\, N_m(b_1, b_2, \ldots, b_m; F_1^m)$$

revealing a positive sensitivity of the compound call C_1 towards r and σ.

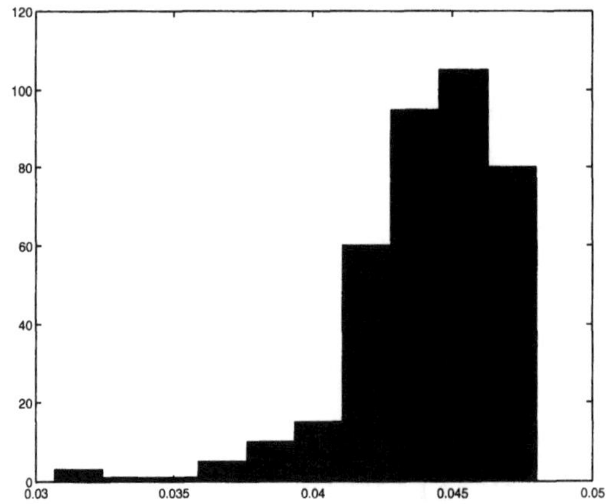

FIGURE 1. Interest rates.

For a particular data set of interest rates (Source: Data stream, interest rates on three month – Belgian Treasury bonds, see Figure 1) the classical average equals 0.0441. In applying some robust statistics, the MCD location estimator (see

Rousseeuw and Van Driessen, 1999) may be considered and provides an estimate of 0.0447 for the same data set. Since we want a 'safe' value for the n-fold compound option from the point of view of the buyer in exploiting the positive sensitivity of the compound option towards the interest rate, the robust estimate has to be preferred.

For an adequate choice of the volatility problems arise. The chosen data set (Daily volatility of Solvay during the period of the last 5 years) does not obey the elliptic condition. The application of the MCD location estimator reveals the presence of a number of outliers. In our search for robust methods which emphasize no assumptions on the distribution, we meet the Hodges-Lehmann location estimator (by Hodges and Lehmann, 1963). An overview of the different estimates gives:

$$\sigma^2_{MI} = 0.0003029498$$

$$\sigma^2_{MCD} = 0.000266752$$

$$\sigma^2_{HL} = 0.000286128.$$

We are confronted with the fact that the MCD location estimator does not provide an estimate for the volatility as expected. In the next table we give a picture of the influence of taking the different estimates for the interest rate and the volatility on the value of the n-fold compound option. We assume a 4-fold compound call option with

- risk free rate of interest $\ln(1 + \frac{r}{100})$
- volatility σ
- exercise prices
  $K_1 = 100\,(1 - e^{-r})$, $K_2 = 100\,(1 - e^{-r})$, $K_3 = 100\,(1 - e^{-r})$, $K_4 = 100$
- maturity dates:
  $t_1 = 21$, $t_2 = 22$, $t_3 = 23$, $t_4 = 24$
- present time $t_0 = 0$
- present value at time t_0: $V = 100$.

The table expresses a strong dependency of the 4-fold towards changes in the interest rate. The changes in the volatility are however too small to elaborate differences in the value of the 4-fold compound option.

TABLE 1. 4-fold compound call evaluated for different estimates of interest rates and volatilities.

σ^2 \ r	0.0441	0.0447
$\sigma^2_{MI} = 0.0003029498$	59.5968	60.0813
$\sigma^2_{MCD} = 0.000266758$	59.5968	60.0813
$\sigma^2_{HL} = 0.000286128$	59.5968	60.0813

We observe from the numerical evaluation of the 4-fold compound option that the robust estimator for the interest rates can be a useful tool for a safer valuation. The differences between the different estimates for the volatility are too small to express a preference between these estimators.

5. Conclusion

In considering a log-normal model for the interest rate a closed-form generalization of the n-fold compound option is introduced and proved.
The numerical evaluation of the n-fold compound option shows that the robust estimators can be used for 'safe' valuations of the n-fold compound option.

Appendix

To integrate a product of an exponential and an n-variate normal CDF, we need the following theorem:

Theorem A.1

$$\int_{-b_1}^{+\infty} \frac{1}{\sqrt{2\pi}} \exp\left(-\frac{q^2}{2}\right) N_{i-1}\left(\frac{fq+b_2}{g_2}, \cdots, \frac{fq+b_i}{g_i}; R\right) dq$$

$$= N_i\left(b_1, \frac{b_2}{\sqrt{f^2+g_2^2}}, \cdots, \frac{b_i}{\sqrt{f^2+g_i^2}}; H\right)$$

with correlation matrices

$$R = (r_{ab})_{a,b=1,2,\ldots,i-1} \quad \textit{such that} \begin{cases} r_{aa} = 1 & \\ r_{ab} = r_{ba} = \dfrac{g_{a+1}}{g_{b+1}} & \textit{for } a < b \end{cases}$$

$$H = (h_{ab})_{a,b=1,2,\ldots,i} \quad \textit{such that} \begin{cases} h_{aa} = 1 & \\ h_{ab} = h_{ba} = \sqrt{\dfrac{f^2+g_a^2}{f^2+g_b^2}} & \textit{for } a < b \end{cases}$$

and where for convenience $g_1 \equiv 0$

Proof. The proof can be found in corollary A.1 of a paper by Thomassen and Van Wouwe (2002).

References

[1] F. Black and M. Scholes, *The Pricing of Options and Corporate Liabilities.* J. Polit. Econ. **81** (1973), 637–654.

[2] R. Geske, *The Valuation of Compound Options.* J. Financial Econ. **7** (1979), 63–81.

[3] J.L. Hodges and E.L. Lehmann, *Estimates of Location Based on Rank Tests.* Ann. Math. Statist. **34** (1963), 598–611.

[4] K.R. Miltersen, K. Sandmann, and D. Sondermann, *Closed Form Solutions for Term Structure Derivatives with Log-Normal Interest Rates.* J. Finance **52** (1997), 409–430.

[5] P.J. Rousseeuw and K. Van Driessen, *A Fast Algorithm for the Minimum Covariance Determinant Estimator.* Technometrics **41** (1999), 212–223.

[6] K. Sandmann and D. Sondermann, *On the Stability of Log-Normal Interest Rate Models.* Working paper, University of Bonn, Bonn, Germany, 1994, available at *http://www.finasto.uni-bonn.de/papers/bonnsfb263.html.*

[7] L. Thomassen and M. Van Wouwe, *The n-fold Compound Option: Theory and Sensitivity.* Submitted, 2002.

[8] L. Thomassen and M. Van Wouwe, *The Influence of a Stochastic Interest Rate on the n-fold Compound Option.* Research Paper 2003-010, UA Faculty of Applied Economics, Antwerp, 2003, available at *http://www.ua.ac.be/tew.*

L. Thomassen and M. Van Wouwe
Faculty of Applied Economics
University of Antwerp
Prinsstraat 13
B-2000 Antwerp
Belgium
e-mail: {liesbeth.thomassen, martine.vanwouwe}@ua.ac.be

Statistics for Industry and Technology, 355–366

Robust Estimations for Multivariate Sinh^{-1}-Normal Distribution

A. Toma

Abstract. In this paper we construct estimators for the parameters of the multivariate $\sinh^{-1}$-normal distribution. We discuss some properties of these estimators, such as consistency, B-robustness and asymptotic normality.

Mathematics Subject Classification (2000). Primary 62G35; Secondary 62H12.

Keywords. Multivariate $\sinh^{-1}$-normal distribution, influence function.

1. Introduction

The problem of robust estimation for the parameters of the multivariate distributions was especially studied in the case of the elliptical distributions and particularly for the multivariate normal distribution. In this paper we consider the multivariate $\sinh^{-1}$-normal distribution. This distribution is generated from the normal distribution by using the inverse hyperbolic sine normal transformation suggested by Johnson (1949) (who generate his system of distributions called Johnson translation system). In using this distribution it can be helpful to specify the mean vector and the covariance structure. This facet of its use has played an important role in discriminant analysis simulation studies (see Johnson, 1987). Several Monte Carlo studies in discriminant analysis have used the members of Johnson translation system, particularly the multivariate $\sinh^{-1}$-normal distribution to model the distributions governing populations. One of the first papers was by Lachenbruch et al. (1973), who studied the estimation of error rates in linear and quadratic discrimination under departures from the multivariate normal distribution. Their study provided the distributional framework for subsequent Monte Carlo work by Koffler and Penfield (1979). One of the specific distributions used from the Johnson system was the ten-variate $\sinh^{-1}$-normal distribution. Also Conover and Iman (1980) used the multivariate $\sinh^{-1}$-normal distribution to compare discriminant analysis procedures. The purpose of this paper is to construct robust estimators for the parameters of the p-variate $\sinh^{-1}$-normal distribution.

For a given p-dimensional $\sinh^{-1}$-normal data set, we define estimators for the mean vector and the covariance matrix, by applying a $\sinh^{-1}$-transform to the data, computing robust estimators for the obtained Gaussian data and transforming those estimators for the $\sinh^{-1}$-normal, using the relationship between the parameters of both distributions. For every choice of robust estimators in the p-variate normal case, other robust estimators for the parameters of the p-variate $\sinh^{-1}$-normal distribution are obtained. The problem of robust estimation in the p-variate normal case has been widely studied, in this sense an overview of some existing estimators of multivariate location and scatter matrix being given in Maronna and Yohai (1998).

In the following we consider those estimators with the property that there exists a functional T defined on a convex set D_T of distributions and valued in the parameter space Θ, such that, if $X_1, X_2, \ldots, X_n$ are random vectors i.i.d. with the distribution $G = G_\theta \in D_T$, then

$$T_n(X_1, X_2, \ldots, X_n) \xrightarrow{P} T(G).$$

The notation G is the same for a probability distribution and its corresponding cumulative distribution function. The functional T is called Fisher consistent if $T(G_\theta) = \theta$, for all $\theta \in \Theta$.

The influence function of the functional T in G measures the effect of an infinitesimal contamination located at a single point x. We thus consider the contaminated distribution $\widetilde{G}_{\varepsilon x} = (1-\varepsilon)G + \varepsilon\delta_x$, where δ_x is the distribution putting all its mass at x. Then the influence function is defined by (see Hampel et al., 1986)

$$IF(x; T, G) = \lim_{\varepsilon \to 0} \frac{T\left(\widetilde{G}_{\varepsilon x}\right) - T(G)}{\varepsilon}.$$

The gross error sensitivity measures approximately the maximum contribution to the estimation error that can be produced by a single outlier and is defined as $\sup_x \|IF(x; T, G)\|$. Whenever the gross error sensitivity is finite, the functional T is called B-robust. It is often said that the estimator T_n associated with the functional T is called B-robust.

Throughout the paper, F denotes the p-variate normal distribution $N_p(\mu, V)$, where $\mu \in \mathbb{R}^p$ and $V \in$SPD(p), with SPD(p) the set of all symmetric and positive definite matrices. F' is the p-variate $\sinh^{-1}$-normal distribution with mean m and covariance matrix S. We will consider estimators t_n and C_n of μ and V, respectively, and will denote by t and C the corresponding statistical functionals. We will suppose that t and C are B-robust and Fisher consistent. Particularly we will take the case when the functionals t and C are affine equivariant meaning that

$$t(AX + b) = At(X) + b$$

$$C(AX + b) = AC(X)A^t,$$

for any $b \in \mathbb{R}^p$ and any $p \times p$ nonsingular matrix A (here the notation $T(Z)$ instead of $T(G)$ means that $Z \sim G$).

The outline of the paper is as follows. In Section 2, we construct estimators for the parameters of the p-variate sinh^{-1}-normal distribution and prove their consistency. In Section 3 we discuss the B-robustness properties, while in Section 4 we give some asymptotic results.

2. The Estimators

Let $X = \left(X^1, \ldots, X^p\right)^t$ be a random vector with normal $N_p(\mu, V)$ distribution and define

$$Y = \left(Y^1, \ldots, Y^p\right)^t = \left[\sinh(X^1), \ldots, \sinh(X^p)\right]^t,$$

where $\sinh(x) = \left(e^x - e^{-x}\right)/2$. The random vector Y has a p-variate sinh^{-1}-normal distribution. For notation let

$$E(Y^i) = m_i,$$

$$\operatorname{cov}\left(Y^i, Y^j\right) = s_{ij},$$

for all $i, j = 1, \ldots, p$ and let μ_i and v_{ij} denote analogous quantities in the $N_p(\mu, V)$ distribution. The following relationships hold:

$$m_i = e^{v_{ii}/2} \sinh(\mu_i), \tag{2.1}$$

$$s_{ii} = \left[e^{v_{ii}} - 1\right] \frac{1 + \cosh(2\mu_i) e^{v_{ii}}}{2} \tag{2.2}$$

$$s_{ij} = e^{(v_{ii}+v_{jj})/2} \left[e^{v_{ij}} \frac{\cosh(\mu_i + \mu_j)}{2} - e^{-v_{ij}} \frac{\cosh(\mu_i - \mu_j)}{2} - \sinh(\mu_i)\sinh(\mu_j)\right] \tag{2.3}$$

for all $i, j = 1, \ldots, p$, where $\cosh(x) = \left(e^x + e^{-x}\right)/2$. These formulae can be found for instance in Johnson (1987).

In the following, for any $p \times p$ symmetric matrix $A = (a_{ij})$, let uvec A denote the $p(p+1)/2$ dimensional column vector $(a_{11}, a_{22}, \ldots, a_{pp}, a_{12}, a_{13}, \ldots, a_{p-1,p})^t$ formed from the elements in the upper triangular half of A, including the diagonal elements.

We consider the functions

$$f : \mathbb{R}^{p+p(p+1)/2} \longrightarrow \mathbb{R}^p \text{ and } g : \mathbb{R}^{p+p(p+1)/2} \longrightarrow \mathbb{R}^{p(p+1)/2}$$

defined respectively by

$$f\begin{pmatrix} x \\ \operatorname{uvec} X \end{pmatrix} = y \text{ and } g\begin{pmatrix} x \\ \operatorname{uvec} X \end{pmatrix} = \operatorname{uvec} Y$$

with

$$y_i = e^{x_{ii}/2} \sinh(x_i) \tag{2.4}$$

$$y_{ij} = e^{(x_{ii}+x_{jj})/2} \left[e^{x_{ij}} \frac{\cosh(x_i + x_j)}{2} - e^{-x_{ij}} \frac{\cosh(x_i - x_j)}{2} - \sinh(x_i)\sinh(x_j)\right] \tag{2.5}$$

for all $i, j = 1, \ldots, p$, where $x = (x_1, \ldots, x_p)^t$, $y = (y_1, \ldots, y_p)^t$ and $X = (x_{ij})$, $Y = (y_{ij})$ are $p \times p$ symmetric matrices. Note that these functions are defined in accordance with the relationship between the parameters of both distributions.

Let $Y_1, \ldots, Y_n$ be a sample drawn from F' which is supposed to be the p-variate $\sinh^{-1}$-normal distribution with mean $m = (m_1, \ldots, m_p)^t$ and covariance matrix $S = (s_{ij})_{1 \leq i,j \leq p}$. We define $m_n(Y_1, \ldots, Y_n)$ and $W_n(Y_1, \ldots, Y_n)$ the estimators of the mean m and of the covariance matrix S, respectively by

$$m_n(Y_1, \ldots, Y_n) = f\begin{pmatrix} t_n(X_1, \ldots, X_n) \\ \operatorname{uvec} C_n(X_1, \ldots, X_n) \end{pmatrix} \tag{2.6}$$

$$\operatorname{uvec} W_n(Y_1, \ldots, Y_n) = g\begin{pmatrix} t_n(X_1, \ldots, X_n) \\ \operatorname{uvec} C_n(X_1, \ldots, X_n) \end{pmatrix} \tag{2.7}$$

where $X_k^l = \ln\left(Y_k^l + \sqrt{(Y_k^l)^2 + 1}\right)$ for all $k = 1, \ldots, n$ and $l = 1, \ldots, p$, X_k^l, Y_k^l being the components of the random vectors X_k and Y_k, respectively.

The problem of robust estimation for the parameters of the multivariate normal distribution has been widely studied. We give here two examples: the M-estimators (see Maronna, 1976) and the S-estimators (see Davies, 1987; Rousseeuw and Leroy, 1987).

In the following, the notation $\|x\|_A^2$ will be used for $x^t A x$ where $x \in \mathbb{R}^p$ and A is a symmetric positive definite matrix. Consider a sample of p-dimensional observations $X_1, \ldots, X_n$. The M-estimators are implicitly defined by

$$\begin{aligned} t_n &= \frac{\sum_{i=1}^n w_1\left(\|X_i - t_n\|_{C_n^{-1}}\right) X_i}{\sum_{i=1}^n w_1\left(\|X_i - t_n\|_{C_n^{-1}}\right)} \\ C_n &= \frac{1}{n}\sum_{i=1}^n w_2\left(\|X_i - t_n\|_{C_n^{-1}}^2\right)(X_i - t_n)(X_i - t_n)^t \end{aligned}$$

where w_1 and w_2 are specified weight functions.

The S-estimators are defined as the solutions (t_n, C_n) to the problem of minimization of $\det(C)$ subject to

$$\frac{1}{n}\sum_{i=1}^n \rho\left(\|X_i - t\|_{C^{-1}}\right) = k_p$$

among all $(t, C) \in \mathbb{R}^p \times \mathrm{SPD}(p)$. The function ρ is nondecreasing, $k_p = E_{F_0}\rho(\|x\|)$, $F_0 = \mathcal{N}_p(0, I_p)$. The S-estimators have excellent robustness properties. Moreover they are fast to compute, in this sense the SURREAL algorithm of Ruppert (1992) is recommended. Theoretical results favor the use of the S-estimators since they combine high efficiency with appealing robustness properties including a smooth influence function.

Remark 2.1. The statistical functionals corresponding to the estimators m_n and W_n are given by

$$m(G) = f\begin{pmatrix} t(G \circ u) \\ \operatorname{uvec} C(G \circ u) \end{pmatrix}, \quad \operatorname{uvec} W(G) = g\begin{pmatrix} t(G \circ u) \\ \operatorname{uvec} C(G \circ u) \end{pmatrix}$$

where u is the function $u(x_1, \ldots, x_p) = [\sinh(x_1), \ldots, \sinh(x_p)]$.

Theorem 2.2. *The estimators m_n and W_n have the properties*

$$m_n \xrightarrow{P} m(F') \text{ and } W_n \xrightarrow{P} W(F')$$

where

$$m(F') = f\begin{pmatrix} t(F' \circ u) \\ \operatorname{uvec} C(F' \circ u) \end{pmatrix}, \quad \operatorname{uvec} W(F') = g\begin{pmatrix} t(F' \circ u) \\ \operatorname{uvec} C(F' \circ u) \end{pmatrix} \tag{2.8}$$

where u is the function from Remark 2.1.

Proof. First let us observe that

$$F = F' \circ u \tag{2.9}$$

where u is the function $u(x_1, \ldots, x_p) = [\sinh(x_1), \ldots, \sinh(x_p)]$.

From the consistency of the estimators t_n and C_n we obtain the consistency of $((t_n)^t, (\operatorname{uvec} C_n)^t)^t$ and then the continuity of the functions f and g imply the consistency of m_n and W_n, respectively. The asymptotical values are obtained using (2.9). □

Let us note the fact that the Fisher consistency of the functionals m and W follows using the Fisher consistency of t and C, on the basis of the relations (2.1) and (2.3).

3. Influence Functions

In this Section we will calculate the influence functions of the functionals m and W and will prove the B-robustness of these functionals. Particularly, we will consider the case when the functionals t and C are affine equivariant.

We recall that for any affine equivariant location functional t possessing an influence function, there exists a function γ_t defined on $[0, \infty[$ and $\mathbb{R}$-valued such that

$$IF(z; t, F) = \gamma_t\left(\|z - \mu\|^2_{V^{-1}}\right)(z - \mu) \tag{3.1}$$

where F is the distribution $N_p(\mu, V)$ (see Toma, 2003).

In Lemma 1 from Croux and Haesbroeck (2000), we see that for any affine equivariant scatter functional C possessing an influence function, there exist two functions α_C and β_C defined on $[0, \infty[$ and $\mathbb{R}$ valued such that

$$IF(z; C, F) = \alpha_C\left(\|z - \mu\|^2_{V^{-1}}\right)(z - \mu)(z - \mu)^t - \beta_C\left(\|z - \mu\|^2_{V^{-1}}\right)V \tag{3.2}$$

where F is the distribution $N_p(\mu, V)$.

Theorem 3.1. *Suppose that t and C are Fisher consistent functionals. The influence function $IF(x;m,F')$ has the components*

$$IF(x;m,F')_i = e^{v_{ii}/2}\left[\cosh(\mu_i)IF(z;t,F)_i + \frac{1}{2}\sinh(\mu_i)IF(z;C,F)_{ii}\right] \quad (3.3)$$

where $z = (z_1,\dots,z_p)^t$, $z_i = \ln\left(x_i + \sqrt{x_i^2+1}\right)$ *for all* $i = 1,\dots,p$.

Proof. First we note that $\widetilde{F}'_{\varepsilon x} \circ u = \widetilde{F}_{\varepsilon z}$, where

$$z = (z_1,\dots,z_p)^t,\ z_i = \ln\left(x_i + \sqrt{x_i^2+1}\right) \quad \text{for all } i = 1,\dots,p,$$

and u is the function from Remark 2.1. Next, using the definition of the influence function, we have

$$\begin{aligned} IF(x;m,F') &= \lim_{\varepsilon\to 0}\frac{m\left(\widetilde{F}'_{\varepsilon x}\right) - m(F')}{\varepsilon} \\ &= \lim_{\varepsilon\to 0}\frac{1}{\varepsilon}\left[f\begin{pmatrix} t\left(\widetilde{F}_{\varepsilon z}\right) \\ \operatorname{uvec} C\left(\widetilde{F}_{\varepsilon z}\right)\end{pmatrix} - f\begin{pmatrix} t(F) \\ \operatorname{uvec} C(F)\end{pmatrix}\right]. \end{aligned}$$

Due to the Fisher consistency of t and C we obtain

$$\begin{aligned} IF(x;m,F')_i &= \lim_{\varepsilon\to 0}\frac{m\left(\widetilde{F}'_{\varepsilon x}\right)_i - m(F')_i}{\varepsilon} \\ &= \lim_{\varepsilon\to 0}\frac{1}{2\varepsilon}\left[e^{t\left(\widetilde{F}_{\varepsilon z}\right)_i + C\left(\widetilde{F}_{\varepsilon z}\right)_{ii}/2} - e^{-t\left(\widetilde{F}_{\varepsilon z}\right)_i + C\left(\widetilde{F}_{\varepsilon z}\right)_{ii}/2}\right. \\ &\quad \left. - e^{\mu_i + v_{ii}/2} + e^{-\mu_i + v_{ii}/2}\right] \\ &= \lim_{\varepsilon\to 0}\frac{1}{\varepsilon}\left\{\frac{e^{\mu_i+v_{ii}/2}}{2}\left[e^{t\left(\widetilde{F}_{\varepsilon z}\right)_i - t(F)_i + \left[C\left(\widetilde{F}_{\varepsilon z}\right)_{ii} - C(F)_{ii}\right]/2} - 1\right]\right. \\ &\quad \left. - \frac{e^{-\mu_i+v_{ii}/2}}{2}\left[e^{-t\left(\widetilde{F}_{\varepsilon z}\right)_i + t(F)_i + \left[C\left(\widetilde{F}_{\varepsilon z}\right)_{ii} - C(F)_{ii}\right]/2} - 1\right]\right\} \\ &= \frac{e^{\mu_i+v_{ii}/2}}{2}\left[IF(z;t,F)_i + \frac{1}{2}IF(z;C,F)_{ii}\right] \\ &\quad - \frac{e^{-\mu_i+v_{ii}/2}}{2}\left[-IF(z;t,F)_i + \frac{1}{2}IF(z;C,F)_{ii}\right] \\ &= e^{v_{ii}/2}\left[\cosh(\mu_i)IF(z;t,F)_i + \frac{1}{2}\sinh(\mu_i)IF(z;C,F)_{ii}\right], \end{aligned}$$

for every $i = 1,\dots,p$, and this completes the proof. □

Corollary 3.2. *Suppose that t and C are Fisher consistent and B-robust. Then the functional m is B-robust.*

The influence function can also be helpful for computing asymptotic variances (cfr. Hampel et al., 1986, page 226). From this point of view it is useful to consider the particular case when t and C are affine equivariant.

Corollary 3.3. *If t and C are affine equivariant, then*

$$IF\left(x; m, F'\right)_i = e^{v_{ii}/2}\Big\{\cosh(\mu_i)\gamma_t\left(\|z-\mu\|^2_{V^{-1}}\right)(z_i-\mu_i)$$
$$+\frac{1}{2}\sinh(\mu_i)\left[\alpha_C\left(\|z-\mu\|^2_{V^{-1}}\right)(z_i-\mu_i)^2-\beta_C\left(\|z-\mu\|^2_{V^{-1}}\right)v_{ii}\right]\Big\} \quad (3.4)$$

for all $i = 1, \ldots, p$.

Figure 1 and Figure 2 represent the norm of the influence function of the mean functional m for the particular case in which $F = \mathcal{N}_2(0, diag(2,1))$. For t_n and C_n both classical maximum likelihood estimators and S-estimators are considered.

In the following, $\operatorname{uvec} W$ denotes the statistical functional defined by

$$(\operatorname{uvec} W)(G) = \operatorname{uvec} W(G). \quad (3.5)$$

Theorem 3.4. *Suppose that t and C are Fisher consistent. The influence function of W in F' is given by*

$$\operatorname{uvec} IF\left(x; W, F'\right) = IF\left(x; \operatorname{uvec} W, F'\right)$$

where the ij component of $IF\left(x; W, F'\right)$ is

$$IF\left(x; W, F'\right)_{ij} = \frac{e^{(v_{ii}+v_{jj})/2}}{2}\left\{\left[e^{v_{ij}}\sinh(\mu_i+\mu_j) - e^{-v_{ij}}\sinh(\mu_i-\mu_j)\right.\right. \quad (3.6)$$
$$- 2\cosh(\mu_i)\sinh(\mu_j)]\, IF(z;t,F)_i + \left[e^{v_{ij}}\sinh(\mu_i+\mu_j) + e^{-v_{ij}}\sinh(\mu_i-\mu_j)\right.$$
$$- 2\sinh(\mu_i)\cosh(\mu_j)]\, IF(z;t,F)_j + \left[e^{v_{ij}}\cosh(\mu_i+\mu_j) - e^{-v_{ij}}\cosh(\mu_i-\mu_j)\right.$$
$$- 2\sinh(\mu_i)\sinh(\mu_j)]\left[IF(z;C,F)_{ii} + IF(z;C,F)_{jj}\right]/2$$
$$\left. + \left[e^{v_{ij}}\cosh(\mu_i+\mu_j) + e^{-v_{ij}}\cosh(\mu_i-\mu_j)\right] IF(z;C,F)_{ij}\right\},$$

for $i, j = 1, \ldots, p$, with $z = (z_1, \ldots, z_p)^t$, $z_i = \ln\left(x_i + \sqrt{x_i^2+1}\right)$ for $i = 1, \ldots, p$.

Proof. As in Theorem 3.1. we use the fact that $\widetilde{F}'_{\varepsilon x} o u = \widetilde{F}_{\varepsilon z}$, where $z = (z_1, \ldots, z_p)^t$, $z_i = \ln\left(x_i + \sqrt{x_i^2+1}\right)$ for all $i = 1, \ldots, p$ and u is the function from Remark 2.1. Then

$$IF\left(x; \operatorname{uvec} W, F'\right) = \lim_{\varepsilon\to 0}\frac{\operatorname{uvec} W\left(\widetilde{F}'_{\varepsilon x}\right) - \operatorname{uvec} W\left(F'\right)}{\varepsilon}$$
$$= \lim_{\varepsilon\to 0}\frac{1}{\varepsilon}\left[g\begin{pmatrix} t\left(\widetilde{F}_{\varepsilon z}\right) \\ \operatorname{uvec} C\left(\widetilde{F}_{\varepsilon z}\right)\end{pmatrix} - g\begin{pmatrix} t(F) \\ \operatorname{uvec} C(F)\end{pmatrix}\right].$$

Using the Fisher consistency of t and C and following an analogous calculus as in Theorem 3.1, we obtain the formula (3.6). □

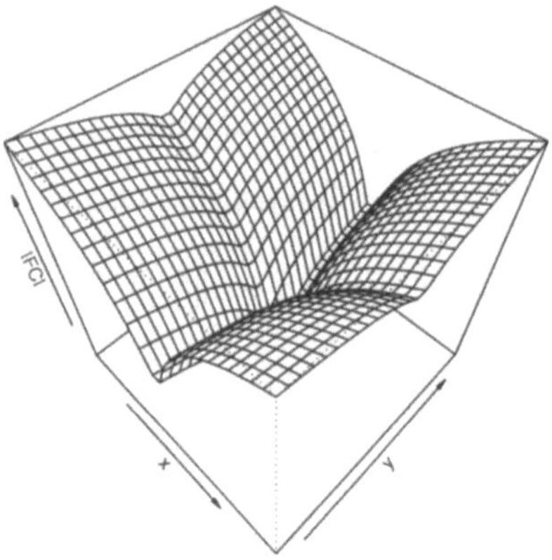

FIGURE 1. Norm of the influence function of the mean for classical MLE at $F = \mathcal{N}_2(0, \mathrm{diag}(2,1))$.

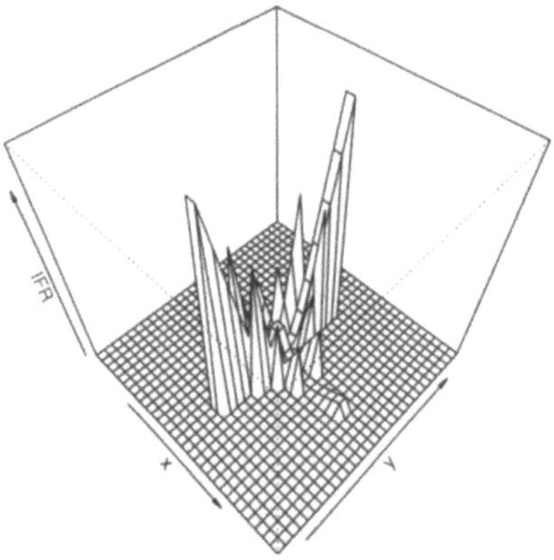

FIGURE 2. Norm of the influence function of the mean for S-estimators at $F = \mathcal{N}_2(0, \mathrm{diag}(2,1))$.

We note the fact that in the particular case when $i = j$, formula (3.6) simplifies to

$$\begin{aligned} IF\,(x; W, F')_{ii} &= e^{v_{ii}}\,(e^{v_{ii}} - 1)\sinh(2\mu_i) IF(z; t, F)_i \\ &\quad + \frac{e^{v_{ii}}}{2}\,(1 - \cosh(2\mu_i) + 2e^{v_{ii}}\cosh(2\mu_i))\, IF(z; C, F)_{ii}. \end{aligned} \tag{3.7}$$

Corollary 3.5. *Suppose that t and C are Fisher consistent and B-robust. Then the functional W is B-robust.*

Particularly, when t and C are affine equivariant, by replacing (3.1) and (3.2) in (3.6) we obtain the ij component of $IF\,(x; W, F')$.

4. Asymptotic Results

In this Section we prove the asymptotic normality of the estimators m_n and W_n and give a result regarding the asymptotic relative efficiency of the couple estimator $\left(m_n^t, (\operatorname{uvec} W_n)^t\right)^t$. The asymptotic normality is obtained through the following lemma, whose proof can be found for example in Mardia et al. (1994).

Lemma 4.1. *Let $Z_1, Z_2, \ldots$ be a sequence of random vectors valued in $\mathbb{R}^K$, asymptotically normal in the sense that*

$$\sqrt{n}\,(Z_n - a) \xrightarrow{\mathcal{L}} N_K(0, V)$$

for $a \in \mathbb{R}^K$ and V a $K \times K$ symmetric and positive definite matrix. Let also h be an application defined on $\mathbb{R}^K$ and $\mathbb{R}^L$ valued which is differentiable in the point a. Then $h(Z_1), h(Z_2), \ldots$ is asymptotically normal. More precisely,

$$\sqrt{n}\,(h(Z_n) - h(a)) \xrightarrow{\mathcal{L}} N_L\left(0, Dh(a)V\,(Dh(a))^t\right)$$

where $Dh(a)$ is the differential of h in a.

Theorem 4.2. *Suppose that $\sqrt{n}\,(t_n - \mu)$ has a limiting normal distribution with zero mean and asymptotic covariance matrix $ASV(t, F)$ and $\sqrt{n}\operatorname{uvec}(C_n - V)$ has a limiting normal distribution with zero mean and asymptotic covariance matrix $ASV(\operatorname{uvec} C, F)$. If t_n and C_n are asymptotically independent, then $\sqrt{n}\,(m_n - m)$ and $\sqrt{n}\operatorname{uvec}(W_n - S)$ have limiting normal distributions with zero means and asymptotic covariance matrices*

$$ASV(m, F') = Df\begin{pmatrix}\mu\\ \operatorname{uvec} V\end{pmatrix}\begin{pmatrix}ASV(t,F) & 0\\ 0 & ASV(\operatorname{uvec} C, F)\end{pmatrix} Df\begin{pmatrix}\mu\\ \operatorname{uvec} V\end{pmatrix}^t \tag{4.1}$$

and

$$ASV(\operatorname{uvec} W, F') = Dg\begin{pmatrix}\mu\\ \operatorname{uvec} V\end{pmatrix}\begin{pmatrix}ASV(t,F) & 0\\ 0 & ASV(\operatorname{uvec} C, F)\end{pmatrix} Dg\begin{pmatrix}\mu\\ \operatorname{uvec} V\end{pmatrix}^t. \tag{4.2}$$

Proof. By the hypothesis conditions we obtain that $\left(t_n^t, (\operatorname{uvec} C_n)^t\right)^t$ is asymptotically normal with mean $\left(\mu^t, (\operatorname{uvec} V)^t\right)^t$ and asymptotic covariance matrix

$$\begin{pmatrix}ASV(t,F) & 0\\ 0 & ASV(\operatorname{uvec} C, F)\end{pmatrix}.$$

Next we apply Lemma 4.1 to $\left(t_n^t, (\operatorname{uvec} C_n)^t\right)^t$ with $h = f$ and find that m_n is asymptotically normal with the mean m and asymptotic covariance matrix given by (4.1).

Likewise, using $h = g$ we obtain the asymptotic normality of $\operatorname{uvec} W_n$. □

As it is known the asymptotic relative efficiency of the estimator T_n of $T(G) = \theta \in \mathbb{R}^s$, could be defined whenever it exists, as

$$\text{Eff}(T,G) = \left(\frac{\det\{ASV(T_{Cl},G)\}}{\det\{ASV(T,G)\}} \right)^{1/s} \tag{4.3}$$

where $\text{ASV}(T_{Cl},G)$ is the asymptotic covariance matrix corresponding to the maximum likelihood estimator (MLE) of θ and $\text{ASV}(T,G)$ is the asymptotic covariance matrix corresponding to T_n.

In the following we suppose that the conditions of Theorem 4.2 are fulfilled. We denote by t_{Cl}, $\text{uvec}\, C_{Cl}$ and U_{Cl} the statistical functionals corresponding to the MLE of μ, $\text{uvec}\, V$ and $(m, (\text{uvec}\, S)^t)^t$, respectively; U will be the statistical functional corresponding to the couple estimator $\left(m_n^t, (\text{uvec}\, W_n)^t\right)^t$.

Theorem 4.3. *The asymptotic relative efficiency of* $\left(m_n^t, (\text{uvec}\, W_n)^t\right)^t$ *is*

$$Eff(U,F') = [Eff(t,F)]^{2/(p+3)} \, [Eff(\text{uvec}\, C,F)]^{(p+1)/(p+3)}, \tag{4.4}$$

where $Eff(t,F)$ *and* $Eff(\text{uvec}\, C,F)$ *are the asymptotic relative efficiencies for* t_n *and* $\text{uvec}\, C_n$*, respectively.*

Proof. The MLE for $\left(m^t, (\text{uvec}\, S)^t\right)^t$ is obtained from the MLE of $\left(\mu^t, (\text{uvec}V)^t\right)^t$ through the transformation h defined on $\mathbb{R}^{p+p(p+1)/2}$ and $\mathbb{R}^{p+p(p+1)/2}$ valued

$$h(x) = \left(f(x)^t, g(x)^t\right)^t,$$

because h is one to one.

Due to the fact that the maximum likelihood estimators for the parameters of the multivariate normal distribution are asymptotically independent, the asymptotic covariance matrix corresponding to the MLE of $\left(m^t, (\text{uvec}\, S)^t\right)^t$ is

$$ASV(U_{Cl},F') = Dh\begin{pmatrix}\mu \\ \text{uvec}\, V\end{pmatrix} \begin{pmatrix} ASV(t_{Cl},F) & 0 \\ 0 & ASV(\text{uvec}\, C_{Cl},F)\end{pmatrix} Dh\begin{pmatrix}\mu \\ \text{uvec}\, V\end{pmatrix}^t \tag{4.5}$$

$\text{ASV}(t_{Cl},F)$ and $\text{ASV}(\text{uvec}\, C_{Cl},F)$ being the asymptotic covariance matrices corresponding to the maximum likelihood estimators of μ and $\text{uvec}\, V$, respectively. On the other hand, the asymptotic covariance matrix corresponding to $\left(m_n^t, (\text{uvec}\, W_n)^t\right)^t$ is

$$ASV(U,F') = Dh\begin{pmatrix}\mu \\ \text{uvec}\, V\end{pmatrix} \begin{pmatrix} ASV(t,F) & 0 \\ 0 & ASV(\text{uvec}\, C,F)\end{pmatrix} Dh\begin{pmatrix}\mu \\ \text{uvec}\, V\end{pmatrix}^t. \tag{4.6}$$

Using (4.3), we obtain

$$\begin{aligned}\mathrm{Eff}(U,F') &= \left(\frac{\det\{ASV(U_{Cl},F')\}}{\det\{ASV(U,F')\}}\right)^{1/(p+p(p+1)/2)} \\ &= \left(\frac{\det\{ASV(t_{Cl},F)\}\det\{ASV(\mathrm{uvec}\,C_{Cl},F)\}}{\det\{ASV(t,F)\}\det\{ASV(\mathrm{uvec}\,C,F)\}}\right)^{2/(p^2+3p)} \\ &= [\mathrm{Eff}(t,F)]^{2/(p+3)}\,[\mathrm{Eff}(\mathrm{uvec}\,C,F)]^{(p+1)/(p+3)},\end{aligned}$$

and this completes the proof. □

5. Conclusions

In this paper we introduced estimators for the mean and covariance matrix of the multivariate sinh^{-1}-normal distribution. We proved the Fisher consistency of the corresponding functionals and derived the influence functions which appear to be bounded. We also obtained the corresponding asymptotic covariance matrices and the asymptotic relative efficiency of the couple estimator.

References

[1] W.J. Conover and R.L. Iman, *The Rank Transformation as a Method of Discrimination with Some Examples.* Comm. Statist. Theory Methods **5** (1980), 465–487.

[2] C. Croux and G. Haesbroeck, *Principal Component Analysis Based on Robust Estimators of the Covariance or Correlation Matrix: Influence Functions and Efficiencies.* Biometrika **87** (2000), 603–618.

[3] P.L. Davies, *Asymptotic Behavior of S-estimators of Multivariate Location Parameters and Dispersion Matrices.* Ann. Statist. **15** (1987), 1269–1292.

[4] F.R. Hampel, E.M. Ronchetti, P.J. Rousseeuw, and W.A. Stahel, *Robust Statistics: the Approach Based on Influence Function.* Wiley, New York, 1986.

[5] M.E. Johnson, *Multivariate Statistical Simulation.* Wiley, New York, 1987.

[6] N.L. Johnson, *System of Frequency Curves Generated by Methods of Translation.* Biometrika **36** (1949), 149–176.

[7] S.L. Koffler and D.A. Penfield, *Nonparametric Discrimination Procedures for Non-normal Distributions.* J. Statist. Comput. Simulation **8** (1979), 281–299.

[8] P.A. Lachenbruch, C. Sneeringer, and L.T. Revo, *Robustness of the Linear and Quadratic Discriminant Function to Certain Types of Non-Normality.* Comm. Statist. Theory Methods **1** (1973), 39–56.

[9] K.V. Mardia, J.T. Kent, and J.M. Bibby, *Multivariate Analysis.* Academic Press, 1994.

[10] R.A. Maronna, *Robust M-estimators of Multivariate Location and Scatter.* Ann. Statist. **4** (1976), 51–67.

[11] R.A. Maronna and V. Yohai, *Robust Estimation of Multivariate Location and Scatter.* In S. Kots, C. Read, and D. Banks, editors, Encyclopedia of Statistical Sciences. Wiley, 1998.

[12] P.J. Rousseeuw and A.M. Leroy, *Robust Regression and Outlier Detection.* Wiley, 1987.

[13] D. Ruppert, *Computing S-estimators for Regression and Multivariate Location/ Dispersion.* J. Comput. Graph. Statist. **1** (1992), 253–270.

[14] A. Toma, *Robust Estimators for the Parameters of Multivariate Lognormal Distributions.* Comm. Statist. Theory Methods **32** (2003), 1405–1417.

A. Toma
Mathematics Department
Academy of Economic Studies
Piata Romana no. 6
Bucharest
Romania
e-mail: aida_toma@yahoo.com

Statistics for Industry and Technology, 367–376

A Robust Estimator of the Tail Index Based on an Exponential Regression Model

B. Vandewalle, J. Beirlant and M. Hubert

Abstract. The objectives of a robust statistical analysis and of an extreme value analysis apparently are contradictory. Where the extreme data are downweighted in robust statistics, these observations receive most attention in an extreme value approach. The most prominent extreme value methods however are constructed on maximum likelihood estimates based on specific parametric models which are fitted to exceedances over large thresholds. So within an extreme value framework some robust algorithms replacing the maximum likelihood part of this methodology can be of use leading to estimates which are less sensitive to few particular observations. This study is motivated by a soil database quality management project, where in the background of Pareto-type tails, automatic identification of suspicious data is needed.

Mathematics Subject Classification (2000). Primary 62G35; Secondary 62G32.

Keywords. Extreme value statistics, robustness.

1. Introduction

In agriculture, a new concept of crop management has emerged, permitting within-field variation of crop techniques as, for instance, the adjustment of fertilizer inputs on the basis of soil sampling and analysis. The development of these techniques has greatly increased the demand for soil data and Laboratories are burdened with large data sets, which inevitably cause concern about outliers and quality of the information. Therefore, automatic outlier detection methods have become a necessity in the database management in order to provide high quality data to Laboratories. The present paper studies Ca and pH records from the Condroz region in Belgium (1505 observations, Goegebeur et al., 2002). The Ca distribution, conditional on pH-level, appears to be right-skewed and long-tailed, resulting in rather frequent large Ca measurements, as can be seen in the scatterplot of Ca versus pH (given in Figure 1). Robust statistical procedures which assume that the regular data points are sampled from a normal distribution will flag too many large observations as outliers. Such long tailed data can be analyzed more efficiently in the framework of extreme value theory. We will present a method which, in the context of a model with heavy tails, allows to point out potential outliers which need to be investigated before further analysis can be done.

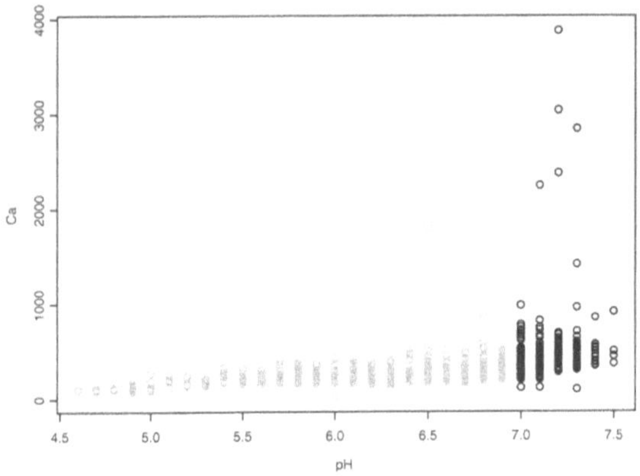

FIGURE 1. Scatterplot of Ca versus pH for one of the communities of the Condroz region in Belgium.

2. Extreme Value Statistics

In extreme value statistics the extreme value index (or tail index), denoted by γ, is used to characterize the tail behavior of a distribution. This real-valued parameter helps to indicate the size and frequency of certain extreme events under a given probability distribution: the larger γ, the heavier the tail.

Consider $X_1, \ldots, X_n$ independent and identically distributed (i.i.d.) random variables with common cumulative distribution function F and quantile function Q. Denote the corresponding order statistics by $X_{1,n} \leq \ldots \leq X_{n,n}$ and suppose that the properly centered and normed sample maxima $X_{n,n} = \max\{X_1, \ldots, X_n\}$ converge in distribution, for $n \to \infty$, to a non-degenerate limit. This limit distribution is necessarily of extreme value type. Indeed, sequences of constants $a_n > 0$ and $b_n \in \mathbb{R}$ can then be found such that

$$\lim_{n\to\infty} P\left(\frac{X_{n,n} - b_n}{a_n} \leq x\right) = H_\gamma(x), \tag{2.1}$$

with

$$H_\gamma(x) = \begin{cases} \exp\Big(-(1+\gamma x)^{-\frac{1}{\gamma}}\Big), & 1+\gamma x > 0, \gamma \neq 0, \\ \exp\Big(-\exp(-x)\Big), & x \in \mathbb{R}, \gamma = 0. \end{cases} \tag{2.2}$$

If (2.1) is satisfied, then F is said to belong to the maximum domain of attraction of H_γ, denoted as $F \in \mathcal{D}(H_\gamma)$.

Distributions F for which $\gamma > 0$ are called Pareto-type (or heavy-tailed) distributions, i.e., $\bar{F}(x) = x^{-1/\gamma} l_F(x)$ with l_F a slowly varying function. The Gumbel class of distributions with $\gamma = 0$ is a quite extensive class of distributions, mainly with exponential decreasing tails. The Weibull class, with $\gamma < 0$ consists of distributions with a finite right endpoint x_+ for which $\bar{F}(x_+ - 1/x) = x^{1/\gamma} l_F(x)$ with l_F again a slowly varying function. A general reference on extreme value statistics is Embrechts et al. (1998).

We will concentrate on Pareto type distributions, i.e., distributions for which there exists a $\gamma > 0$ such that $\bar{F}(x) = x^{-1/\gamma} l_F(x)$ or, $U(x) = x^{\gamma} l_U(x)$ with l_F, l_U slowly varying functions and $U(x) = Q(1 - 1/x)$ with Q the quantile function of F. As for slowly varying functions $l_F(tx)/l_F(t) \to 1$ for all $x > 0$ and $t \to \infty$, the conditional distribution of relative excesses $P(\frac{X}{t} > x \mid X > t)$ converges to $x^{-1/\gamma}$ for all $x > 1$.

A graphical tool for checking Pareto type behavior is the Pareto quantile plot. As log-transformed Pareto distributed random variables are exponentially distributed, the hypothesis of strict Pareto behavior can be verified by looking at an exponential quantile plot based on the log-transformed data, leading to a Pareto quantile plot

$$\left(\log\left(\frac{n+1}{j}\right), \log x_{n-j+1,n}\right), \qquad j = 1, \ldots, n. \tag{2.3}$$

For Pareto type distributions, since $U(x) = x^{\gamma} l_U(x)$, it follows that

$$\log U(x) = \gamma \log x + \log l_U(x). \tag{2.4}$$

Since $\log l_U(x)/\log x \to 0$ as $x \to \infty$ we have that $\log U(x) \sim \gamma \log x$ as $x \to \infty$, which implies that for Pareto type distributed data, the Pareto quantile plot, ultimately for the smaller j-values, shows a linear behavior, with slope γ.

In Figure 2(a), the Pareto quantile plot is shown for the variable Ca for one of the communities in the Condroz database. The last seven observations that do not follow the ultimate linearity of the rest of the Pareto quantile plot can be considered as outliers with respect to the Pareto model. As can be seen from the Ca versus pH scatterplot given in Figure 1, extreme Ca measurements tend to occur more often at higher pH levels, justifying a tail analysis conditional on higher pH levels. In Figure 2(b), the Pareto quantile plot is shown for the variable Ca, conditional on pH-levels from 7 up to 7.5 (428 observations). Now, six observations can be indicated as possible outliers with respect to the Pareto model. We will use the data set with pH-levels from 7 up to 7.5 in the sequel.

The problem of estimation of the extreme value index, and of extreme quantiles and small exceedance probabilities, in case the distribution is of Pareto type, has been studied in great detail in the recent literature. Hill (1975) introduced the estimator

$$\hat{\gamma}_{k,H} = \frac{1}{k} \sum_{j=1}^{k} \log x_{n-j+1,n} - \log x_{n-k,n} \tag{2.5}$$

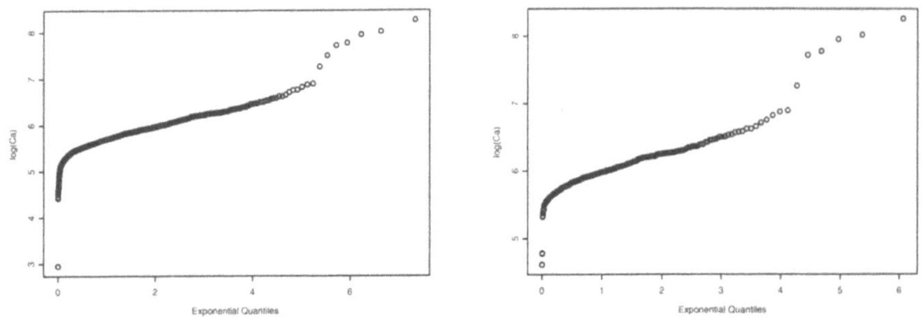

(a) (b)

FIGURE 2. Pareto quantile plot of Ca for the Condroz data (a) not taking into account pH and (b) conditional on pH-levels from 7 up to 7.5.

or, equivalently

$$\hat{\gamma}_{k,H} = \frac{1}{k}\sum_{j=1}^{k} j(\log x_{n-j+1,n} - \log x_{n-j,n}). \tag{2.6}$$

This estimator has received much attention in the literature. As the Hill estimator measures the average increase of the Pareto quantile plot above an anchor point $(\log(\frac{n+1}{k+1}), \log x_{n-k,n})$ it can be interpreted as a slope estimator of the linear part of the Pareto quantile plot. More recently several authors have recognized and exploited the potential of quantile plots in estimating $\gamma > 0$ (Beirlant et al., 1996, Schultze and Steinebach, 1996, and Kratz and Resnick, 1996).

Recently, in Beirlant et al. (1999), Feuerverger and Hall (1999) and Beirlant et al. (2002), it was proven that under some suitable conditions on the slowly varying function l_F and for k relatively small with respect to n (i.e., $k/n = o(1)$ as $k, n \to \infty$), the scaled log-spacings

$$y_j = j\ (\log x_{n-j+1,n} - \log x_{n-j,n}), \qquad 1 \le j \le k \le n, \tag{2.7}$$

approximately follow an exponential regression model

$$y_j {\sim}_d \left(\gamma + b_{n,k}\left(\frac{j}{k+1}\right)^{-\rho}\right) g_j, \tag{2.8}$$

with the g_j denoting i.i.d. standard exponential random variables, $b_{n,k} \to 0$ as $k, n \to \infty$, and $\rho < 0$. From this model γ can be estimated jointly with $b_{n,k}$ and ρ using the maximum likelihood method. In comparison to the Hill estimator, the maximum likelihood estimator for γ based on (2.8) is typically more stable over k and performs better with respect to bias.

Figure 3(a) shows the Hill (solid line) and maximum likelihood (broken line) estimates for the tail index of the conditional distribution of Ca as a function of k. On this plot, we clearly see the influence of the outlying Ca measurements on the γ-estimates. Considering decreasing k, then when k becomes smaller than 100 the estimates first increase drastically and then suddenly decrease for the smallest k-values. Figure 3(b) shows estimates of the asymptotic mean squared error of the Hill estimator

$$\widehat{AMSE}(\hat{\gamma}_{k,H}) = \frac{\hat{\gamma}_{k,ML}^2}{k} + \left(\frac{\hat{b}_{n,k,ML}}{1-\hat{\rho}_{k,ML}}\right)^2, \tag{2.9}$$

which, as in Beirlant et al. (1999), can be used as a criterion to find an optimal sample fraction k_{opt} for the Hill estimator. Here, k_{opt} is found to be 249 (vertical line) leading to an estimate $\hat{\gamma}_{k_{opt},H} = 0.298$.

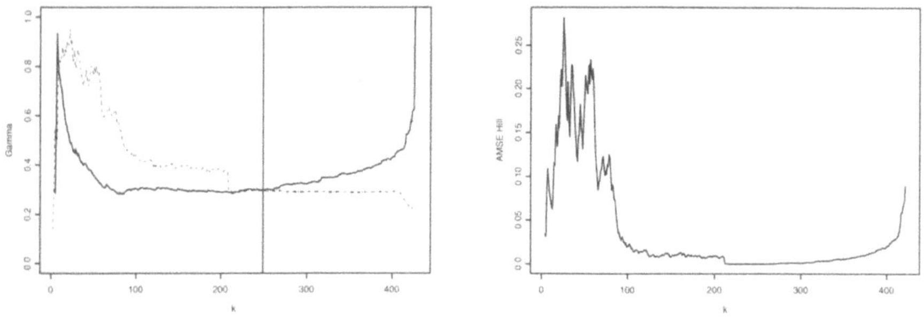

(a) (b)

FIGURE 3. (a) Hill (solid line) and maximum likelihood (broken line) estimates of the conditional Ca measurements and (b) AMSE estimates for the Hill estimator as a function of k.

3. Robust Estimation of the Tail Index

To obtain a robust estimate of γ, we start by transforming the exponential regression model (2.8) to its linearized form, given by

$$y_j \sim_d \gamma + b_{n,k}\left(\frac{j}{k+1}\right)^{-\rho} + \gamma e_j, \qquad 1 \le j \le k \le n, \tag{3.1}$$

where $e_j = g_j - 1$. Further we specify a canonical value for ρ. In Matthys and Beirlant (2000) it is shown that for most applications a ρ value between -2 and 0 is most appropriate. Moreover, estimates of γ and $b_{n,k}$ are not highly influenced by a specific choice of ρ. Hence, they recommend to use $\rho = -1$ or $\rho = -0.5$. For a fixed value of k, expression (3.1) becomes a linear regression model with asymmetric errors of the form

$$y_j = \theta_1 + \theta_2 t_j + \sigma e_j, \qquad j = 1, \ldots, k, \tag{3.2}$$

with $t_j = \left(\frac{j}{k+1}\right)^{-\rho}$, $e_j = g_j - 1$, $\theta_1 = \gamma$, $\theta_2 = b_{n,k}$ and $\sigma = \gamma$.

In Marazzi and Yohai (2004) high efficiency and high breakdown point estimators were proposed for this type of regression models. In general, consider k observations $(\mathbf{t}_j, y_j)$ with $j = 1, \dots, k$ and $\mathbf{t}_j = (1, t_{2j}, \dots, t_{pj})'$ that satisfy the linear relationship

$$y_j = \theta_1 + \theta_2 t_{2j} + \ldots + \theta_p t_{pj} + \sigma e_j, \qquad j = 1, \dots, k, \tag{3.3}$$

where $\boldsymbol{\theta} = (\theta_1, \ldots \theta_p)' \in \mathbf{R}^p$ is a vector of regression parameters and σ is a scale parameter. Here, the error terms e_j are assumed to be i.i.d. as a random variable e with cdf $F_{0,1}$, which is the standard element of a parametric location-scale family of asymmetric distributions with density function $f_{\lambda,\sigma}$ and cdf $F_{\lambda,\sigma}(z) = F_{0,1}((z - \lambda)/\sigma)$.

As a robust initial estimator for the general model (3.3) with asymmetric errors, a corrected S-estimator for regression is suggested. The biweight S-estimate $(\hat{\boldsymbol{\theta}}^*, \hat{\sigma}^*)$ of $(\boldsymbol{\theta}, \sigma)$ with 50% breakdown value is defined by $\hat{\boldsymbol{\theta}}^* = \arg\min_{\theta} S_{m_0}(\boldsymbol{\theta})$ and $\hat{\sigma}^* = S_{m_0}(\hat{\boldsymbol{\theta}}^*)$. For a given $\boldsymbol{\theta}$, $S_{m_0}(\boldsymbol{\theta})$ is an M-estimator of scale of the residuals, as the solution of

$$\frac{1}{k-p} \sum_{j=1}^{k} \chi_{m_0}\left((y_j - \mathbf{t}_j'\boldsymbol{\theta})/S\right) = 0.5 \tag{3.4}$$

with respect to S (Rousseeuw and Yohai, 1984). The function χ_m and the constants a_0 and m_0 are defined by

$$\chi_m(z) = \begin{cases} 3(z/m)^2 - 3(z/m)^4 + (z/m)^6 & \text{if } \mid z \mid \leq m, \\ 1 & \text{if } \mid z \mid > m, \end{cases} \tag{3.5}$$

and satisfy $\int \psi_{m_0}(z - a_0) f_{0,1}(z) dz = 0$ and $\int \chi_{m_0}(z - a_0) f_{0,1}(z) dz = 0.5$, where $\psi_m = \frac{d}{dz}\chi_m$. A consistently corrected S-estimate for $\boldsymbol{\theta}$ and σ is then given by $\hat{\theta}_1^0 = \hat{\theta}_1^* - \hat{\sigma}^* a_0$, $\hat{\theta}_j^0 = \hat{\theta}_j^*$ for $j = 2, \dots, p$ and $\hat{\sigma}^0 = \hat{\sigma}^*$.

Next, a reweighting step is applied, as proposed in Rousseeuw and Leroy (1987). Observations whose standardized residual $r_j = \left(y_j - \mathbf{t}_j'\hat{\boldsymbol{\theta}}^0\right)/\hat{\sigma}^0$ has a large negative log-likelihood $\rho_j = \rho(r_j) = -\log(f_{0,1}(r_j))$ have a small likelihood under the model $F_{0,1}$ and can be flagged as outliers. They are given a weight $w_j = I_{(\rho_j < \eta)}$, where I denotes the indicator function, and η is a large quantile of the cdf of $\rho(e)$. Finally, the maximum likelihood estimates of $(\boldsymbol{\theta}, \sigma)$ in model (3.3) are computed on the data points with $w_j = 1$.

When we apply this robust procedure to (3.2) we obtain preliminary estimates $(\hat{\theta}_1^0, \hat{\theta}_2^0), \hat{\sigma}_1^0$ and standardized residuals

$$r_j = \frac{\left(y_j - \hat{\theta}_1^0 - \hat{\theta}_2^0 \left(\frac{j}{k+1}\right)^{-\rho}\right)}{\hat{\sigma}^0}. \tag{3.6}$$

Note that the constants a_0 and m_0 that are used in the definition of the S-estimators are found to be $a_0 = -0.5700$ and $k_0 = 0.9466$. From the model cdf $F_{0,1}(z) = 1 - e^{-(z+1)}$ for $z \geq -1$, it follows that $\rho(e)$ is standard exponentially distributed, hence we set $\eta = -\log(1-0.95)$, the 0.95-quantile of the standard exponential distribution. Of course, also other quantiles could be considered as well. We now indicate the zero weight log-spacings y_j as outliers and remove the corresponding x_i from our data set. More precisely, let y_J be the outlying log-spacing with largest index, then all x_i with $n-J+1 \leq i \leq n$ are flagged as outliers under the Pareto model.

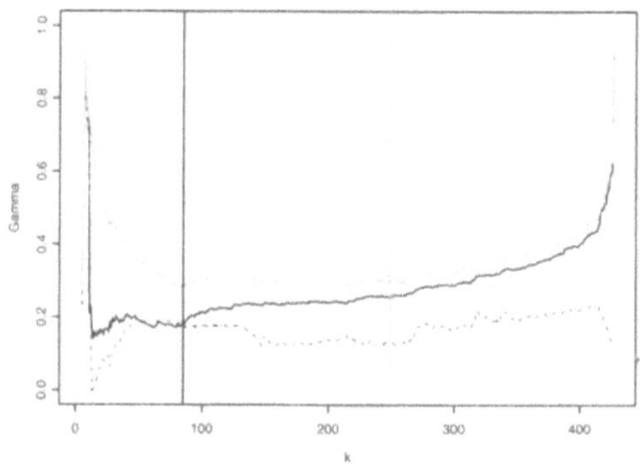

FIGURE 4. Hill (solid line) and maximum likelihood (broken line) estimates of the conditional Ca measurements after rejection of the outliers found for each k, together with the Hill (solid grey line) and maximum likelihood (broken grey line) estimates before rejection.

4. Application

The results after applying our proposal to the Condroz data, are depicted in Figure 4. Note that we have set $\rho = -1$ in (3.1). We have also tried some other values of ρ between -2 and 0 but this did not change the result. The broken black line shows the maximum likelihood estimates after removal of the outliers that were found for each k, whereas the solid black line exposes the Hill estimator on the same reduced data sets. On this plot we have also superimposed (in grey) the ML and Hill curves of Figure 3 that are based on the full data set. We see that both robust curves yield much lower estimates for the tail index. Moreover they are rather constant for intermediate values of k, which gives support to the Pareto model assumption. For very small values of k, the estimates are however still not stable. This is due to the small number of observations in the regression model (3.2), where even an estimator with 50% breakdown is biased in the presence of outliers.

It is observed that our method found six outliers for most k values, except for the smaller ones. The dark vertical line in Figure 4 is set at $k_{opt} = 85$. To obtain this optimal sample fraction for the Hill estimator, we computed its AMSE as in

(2.9). Since not all the ML estimates were based on the same data set for all k, here the ML estimators were computed on the data set after the six largest observations were removed. The AMSE curve, shown in Figure 5, attains its minimum at $k_{\text{opt}} = 85$, from which $\hat{\gamma}_{k_{\text{opt}},H} = 0.177$ follows.

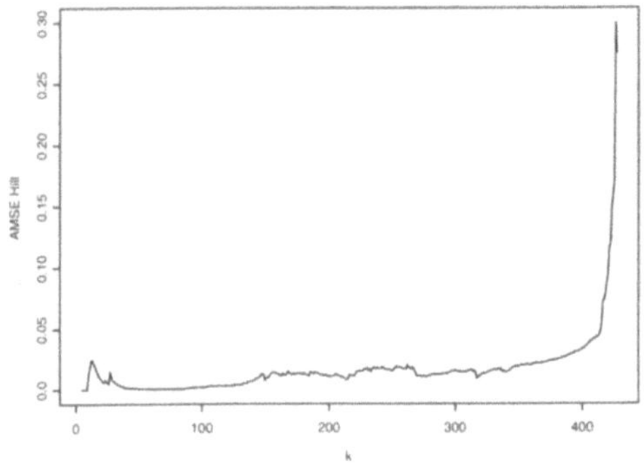

FIGURE 5. AMSE estimates for the Hill estimator after rejection of the six largest observations.

Figure 6 shows the $\rho_j = \rho(r_j)$ for the regression data (t_j, y_j) from model (3.2) for $k = k_{\text{opt}} = 85$, together with the cut-off line that separates the outliers from the regular observations. We see that t_6 is the observation with largest index that exceeds the cut-off, hence six observations are flagged out as being unlikely under the Pareto model. This confirms our findings from the Pareto quantile plot in Figure 2(b).

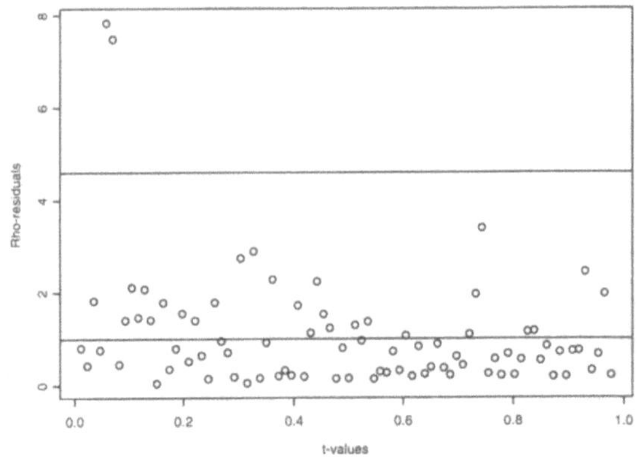

FIGURE 6. The $\rho(r_j)$ for the regression data (t_j, y_j) from model (3.2) with $k = 85$ together with the expected value and cut-off line that separates the outliers from the regular observations.

5. Conclusions and Outlook

In this paper we have introduced a new robust estimator of the tail index of Pareto type distributions. It is obtained by applying a robust regression estimator for a linear model with asymmetric errors to scaled log-spacings. When we applied this method to Ca and pH measurements from the Condroz region in Belgium, we could easily identify the outliers that were also seen on a Pareto quantile plot, and we obtained a much lower estimate of the tail index.

In our further research we will study this robust estimator in more detail. We will investigate its breakdown value, its influence function and its performance at simulated data sets.

As a different approach, we will also study the application of a robust regression estimator to the Pareto quantile plot data (2.3). In particular we will consider the deepest regression method (Rousseeuw and Hubert, 1999) as it yields consistent estimates under heteroscedastic error distributions. Nonconstant variances are likely to occur in quantile plots.

References

[1] J. Beirlant, G. Dierckx, Y. Goegebeur, and G. Matthys, *Tail Index Estimation and an Exponential Regression Model.* Extremes **2** (1999), 177–200.

[2] J. Beirlant, G. Dierckx, A. Guillou, and C. Stărică, *On Exponential Representations of Log-Spacings of Extreme Order Statistics.* Extremes **5** (2002), 157–180.

[3] J. Beirlant, P. Vynckier, and J.L. Teugels, *Tail Index Estimation, Pareto Quantile Plots, and Regression Diagnostics.* J. Amer. Statist. Assoc. **91** (1996), 1659–1667.

[4] P. Embrechts, C. Klüppelberg, and T. Mikosch, *Modelling Extremal Events for Insurance and Finance.* Springer, New York, 1998.

[5] A. Feuerverger and P. Hall, *Estimating a Tail Exponent by Modelling Departure from a Pareto Distribution.* Ann. Statist. **27** (1999), 760–781.

[6] Y. Goegebeur, V. Planchon, J. Beirlant, and R. Oger, *Quality Assessment of Pedochemical Data Using Extreme Value Methodology.* Technical Report, 2002, available at *http://www.kuleuven.ac.be/ucs/technical_reports.htm.*

[7] B.M. Hill, *A Simple General Approach to Inference About the Tail of a Distribution.* Ann. Statist. **3** (1975), 1163–1174.

[8] M. Kratz and S. Resnick, *The qq-Estimator of the Index of Regular Variation.* Comm. Statist. Stoch. Mod. **12** (1996), 699–724.

[9] A. Marazzi and V. Yohai, *Adaptively Truncated Maximum Likelihood Regression With Asymmetric Errors.* J. Statist. Plann. Inference **122** (2004), 271–291.

[10] G. Matthys and J. Beirlant, *Adaptive Threshold Selection in Tail Index Estimation.* In P. Embrechts, editor, Extremes and Integrated Risk Management, pages 37–49. Risk Books, UBS Warburg, 2000.

[11] P.J. Rousseeuw and M. Hubert, *Regression Depth.* J. Amer. Statist. Assoc. **94** (1999), 388–402.

[12] P.J. Rousseeuw and A. Leroy, *Robust Regression and Outlier Detection.* John Wiley, New York, 1987.

[13] P.J. Rousseeuw and V. Yohai, *Robust Regression by Means of S Estimators.* In J. Franke, W. Härdle and D. Martin, editors, Robust and Nonlinear Time Series Analysis, Lecture Notes in Statist. **26**, pages 256–272. Springer, New York, 1984.

[14] J. Schultze and J. Steinebach, *On Least Squares Estimates of an Exponential Tail Coefficient.* Statist. Decisions **14** (1996), 353–372.

B. Vandewalle, J. Beirlant and M. Hubert
Department of Mathematics and University Center of Statistics
Katholieke Universiteit Leuven
de Croylaan 54
B-3001 Heverlee
Belgium
e-mail: {Bjorn.Vandewalle, Jan.Beirlant, Mia.Hubert}@wis.kuleuven.ac.be

Statistics for Industry and Technology, 377–385

Robust Processing of Mechanical Vibration Measurements

S. Vanlanduit and P. Guillaume

Abstract. When performing mechanical vibration measurements, it is often assumed that the disturbing measurement noise behaves Gaussian. In reality, this is not always the case, and therefore classical least-squares procedures can give poor results when processing the measurements. In this paper, several robust statistical procedures will be used in four selected problems in vibration measurement and analysis. It will be shown that the procedures improve the results compared to classical least-squares methods. Because fast processing is required, a tradeoff between robustness of the methods and computation speed is made. In particular the following steps in vibration engineering were robustified: (1) positional calibration, (2) measurement post-processing, (3) system identification and (4) data classification for damage detection.

Mathematics Subject Classification (2000). Primary 74H45; Secondary 93E12.

Keywords. Robust, vibration analysis, signal processing.

1. Motivation

An important research domain in the field of mechanical engineering, is the analysis of vibrating systems. This so-called 'modal analysis' is an important tool for car manufacturers who want to determine the sound field inside a vehicle. Also, modal analysis offers the capability of detecting damage in structures (like for instance bridges or airplane components).

A modal analysis typically requires three steps. Firstly, the structure is excited and measurement data is collected from vibration sensors. Secondly, – after pre-processing – this data is processed using parameter estimation methods in order to obtain a parametric model of the structure under test. In a third step, this model can be used to draw conclusions on the structural behavior (acoustic radiation, presence of damage,...). Alternatively, this post-processing step can also be performed using nonparametric data (as for instance the so-called Frequency Response Functions or FRFs).

In general, the measurement noise on the vibration data is assumed to be normally distributed. However, in reality this is not always the case. Indeed, some of the sensors can fail, giving rise to outliers in the measurements. In addition, even when the sensor works correctly, disturbing noise with a large amplitude (e.g., from neighboring machines) can seriously hamper the successful processing of data using classical (least-squares) techniques.

In this article, an overview of potential applications of robust processing techniques in vibration engineering is given based on several case studies. For each of the considered case studies, a classical least-squares and a proposed robust solution is given. The case studies represent different steps typically executed in vibration engineering:

- Calibration of measurement equipment: a robust technique for the calibration of an optical scanning vibrometer (Section 2).
- Robust pre-processing of measured vibration spectra (Section 3): elimination of measurement dropouts.
- Robust model identification (Section 4).
- Data classification for structural damage detection using Principal Component Analysis (Section 5).

2. Laser Position Calibration

A particular measurement instrument which is used a lot when the vibration response at a large number of locations is desired is the scanning Doppler laser vibrometer (SLDV) (Halliwell, 1979). This device discretely scans spatial locations with a laser beam, which is moved horizontally and vertically with the aid of 2 rotating mirrors. Before the measurements can be done, a calibration between scanning mirror angles (θ_i, ϕ_i) and the coordinates of the pixels (p_i, q_i) (corresponding to the location of the laser beam on the object under test) has to be performed. When the laser beam is oriented at randomly selected angles (θ_i, ϕ_i), the relation between the mirror angles and the coordinates is given by:

$$p_i = p_0 + k_1 d \tan\theta_i \tag{2.1}$$

$$q_i = q_0 + k_2 d \frac{\tan\phi_i}{\cos\theta_i} \tag{2.2}$$

where p_0, q_0, k_1, k_2 and d depend on the position of the SLDV instrument (with respect to the position of the object) and the camera zoom. When considering N_p random angles (θ_i, ϕ_i) Equations (2.2) and (2.1) can be written in matrix form:

$$\mathbf{Ac} = \mathbf{b} \tag{2.3}$$

where $\mathbf{A}$ is an $2N_p$ by 4 matrix, $\mathbf{b}$ a $2N_p$ length vector containing the pixel coordinates $p_1, \ldots, p_{N_p}, q_1, \ldots, q_{N_p}$ and $\mathbf{c} = [p_0, q_0, kd_1, kd_2]$. The calibration is performed by solving the system of equations in Equation 2.3. Unfortunately, some of the coordinates p_i, q_i (i.e. the elements of $\mathbf{b}$) have errors orders of magnitude larger than the average error. This means that when solving Equation 2.3 using a

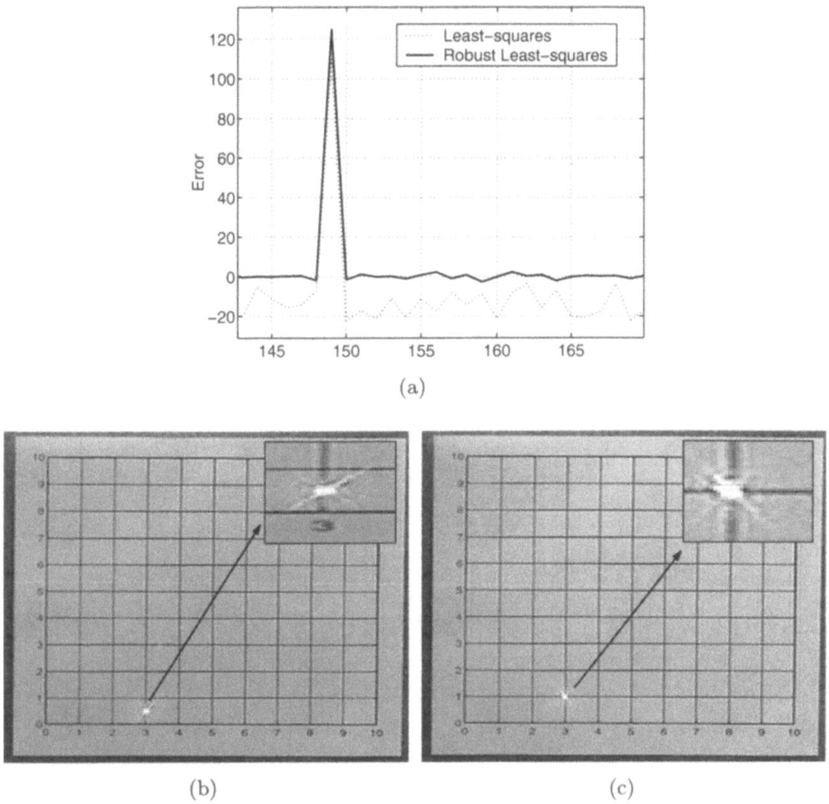

FIGURE 1. (a): Residual errors $\mathbf{A}\hat{\mathbf{c}} - \mathbf{b}$ for the Least-squares and robust Least-squares solution of the laser position calibration problem. (b-c): Validation experiment for least-squares (b) and least trimmed squares (c) solution: target position on location (3,1).

least-squares regression, an incorrect calibration is performed. This can be clearly seen in the LS residual errors $\mathbf{A}\hat{\mathbf{c}} - \mathbf{b}$ in Figure 1a. Using a robust regression algorithm – like the least trimmed squares (LTS) (Rousseeuw and Van Driessen, 2000) – a much better result can be obtained (see Figure 1a). When the parameters $\mathbf{c}$ are estimated, the angles (θ, ϕ) which are needed to place the laser beam at a certain validation location (p, q) can be computed. In the validation experiment shown in Figures 1b and 1c the laser beam is targeted onto position (3,1) of a grid on a paper. It is clear that using the LS calibration it is not possible to hit the correct location (3,1), while the robust LTS solution exactly aims at the target location. More details on the problem can be found in Vanlanduit et al. (2003).

3. Pre-processing of Measurement Spectra

When using optical vibration measurement sensors on dark structures, the reflected laser light can suddenly drop out, resulting in outliers (see Figure 2a). Instead of using time-domain measurements $x(t)$ sampled at time instances $\delta t, \ldots, N_t \delta t$, in vibration analysis, the data is usually transformed into the Fourier domain using a Discrete Fourier Transform (DFT):

$$X(\omega) = DFT(x(t)) \tag{3.1}$$

the result is that the error from outliers is smeared out over the full frequency band in the spectra $X(\omega)$ (as can be seen clearly in Figure 2b). The DFT transforma-

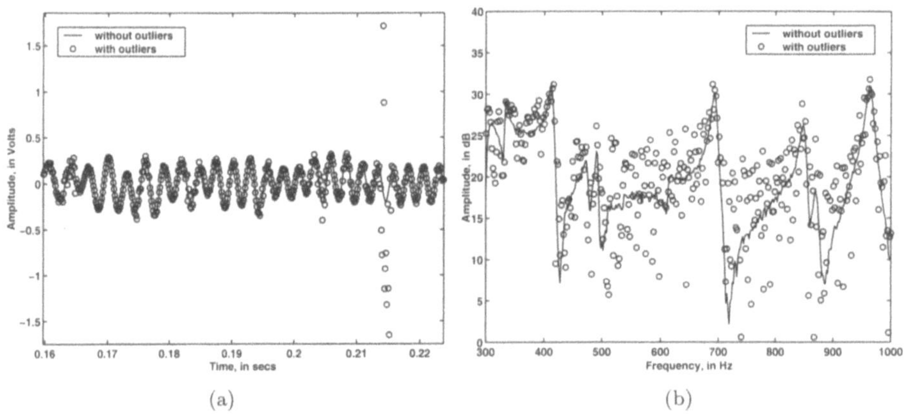

FIGURE 2. Velocity measurement without and with outliers. Left: time domain, right: frequency domain.

tion in Equation 3.1 can equivalently be expressed using a matrix transformation (although in practice a faster transformation – the Fast Fourier Transform or FFT – algorithm is used):

$$\mathbf{X} = \mathbf{W}\mathbf{x} \tag{3.2}$$

with $\mathbf{X}$ and $\mathbf{x}$ length N_f and N_t vectors respectively (with N_t and N_f the number of measured time samples and excited frequencies respectively) and $\mathbf{W}$ a known N_f by N_t matrix (typically $N_t = 1024$ and $N_f = 512$). In principal, existing robust regression methods could be used to solve Equation 3.2 in order to be resistent to outliers. However, for high spatial resolution measurements, thousands of spectral measurements have to be performed. Therefore, the processing time should be limited to less than a second, excluding all robust regression methods known to the authors.

Instead, the following quasi-robust – but fast – procedure is proposed:

Algorithm: Iterative two step LS

1. Apply a median filter to $[x(\delta t), \ldots, x(N_t \delta t)]$ to obtain $[\widehat{x}(\delta t), \ldots, \widehat{x}(N_t \delta t)]$
2. Compute the Median Absolute deviation (MAD) of the residual $e = x - \widehat{x}$:
$$MAD(e) = \text{median}(|e - \text{median}(e)|) \tag{3.3}$$
When divided by 0.6745 the MAD is a robust estimator of the scale which is consistent with the normal distribution (Huber, 1981).
3. Compute the set I of timesamples which are outliers:
$I = \{i \mid |e(i\delta t)| > 5 \times MAD\}$. Put $\widetilde{x} = x(I)$. Remark that for a normal distribution with outliers it would be more natural to take e.g. 1.96*MAD (to obtain the $\alpha = 5\%$ confidence level). From experimental investigations it was observed that the tail of the measurement noise distribution (without any outliers included) is much heavier than for a normal distribution, and therefore a higher value was needed. The value of 5*MAD was chosen because it worked well on all investigated experimental data.
4. Compute a cubic spline interpolant $\widehat{x}$ of $\widetilde{x}$ in the time samples $\delta t, \ldots, N_t \delta t$ (Dierckx, 1993).
5. Put $x(I) = \widehat{x}(I)$.
6. Compute the low-pass filtered $\overline{x}$ of x.
7. Put $x(I) = \overline{x}(I)$.
8. Repeat Steps 6. to 7. N_{iter} times.
9. The processed data is equal to: $\widehat{x} = \overline{x}$.

In Vanlanduit et al. (2002) it is demonstrated that the Iterative two step LS solution converges to the global solution. In Figure 3 the results of the quasi-robust processing of vibration measurements with a dropout are given. It is clear, that the results using the introduced method improve drastically (the two vibration modes around 470Hz and 540Hz are much more clear in Figure 3d than in Figure 3c). Less than a tenth of a second of processing time is needed for $N_{iter} = 100$ iterations on a Pentium 4 1.4GHz processor.

In the next section the implementation of a robust modelling technique for vibration measurements will be illustrated.

4. Modal Parameter Estimation

In vibration engineering, the mechanical structure is excited with a force $x(t)$ and the resulting response $y(t)$ is measured. From these measurements $x(t)$ and $y(t)$ the spectra $X(\omega)$ and $Y(\omega)$ are computed (using a DFT) which in turn lead to the computation of the so-called Frequency Response Function (FRF) H: $H = \frac{Y}{X}$. Each vibrating system can be modelled as a fraction of polynomials in $\sqrt{-1}\omega$, and the measured FRF H is used to identify the polynomial's coefficients, which represent the system parameters. this can be done by minimizing the error E:

$$E(\omega) = H(\omega) - \frac{B(\omega)}{A(\omega)} \tag{4.1}$$

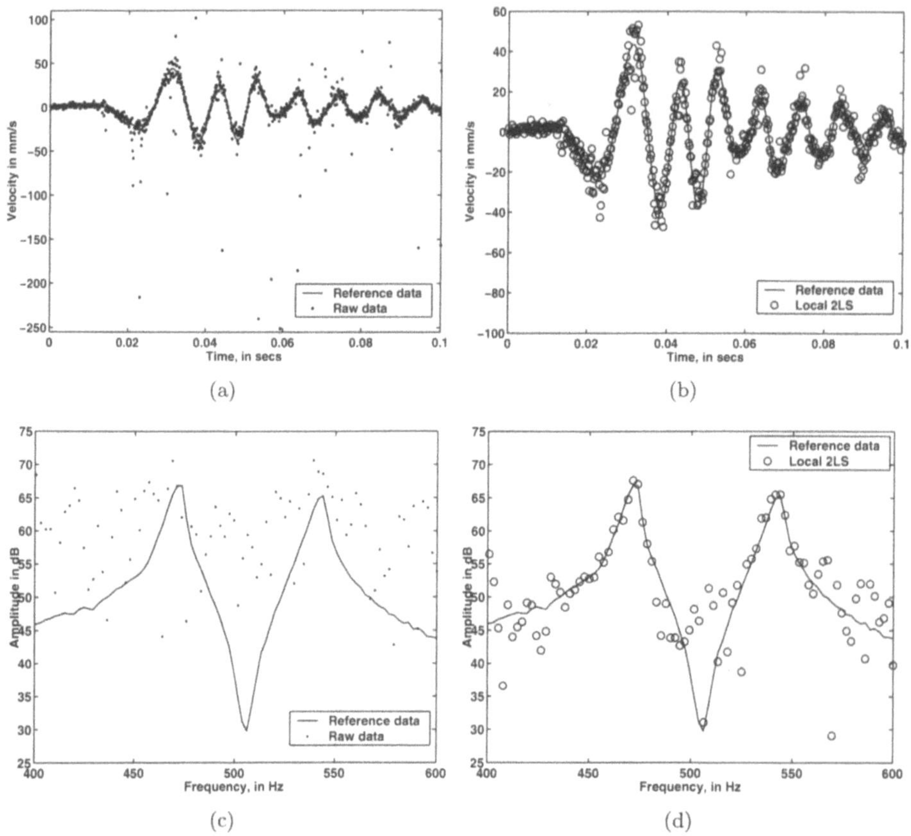

FIGURE 3. (a) Raw time data x and reference data without outlier, (b) quasi-robust pre-processed time data $\widehat{x}$ (see Iterative 2 step LS algorithm), (c) raw frequency domain data X and reference data without outlier, (d) quasi-robust pre-processed frequency domain data FFT($\widehat{x}$). Two vibration modes are visible by the peaks around 470Hz and 540Hz.

where A and B are polynomials in $\sqrt{-1}\omega$. A first possibility to estimate the model parameters is by minimize a quadratic cost function κ_2:

$$\kappa_2(\theta) = \sum_{k=1}^{N_f} \left| H(\omega_k) - \frac{B(\omega_k, \theta)}{A(\omega_k, \theta)} \right|^2 . \tag{4.2}$$

This cost function gives high weights to outlying values in the error. A much more robust alternative is the use of a logarithmic cost function κ_{log} (Guillaume et al.,

1995):

$$\kappa_{\log}(\theta) = \sum_{k=1}^{N_f} \left| \log \left(\frac{B(\omega_k, \theta)}{H(\omega_k) A(\omega_k, \theta)} \right) \right|^2 . \tag{4.3}$$

Both cost functions are nonlinear in the parameters and therefore their estimation is quite involved (an iterative scheme like the Levenberg-Marquard optimization algorithm is often used to perform this task). The logarithmic cost function is much more robust than the least-squares solution (indeed, the logarithmic transformation gives a lower weight to outliers in the error E) Remark that the solution is not truly robust, since the influence of outliers is not bounded. Moreover, it was shown in Guillaume et al. (1995) that the logarithmic cost solution leads to nearly efficient estimates.

In Figure 4, an FRF with outliers due to disturbing 50Hz and harmonics components is shown. It can be seen that the logarithmic cost function (Figure 4b) is able to identify the vibration mode at 120Hz (although some bias is present), while the quadratic cost function (Figure 4a) gives completely erroneous results.

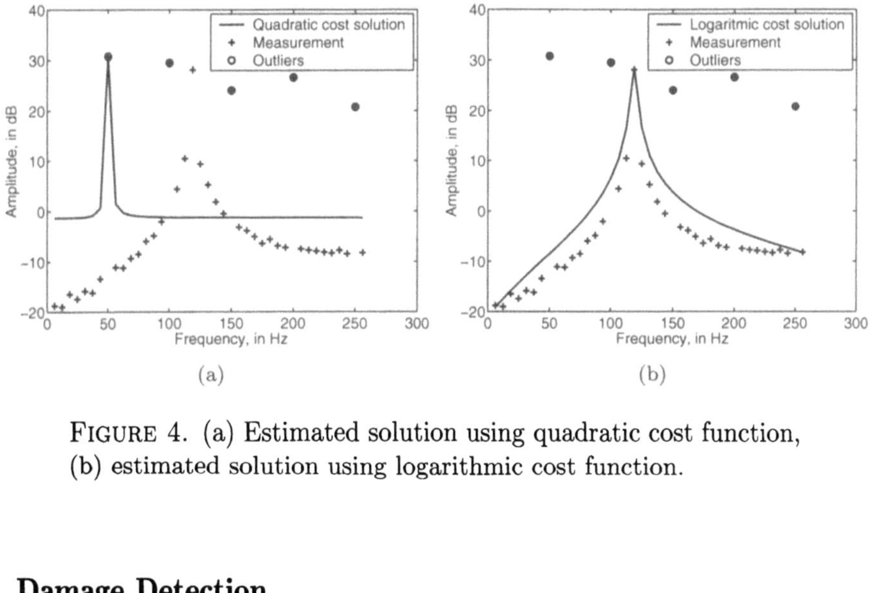

(a) (b)

FIGURE 4. (a) Estimated solution using quadratic cost function, (b) estimated solution using logarithmic cost function.

5. Damage Detection

When a structure is damaged (cracks, delaminations, loose parts) its vibration behavior changes. This can be observed through the FRFS $H(\omega)$ of the structure: the vibration modes (peaks in the FRFs) are shifted towards lower frequencies.

Consider a matrix $\mathbf{H} = [|H_1|, \ldots, |H_n|]^t$, where each column of $\mathbf{H}$ contains an FRF of a structure measured in a certain condition. The goal is to identify the outliers between the n conditions (these outliers then represent the damaged specimen). Because FRF of an intact structure form a subspace, the outlier detection can be done using Principal Component Analysis (PCA). However, FRFs of a damaged structure do not belong to the subspace of the intact structure. By plotting the distance of the n different objects to the subspace it is possible to detect damaged specimen. However, by using classical PCA the correct subspace is not found back because of the presence of the damaged FRFs which are outliers. Therefore, a robust PCA should be used. In this paper the RAPCA algorithm (Hubert et al., 2002) was used to perform this task. In Figure 5a 27 FRFs, measured at different time periods and in different conditions (21 intact and 6 damaged) are shown. From Figure 5b it can be seen that the 6 damaged beams can be identified from the robust PCA, while the classical PCA erroneously detects the cracked beams (1 intact and only 3 damaged beams are marked as an outlier).

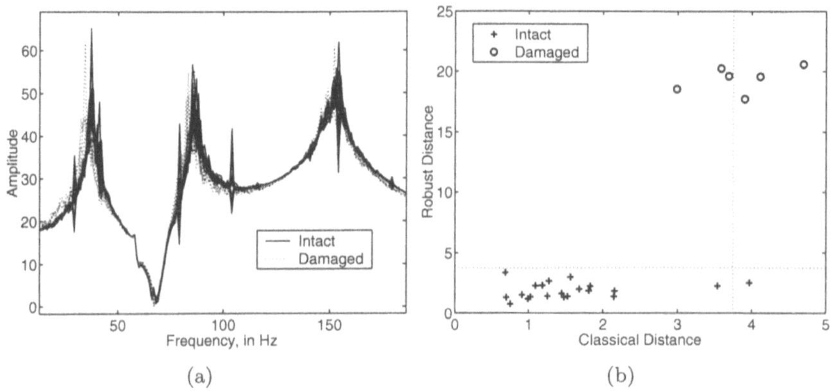

FIGURE 5. (a) Frequency Response Functions of a steel beam for 27 conditions (21 intact and 6 cracked), (b) Robust versus classical PCA distance for the 27 FRFs in (a).

6. Conclusions

From the examples in the paper, it is clear that there is an important need for robust procedures to process vibration measurement data. This is in particular the case when optical vibration measurement instrumentation is used. On the other hand, because of the high amount of data that should be processed (typically thousands of measurement records of a few thousands of time samples) a tradeoff between robustness and computational efficiency has to be made. In the current paper four adapted robust processing techniques were proposed for the application during the different steps of the vibration engineering procedure: calibration, pre-processing, model identification and damage detection. It was shown on experimental data that the results were much better than the classical least-squares solutions.

References

[1] P. Dierckx, *Curve and Surface Fitting With Splines.* Oxford University Press, Oxford, 1993.

[2] P. Guillaume, R. Pintelon, and J. Schoukens, *Robust Parametric Transfer Function Estimation Using Complex Logarithmic Frequency Response Data.* IEEE Trans. Autom. Control **40** (1995), 1180–1190.

[3] N.A. Halliwell, *Laser Doppler Measurement of Vibrating Surfaces: a Portable Instrument.* J. Sound Vibrat. **62** (1979), 312–315.

[4] P. Huber, *Robust Statistics.* John Wiley, New York, 1981.

[5] M. Hubert, P.J. Rousseeuw, and S. Verboven, *A Fast Method For Robust Principal Components With Applications to Chemometrics*, Chemometr. Intell. Lab. Syst. **60** (2002), 101–111.

[6] P.J. Rousseeuw and K. Van Driesen, *Computing LTS Regression for Large Data Sets.* Estadistica, to appear, 2004.

[7] S. Vanlanduit, P. Guillaume, B. Cauberghe, and P. Verboven, *An Automatic Position Calibration Method For the Laser Scanning Doppler Vibrometer.* Meas. Sci. Technol. **14** (2003), 1496–1476.

[8] S. Vanlanduit, P. Guillaume, and J. Schoukens, *On-Line Robust Processing Techniques For Measurement Dropout Elimination.* Meas. Sci. Technol. **13** (2002), 1183–1190.

S. Vanlanduit and P. Guillaume
Vrije Universiteit Brussel
Pleinlaan 2
B-1050 Brussel
Belgium
e-mail: {steve.vanlanduit, Patrick.Guillaume}@vub.ac.be

Statistics for Industry and Technology, 387–400

Quadratic Mixed Integer Programming Models in Minimax Robust Regression Estimators

G. Zioutas

Abstract. The robust estimation of regression parameters is formulated in terms of mixed integer-quadratic programming problem. The main contribution of this technique is that it improves the estimator efficiency by down-weighting only bad influential points, either y-outliers or x-outliers. We follow the minimax strategy where the objective function of our mathematical programming formulation is mainly a Huber loss function, and bad influential outliers pulled towards the regression line with low cost. This penalized pulling cost is a function of Mallows type weights, and in the modified data a GM estimator (Schweppe type) could be defined. The main advantage of the proposed technique is that data points are not down-weighted, unless they have increased substantially the square residuals. Previously published mixed integer formulations withdraw data points, the most influential even if they are not bad influential points. GM estimators are compared to our proposal via simulated experiments, the robust estimator obtained by quadratic programming is reasonable.

Mathematics Subject Classification (2000). Primary 05C38, 15A15; Secondary 05A15, 15A18.

Keywords. Bounded influence, outliers, quadratic programming, mixed integer programming, robust regression.

1. Introduction

It is well known that least squares estimators are very sensitive to departures from the normality assumption on the error distribution (Hampel et al., 1986 and Huber, 1981). These deviations are usually referred to as outliers in statistics and affect the estimators. Different kinds of outliers or influential points can occur for many reasons, but in this paper we consider the estimation of regression parameters mainly in the presence of x-space outliers (high leverage points) in the data which could be distinguished in "bad" leverage points and "good" leverage points.

- "Bad" leverage points are regression outliers, which do not lie in the pattern of the bulk of the data and are highly influential. It is possible that only a single "bad" leverage point can render least squares regression coefficients meaningless. Therefore, their influence should be limited.
- "Good" leverage points are high leverage points, which lie in the pattern of the bulk of the data. Usually, "good" leverage points do not correspond to large residuals. They carry valuable information about the regression line and contribute to the precision of the parameter estimation. Therefore, such points should not be downweighted, otherwise significant information will be lost (Krasker and Welsch, 1982).

A group of robust methods has been developed where the estimators are less sensitive to outliers. Maximum Likelihood type estimation (M-estimation), and General Maximum Likelihood type estimation (GM-estimates), with Mallows weights or cutting big residuals with Schweppe forms are few among the popular robust techniques. In this paper we will be concerned with bounded influence estimates (GM), by combining Huber loss function, Mallows weights and Schweppe residual size.

It is considered whether "good" leverage points could be weighted appropriately (reducing down-weight) so as to increase estimator precision. Our basic idea was to restrict the down-weighting activation only for the largest residuals.

In the presented approach, the influence of every outlier is bounded by following the general robust procedure, where an observation point (x_i, y_i) could be down-weighted by pulling y_i-values towards its fitted values. The pulling distance for each y_i is penalized proportionally to given weights $w(x_i)$, which have been proposed by Hampel for the Mallows-type estimator (Hampel, 1978, 1986). The proposed estimator is defined by minimizing an objective function similar to Huber's "loss" function, which is the sum of squared winsorized residuals and the penalty cost of the overall pulling.

The robust procedure has been formulated by Camarinopoulos and Zioutas (2002) as an optimum allocation problem, where the objective function ("loss" function) is minimized subject to a given total distance of overall pulling. Thus, the available pulling resource is optimally allocated to outlying points, with respect to both the residual size and the pulling cost. Since the amount of overall pulling is limited, it may not be sufficient to pull all the high leverage points, which correspond to a low penalty cost. Therefore, the down-weighting of "good" leverage points is reduced. A solution to the optimum allocation problem is obtained by using mathematical programming, an optimization technique suitable to many statistical procedures (Arthanari and Dodge, 1993).

Alternatively, to down-weigh high influential points which corresponds only to big residuals, an improved mathematical programming formulation is developed in this work. Specifically, a quadratic mixed integer programming formulation is used with objective function a Huber type loss function. The penalty cost of pulling high leverage points is very low (Mallows's weights), but the pulling activation starts only for points which correspond to big residuals, suggested by Schweppe or other GM-estimators.

The proposed technique has also the flexibility to establish constraints for the regression parameters by adapting the mathematical programming formula. Also, a sensitivity analysis, by fluctuating the pulling resource, may potentially provide insight regarding the most catastrophic outlier detection.

The new estimator, called QMIP, is defined by the minimization of a convex function, leading to a unique solution. Also, the bad influence of outlying points is bounded, consuming the least possible amount of overall pulling. These factors contribute to good robust properties (Huber, 1981).

In the next section, we briefly review the robust algorithm for M and GM estimators. Section 3 describes the proposed QMIP estimator and the mathematical programming formulation. The Monte Carlo design and results are presented in Sections 4 and 5.

2. Robust Methodology

Consider the linear regression model,

$$y = x^T \beta + e \tag{2.1}$$

where y is the response variable, x is a $p \times 1$ vector of explanatory variables, β is a $p \times 1$ vector of unknown parameters, and e is a random error with expectation zero and variance σ^2. We observe a sample $(x_1, y_1), \ldots, (x_n, y_n)$ and we wish to construct a robust estimator in the sense that the influence of any observation (x_i, y_i) on the sample estimator is bounded. Of course, the original least squares estimate of β does not satisfy the robustness requirement.

2.1. "Loss" Function

For known scale parameter σ, an M-estimator of Huber-type (Huber, 1973, 1981) could be defined by minimizing the sum of less rapidly increasing functions of the residuals,

$$\underset{\beta}{\text{minimize}} \sum_{i=1}^{n} \rho_c(u_i/\sigma)\sigma^2 \tag{2.2}$$

where ρ_c is a less rapidly increasing function known as "loss" function of the residuals

$$u_i = y_i - x_i^T \beta$$

$$\rho_c(u_i/\sigma) = \begin{cases} \frac{1}{2}(u_i/\sigma)^2 & \text{for } |u_i/\sigma| \le c \\ c(u_i/\sigma) - c^2/2 & \text{for } |u_i/\sigma| > c. \end{cases} \tag{2.3}$$

Differentiating the expression (2.2) with respect to the regression coefficient β yields

$$\sum_{i=1}^{n} \psi_c(u_i/\sigma)x_{ij} = 0 \quad \text{for } j = 1, \ldots, p \tag{2.4}$$

where ψ_c is the derivative of ρ and from (2.3) we obtain

$$\psi_c(u_i/\sigma) = \begin{cases} u_i/\sigma & \text{for } |u_i/\sigma| \leq c \\ c & \text{for } |u_i/\sigma| > c. \end{cases} \tag{2.5}$$

Even though $\psi_c()$ limits the effect of large residuals, the influence of an observation (y_i, x_i) can be arbitrary large because x_i multiplies $\psi_c()$. Thus, the M-estimators do not bound the influence of outliers in the covariate space (regression outliers).

In 1975 (Handschin et al., 1975), Schweppe modified (2.2), minimizing with respect to β, a similar form of "loss" function, as follows:

$$\sum_{i=1}^{n} \sigma^2 v(x_i)^2 \rho_c(u_i/(\sigma v(x_i))) \tag{2.6}$$

where $v(x_i) = \sqrt{1-h_i}$ and h_i are the diagonal elements of the matrix $H = X(X^TX)^{-1}X^T$. If the ith observation is an outlier in x-space, it is indicated by a large value of h_i and the corresponding small value of $v(x_i)$ contributes to more drastic down-weighting. Thus, the solution of (2.6) could provide bounded influence estimator but with the additional property that if $u_i/v(x_i)\sigma$ is small, the effect of $v(x_i)$ will be cancelled out.

Mallows has proposed a bounded-influence estimator by minimizing

$$\sum_{i=1}^{n} \sigma^2 w(x_i) \rho_c(u_i/\sigma) \tag{2.7}$$

where $w(x_i)$ are certain weights that depend on the distance of x_i from the center of the multivariate regression space cloud. This estimator bounds the influence of an outlying x_i but there is one problem that is immediately apparent. Any down-weighting of high leverage points regardless of the underlying residual cannot be efficient.

Schweppe-type estimator has the potential to help overcome some of the efficiency problems noted for the Mallows estimator. Hill (1977) suggested using Schweppe forms with Mallows weights for more drastic down-weighting, and solving

$$\sum_{i=1}^{n} w(x_i)\psi_c(u_i/(w(x_i)\sigma))x_{ij} = 0 \qquad \text{for } j = 1, \ldots, p \tag{2.8}$$

found that this last estimator has an advantage over the above estimators. Therefore, the Schweppe objective function is typically preferred in most GM-estimators as well as in most compound estimators (Simpson and Montogomery, 1998). But, there is still a small problem when the (y_i, x_i) is a "good" leverage point, corresponding to small residual u_i and very small $w(x_i)$ such that $u_i/(w(x_i)\sigma) > c$. By down-weighting such points, some information may be lost.

In 1977 Welsch obtained a bounded influence estimator by solving the above minimization problem where $v(x_i) = (1-h_i)/h_i^2$. In order to bound the influence of "gross" errors in the independent variables, Krasker and Welsch (1982) proposed a robust estimator (KW) by limiting the sensitivity of the estimators in an efficient way. Their choice of $v(x_i)$ is similar to the last mentioned one.

It is remarkable that most of the above estimators can be defined by solving the problem (2.2), but differ only in the choice of the tuning constant c (Huber, 1983). Different values of the tuning parameter c, $c = c_i$, lead to different bounded influence estimators (Krasker and Welsch, 1982). For example, if we wish to define a Schweppe-type estimator we minimize Huber's loss function (2.2) or solve the equation system of (2.4) using $c = c_i = \sqrt{1 - h_i}$

2.2. General Robust Procedure

Multiplying both sides of (2.3) by $2\sigma^2$ and after simple calculus on the right-hand side we obtain the following equivalent "loss" function,

$$2\sigma^2 \rho_{c_i\sigma}(u_i) = \begin{cases} {u_i}^2 & \text{for } |u_i| \le c_i\sigma \\ c_i^2\sigma^2 + 2c_i\sigma\varepsilon_i & \text{for } |u_i| > c_i\sigma \end{cases} \tag{2.9}$$

where ε_i indicates the size of shorten big residuals u_i following the manner:

$$\varepsilon_i = \begin{cases} |u_i| - c_i\sigma & \text{for } |u_i| > c_i\sigma \\ 0 & \text{for } |u_i| \le c_i\sigma \end{cases} \tag{2.10}$$

or equivalently, ε_i could be interpreted as the distance of pulling y_i towards its fitted value and from now on we shall call this "pulling" distance.

Combining equations (2.9) and (2.10), an alternative expression of the "loss" function (2.2) is obtained and the robust estimator can be defined by solving an equivalent problem,

$$\underset{\beta}{\text{minimize}} \sum_{i=1}^{n} ({u_i^*}^2 + 2c_i\sigma\varepsilon_i) \tag{2.11}$$

where ${u_i^*}^2$ are the modified residuals, which we call metrically Winsorized residuals,

$$u_i^* = \begin{cases} u_i & \text{for } |u_i| \le c_i\sigma \\ c_i\sigma = u_i - \varepsilon_i & \text{for } u_i > c_i\sigma \\ c_i\sigma = u_i + \varepsilon_i & \text{for } u_i < -c_i\sigma \end{cases} \tag{2.12}$$

and $2c_i\sigma\varepsilon_i$ can be considered as the penalty cost of pulling y_i towards its fitted value for a distance ε_i. Obviously, for influential points with large value of h_i the corresponding cutting parameter c_i is small leading to low pulling cost. Therefore, solving (2.11), the y_i of high leverage points may be pulled for longer distances yielding more drastic down-weights. An important interpretation of the problem (2.11) is that the estimator is defined by modifying the observations y_i with the smallest possible amount of overall pulling distances $E = \sum_{i=1}^{n} \varepsilon_i$ and this leads to minimax robust estimator (minimax theory, Huber, 1981).

3. New Robust Approach

The scope of the proposed quadratic programming formulation (Camarinopoulos and Zioutas, 2002) was based on the idea to restrict the down-weighting resource or equivalently to constrain the total pulling distance. Then, the estimator selectively down-weights "bad" leverage points by allocating optimally the pulling resource. As optimality criteria, the leverage measure h_i and the residual u_i were considered. In case of unconstraint pulling resource, the proposed estimator QP leads to a Schweppe-type estimator where Mallows weights used as an argument in the ψ-function, (Rousseeuw and Leroy, 1987; Coakley and Hettmansperger, 1993; Simpson and Montgomery, 1998).

The proposed robust procedure has as starting point:

- the weight proposals for Mallows-type estimator $w_i = w(x_i)$,
- the total pulling distance E, obtained empirically from Mallows-type estimator.

The robust estimator QP is defined by optimally allocating the total distance as in the following problem:

$$\underset{\beta}{\text{minimize}} \sum_{i=1}^{n} (u_i^{*^2} + 2cw_i\sigma\varepsilon_i)$$

$$\text{subject to:}$$

$$\sum_{i=1}^{n} \varepsilon_i \le E. \tag{3.1}$$

In the above objective function the penalty cost, $2cw_i\sigma\varepsilon_i$, of pulling y_i towards its fitted value depends on weight $w_i = w(x_i)$, which is based on Hampel's leverage measure. For an outlying observation in the x-space the corresponding penalty cost is very small and thus the y_i could be pulled easily towards the regression line, resulting in excessive down-weighting. Therefore, the proposed estimator QP is a bounded influence estimator.

The above approach has the potential to restrict the total down-weighting resource E. In an optimum solution of the problem (3.1), most of the down-weighting resource is allocated preferably to catastrophic outliers, like the "bad" leverage points, since the restricted resource E may not be sufficient for down-weighting every leverage point.

But, a difficulty arises in the formulation (3.1) with the optimum size of overall pulling, E. The source of down-weighting is a significant factor for the robustness and effectiveness and the optimum size of this given constant has to be searched.

In the following proposed approach the source of pulling is not restricted, but a new restriction inserts into the model instead. A point is modified only if its residual u is greater than the specified upper limit in Schweppe or other GM-estimator. In this section first two alternative models are described for the new robust approach and then the estimates are obtained by solving the corresponding mathematical programming problem.

3.1. The Proposed QMIP Estimators

In the new proposal, the QMIP1 estimator is defined by minimizing the same loss function as in (3.1), under a new constraint, concerning the starting rule of pulling activation. This constraint allows the shortening of big residuals according to a Schweppe form,

$$\underset{\beta}{\text{minimize}} \sum_{i=1}^{n} ({u_i^*}^2 + 2cw_i\sigma\varepsilon_i)$$

$$\text{subject to:}$$

$$|u_i^*| = \begin{cases} |u_i| - \varepsilon_i & \text{when } |u_i| > c(\sqrt{1-h_i})\sigma \\ |u_i| & \text{when } |u_i| \leq c(\sqrt{1-h_i})\sigma. \end{cases} \tag{3.2}$$

In the above objective function the penalty cost, $2cw_i\sigma\varepsilon_i$, of pulling y_i towards its fitted value depends on Mallows's weight $w_i = w(x_i)$, which is based on leverage measure. For an outlying observation in the x-space the corresponding penalty cost is very small and thus the y_i could be pulled easily towards the regression line, resulting in excessive down-weighting. Therefore, the proposed estimator is a bounded influence estimator.

It should be noted here that:

- the pulling cost is a significant factor for the degree of robustness,
- the constraint concerning the residual size is a significant factor for the effectiveness of the proposed estimator, indicating which data points must be down-weighted.

An alternative formulation of the above approach leads to a new robust estimator QMIP2. By reducing the pulling cost to zero, large outliers (x-outliers, y-outliers) can be discarded from the sample. The QMIP2 estimator can be defined by solving the problem

$$\underset{\beta}{\text{minimize}} \sum_{i=1}^{n} ({u_i^*}^2 + 0\varepsilon_i)$$

$$\text{subject to:}$$

$$|u_i^*| = \begin{cases} 3c(\sqrt{1-h_i})\sigma & \text{when } |u_i| > 3c(\sqrt{1-h_i})\sigma \\ u_i & \text{when } |u_i| \leq 3c(\sqrt{1-h_i})\sigma. \end{cases} \tag{3.3}$$

The residual constraint determines which data points should be removed. In the robust procedure (3.3), a data point is pulled free towards the fitted value if its residual is bigger than $3c\sigma\sqrt{1-h_i}$. Free pulling means that the corresponding point does not influence the estimation, this is equivalent with discarding it.

An interpretation of the residual constraint is the fixed penalty cost, $(3c\sigma\sqrt{1-h_i})^2$, for discarding a data point (x_i, y_i). Equivalently, this discarding cost can be included in the loss function as is shown in the mathematical formula (3.5) in the next paragraph. A significant advantage of this discarding cost is the effectiveness of the resulting estimator, since only catastrophic outliers are deleted from the sample. Also, the pulling cost for down-weighting "smaller" outliers could be included in the same loss function.

For the proposed estimator QMIP2 defined by solving the problem (3.3), a better high break down point than for the above mentioned GM-estimators is expected. This is a consequence of the free pulling or equivalently discarding the bad outliers. In the proposed robust procedure, the loss function increases only by a fixed cost for discarding a bad outlier, it does not take into account how far it is. However, the break down property of the last QMIP2 estimator has to be searched.

The objective functions in (3.2) and (3.3) are convex and therefore global optimum solutions exist. The new estimators QMIP1 and QMIP2 have the bounded influence property. Furthermore, influential points are down-weighted with the smallest possible amount of overall pulling and this gives good robust properties to the proposed estimators (Huber, 1981).

The solutions of the above problems, (3.2) and (3.3), are obtained via quadratic and mixed integer programming as described in the following subsection.

3.2. Formulation of Quadratic and Mixed Integer Programming Problem

One technique that has the potential to increase the scope for application of efficient statistical methodology is mathematical programming. Barrodale and Roberts (1970, 1973) proposed a mathematical formulation for a regression model. Convex-programming theory has been developed for solving restricted least-squares problems (Arthanari and Dodge, 1993).

So far, mathematical programming techniques have been used in regression models for various reasons. Normally, linear programming is employed in the context of the absolute error criterion.

In robust regression, Musicant and Mangasarian (2000) presented a new approach with a convex quadratic program, in order to solve Huber regression. They found their approach simpler and faster.

We use a modification of the proposed quadratic programming formulation (Camarinopoulos and Zioutas, 2002). Thus, the minimization problem (3.2) for defining the robust QMIP1 estimate of the parameter β is expressed as a quadratic mixed integer programming problem as follows,

$$\underset{\beta,\ \varepsilon_i, u_i^*,\ \delta_i}{\text{minimize}} \sum_{i=1}^{n} ({u_i^*}^2 + 2cw_i\sigma\varepsilon_i)$$

subject to:

$$
\begin{aligned}
x_i^T\beta_1 - x_i^T\beta_2 + u_i^* + \varepsilon_i &\geqslant y_i\\
x_i^T\beta_1 - x_i^T\beta_2 - u_i^* - \varepsilon_i &\leq y_i\\
\varepsilon_i &\leq \delta_i 1000\\
u_i^* &\geq \delta_i c(\sqrt{1-h_i})\sigma\\
\beta_1, \beta_2\ \varepsilon_i, u_i^*,\ \delta_i &\geq 0\\
\delta_i &: \text{zero-one variable}\\
\text{for}\ \ i &= 1, 2, 3, \ldots, n
\end{aligned}
\tag{3.4}
$$

where δ_i is a an integer decision variable (zero-one), indicating which points have to be pulled. The first two constraints determine the regression line and the resulting residuals. If $\delta_i = 1$, due to the third constraint the pulling distance ε_i can take positive value, and from the fourth constraint the modified residual could not be less than the specified size, since $\delta_i \neq 0$.

The regression coefficients $\beta_1^T = (\beta_{11}, \ldots, \beta_{1p})$, $\beta_2^T = (\beta_{21}, \ldots, \beta_{2p})$, are non-negative vector variables. An estimation of parameter β is obtained by solving (3.4) and setting $\beta = \beta_1 - \beta_2$.

The quadratic programming problem (3.4) has a continuous and convex objective function and the constraints form a closed convex set. Therefore, during the minimization process the simplex method searches the feasible solutions on the extreme points or corners of the convex region. The iterative method stops at the extreme point, $(\beta_1, \beta_2, u^*, \varepsilon, \delta)$, which is the unique global minimum, since the objective function is continuous and convex.

A final interpretation of the values of the decision variables $(\beta_1, \beta_2, u^*, \varepsilon, \delta)$ in the basic optimum solution of the above problem is that the objective function is minimized and all the constraints are satisfied. Especially, for $\delta_i = 1$:

- the corresponding point has to be modified by pulling it for a distance ε_i,
- the corresponding residual u_i is modified to so as $u_i^* = u_i - \varepsilon_i = c(\sqrt{1-h_i})\sigma$
- the corresponding penalty cost is $2cw_i\sigma\varepsilon_i$.

The mathematical formula for the QMIP2 estimator is very similar with the formula (3.4), as follows:

$$
\underset{\beta, \varepsilon_i, e_i, u_i^*, \delta_i}{\text{minimize}} \sum_{i=1}^{n} ({u_i^*}^2 + \delta_i(3c(\sqrt{1-h_i})\sigma)^2 + 2c_i\sigma e_i)
$$

subject to:

$$
\begin{aligned}
x_i^T\beta_1 - x_i^T\beta_2 + u_i^* + \varepsilon_i + e_i &\geqslant y_i\\
x_i^T\beta_1 - x_i^T\beta_2 - u_i^* - \varepsilon_i - e_i &\leq y_i\\
\varepsilon_i &\leq \delta_i 10000\\
\beta_1, \beta_2\ \varepsilon_i, u_i^*,\ \delta_i &\geq 0\\
\delta_i &: \text{zero-one variable}\\
\text{for}\ i &= 1, 2, 3, \ldots, n.
\end{aligned}
\tag{3.5}
$$

When the binary decision variable δ_i takes the value one, $\delta_i = 1$, then from the third constraint ε_i takes a positive value. Since there is no any pulling cost the bad outlier can be pulled as long as it is required for a better fitting. In other words, the influence of a bad outlier is eliminated. For each such decision there is a fixed penalty cost in the loss function. The size of this fixed discarding cost could be a function of the Schweppe critical value.

If it is desired, a second weight function could cooperate simultaneously in order to down-weight smaller outliers. Thus, a new decision variable e, presenting the pulling distance of the second weight function, inserts into the formula, and the corresponding pulling cost is included in the loss function.

The quadratic programming still has a continue and convex objective function, and therefore, a global optimum solution could be provided by the simplex method.

The robust properties of the proposed estimators are illustrated numerically with Monte Carlo simulation, which is discussed in the following section.

4. Simulation Design

To evaluate the performance of the new robust approach, we carried out a simulation study that compared it to well-known robust methods. To carry out one simulation run, we proceeded as follows. The distributions of independent variables and the values of parameters were given. Errors were generated according to an error distribution and observations, y_i, were obtained following the regression model (2.1). In this study we considered five factors:

1) number of regression outliers,
2) number of "good" leverage points,
3) tuning constants,
4) true model,
5) type of estimate.

$N = 100$ replications have been chosen for this Monte Carlo study in order to obtain a relative error $(\widehat{\beta}_j - \beta_j)/\beta_j < 10\%$ for all $j = 1, 2, 3, \ldots, p$, with a reasonable confidence level of at least 90% for all the simulation estimates.

All the constants of the robust estimates are tuned so that all the resulting average weights are the same. Thus, for all methods we down-weigh approximately the same fraction of data. According to the robustness literature any other choice for the scale parameter does not change the performance of our proposed robust estimator.

5. Results

After 100 replications, using samples of size $n = 25$ for a Gaussian regression model contaminated by regression outliers and "good" leverage points, Monte Carlo results have been obtained. The three Tables presented correspond to QP,

QMIP1 and QMIP2 estimators and the results are based on lightly different groups of data contaminated by regression outliers and "good" leverage points.

The performance of the robust estimators is measured by Monte Carlo estimates (actual calculations) of the following criteria: Mean Square Error, Mean Absolute Error, Norm of the Bias, and Trace of the Covariance matrix.

All of the following conclusions were supported by careful examination of the individual estimates. Taking account the four performance criteria mentioned above, as it was expected, the QP, QMIP1 and QMIP2 estimators had the best performance, in the presence of both regression outliers and "good" leverage points, Table 1, Table 2 and Table 3.

TABLE 1. Number of regression outliers 3, "good" leverage points 3. $\beta_1 = 1.20$, $\beta_2 = -0.80$.

Est.	Var. of β_1	Var. of β_2	Trace of Covar	Mean Absol. Err. β_1	Mean Absol. Err. β_2	Norm of Bias β	Mean Squar. Fit. Error
MLL	.021	.008	.028	.184	.122	.230	277.8
KW	.029	.011	.039	.154	.104	.198	277.9
QP	**.021**	**.008**	**.028**	**.139**	**.098**	**.180**	**272.8**
MH	.028	.010	.038	.176	.116	.222	278.4

TABLE 2. Number of regression outliers 3, "good" leverage points 3. $\beta_1 = 1.20$, $\beta_2 = -0.80$.

Estimators.	Mean. of β_1	Mean. of β_2	Trace of β	Mean Square Error of β	Mean Squar. Fit. Error
QMIP1	**1.169**	**-0.802**	**0.009**	**0.010**	**261**
Schweppe-type, Mallows weights	1.148	-0.787	0.020	0.023	265

TABLE 3. Number of regression outliers 3, "good" leverage points 3. $\beta_1 = 1.20$, $\beta_2 = -0.80$.

Estimators.	Mean. of β_1	Mean. of β_2	Trace of β	Mean Square Error of β	Mean Squar. Fit. Error
QMIP2	**1.145**	**-0.806**	**0.024**	**0.022**	**264**
Schweppe-type, Mallows weights	1.088	-0.754	0.054	0.068	275

5.0.1. Computation Time. All experiments ran on a 1200 Mhz Athlon AMD Processor. All computations of robust estimators were carried out using Fortran library ROBETH (1993), provided by Marazzi. A new optimization algorithm "solver FortMP/QMIP"(CARISMA, Brunel University) which has been developed for large scale mathematical programming problems can reduce the computational time of the solution of the above QP-Mixed Integer problems, making the new procedure reasonable and promising.

For the QP estimator the computational time is competitive to the ordinary statistical procedures

For the QMIP1 estimator the computation is slower than the standard statistical methods. But, for a medium size sample the computation is still fast. Therefore, the proposed approach is reasonable.

For the QMIP2 estimator, generally, the computation is slow. For a sample size up to 50 or even 100 observations the solution is obtained quickly. Suitable algorithms or alternative formulations have to be searched for larger problems.

6. Final Remarks

The basic aim of this work was to develop useful mathematical programming formulas for robust estimation. The quadratic mixed integer programming method offers a systematic approach to improve the sample behavior of bounded influence type estimator. Since regression outliers are the most dangerous and often "good" leverage points exist in the data, we recommend the use of the QP and QMIP estimator, which provides a good protection against bias while keeping good efficiency.

Based on the above optimality criteria and results, we conclude that the QMIP estimator works well in many circumstances and is reasonable for all x-outliers including "good" leverage points. From this study, one can gain by using quadratic mixed integer programming for bounded influence regression estimators.

Also, the mathematical programming technique in robust estimation offers flexibility in being able to

- constraints for parameter vector β when it is required, for example in a outlier-multicollinearity problem (Simpson and Montogomery, 1996),
- identify the most catastrophic outliers in a sample of observations,
- combination of two or more weight functions, in order to bound the "position" weight function for points with small h, a desirability which has been discussed by Huber and Mallows (Huber, 1983).

Further research is needed to determine possible better choices of the combined weight functions and the modified residual size.

References

[1] T.S. Arthanari and Y. Dodge, *Mathematical Programming in Statistics.* John Wiley& Sons, Inc, 1993.

[2] I. Barrodale and D.K. Roberts, *Applications of Mathematical Programming to lp-Approximation.* In J.B. Rosen, O.L. Mangasarian, and K. Ritter, editors, Nonlinear Programming, pages 447–464. Academic Press, New York, 1970.

[3] I. Barrodale and D.K. Roberts, *An Improved Algorithm for Discrete l1 Linear Approximation.* SIAM J. Numer. Anal. **10** (1973), 839–848.

[4] L. Camarinopoulos and G. Zioutas, *Formulating Robust Regression Estimation as an Optimum Allocation Problem.* J. Statist. Comput. Simulation **72** (2002), 687–705.

[5] C.W. Coakley and T.P. Hettmansperger, *A Bounded Influence, High Breakdown, Efficient Regression Estimator.* J. Amer. Statist. Assoc. **88** (1993), 872–880.

[6] F.R. Hampel, *Optimally Bounding the Gross Error Sensitivity and Influence of Position in Factor Space.* In Proceedings of the ASA Statistical Computing Section, pages 59–64. American Statistical Association, Washington, D.C., 1978.

[7] F.R. Hampel, E.M. Ronchetti, P.J. Rousseeuw, and W.A. Stahel, *Robust Statistics, the Approach Based on Influence Functions.* John Wiley, New York, 1986.

[8] E. Handschin, J. Kohlas, A. Flechter, and F. Schweppe, *Bad Data Analysis for Power System Estimation.* IEEE Transactions Power Appar. Syst. **2** (1975), 329–337.

[9] R.W. Hill, *Robust Regression When There Are Outliers in the Carriers.* Ph.D. Dissertation, Harvard University, Boston, MA, 1977.

[10] P.J. Huber, *Robust Regression : Asymptotics, Conjectures, and Monte Carlo.* Ann. Statist. **1** (1973), 799–821.

[11] P.J. Huber, *Robust Statistics.* John Wiley, New York, 1981.

[12] P.J. Huber, *Minimax Aspects of Bounded-Influence Regression.* J. Amer. Statist. Assoc. **78** (1983), 66–72.

[13] W.S. Krasker and R.E. Welsch, *Efficient Bounded-Influence Regression Estimation.* J. Amer. Statist. Assoc. **77** (1982), 595–604.

[14] J.W. McKean, S.J. Sheather, and T.P. Hettmansperger, *Robust and High-Breakdown Fits of Polynomial Models.* Technometrics **36** (1994), 409–415.

[15] C.L. Mallows, *Discussion on Minimax Aspects of Bounded-Influence Regression.* J. Amer. Statist. Assoc. **78** (1983), 66–80.

[16] A. Marazzi, *Algorithms, Routines, and S Functions for Robust Statistics.* Wadsworth and Brooks/Cole, Pacific Grove, California, 1993.

[17] D. Musicant and O.L. Mangasarian, *Support Vector Regression.* International Symposium on Mathematical Programming, August 10, 2000.

[18] P.J. Rousseeuw and A.M. Leroy, *Robust Regression and Outlier Detection.* Wiley, New York, 1987.

[19] J.R. Simpson and D.C. Montogomery, *A Biased-Robust Regression Technique for the Combined Outlier-Multicollinearity problem.* J. Statist. Comput. Simulation **56** (1996), 1–22.

[20] J.R. Simpson and D.C. Montogomery, *A Robust Regression Technique Using Compound Estimation.* Naval Res. Logist. **45** (1998), 125–139.

[21] R.E. Welsch, *Regression Sensitivity Analysis and Bounded-Influence Estimation.* In J. Kmenta and J.B. Ramsey, editors, Evaluation of Econometric Models, pages 153–167. Academic Press, New York, 1980.

G. Zioutas
General Department
Faculty of Engineering
Aristotle University of Thessaloniki
P.O. Box 464
GR 54124 Thessaloniki
Greece
e-mail: `zioutas@eng.auth.gr`

MIX
Papier aus verantwortungsvollen Quellen
Paper from responsible sources
FSC® C105338

If you have any concerns about our products,
you can contact us on
ProductSafety@springernature.com

In case Publisher is established outside the EU,
the EU authorized representative is:
Springer Nature Customer Service Center GmbH
Europaplatz 3, 69115 Heidelberg, Germany

Printed by Libri Plureos GmbH
in Hamburg, Germany